D0414761

Engineering and Construction Law and Contracts

J. K. Yates, Ph.D

Department Head and the Joe W. Kimmel Distinguished Professor of Construction Management
Western Carolina University

Prentice Hall

Boston Columbus Indianapolis New York San Francisco Upper Saddle River
Amsterdam Cape Town Dubai London Madrid Milan Munich Paris Montreal Toronto
Delhi Mexico City Sao Paulo Sydney Hong Kong Seoul Singapore Taipei Tokyo

Editorial Director: Vernon R. Anthony
Acquisitions Editor: David Ploskonka
Editorial Assistant: Nancy Kesterson
Director of Marketing: David Gesell
Marketing Manager: Derril Trakalo
Senior Marketing Coordinator: Alicia Wozniak
Marketing Assistant: Les Roberts
Managing Editor: Alexandrina Benedicto Wolf
Inhouse Production Liaison: Louise Sette
Senior Operations Supervisor: Patricia Tonneman
Operations Specialist: Laura Weaver
Senior Art Director: Jayne Conte
Cover Designer: Karen Salzbach
Cover Art: Superstock
Full-Service Project Management: Aparna Yellai/PremediaGlobal
Composition: PremediaGlobal
Printer/Binder: Edwards Brothers
Cover Printer: Lehigh-Phoenix Color/ Hagerstown
Text Font: 10/12 Minion

Credits and acknowledgments borrowed from other sources and reproduced, with permission, in this textbook appear on appropriate page within text.

Library of Congress Cataloging-in-Publication Data

Yates, J. K.
Engineering and construction law and contracts / J.K. Yates.
p. cm.
Includes bibliographical references and index.
ISBN-13: 978-0-13-503352-4 (alk. paper)
ISBN-10: 0-13-503352-7 (alk. paper)
1. Construction contracts—Popular works. 2. Engineering contracts—Popular works. 3. Construction industry—Law and legislation—Popular works. 4. Engineering law—Popular works. 5. Construction contracts—United States—Popular works. 6. Engineering contracts—United States—Popular works. 7. Construction industry—Law and legislation—United States—Popular works. 8. Engineering law—United States—Popular works. I. Title.
K891.B8Y38 2011
343′.07869—dc22

2010029035

10 9 8 7 6 5 4 3 2 1

Prentice Hall
is an imprint of

www.pearsonhighered.com

ISBN 13: 978-0-13-503352-4
ISBN 10: 0-13-503352-7

To all of the engineering and construction students who helped the author understand that the subjects of law and contracts are foreign to them and that they have to be translated into their language.

To the 90,000 volunteers of the Red River Flood Fight of 2009 who saved the city of Fargo, North Dakota. The Flood Fight of 2009 was a massive civil engineering and construction project that was completed in seven days and was led by the mayor of Fargo, who is a civil engineer. The author of this book was trapped in one of the flooded homes for three weeks, providing uninterrupted time to complete this book.

PREFACE

This book was written to provide an up-to-date compendium of information on engineering and construction law and contracts. The material is presented from the perspective of how legal issues affect engineering and construction professionals on a daily basis in their careers. Examples and case studies illustrate how engineers and constructors could become involved in the legal affairs of their profession early in their careers.

Although there are numerous books available that cover the legal aspects of engineering and construction, many of them mainly address construction contracts rather than focusing on legal issues that affect engineers and constructors. This book is unique because it was written from the perspective of practicing engineers and constructors with minimal use of legal terminology. If legal terms are used, they are explained using common vernacular rather than legal textbook definitions.

The author has taught law courses at both the undergraduate and graduate level and adapted the material to reflect the background of students majoring in engineering and construction rather than law. Each year at least one of the engineering or construction students chose to pursue a degree in law after completing their engineering law courses; they are now practicing lawyers or owners of engineering, construction, or legal firms.

This book is divided into four parts. Part One has eleven chapters that focus on engineering and construction laws as they pertain to practicing engineers and constructors. These chapters cover court proceedings; labor and environmental laws that affect engineering and construction projects, including Leadership in Energy and the Environment (LEED) and sustainability practices and legislation; professional ethics; whistle-blowing; professional registration; agency relationships; negligence and tort liability; real property, including forms of ownership of real estate; and forms of ownership of companies.

Part Two includes five chapters that focus on engineering and construction contracts. These chapters address forming engineering and construction contracts; the components of contracts; contract conditions from the perspective of which clauses help firms avoid breach of contract or the clauses that could be used as defenses for breach of contract; specifications; and government contracting.

Part Three focuses on change orders, claims, consequential damages, and documentation to support claims. It discusses traditional dispute resolution techniques, but it also includes alternative dispute resolution methods such as dispute review boards, early neutral evaluation, minitrials, rent-a-judge, court-annexed arbitration, and summary jury trials. Chapter 18 includes a discussion on information technology, the electronic marketplace, Building Information Modeling (BIM), and the legal effects these areas are having on contracts and engineering law. The other topics addressed include the admissibility of e-mails in dispute resolution proceedings, confidentiality issues, liability for computer-aided design (CAD) software format conversions, and discrepancies between electronic and paper documentation. Chapter 19 discusses risk management and construction insurance.

Part Four covers international engineering and construction laws, contracts, and arbitration. An American firm purchasing any materials or products from overseas is involved in international contracts. Chapter 20 provides information on the obligations and ramifications of international contracts and how to settle disputes that arise during the execution of international projects. International engineering and construction laws are also discussed in the context of how particular laws affect engineers or constructors when they are working on foreign projects. Chapter 21 summarizes the ramifications of engineers and constructors working in the twenty-first century global legal environment.

Sample engineering and construction contracts are included in the Appendices along with the American Arbitration Association Construction Industry Arbitration

Rules and Mediation Procedures, since the majority of construction disputes are now settled through negotiation, mediation, or arbitration rather than litigation.

Along with this book, instructional materials are available for educators, including an instructor's manual containing solutions to the discussion questions at the end of each chapter and sample examination questions.

This book is a resource that helps prepare engineering and construction professionals to work within the legal framework of their profession and to be able to address the daily legal issues that arise during the design and construction of projects.

Download Instructor Resources from the Instructor Resource Center

To access supplementary materials online, instructors need to request an instructor access code. Go to www.pearsonhighered.com/irc to register for an instructor access code. Within 48 hours of registering, you will receive a confirming e-mail including an instructor access code. Once you have received your code, locate your text in the online catalog and click on the Instructor Resources button on the left side of the catalog product page. Select a supplement, and a login page will appear. Once you have logged in, you can access instructor material for all Prentice Hall textbooks. If you have any difficulties accessing the site or downloading a supplement, please contact Customer Service at http://247.prenhall.com.

ACKNOWLEDGMENTS

The genesis of this book occurred when the author first taught a course on engineering and construction law and contracts. Three weeks into the semester a student raised his hand and said, "Dr. Yates, everything you have been teaching us is really interesting but unfortunately we do not understand a word of it. Could you please translate it into plain English." The rest of that semester the author was translating legal phrases, legal concepts, laws, and contracts into simple phrases and terminology. If the students still did not understand something, it would be explained in two or three different ways. Over the years the author has taught contract and law courses at both the undergraduate and graduate level. Each year one of the civil engineering students in these courses has had more than a passing interest in laws and contracts and continued on to graduate from law school. This book was written for all of the students who have taken contracts and law courses from the author, because they provided the author with insight into how legal topics should be presented to engineering and construction professionals.

The subjects of engineering and law are taught differently in that engineering requires analytical reasoning of new concepts and law requires an excellent memory and a reliance on precedent legal cases. Yet the importance of having a basic understanding of legal issues and contracts increases every year, as the number of lawsuits escalates in the United States. The former civil engineering students who graduated from law school demonstrate that engineers are able to master legal concepts if they are introduced to them in a manner that is relevant to their particular profession. I thank these students for showing me this by their success in the legal profession.

Eric Krassow, my first editor at Prentice Hall/Pearson Education, asked me what other book I could write for his company just as I was concluding negotiations on a contract for a book on Engineering and Construction Productivity Improvement. At first I thought he was joking but he was not. A few weeks later I provided him with a prospectus for this book. He probably still does not believe that I really like doing the index for books. I would like to thank him for asking me to write this book. It will provide a much wider audience of engineers and constructors with a resource that will help them understand legal concepts and contracts, and they will be able to utilize this knowledge in their careers. I also appreciate his patience and guidance on both of my books published by his firm.

David Ploskonka took over as my second editor after Eric Krassow was promoted, and he had to quickly grasp the intention of this book and help shepherd it through to completion. That was no easy task; I thank him for this time, effort, and support. I also thank Lucy Palmieri and Aparna Yellai for their production assistance, which occured in the middle of my move to North Carolina.

Several of my former graduate students contributed material that was included in this book. I would like to thank them and acknowledge their contributions. Joseph Smith conducted research and helped write Chapter 20, which is on international contracts and arbitration. His interest in international law has led him to working in Shanghai, China. Alan Epstein contributed material included in Chapter 17, on construction claims. He is one of my former students who continued on to law school and now owns a highly successful engineering firm in New York City. Juan Duran, who was the vice president of operations in Latin America for a major engineering and construction firm, provided the material in Chapter 18 on Dispute Resolution Boards. Hala El-Nagar conducted research on construction documentation used as indicators of delays; Table 17.1 summarizes her research on the different types of construction documentation that helps to justify delays. She is currently a professor at Abu Dhabi University.

I would like to acknowledge and thank Matthew Erickson, who is an undergraduate student at North Dakota State University, for drawing Figure 6.2. Carlos Sotelo contributed the drawings in Chapter 13 of the recreation center for the moon; he is now working in Peru an as engineer. I would like to thank Dr. Rita Meyninger for contributing the case study in Appendix B on Love Canal. She was a regional director for the Federal Emergency Management Administration (FEMA) under presidents Carter and Clinton.

I would like to thank Dean Gary Smith, the dean of the College of Engineering and Architecture at North Dakota State University, for providing a month of summer support while I completed the first draft of this book.

I would like to acknowledge the reviewers of this book: Phil Lewis, North Carolina State University, James Butler, Georgia Institute of Technology and Richard Storrs, Georgia State University. I would also like to acknowledge the lawyers I worked with as an expert witness. They provided me with additional insight into the inner workings of the legal system and exposure to legal cases in the construction industry.

ABOUT THE AUTHOR

Dr. J. K. Yates is the Department Head and the Joe W. Kimmel Distinguished Professor of Construction Management in the Department of Construction Management at Western Carolina University in Cullowhee, North Carolina. Dr. Yates was previously a professor at North Dakota State University; in charge of the Construction Engineering and Management focus area at Ohio University; and the program coordinator for the Construction Engineering program in the Civil and Environmental Engineering Department at San Jose State University in California. Dr. Yates was also a professor at New York University (formerly Polytechnic University) and at Iowa State University, as well as a visiting professor at the University of Colorado.

Dr. Yates received a Bachelor of Science degree in civil engineering from the University of Washington and a Doctor of Philosophy (Ph.D.) degree in civil engineering from Texas A&M University, with minors in global finance and management, global political science, business analysis, construction science, and archeology/anthropology.

Dr. Yates has worked for several domestic and global engineering and construction firms and as a consultant during legal cases and on international contracts. Dr. Yates is the author of eight books and numerous refereed journal articles. She is a member of the American Society of Civil Engineers, the Project Management Institute, and the American Association of Cost Engineers International. Dr. Yates received the *Distinguished Professor* award from the Construction Industry (CII) in 2010 and for Polytechnic University in 1994, was the Associated General Contractor's *Outstanding Construction Professor in America* in 1997, was one of the Engineering News Records *Those Who Made Marks on the Construction Industry* in 1991, and received the Ron Brown award for industry/academic collaborations with the Hewlett Packard Foundation from President Clinton in 2001. Dr. Yates has traveled for work and pleasure to thirty-one countries and worked in Bontang, East Kalimantan on the Island of Borneo in Indonesia.

CONTENTS

Chapter 1

INTRODUCTION 1

1.1 The Importance of Understanding Engineering and Construction Laws and Contracts 1

1.2 How Contracts and Laws Are Used in the Engineering and Construction Industry 3

1.3 Writing Engineering and Construction Contracts 3

1.4 Interpreting Engineering and Construction Contracts 4

1.5 Administering Engineering and Construction Contracts 4

1.6 Avoiding Breach of Contract 4

1.7 Tort Liability of Engineers and Constructors 5

Chapter Summary 5

Key Terms 5

Discussion Questions 5

Chapter 2

THE UNITED STATES AND OTHER LEGAL SYSTEMS 6

2.1 Introduction 6

2.2 Civil Law, Common Law, Shari'a Law, and Asian Legal Systems 6

2.3 The United States Legal System 10

2.4 The Purpose of Constitutions 10

2.5 The Federal and State Systems of Government 10

2.6 Exclusive Jurisdiction of the Federal Courts 11

2.7 Exclusive Jurisdiction of the State Courts 12

2.8 The Responsibility of Engineers and Constructors for Implementing Laws 12

Chapter Summary 12

Key Terms 12

Discussion Questions 13

References 13

Chapter 

THE UNITED STATES FEDERAL AND STATE COURT SYSTEMS 14

3.1 Introduction 14

3.2 The Federal Court System 14

3.3 State Court Systems 16

3.4 Lawsuits 17

Chapter Summary 19

Key Terms 19

Discussion Questions 19

References 20

Chapter 4
UNITED STATES LABOR LAWS THAT AFFECT ENGINEERS AND CONSTRUCTORS 21

4.1 Introduction 21
4.2 Federal Labor Laws 21
4.3 The Occupational Safety and Health Act of 1970 25
4.4 Unions and Union Labor 26
4.5 Union Structures 27
4.6 Jurisdictional Disputes 27
4.7 Open Shop (Nonunion) Environments 28
4.8 Double Breasting 29
Chapter Summary 29
Key Terms 29
Discussion Questions 29
References 30

Chapter 5
ENVIRONMENTAL AND SUSTAINABILITY LAWS THAT AFFECT ENGINEERS AND CONSTRUCTORS 32

5.1 Introduction 32
5.2 The Council on Environmental Quality (CEQ) 32
5.3 The Environmental Protection Agency and Environmental Impact Statements 33
5.4 Federal Environmental Laws of Concern to Engineers and Constructors 33
5.5 Sustainability Considerations in Engineering and Construction 37
5.6 Environmental Protection Agency Sustainability (EPA) Laws 40
5.7 United States Government Sustainability Laws 40
5.8 Global Environmental Management Standards 42
5.9 Global Sustainability—The United Nations Framework Convention on Climate Change (UNFCCC) 42
Chapter Summary 45
Key Terms 46
Discussion Questions 46
References 47

Chapter 6
ENGINEERING PROFESSIONAL ETHICS 49

6.1 Introduction 49
6.2 The Definition of Ethics and Morals 51
6.3 The Role of Engineering Professional Societies in Ethics 52
6.4 The Purpose of Professional Codes of Ethics 52
6.5 Enforcing Engineering Codes of Ethics 54
6.6 Conflicts Created by Engineering Professional Codes of Ethics 54
6.7 Whistle-Blowing in the Engineering Profession 55
6.8 Deciding When to Blow the Whistle 55
6.9 Methods for Whistle-Blowing 56
6.10 Whistleblower Protection Laws 57
Chapter Summary 57
Key Terms 57
Discussion Questions 57
References 58

Chapter 7
ENGINEERING PROFESSIONAL REGISTRATION AND THE CONTRACTOR LICENSING PROCESS 59

7.1 Introduction 59
7.2 Justification for Engineering Professional Registration 59
7.3 The Professional Engineering Licensing Process 59
7.4 Out-of-State Engineering Practice 61
7.5 Criminal Violations of Engineering Licensing Laws 61
7.6 Moonlighting as an Engineer—Unregistered Design Professionals 61
7.7 Contractor Licensing Requirements 62

7.8 Depositions Provided by Engineers and Contractors 63
7.9 Expert Witness Testimony 63
Chapter Summary 64
Key Terms 64
Discussion Questions 64
References 65

Chapter 8
AGENCY RELATIONSHIPS 66
8.1 Introduction 66
8.2 Definition of Agency 66
8.3 Creating Agency Relationships 66
8.4 Terminating Agency Contracts 67
8.5 Disputes Between Third Parties and Agents 67
Chapter Summary 68
Key Terms 68
Discussion Questions 68
References 68

Chapter 9
TORT LAW 69
9.1 Introduction—Definition of Tort Law 69
9.2 Negligence 69
9.3 Negligence of Engineers and Contractors 71
9.4 Defenses to the Liability of Engineers or Contractors 72
9.5 Tort Liability Compensation 73
9.6 Indemnification Clauses 73
Chapter Summary 74
Key Terms 74
Discussion Questions 74
References 74

Chapter 10
LEGAL ISSUES RELATED TO REAL PROPERTY 75
10.1 Introduction 75
10.2 Definition of Real Property 75
10.3 Options for Ownership of Real Property 76
10.4 Mortgages 77
10.5 Encumbrances 77
10.6 Servitudes 77
10.7 Doctrine of Support 78
10.8 Rental Property 78
10.9 Liens 78
10.10 Eminent Domain 78
10.11 Right-of-Way 79
10.12 Zoning 79
Chapter Summary 79
Key Terms 79
Discussion Questions 79
References 80

Chapter 11
FORMS OF OWNERSHIP OF FIRMS 81
11.1 Introduction 81
11.2 Sole Proprietorship 81
11.3 Partnerships 81
11.4 Corporations 82
11.5 Joint Ventures 82
Chapter Summary 83
Key Terms 83
Discussion Questions 83
References 83

Chapter 12
FORMING ENGINEERING AND CONSTRUCTION CONTRACTS 84
12.1 Introduction 84
12.2 The Definition of Contracts 84
12.3 Construction Industry Project Team Members 85
12.4 Construction Project Development 86
12.5 Contracts Used during the Phases of Construction Projects 86
12.6 Project Contract Administration Models 88

12.7 Formation Principles of Contracts 90
12.8 Forming Contracts for Engineering and Construction Services 94
12.9 Engineering Contracts 96
12.10 Construction Contracts 96
Chapter Summary 99
Key Terms 99
Discussion Questions 99
References 100

Chapter 13
CONTRACTS FOR ENGINEERING AND CONSTRUCTION SERVICES 101
13.1 Introduction 101
13.2 Construction Bid Proposals 101
13.3 The Main Components of Construction Contracts 103
13.4 Components of Engineering Contracts 112
13.5 Construction Management Contracts 115
13.6 Other Contract Documents 115
Chapter Summary 116
Key Terms 116
Discussion Questions 116
References 116

Chapter 14
CONTRACT TERMS AND CONDITIONS 117
14.1 Introduction 117
14.2 Duties and Responsibilities of the Architect/Engineer (A/E) 117
14.3 Topics Frequently Covered or Avoided in Construction Contracts 118
14.4 Construction Contract Clauses Requiring Interpretation 120
14.5 Performance and Breach of Contract 124
14.6 Types of Breach of Contract 127
14.7 Ways to Terminate Contracts 127
Chapter Summary 128
Key Terms 128
Discussion Questions 129
References 129

Chapter 15
SPECIFICATIONS 130
15.1 Introduction—Definition of Specifications 130
15.2 Types of Specifications 130
15.3 Standard Specifications 132
15.4 International Standards and the International Organization for Standardization 134
15.5 International Technical Standards 138
15.6 Opportunities for the Misinterpretation of Specifications 139
15.7 The Construction Specification Institute (CSI) Master Format 141
Chapter Summary 142
Key Terms 142
Discussion Questions 143
References 143

Chapter 16
GOVERNMENT CONTRACTING 145
16.1 Introduction 145
16.2 Government Bonding Requirements 145
16.3 Bonding Procedures 146
16.4 Types of Bonds 147
16.5 Advantages and Disadvantages of Bonding 150
16.6 Government Construction Contracts 151
16.7 What Constitutes a Construction Claim Against the Government 152
16.8 Stop Notices 153
16.9 Mechanic's Liens 153
Chapter Summary 156
Key Terms 156
Discussion Questions 156
References 156

Chapter 17

CHANGE ORDERS AND CLAIMS 158

17.1 Introduction 158
17.2 Definition of Changes 158
17.3 Change Orders 159
17.4 Addenda 161
17.5 Construction Claims 162
17.6 Construction Documentation to Support Claims 162
17.7 Construction Delay Claims 163
17.8 Utilizing Critical Path Method Schedules for Delay Analysis 165
17.9 Quantification of contracter Delay Claims 168
17.10 Owner Damages for Delay 168
17.11 Minimizing Delay Claims—Appropriate Clauses in Specifications 168
Chapter Summary 169
Key Terms 169
Discussion Questions 169
References 170
Legal Cases 170
Board of Contract Appeals Decisions 171

Chapter 18

CONTRACT DISPUTE RESOLUTION TECHNIQUES 172

18.1 Introduction 172
18.2 Contract Claims 172
18.3 Information Technology and the Electronic Marketplace 173
18.4 Building Information Modeling (BIM) 174
18.5 Contract Negotiations 175
18.6 Contract Mediation 176
18.7 Arbitration 176
18.8 Litigation 177
18.9 Dispute Review Boards 178
18.10 Alternative Dispute Resolution Techniques 179
Chapter Summary 180
Key Terms 180
Discussion Questions 181
References 181

Chapter 19

RISK MANAGEMENT AND CONSTRUCTION INSURANCE 182

19.1 Introduction 182
19.2 Indemnity 182
19.3 Professional Liability Insurance 182
19.4 Commercial Insurance 183
19.5 Green Building Technology Liability 185
Chapter Summary 185
Key Terms 186
Discussion Questions 186
References 186

Chapter 20

INTERNATIONAL LAW, CONTRACTS, AND ARBITRATION 187

20.1 Introduction 187
20.2 International Conventions 187
20.3 Regional Legal Jurisdictions 187
20.4 International Engineering and Construction Contracts 187
20.5 International Construction Contract Clauses 188
20.6 Claims and Change Orders 192
20.7 Global Contract Dispute Resolution Techniques 193
20.8 International Arbitration 193
20.9 Anticorruption Legislation 194
20.10 Kidnapping and Ransom Insurance 194
20.11 Changing Governments 195
20.12 Liability Issues 195
Chapter Summary 197
Key Terms 197
Discussion Questions 198
References 198

Chapter 21
INTERNATIONAL SUMMARY 200

References 201

Appendix A UNITED STATES CONSTITUTION 202

Appendix B THE ENGINEERING AND HEALTH HISTORY OF LOVE CANAL 211

Appendix C AMERICAN SOCIETY OF CIVIL ENGINEERS CODE OF ETHICS 216

Appendix D CODES OF ETHICS OF ENGINEERS 219

American Society of Agricultural and Biological Engineers Code of Ethics 219

American Institute of Chemical Engineers Code of Ethics 219

American Society of Mechanical Engineers Code of Ethics 220

Institute of Electronic and Electrical Engineers Code of Ethics 220

Institute of Industrial Engineers Code of Ethics 221

Appendix E 222

Suggested Form of Agreement Between Owner and Contractor for Construction Contract (Stipulated Price) 224

Appendix F 236

Agreement Between Owner and Engineer for Professional Services 238

Appendix G 327

Short Form of Agreement Between Owner and Engineer for Professional Services 329

Appendix H 339

Standard General Conditions of the Construction Contract 341

Appendix I 409

Guide to the Preparation of Supplementary Conditions 411

Appendix J CONSTRUCTION INDUSTRY ARBITRATION RULES AND MEDIATION PROCEDURES 437

INDEX 454

INTRODUCTION

1.1 THE IMPORTANCE OF UNDERSTANDING ENGINEERING AND CONSTRUCTION LAWS AND CONTRACTS

Throughout the world, the **rule of law** is relied upon to help keep peace, foster economic development or expansion, promote commerce, stabilize communities, and allow governments a means for controlling society. Without laws there is a potential for anarchy or chaos. In some parts of the world, such as Afghanistan where some areas are governed by Shara'i law and others by government laws, there is no agreement as to what source of laws will be used to govern the country. Fortunately, in the United States there is only one constitution, the **United States (U.S.) Constitution**, that is used as the basis for the U.S. legal system and for framing the government of the United States.

When someone decides to become an engineering or construction professional, they are responsible for following not only the laws that guide society but they have to also implement the laws that have a direct impact on their profession. Engineers in particular have the added responsibility of being the guardians of society through their ethical obligation to protect the welfare of the public, as stated in engineering professional codes of ethics.

This book provides an introduction to engineering and construction **contracts** and to the laws that are relevant to the engineering and construction professions in a manner that reduces complicated legal concepts to a basic level in order to allow engineering and construction professionals to understand them. The topics included in this book are:

Part I

1. Introduction
2. The United States and Other Legal Systems
3. The United States Federal and State Court Systems
4. United States Labor Laws that Affect Engineers and Constructors
5. Environmental and Sustainability Laws that Affect Engineers and Constructors
6. Engineering Professional Ethics
7. Engineering Professional Registration and the Contractor Licensing Process
8. Agency Relationships
9. Tort Law
10. Legal Issues Related to Real Property
11. Forms of Ownership of Firms

Part II

12. Forming Engineering and Construction Contracts
13. Contracts for Engineering and Construction Services
14. Contract Terms and Conditions
15. Specifications
16. Government Contracting

Part III

17. Change Orders and Claims
18. Contract Dispute Resolution Techniques
19. Risk Management and Construction Insurance

Part IV

20. International Law, Contracts, and Arbitration
21. International Summary

It is crucial that engineers and constructors understand construction contracts, how construction contracts are used in the engineering and construction industry, and how the legal system interprets contracts because they are the ones that will either write or interpret contracts for construction projects.

In the United States, in the twenty-first century, many of the activities that people engage in require contracts. Samples of areas where contracts are common are listed in Table 1.1.

Table 1.1. Samples of Common Contracts

Adopting animals	Marriage
Airline tickets	Medical insurance
Athletic center memberships	Memberships in associations and organizations
Automobile repairs	Opening checking and savings accounts
Club memberships	Opening credit and charge card accounts
Day care	Ordering merchandise using computers
Digital video rentals	Passports
Downloading music and videos	Pest control
Driver's license	Purchasing items that cost over $500.00
Grocery club cards	Procuring engineering services
Hiring contractors	Professional society memberships
Hiring employees or laborers	Renting or owning a home, apartment, co-op, or condominium
Homesteading a house	Telephone and cell phone service
Hospital admission	Selling a home, co-op, or condominium
House cleaning	Software licenses
Hunting and fishing licenses	Social Security
Insurance—house, automobile, boat, liability, property, and others	Storage units
Internet accounts	Union memberships
Internet auction sites	Utilities such as water, electric, cable, satellite television, natural gas, and trash collection
Internet service	Vehicle registration
IRA accounts	Warranties
Leasing or buying an automobile	Writing a will or a living trust
Library cards .	
Magazine subscriptions	

The global financial crisis that devastated the U.S. economy in the fall of 2008 was precipitated by large numbers of defaults on home loans that were caused by homeowners not fully understanding the terms of their home **mortgage** (loan) contracts. Homeowners were provided with contracts that established house payments at artificially low starting rates in order for homebuyers to qualify for their monthly mortgage payments. At a future point in time, which usually is three to five years, the mortgage payments are adjusted to a higher, more realistic rate that reflects the actual cost of borrowing the money. When the interest rates on the home loans that were issued in the early part of the first decade of the twenty-first century were drastically increased, after three years of mortgage payments, some homeowners could not afford to pay their mortgage each month. This resulted in many homeowners defaulting on their mortgages and the banks foreclosing on the homes.

If the conditions listed in the original mortgage contracts had been understood by the purchasers of homes, they might not have committed to the terms of the loans. If the original loans had utilized the actual current interest rates, rather than artificially low interest rates, some homebuyers would not have been able to qualify for their home loans. These types of loans, called **adjustable rate mortgages** (ARMs), were being used to qualify home buyers for more expensive homes than they would have been eligible for undertraditional fixed-rate mortgages, where the interest rate charged on a loan remains the same for the entire term of the loan.

A second factor that contributed to the mortgage crisis of 2008 was that mortgages were bundled together into **collateralized debt obligations** (CDOs) and sold to investors throughout the world. When foreclosures became more prevalent, the investment instruments that contained mortgages in default lost their value, thus leaving investment firms and individual investors with worthless investments.

The third factor that also caused banking and investment institutions to either declare bankruptcy or require massive government loans to bail them out was the repeal of the **Glass-Steagall Act** in the 1990s, which required commercial banks to operate separately from investment banks. Once this law was repealed, commercial banks were no longer restricted from bundling loans and selling them to investors or other banks. By the time investment firms or banks purchased the bundled mortgages, which might have been sold previously to any number of other banks or investment firms, there was no way of determining the quality of the mortgages. When mortgages became toxic, and banks started foreclosure proceedings, the effects were felt throughout the world by investors who held bundled mortgages. The mortgage crisis of 2008 is one example of why it is important to fully understand the terms and conditions of contracts before signing them.

Another example that helps to illustrate the importance of understanding the terms and conditions of contracts is when one party is in default of the terms of their contract. If someone defaults on a mortgage contract, they assume that the consequences of the default are that they will forfeit their down payment that they paid to secure the mortgage and that the mortgage holder will take possession of the house or other structure. Unfortunately, many people do not realize that their legal liability could continue even after the mortgage holder takes possession of the property. The mortgage holder will attempt to sell the property through an auction, and if it is sold for less than the value of the original mortgage, the original owner of the property could be legally liable for the difference between the original mortgage value and the amount the property is sold for at auction. If the mortgage holder incurs any legal costs during the resale of the property, the original owner is also liable for those costs plus any interest incurred while the legal transactions are transpiring during the resale of the property.

For other types of contracts there is also legal recourse against the party that defaults on the contract. The party that does not default is entitled to compensation to reimburse them for the defaulting party not fulfilling their legal obligation by adhering to the terms of the contract. The nondefaulting party may also be entitled to additional funds to compensate them for the legal costs associated with proving that the other party did not fulfill the contract.

1.2 HOW CONTRACTS AND LAWS ARE USED IN THE ENGINEERING AND CONSTRUCTION INDUSTRY

Contracts are required in order to set forth the terms and conditions of agreements, protect whoever is a party to a contract from being taken advantage of by the other signatories to the contract, and protect parties in case the other signatories to a contract default on their legal obligations, as set forth in the contract. From a legal perspective, the party who understands the terms and conditions of their contract has a better chance of prevailing in a court of law should issues arise related to the terms of a contract. Contracts have been evolving into complicated legal instruments that only lawyers are able to interpret. This is creating problems for members of the engineering and construction professions because most lawyers understand the legal aspects of contracts, but not the inner workings of the engineering and construction professions. Therefore, engineers and constructors are now required to have an understanding of the basics of contract law in order to be able to determine whether a contract addresses the requirements of an engineering or construction project, and whether a contract favors an owner to the detriment of the engineer or the constructor.

Similar to the professions of engineering and construction, the legal profession has its own terminology that is unique to the profession and might not be understood by people outside the profession. It is the terminology that hinders others from understanding legally binding contracts. This book was written to try to demystify not only the terminology but also the engineering and construction contracting process. The focus of this book is on providing engineering and construction professionals with knowledge that allows them to determine whether a contract is written in such a manner that it does not favor only one party and with the ability to know which contract clauses they would be able to use as a defense if a **breach of contract** (contract default) occurs during the execution of a contract.

Contracts are also used to prevent one party from benefiting at the expense of another party. If a contract does not explicitly state what is to be performed in order to meet the conditions of the contract, then a party might end up performing more work than was originally intended in the contract. The opposite could also occur when one party performs less than what was intended in a contract.

If a contract is fulfilled by all the parties to the contract, then the terms and conditions of the contract are used only to stipulate the requirements for performance of the contract; but if a contract is not successfully completed, then the conditions of the contract are utilized to determine who is responsible for a default. The more detailed a contract is, the more it could increase the cost of implementing the contract, because it could make a contract harder to perform or more difficult for someone to meet the contract requirements.

Laws that Influence the Engineering and Construction Professions

Engineers and constructors should have a basic understanding of the laws that influence their work. In addition to contact law, engineers and constructors have to comply with labor laws, environmental laws, safety regulations, and **negligence (tort)** laws. Building codes also have to be followed along with any local laws or ordinances that affect engineering and construction. Laws are not only written to protect society and consumers also protect engineers and constructors if they understand how to use them to their advantage. This book discusses some of the laws that affect engineers and constructors and provides insight into how laws could be used by engineering and construction professionals to help ensure that they are not liable for violating laws.

1.3 WRITING ENGINEERING AND CONSTRUCTION CONTRACTS

Sometimes, when engineering and construction professionals are college students, they may have a misconceived notion that lawyers write the engineering and construction contracts; the firms they work for will use them for projects. In reality, most engineering and construction firms either write their own contracts or they purchase standard engineering and construction contracts through professional organizations or trade associations such as the **American Institute of Architects**, the **Joint Engineers Construction Document Committee**, or the **Associated General Contractors**. Standard form contracts require modification in order to adapt them to the specific requirements for each individual engineering or construction project.

Since there are similar types of projects within a particular segment of the construction industry, the general conditions contained in the standard form contracts that are purchased from professional or trade organizations could be used on construction projects with only minor modifications. Most construction projects also require detailed information that is specific to each project; this information is customized for each project and included in the supplementary conditions and the specification sections of construction contracts.

In order to be able to effectively write construction contracts, engineering and construction professionals should understand what the potential impact is for each type of contract clause. This book includes a discussion on standard contract clauses and additional information on the potential impacts of construction clauses, to help engineering and construction professionals learn to draft construction contracts that will help protect their firm from legal entanglements or that will provide legal defenses in the event of a lawsuit.

1.4 INTERPRETING ENGINEERING AND CONSTRUCTION CONTRACTS

In order to be able to interpret contracts, engineering and construction professionals should have a basic understanding of contract law, along with being familiar with the laws that impact engineering and construction. Depending upon how engineering and construction contracts are written, they could be interpreted in a variety of ways; therefore, it is essential to be aware of the different potential interpretations of clauses in case legal proceedings are required to clarify contract provisions. Sometimes contracts are intentionally written in vague terms to allow for latitude in the manner in which they are interpreted by engineering and construction professionals who are working on construction projects. If contracts with vague clauses are used on construction projects, it could increase the likelihood that lawyers will be required to help interpret clauses if a lawsuit is filed during a construction project or after its completion. Engineering and construction professionals should learn how to interpret construction clauses because not only it provides them with an ability to not only be able to defend the position of their firm relative to the contract but it also helps them determine which clauses should be included in contracts and which clauses to avoid including in contracts.

The importance of including clear and concise construction clauses is not fully realized until something goes wrong on a construction project. If all of the parties to a contract perform their legal obligations according to their contract, then there will not be any legal ramifications of individual contract clauses. Individual contract clauses become important when one of the parties to a contract does not perform as specified in the contract. It is at this point that the parties to the contract will examine individual clauses to try to locate a clause or clauses that will support a defense for their actions or a clause that will help indicate inaction or illegal action by the other party to the contract. This book provides key insights into the interpretation of engineering and construction contract clauses from the perspective of providing legal defenses or for locating clauses that could be used to prosecute an offending party.

1.5 ADMINISTERING ENGINEERING AND CONSTRUCTION CONTRACTS

After a construction contract is written, the contract is let out for bid, the bid proposals are evaluated, the contract is awarded, and then one or more parties will be responsible for the administration of the contract. Normally an architect, an engineer, or a construction professional will be assigned responsibility for administering construction contracts. **Contract administration** requires that the responsible party be intimately familiar with all aspects of the contract. Some engineering and construction firms will assign responsibility for contract administration to an engineering or construction professional who worked on the bid estimate proposal because he or she would be familiar with all aspects of the contract after having been involved in evaluating and analyzing it while they were working on the bid proposal estimate. In some situations, the engineer who wrote a construction contract will be assigned to administer it during construction as a representative of the owner.

Construction administration requires that an individual, or individuals, be able to determine whether a project is being built according to the contract and specifications. The contract administrator is responsible for evaluating whether work meets the contract and specification requirements, which requires familiarity with contracts and specifications, as well as construction operations, methods, and materials.

1.6 AVOIDING BREACH OF CONTRACT

One of the goals of any firm that is a party to an engineering or construction contract is to avoid being in breach of contract. Breach of contract occurs when one party to a contract does not fulfill their legal obligations, as stated in the contract. Breach could either be partial or full, depending on whether only a small portion of the contract is not completed or the entire obligation is not performed according to the contract documents. There are also other levels of breach; they are discussed in Chapter 14.

In order to avoid being in breach of contract, members of firms should ensure that all aspects of a contract are implemented exactly as they are described in a contact. This requires that contracts be evaluated and analyzed in order to determine how every item listed in them has to be constructed and what materials and quality of workmanship are necessary to avoid breaching the contract.

1.7 TORT LIABLITY OF ENGINEERS AND CONSTRUCTORS

In addition to understanding engineering and construction contracts, engineers and constructors should also be aware of the legal liability they assume while working on designing and constructing construction projects. The area of liability with the greatest impact on engineering and construction professionals is the area of **tort liability**, which is referred to as negligence. Tort liability is the area of law that deals with intentional or unintentional acts that result in an individual, or individuals, being injured or killed or property being damaged or destroyed, as a result of an action or inaction of an individual or individuals.

One example of tort liability is when an engineer is aware that there is an unsafe condition at a job site and he or she does not take remedial action to eliminate the dangerous situation and someone is injured or killed as a result of the unsafe condition. Chapter 9 provides detailed information on the legal ramifications of tort liability and how it affects engineering and construction professionals who design and construct structures.

Chapter Summary

This chapter provided an introduction to some of the concepts related to engineering and construction law and contracts that affect engineering and construction professionals if they are involved in either designing or constructing projects. The legal aspects mentioned in this chapter that are important to engineers and constructors are how laws are used in the engineering and construction industry, writing and interpreting contracts, administering contracts, avoiding breach of contract, and preventing tort liability. In-depth coverage of each of these topics is provided throughout this book along with additional information that is useful to engineering and construction professionals interested in learning more about engineering and construction law and contracts.

Key Terms

adjustable rate mortgages
American Institute of Architects
Associated General Contractors
breach of contract
collateralized debt obligations
contract administration
contracts
Glass-Steagall Act
Joint Engineers Construction Document Committee
mortgage
negligence
rule of law
tort
tort liability
United States Constitution

Discussion Questions

1.1 Why is it is important for engineers to have a basic understanding of contract law?

1.2 Why is it important for constructors to have a basic understanding of contract law?

1.3 Why would it be difficult for an engineer or a constructor who is not familiar with construction law to administer a contract as the owner's representative at a construction job site during construction?

1.4 Why is breach of contract the central focus of all of the parties to a contract?

1.5 Why do engineers and construction professionals need to understand what tort liability is, and how they might be affected by tort liability, while they are working at construction job sites?

1.6 Why should engineers or constructors know about the different clauses in contracts?

1.7 Why is it easier for someone who has worked on a bid estimate proposal to administer a construction contract rather than someone who has just been assigned to work on a construction project?

1.8 Do contracts have to be written by lawyers or could they be written by engineering or construction professionals? Explain why or why not.

1.9 Why is it important for all types of individuals to understand the basics of contract law?

1.10 Explain why tort liability is of concern to engineers and constructors.

CHAPTER **TWO**

THE UNITED STATES AND OTHER LEGAL SYSTEMS

2.1 INTRODUCTION

This chapter discusses the U.S. legal system and introduces civil law, common law, Shari'a law, and Asian law legal systems. A discussion is provided on how the laws that are used in the United States are formed, the purpose of having a constitution, the difference between federal and state systems of government, the three main branches of the U.S. government, administrative agencies, how administrative agencies administer laws, and the responsibility of engineering and construction professionals for implementing laws when they work for government agencies.

This chapter also addresses how federal and state governments and their judicial systems affect construction contracting and the legal aspects of designing and building construction projects. Engineering and construction professionals interact with personnel from government administrative agencies while they are designing or building construction projects. Examples of the administrative agencies that they interact with follow:

- **Federal Highway Administration (FHA)**
- **Government Services Administration (GSA)**
- **Occupational Health and Safety Administration (OSHA)**
- **State Departments of Transportation**
- **Permit Departments**
- **Surveying Departments**
- **Inspection Agencies**
- **Plan Approval Departments**

Having a basic understanding of how federal and state governments function helps engineers and constructors to operate more effectively within the confines of the U.S. federal and state legal systems.

2.2 CIVIL LAW, COMMON LAW, SHARI'A LAW, AND ASIAN LEGAL SYSTEMS

Knowing the origins of legal systems when trying to settle disputes with firms from foreign countries provides an advantage for engineering and construction personnel when they are involved in legal disputes, but even within one nation there is a high degree of variability in the outcomes of legal cases (Stokes 1980; Knutson 2005).

The most frequently used legal systems are **civil law** and **common law**. Civil law is derived from the ancient **Roman legal system** (**Romano-Germanic**), and common law legal systems evolved during the **Norman Conquest** to unify England (Katz 1986). The **Germanic Code**, based on Romano-Germanic law, was adopted in 1896; this code relies on some procedures from the civil law legal system. There are also governments in some Asian countries that do not follow either common law or civil law legal systems. In some Islamic countries, the governments enforce **Shari'a law**, which is based on the holy book of the Islamic faith, the **Koran (Quran)**. The following sections discuss these four legal systems.

Civil Law Jurisdictions

Civil law legal systems are the most prevalent legal systems in the world. Civil law jurisdictions include Europe, South America, Scotland, Quebec, Louisiana, and the former French colonies including South Korea (Calvi and Coleman 1997). Elements of civil law legal systems are also present in some of the former Eastern Bloc countries and other communist countries. Many aspects of civil law are still based on the original Roman precepts of legislation, administrative edicts, and judicial reasoning. In AD 528, the emperor of Constantinople Justinian published all the laws of the empire as the **Code of Justinian** or **Corpus Juris Civilis**. Many civil law legal systems are still based on this code, and its precepts are also used in some of the common law legal systems. When the code was revived in the Middle Ages, it became known as Romano-Germanic law and as "a rule of conduct intimately linked to ideas of justice and morality" (Calvi and Coleman 1997, 23).

In civil law jurisdictions, judges are more involved in the litigation process than they are in common law proceedings because they are responsible for promoting the discovery of all the facts related to cases. Civil law does not rely on the use of previous cases (**precedent law**) to plead existing cases; rather, when cases are heard in civil law courts, they are evaluated based on existing laws (Stokes 1980). Civil law relies on decisions that are based on legal codes rather than precedent cases; it does not have rigid rules of evidence or stringent procedures such as the ones used in common law proceedings.

It is easier to prove that a liquidated damages clause is valid in civil law legal systems than in common law legal systems because legal cases are evaluated based solely on the merits of the liquidated damage clause (Knutson 2005). In civil law legal systems, awards for **liquidated damages** (damages that contractors pay on a per diem basis if a project is not completed on time and an owner is able to prove a financial loss due to a delay) are normally based on actual damages (Knutson 2005, xxi).

Common Law Jurisdictions

Examples of common law jurisdictions include Great Britain and the former colonies of Great Britain such as the United States, India, and Malaysia (Knutson 2005). The United States is predominately a common law country, but the State of Louisiana utilizes elements of civil law because it was originally settled as a colony of France (Katz 1986). Common law is based on the legal system that developed in the United Kingdom when the kings consolidated their judicial power, formed uniform tribunals, and no longer allowed local courts. Common law is also referred to as **judge-made laws** because the judges formalized court rules and applied them throughout the kingdom. Judicial decisions were recorded by the court system and used as precedents for future legal cases. These changes to the legal system made the law common to the entire nation (Calvi and Coleman 1997).

Litigation in common law legal systems is conducted by opposing attorneys who plead their case before a judge or a jury (Stokes 1980). Common law jurisdictions typically emphasize the intent of contracts, and precedent laws (previous cases) are referenced to support cases, whereas civil law legal systems focus on literal interpretations of contract clauses (Knutson 2005; Stokes 1980). Liquidated damages have to be actual losses incurred by an owner, or they will not be awarded in common law legal systems (Knutson 2005).

Some of the terminology used in the **British legal system** is different than the terminology used in the **U.S. legal system**, even though they are both common law legal systems. In the United States, **attorneys** investigate legal cases, and when they are representing clients during court proceedings, they are referred to as **lawyers** for the plaintiff and lawyers for the defense. In the British legal system, attorneys who gather facts for a case are called **solicitors** and the attorney who delivers the argument in court is called a **barrister**. In the United States, the burden of proof is on the person who files a lawsuit; in the United Kingdom the burden of proof is on defendants.

In the United States, either six or twelve jurors and one or two alternates are selected from hundreds of potential jurors after the jurors are questioned by lawyers. In Britain, the names of potential jurors are written on pieces of paper, put into a box, and then the names of twelve jurors and one or two alternates are selected from the box.

One legal anomaly in the United Kingdom (U.K.) is that an **enemy alien** is not allowed to sue in an English court; however, a plaintiff may sue an enemy alien. An enemy alien is defined as someone, or an organization, from a country that is at war with the United Kingdom. A citizen of a foreign nation who is legally living in the United Kingdom is not considered to be an enemy alien (Hill 1998).

The legal system in Canada is unique because Quebec has a civil law system and a common law system is used in all of the remaining nine provinces. French is the official language in Quebec, but English is the official language in the other provinces.

Within the U.S. common law legal system there are several different types of law including:

- **Criminal**—crimes against the state relating to activities that are forbidden by the government
- **Constitutional**—laws set by the Constitution
- **Statutory**—laws based on the collective majority of the will of the people
- **Administrative**—Congress regulates activities that affect the lives of citizens on a daily basis
- **Case**—utilizing precedent law cases
- **Common**—common law rules that are set in cases and followed in future cases (Calvi and Coleman 1997, 8–9)

Asian Legal Systems

Many Asian countries are examples of jurisdictions that are neither common law nor civil law legal systems. The legal system in Japan is derived from ancient **Japanese law**, but it was influenced by French civil law. Japanese citizens are extremely averse to litigation, as is demonstrated by the fact that the per capita litigation rate in Japan is one-twentieth the per capita litigation rate of the United States (Katz 1986). In Japan, there are eight engineers for every lawyer, and in the United States there are eight lawyers for every engineer.

The process of making decisions by consensus used in Japanese firms is called **nemawashi (root-binding)**. Before a decision is made, everyone is consulted and everyone in the group has to accept the decision before it is implemented, which could delay decisions for years.

Lawsuits are not common in Japan due to the legal system taking two to twenty years to settle cases. The main purpose of lawsuits is to force the opposing party to take moral responsibility for their action. Contracts are regarded with suspicion by Japanese, because they feel that relationships should change with the circumstances. Even though elaborate contracts are not used, the Japanese are honorable in their business dealings and they use the term **seig**, which means **right principles** in reference to legal matters.

The **Ministry of International Trade and Industry** regulates the economy, and government central planning includes the consensus process where private enterprises also provide input. The government of Japan is a **bicameral parliament** that includes a **Diet** of two houses—an upper house that performs ceremonial functions and a lower house that holds the real power. The prime minister is selected by the parliament rather than by popular vote. Japan was ruled by the **Liberal Democratic Party** (LDP) from 1955 until 2009.

In the **People's Republic of China, conciliation** is used more frequently than litigation because "the basic goal of Chinese social philosophy is to attain harmony and mediation (conciliation) is compatible with this Confucian ideal" (Redfern and Hunter 1986; Katz 1986, 246). In the People's Republic of China, the law codes of the **Ch'in** have been in use since 220 BC, and they were not modified until the twentieth century. The current legal system in the People's Republic of China resembles socialist systems as described in their constitution—"socialist country under the people's democratic dictatorship and led by the proletariat on the basis of a worker-peasant alliance. Socialism is the political system of the country where all power belongs to the people" (Xichuan and Lingyuan 1990, 214).

Laws are created by the **National People's Congress** (NPC), which also enforces laws, interprets laws, and determines penalties for violations of laws. "In China, the government is divided into central, provincial, and local levels. At times these different levels of government may contradict each other or apply the law inconsistently" (Ling and Low 2007, 240). When criminal laws are violated, citizens are required to participate in ideological reform through reeducation programs and forced labor (Xichuan and Lingyuan 1990).

In the People's Republic of China, construction is the third largest business sector behind manufacturing and agriculture. Construction firms are either **Foreign Invested Construction Enterprises** (FICE), which include **Sino-Foreign** (Asian-foreign) equity construction joint ventures and Sino-Foreign cooperative construction firms, or **Wholly Foreign Invested Construction Enterprises** (WFICE). For a firm to be classified as an FICE, its overseas personnel have to reside in the People's Republic of China for three months each year. Members of FICEs that want to upgrade their business classification have to meet government requirements based on a minimum number of years performing work in China, previous work experience, registered capital, technical capabilities, and number of personnel (Ling and Low 2007). According to the article, *Awarding Construction Contracts on Multicriteria Basis in China*, by Shen, Drew, and Shen (2004, 386):

> Contractors operating in China must be registered under the Business License and Qualification Scheme (Ministry of Construction, 2002). Registered contractors can only conduct their businesses within approved types of works based on their registration grades. The criteria for assessing the level of registration grade include the amount of paid-up capital, number of staff, technological capability, and previous track record. Unfortunately, only projects in China are counted as a legitimate track record. This is a problem for foreign AEC firms that have not undertaken projects in China.

Having a higher grade classification allows construction firms to work on a larger variety of construction projects. Government regulations for construction are located in the **Regulation on Administration of Foreign-Invested Construction Enterprises.** Engineers and constructors use the **International Federation of Consulting Engineers** contracts or contract forms that are issued by the **Ministry of Construction and National Commerce and Administration Bureau**. Quantities are measured using the **Royal Institute of Chartered Surveyors** measurement methods. Construction is subject to the **Regulation on Examining Occupational Safety and Health Management System** issued by the **National Economic and Trade Commission** (Commerce Clearing House, Asia Pte Limited 2006).

Construction is regulated by the **Construction Law of the People's Republic of China**, which includes eighty-five articles in eight chapters that cover (Commerce Clearing House, Asia Pte Limited 2006, 79):

1. General provisions
2. Construction licensing
3. Contract awarding and contracting of construction projects
4. Supervision of construction projects
5. Management of construction safety and operation
6. Management of construction project quality
7. Legal responsibilities and appendices

Companies are required to work through a state agency to do business in the country and to have local partners. The required amount of local participation on projects depends on the government classification of a foreign company. There are classifications of special grade and grades I, II, and III, which include primary contractors that are allowed to subcontract some of the work and use specialty contractors. Labor contractors are classified as grades I or II. Classifications are based on management expertise, number of engineers, the amount of fixed assets and circulating capital, mechanical equipment, and total output per year. As of 2004, there were over 100,000 registered construction companies, and less than 3,000 were classified as grade I, which are the firms that are able to engage in any type of construction (Shen, Drew, and Shen 2004).

There are over a thousand **joint venture** construction companies in China. Foreign companies are able to enter the construction market through the following:

- Joint ventures or cooperative engineering and construction companies
- Real estate development companies
- Providing only construction supervision
- A construction consulting firm
- Bidding on projects as a joint general contractor
- Supervising projects funded by foreign companies or overseas loans
- Setting up representative offices in the country (Shen, Drew, and Shen 2004)

The **Foreign Ministry** assists foreigners, and there is a **Ministry of Construction**, which is a department of the **State Council**. There is a construction committee or construction department in every province. The **Information Department** provides official information about the country. There are four levels of government that develop regulations—the republic, provinces, regions, and local agencies.

In the **Republic of Korea** (South Korea), the country was closed to foreign business until 1994. The November Declaration of the **Fourth World Trade Organization Ministerial Conference** opened up South Korea to more foreign investment. Foreign firms that obtained construction licenses in the late 1990s returned their licenses and started focusing exclusively on construction management rather than construction because they could not secure enough work. The other Asian countries have legal systems that are similar to the legal systems of their historical colonizers such as Great Britain, France, or Spain.

Shari'a Law Jurisdictions

Some Islamic countries in the world are governed by Shari'a law. In Shari'a law jurists try to deduce rulings from the legislation issued from God. "A Shari'a ruling is an instance of legislation issuing from God, the Exalted, with the purpose of organizing the life of human beings. The revealed addressed (*al-khitabat*) in the Koran (Quran) and the Sunnah are not considered to be religious rulings in themselves, although they disclose and shed light on religious rulings" (Al-Sadr 2003, 54). Religious rulings in the Shari'a are divided into two types:

> 1. First, rulings which are related to the actions of human beings and directly address their conduct in different aspects of personal, devotional, family and social life, all of which are dealt with and organized by the Shari'a. These include such rulings as the unlawfulness of consuming alcoholic beverages, the obligatoriness of prayer, the obligation to give financial support to certain close relatives, the permissibility of reviving neglected land and the obligation of rulers to act with justice.
> 2. Second, rulings that do not address human actions and conduct directly, but legislate for certain situations that have an indirect effect on human conduct, such as the rulings that regulate matrimonial relations. . . .Rulings of this type are called situational rulings (*al-ahkam al-wad'iyyah*). There is a strong connection between situational rulings and rulings of general obligation (*al-ahkam al-taklifiyyah*), since no situational ruling exists without one of general obligation existing alongside it. (Al-Sadr 2003, 55)

There are five types of rulings of general obligations:

1. Obligations (*al-wujub*): A ruling that impels a person toward the matter in question at the level of absolute requirement.
2. Recommendations (*al-istihbab*): A ruling the impels a person toward the matter in question at a level lower than obligations.
3. Prohibition (*al-hurmah*): A ruling that restrains a person from the matter in question at the level of absolute obligation.
4. Reprehensibility (*al-kirahah*): A ruling that restrains a person from the matter in question at a level not reaching that of absolute obligation.
5. Permissibility (*al-ibuhah*): A ruling that arises whenever the legislator grants those held accountable by law the opportunity to choose the stand he wished to adopt with respect to a particular matter. The result is freedom to perform or omit a particular act. (Al-Sadr 2003, 55)

"Modern nations in the Middle East often seek to synthesize Islamic laws and western legal systems. The religious law prevails in regard to family, inheritance, and criminal matters, while the western influence is frequently stronger in the area of commercial matters" (Calvi and Coleman 1997, 26). Islamic law has adopted portions of the **Napoleonic Code**, and some of the Northern African nations use elements of the Napoleonic Code at the national level, and tribal and local laws are followed at the local level.

The three areas that are different in some of the Islamic countries and that directly affect contract law are the areas of bonding, project records, and interest on loans. Performance bonds are rarely used because owners do not want to have to sue a bonding company if it is in a foreign nation; therefore, owners withhold a percentage of a contract until it is satisfactorily completed and signed off. Contracts formed under Islamic law with engineering and construction firms all require project records to be in Arabic, and the records have to be physically located in the host country. The charging of interest on loans is illegal under Shari'a law, but investments could provide products or services in return for loans.

Corporate taxes in some Middle Eastern countries are between 30 and 50 percent. If a company is in a joint venture with a Saudi Arabian firm, their corporate taxes could be exempt for up to five years and they are eligible for loans from the **Saudi Industrial Development Loan Fund** (Stokes 1980). Government projects in Saudi Arabia are required to follow the **Government Tender Law of Saudi Arabia**. The government of Saudi Arabia does not allow labor unions or strikes (Al-Jarallah 1983; Stokes 1980).

Other Religious Laws

In addition to Shari'a law, there are other religious laws that mostly govern the moral behavior of human beings and that directly address their conduct in different aspects of personal, devotional, family, and social life, including **Buddhism**, **Hinduism**, and the **Jewish faith**, but the laws prescribed by these religions are not used to govern countries.

2.3 THE UNITED STATES LEGAL SYSTEM

In the U.S. engineering and construction industry, projects are not feasible unless the parties involved in their development and implementation know that they have legal recourse against the opposing party when work is not performed properly, when they are not paid to perform work, or when they have their property forcibly removed from them. All three of these situations are part of the foundation of the U.S. Constitution and legal system.

In addition to the U.S. government creating, monitoring, and enforcing laws that protect its citizens, it also established a system to ensure that there are checks and balances on each branch of the government. In the United States, power is shared between the branches of the federal government and between the federal and state governments.

By allowing state governments to share power, a system was created that permits each state government to pursue the ideals of its citizens and that also accounts for differences in political, social, or economic viewpoints (Bockroth 2000). State governments are also able to enact their own laws related to capital punishment, social services, contracts, professional registration, and other areas not governed by the federal government. State governments have their own senators and congressional representatives, in addition to their federal senators and representatives, and this makes it easier for citizens to have access to the people who are enacting legislation that affects the citizens of each state.

Contracts are administered and interpreted by state judicial systems. State laws determine how contracts are enforced, the remedies prescribed for breach of contract, the responsibilities required for preventing tort liability, and the laws that govern the ownership of property.

When construction projects are being built, they are subject to the laws of the state where they are located regardless of where the owner of the project resides, or the location of the design or construction firm. The reason for this legal requirement is that it is the citizens of the state where a project is built who are affected by the operation of a structure or its presence within their state.

2.4 THE PURPOSE OF CONSTITUTIONS

When any type of a group or organization is being formed, one of the first items of business is to write a constitution. Constitutions are written to set forth a set of rules, guidelines, or principles to help regulate how the organization will be operated and to explain the operations of the organization to new or potential members. The United States created its Constitution in 1776, and there were only seven articles in the original Constitution. Since 1776, there have been twenty-seven amendments to the Constitution. The **Bill of Rights** is contained in the first ten amendments.

The original drafters of the U.S. Constitution did not want it to prescribe all of the details on how the government would be operated but rather to set forth the responsibilities and duties of the government officials in charge of administering the government and regulating the power of the federal and state governments. The U.S. Constitution is limited in the scope of its coverage, and it requires members of the federal Supreme Court to interpret whether new laws contradict the U.S. Constitution and also the legality of new laws, as they pertain to the Constitution (United States Government 1776). A copy of the U.S. Constitution is included in Appendix A.

Since construction contracts are administered by state governments, they are not affected by the federal Constitution unless new articles pertain to workers such as the federal **Civil Rights Act of 1964**, which guarantees equal employment opportunities and outlaws discrimination on the basis of race, religion, national origin, or sexual orientation.

All of the fifty states have their own individual state constitutions containing articles that pertain to the administration of their state government and that explain how the power is dispersed between the legislative, executive, and judicial branches.

2.5 THE FEDERAL AND STATE SYSTEMS OF GOVERNMENT

The three main branches of the federal government are the (1) **executive branch**, (2) **legislative branch**, and (3) **judicial branch**. **Administrative agencies** operate as an unofficial fourth branch of the government.

The Executive Branch

The executive branch of the government includes the president at the federal level, governors at the state level, and mayors at the municipal level. Members of the executive branch are only able to enact legislation by issuing executive orders since enacting laws is the responsibility of the legislative branch. The executive branch has the power to veto legislation, but the legislature may override a veto with a two-thirds majority vote of the legislature.

The Legislative Branch

The legislative branch at the federal level includes both the **Senate** and the **House of Representatives**. The Senate includes two senators from each state, and the House of Representatives has 435 members (2009 data). Each state has a different number of representatives in proportion to the state population. Examples are North Dakota with one representative at large for a population of 639,715 and New York with twenty-nine representatives for a population of 19,297,729 (U.S. Census 2009). At the state level, each state has a different number of senators and representatives.

At both the federal and state levels, the legislature proposes laws, forms committees to review and modify proposed legislation, and passes legislation that is then implemented by members of administrative agencies. Laws are voted on first in the House of Representatives and then they are reviewed and voted on in the Senate. After laws are approved, they are sent to the president for his or her signature. The president may veto a proposed law; it requires a two-thirds majority of the legislature to override a presidential veto. A similar process takes place at the state level with governors having final approval of all proposed laws.

The Judicial Branch

The judicial branch is responsible for judicial review of court cases and for determining whether laws violate the tenants of the Constitution. The judicial branch also interprets legislation and administrative regulations and provides remedies for violations of laws and regulations. At the federal level, the judicial branch includes the federal court system and the U.S. Supreme Court. Each of the fifty states also has a supreme court and lower level courts.

The judicial branch has to follow the U.S. Constitution, which guarantees *due process of law*, which requires that when someone is sued, he or she has a legal right to be told what the lawsuit is about, who has filed the lawsuit, and when and where a trial will be held to decide the outcome of lawsuits.

Administrative Agencies

Administrative agencies are established because it is important for industries to be regulated by individuals who have some expertise related to the industry. Administrative agencies are also established because the people involved in creating laws do not have the time, nor the expertise, to be directly involved in the implementation and regulation of laws.

When new legislation is passed, it may include a provision for establishing an administrative agency to implement and regulate the new law. The president of the United States is responsible for appointing the heads of the major administrative agencies; therefore, they are political appointments. Members of Congress hold committee hearings in which members of administrative agencies testify on the operations of the agencies, and this provides legislative oversight of administrative agencies. Administrative agencies enforce regulations by imposing disciplinary actions for violations of their regulations. Examples of administrative agencies include:

- **Federal and state licensing commissions (boards)**
- **The Federal Bureau of Investigation (FBI)**
- **The Central Intelligence Agency (CIA)**
- **The National Security Administration (NSA)**
- **The Federal Emergency Management Administration (FEMA)**
- **The Housing and Urban Development Agency (HUD)**
- **The Federal Highway Administration (FHA)**
- **State Departments of Transportation (DOTs)**

State licensing commissions create the examinations used in licensing processes; they also determine how the licensing and registration process will be implemented for each profession. Engineers and contractors deal with licensing boards in order to become registered professional engineers and licensed contractors. Each state sets licensing requirements including the types of tests that are required and the experience requirements such as two or four years experience before someone is eligible to take the professional engineers licensing examination.

2.6 EXCLUSIVE JURISDICTION OF THE FEDERAL COURTS

The federal court system was established to review legal cases that relate to situations in which the entire United States would be affected by a decision. The areas where the federal courts have exclusive jurisdiction are:

1. **Admiralty** including crimes committed on the seas
2. **Bankruptcy** where creditors from multiple states are affected
3. **Currency** issues
4. **Patents** and **copyrights** that affect all of the citizens of the United States
5. Actions against the United States brought by citizens or governments of other countries
6. Violations of federal criminal laws

Federal courthouses in each state review cases that are related to matters of the federal Constitution or federal laws. Federal courts also review cases that involve citizens from different states under **diversity of citizenship** laws if the cases involve disputes that are over $75,000.

2.7 EXCLUSIVE JURISDICTION OF THE STATE COURTS

State court systems were established to rule on matters related to issues that occur within each state. Contract law resides in the jurisdiction of state courts, as do violations of state criminal statutes. States also preside over matters related to state constitutions, criminal law, real estate, insurance, education, corporations, tort liability, and negligence. Another area under state jurisdiction is **family matters** such as divorce, child custody, and adoption proceedings. States also have jurisdiction in cases that involve probate matters and the settling of estates after someone passes away. Any other areas not covered by the federal court system are relegated to state jurisdiction.

State legislation governs engineering and construction including the administration of contracts and incorporation laws. If construction projects are publically financed projects, then they require public bonds that are voted on by citizens of a state at either the state or local level. Individual states also administer contractor licensing examinations and the professional registration of engineers.

2.8 THE RESPONSIBILITY OF ENGINEERS AND CONSTRUCTORS FOR IMPLEMENTING LAWS

When engineers and construction management students graduate from college, they will either be employed by private firms or work for one of the many government agencies that implement and regulate laws. As explained in Section 2.4.4, administrative agencies are responsible for developing and implementing the regulations that help the government enforce laws. When engineers and constructors are employed by administrative agencies, they assist in interpreting laws and developing and enforcing government regulations.

In both the public and private sectors, engineers are responsible for the safety and welfare of citizens. The manner in which this is achieved is through the safe design of structures, complete and accurate contract documents, and the use of proper construction materials and processes. Chapter 6 provides detailed information on the ethical responsibilities of engineers as set forth in the engineering codes of ethics.

Chapter Summary

The purpose of this chapter was to provide background information on the elements of the United States and other legal systems. The first part of the chapter covered the different types of legal systems including common law, civil law, Asian law, and Shari'a law. Information was included on how laws are formed in the United States and how government entities are involved in the development and implementation of laws. The U.S. Constitution was discussed in terms of its role in the operation of federal and state governments.

The three main branches of the federal government—the (1) executive, (2) legislative, and (3) judicial—were covered in this chapter along with how administrative agencies help implement and regulate laws. This chapter briefly discussed how engineers and contractors become involved in implementing laws.

Key Terms

administrative agencies
administrative law
admiralty
attorneys
bankruptcy
barristers
bicameral parliament
Bill of Rights
British legal system
Buddhism
case law
Central Intelligence Agency (CIA)
Ch'in
civil law
Civil Rights Act of 1964
Code of Justinian
common law
conciliation
constitutional law
Construction Law of the People's Republic of China
copyright
Corpus Juris Civilis
criminal law
currency
Department of Transportation
Diet
diversity of citizenship
enemy alien
executive branch
family matters
federal and state licensing commissions/boards
Federal Bureau of Investigation (FBI)
Federal Emergency Management Administration (FEMA)
Federal Highway Administration (FHA)
Foreign Invested Construction Enterprise
Foreign Ministry
Fourth World Trade Organization Ministerial Conference
Germanic Code
Government Services Administration (GSA)
Government Tender Law of Saudi Arabia
Hinduism
House of Representatives
Housing and Urban Development (HUD)
Information Department
Inspection Agencies
International Federation of Consulting Engineers
Japanese law
Jewish faith
joint venture
judge-made laws
judicial branch
Koran
lawyers
legislative branch
Liberal Democratic Party
liquidated damages
litigation
Ministry of Construction
Ministry of Construction and National Committee and Administration Bureau
Ministry of International Trade and Industry
Napoleonic Code

National Economic and Trade Commission
National People's Congress
National Security Administration (NSA)
nemawashi (root-binding)
Norman Conquest
Occupational Safety and Health Administration (OSHA)
patents
People's Republic of China
Permit Departments
Plan Approval Departments
precedent law
probate
Quran
Regulation on Administration of Foreign-Invested Construction Enterprises
Regulation on Examining Occupational Safety and Health Management System
Republic of Korea
Roman legal system
Romano-Germanic
Royal Institute of Chartered Surveyors
Saudi Industrial Development Loan Fund
seig (right principles)
Senate
Shari'a law
Sino-Foreign
solicitors
State Council
State Departments of Transportation (DOTs)
statutory law
Surveying Departments
U.S. legal system
Wholly Foreign Invested Construction Enterprises

Discussion Questions

2.1 What is the purpose of constitutions, and why would the members of a firm adopt a constitution?

2.2 How are engineers and constructors responsible for implementing laws?

2.3 What areas of the legal system are the exclusive jurisdiction of states?

2.4 What areas of the legal system are the exclusive jurisdiction of the federal government?

2.5 Which branch of the government is responsible for interpreting laws? How are they able to do it?

2.6 Who is the head of the executive branch at the federal, state, and municipal levels?

2.7 How are the individuals who are in charge of administrative agencies selected and appointed to their positions?

2.8 How are laws created in the United States at the federal level?

2.9 Why was the U.S. system of law established so that power is shared between the federal and state governments?

2.10 Who has jurisdiction over construction contracts and why?

References

Al-Jarallah, M. 1983. Construction industry in Saudi Arabia. *American Society of Civil Engineers Journal of Construction Engineering and Management,* 109 (4): 355–368.

Al-Sadr, M. B. 2003. *Principles of Islamic Jurisprudence.* North Haledon, NJ: Islamic Publications International.

Bockroth, J. T. 2000. *Contracts and the Legal Environment for Engineers and Architects.* New York: McGraw-Hill Publishers.

Calvi, J., and S. Coleman. 1997. *American Law and Legal Systems.* Upper Saddle River, NJ: Prentice Hall Publishers.

Commerce Clearing House, Asia Pte Limited. 2006. *PRC Construction Law—A Guide for Foreign Companies*. Beijing, China: Author.

Hill, J. 1998. *International Commercial Disputes.* London: LLP Limited.

Katz, A. N. 1986. *Legal Traditions and Systems: An International Handbook.* Westport, CT: Greenwood Press.

Knutson, R. 2005. *FIDIC: An Analysis of International Construction Contracts.* The Hague: Kluwer Law International.

Ling, F., and S. Low. 2007. Legal risks faced by foreign architectural, engineering, and construction firms in china. *American Society of Civil Engineers, Journal of Professional Issues in Engineering Education and Practice,* 133 (3): 238–245.

Redfern, A., and M. Hunter. 1986. *Law and Practice of International Commercial Arbitration.* London: Sweet and Maxwell.

Shen, L., Q. Li, D. Drew, and Q. Shen. 2004. Awarding construction contracts on multicriteria basis in China. *American Society of Civil Engineers, Journal of Construction Engineering and Management,* 130 (3): 385–393.

Stokes, M. 1980. *International Construction Contracts.* New York: McGraw-Hill.

U.S. Census. 2009. www.quickfacts.census.gov/qfd/state (accessed April 20, 2009).

U.S. Government. United States Constitution. Washington, DC: United States Government Printing Office.

Xichuan, D., and Z. Lingyuan. 1990. *China's Legal System.* Beijing: New World Press.

CHAPTER THREE

THE UNITED STATES FEDERAL AND STATE COURT SYSTEMS

3.1 INTRODUCTION

This chapter explains how the U.S. federal and state court systems administer and process legal cases. The first part of the chapter discusses the federal and state court systems and the different court venues that review each type of legal case. The second half of the chapter covers how lawsuits are initiated, pretrial activities including discovery, statute of limitations, legal presentation, prejudgment remedies, preliminary injunctions, jury trials, lawsuit judgments, and the appeals process.

3.2 THE FEDERAL COURT SYSTEM

This section provides information on the federal court system, how it was established, how it functions, and its different levels of legal venues.

United States Constitution Articles and Amendments

This section discusses the articles and amendments in the U.S. Constitution that directly impact the U.S. legal system. Many of the other articles and amendments not mentioned in this section pertain to the establishment and management of the government rather than legal processes that affect the federal court system.

Article III of the **U.S. Constitution** empowers the **U.S. Supreme Court**, and lower federal courts, with judicial power for the United States. The Fifth and Sixth Amendments require that every citizen be afforded **due process of law** if they are accused of a crime. Due process of law means that anyone accused of a crime has the following rights:

- The right to be informed of the nature and cause of an accusation
- The right to a speedy and public trial
- The right to remain silent before his or her accusers
- The right to a trial by an impartial jury in the state where the alleged crime was committed
- The right to be confronted with the witnesses against him
- The right to have a compulsory process for obtaining witnesses in his favor
- The right to have counsel to assist with his defense (U.S. Constitution art. V and VI)

The Eighth Amendment prohibits excessive bail or fines and protects individuals from cruel and unusual punishment resulting from a criminal conviction. The Tenth Amendment states that all the powers that are not delegated to the United States are delegated to the individual states. The Eleventh Amendment to the U.S. Constitution prohibits the federal judicial system from hearing cases that involve individuals from one state against individuals from another state or foreign entity; therefore, these types of cases are addressed at the state level unless they exceed $75,000.

Federal Level Courts

The federal level trial courts are involved in settling cases that involve dispute resolution and the federal appellate courts create rulings that lead to legal precedents that are used in future court cases. The U.S. Supreme Court mainly addresses cases that require interpretation of their constitutionality or the constitutionality of federal legislation. Figure 3.1 provides a diagram of the federal court system.

The scope of the jurisdiction of the federal court system is limited by the power invested in it by the U.S. Constitution that limits it to the following:

1. All cases in law and equity arising under the Constitution, laws, and treaties of the United States
2. All cases affecting ambassadors, other public ministers, and consuls
3. Controversies between two or more states

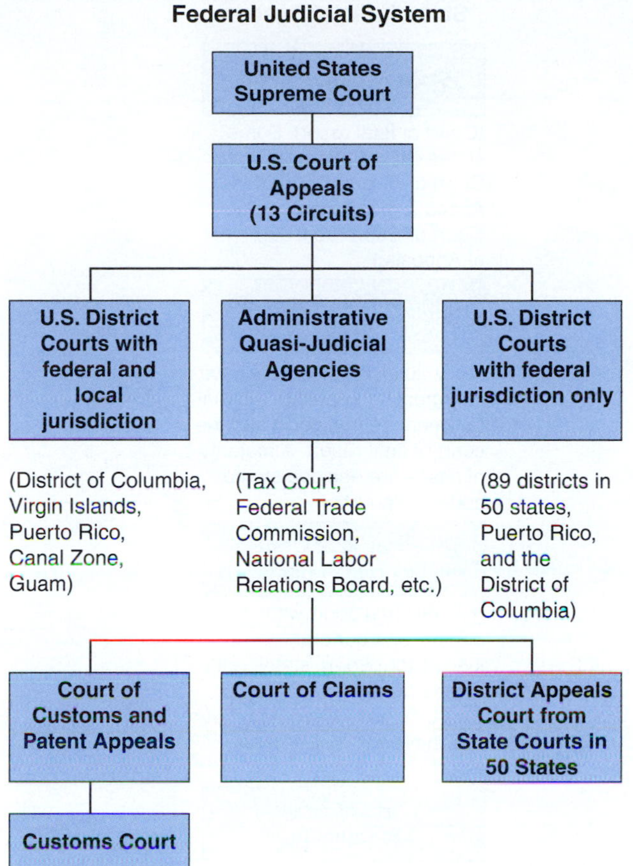

FIGURE 3.1 The Federal Judicial System.

Source: *Congressional Quarterly*, 3rd ed., 1993, Washington D.C., p. 27.

4. Controversies to which the United States is a party
5. Controversies between a state and citizens of another state
6. Controversies between citizens of different states
7. Controversies between citizens of the same state claiming land grants of different states
8. Controversies between a state, or its citizens, and foreign states, citizens, or subjects (Ball 1987, 2)

The requirements for having a case heard in the federal legal system include the following:

1. Having personally suffered a real injury
2. Where a right or obligation allegedly deprived or denied has been protected by statute or constitution
3. For which the federal courts can provide a remedy
4. That the plaintiff has followed the appropriate channels in his or her pursuit of justice
5. Where the issue has not resolved itself prior to judicial examination (Ball 1987, 133)

The first level of federal courts includes the **Court of Claims**, **Court of Customs and Patent Appeals**, and the **U.S. Military Appeals Court**. The Court of Claims hears cases such as public contract disputes, back pay, refunds of taxes, contract disputes, and tax liability. The U.S. Court of Customs and Patent Appeals addresses cases involving customs duties, the value of imported goods, and the exclusion of merchandise from the United States. The U.S. Military Appeals Court only hears cases involving military court marshals (Ball 1987).

The next level of federal courts is the **U.S. District Courts** and they hear cases such as antitrust suits, social security, truth in lending, civil rights, bankruptcy, passports, naturalization, racial integration, sexual discrimination in employment, and occupational safety and health matters. U.S. District Courts also hear **diversity of citizenship** cases that involve lawsuits between citizens of different states if the amount being disputed exceeds $75,000 (Ball 1987).

The federal district courts have jurisdiction over both civil and criminal cases. The federal criminal courts hear cases such as organized crime, the armed forces, interstate crimes, embezzlement, immigration law, burglary, homicide, robbery, draft violations, narcotics, and immigration violations (Ball 1987).

The **U.S. Court of Appeals** has thirteen federal circuit courts and they hear appeals from federal district courts. In addition, they also review cases involving federal administrative boards and commissions and the decisions of boards such as the National Labor Relations Board, the Immigration and Naturalization Service, the Internal Revenue Service, and the Interstate Commerce Commission (Ball 1987).

The highest federal court level is the U.S. Supreme Court, which reviews lower level court cases that involve questions about the constitutionality of legislation and cases that involve violations of the U.S. Constitution. In order for the U.S. Supreme Court to review a case from a state court system it has to involve a federal statute or include a situation in which a state constitution or statute is in conflict with the U.S. Constitution. There are three types of reviews that the U.S. Supreme Court considers:

1. **Writ of Certiorari**—This is an order from a higher court to a lower court to send them the record of a case so that the higher court may review it. At least four of the nine judges have to agree that a case merits the attention of the court before it will issue the writ.
2. **Appeal as a Right**—If a federal court of appeal rules that a state statute conflicts with the U.S. Constitution or treaties, it could be appealed to the U.S. Supreme Court.
3. **Certification**—This occurs when a lower federal court asks the U.S. Supreme Court about a question of law. (Calvi and Coleman 1997, 85; Ball 1987, 120–121)

U.S. Supreme Court judges are selected by the president of the United States and confirmed by the legislature. Their appointment is for life unless they choose to retire. The only means for removing a U.S. Supreme Court judge is through impeachment proceedings, which require a majority vote in the House of Representatives and a trial in the Senate and a two-thirds vote of the Senate.

Other judges are appointed for a definite term such as the U.S. Claims Court judges, who serve fifteen-year terms, bankruptcy judges serve fourteen-year terms, and the U.S. Court of Military Appeals judges serve for fifteen years.

Class Action Lawsuits

A **class action lawsuit** is a lawsuit where a plaintiff is permitted to file a lawsuit on the behalf of other individuals who all have been harmed in a similar manner. The lawsuit has to be specific rather than abstract. The lawsuit should involve as many members of the group who have been harmed that could be identified through reasonable procedures such as sending out letters soliciting individuals who have been harmed in a similar manner by the same organization. One example of class action lawsuits is lawsuits against automobile manufacturers when a defect in a particular automobile, or a class of automobiles, causes similar injuries to drivers or passengers of the vehicles. Another example is when there are toxic chemicals in a local water system and a direct causal connection could be demonstrated between the chemicals in the water and harm to local residents.

3.3 STATE COURT SYSTEMS

Individual states hear cases that are not included in the jurisdiction of the federal courts including cases related to criminal law, civil codes, state constitutions, marriage, corporations, real estate, insurance, and education. At the state criminal court level, there are **misdemeanor** and **felony** crimes. A felony is a crime that would receive a penalty that would exceed one year in jail and be served in a state penitentiary such as burglary and murder. Only a small proportion of criminal cases, between 8 and 10 percent, actually go to trial. Most criminal cases are settled through a process called **plea bargaining** where the accused pleads guilty to the charges, or to lesser charges, in exchange for reduced jail time. Figure 3.2 shows a diagram of state court systems.

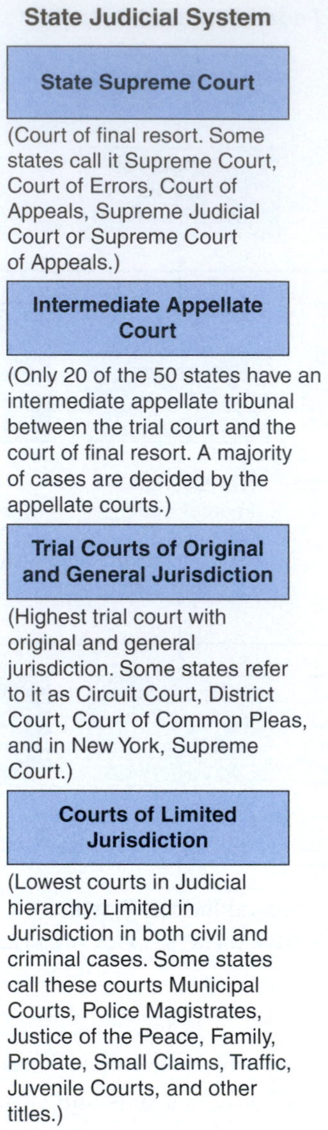

State Judicial System

State Supreme Court

(Court of final resort. Some states call it Supreme Court, Court of Errors, Court of Appeals, Supreme Judicial Court or Supreme Court of Appeals.)

Intermediate Appellate Court

(Only 20 of the 50 states have an intermediate appellate tribunal between the trial court and the court of final resort. A majority of cases are decided by the appellate courts.)

Trial Courts of Original and General Jurisdiction

(Highest trial court with original and general jurisdiction. Some states refer to it as Circuit Court, District Court, Court of Common Pleas, and in New York, Supreme Court.)

Courts of Limited Jurisdiction

(Lowest courts in Judicial hierarchy. Limited in Jurisdiction in both civil and criminal cases. Some states call these courts Municipal Courts, Police Magistrates, Justice of the Peace, Family, Probate, Small Claims, Traffic, Juvenile Courts, and other titles.)

FIGURE 3.2 State Judicial System.
Source: *Congressional Quarterly*, 3rd ed., 1993, Washington D.C., p. 27.

In state court systems, there are multiple layers of courts. Some states have **small claims courts** that hear cases that involve disputes under a certain value such as $500 or $5,000, depending on the size of the jurisdiction. The cases heard in small claims court usually do not involve lawyers and they are settled quickly, sometimes in a manner of minutes.

Another state court level is **trial courts**. Trial courts are referred to by different names throughout the United States such as **superior courts**, **district courts**, or **circuit courts** (Calvi and Coleman 1997). If a metropolitan area is large enough, it may have separate courts to hear cases for criminal offenses; personal injury (tort); commercial disputes; domestic grievances including divorce, child custody, and adoption; and probate, which addresses the transfer of property when someone passes away.

Trial courts also hear cases related to actions that result from administrative agencies regulating particular industries such as hazardous waste mitigation, environmental degradation, and employment injuries. Criminal courts differ from trial courts because in criminal cases the state presses charges against the plaintiff rather than private citizens filing lawsuits against other individuals, as is done in the other types of trial courts. Criminal courts hear cases that are about violations of the laws that were formed by the government to protect members of society.

In state court systems, cases may be heard in an appellate court if there is justification and some of the larger states have intermediate appellate courts. In a few exceptional cases, a lawsuit may be heard by the state supreme court after it moves through the appellate court system. The state supreme courts hear cases that pertain to the determination of the constitutionality of the subject of cases.

Judges in the state appellate and supreme court systems are appointed by governors for an indefinite term and state lower court judges are elected by the general population. Some states have nominating committees composed of lawyers, private citizens, and incumbent judges, who screen candidates and then submit three names to the governor. The governor then appoints one of the recommended judges for a trial period of one year before their name is submitted to the voters for their approval. The term of a judge ends when they choose to retire, they pass away while serving on the bench, or there is a term limit.

Judges at the state level could be removed for senility, being incompetent, committing a crime, or abusing their power. There are four methods for removing state judges. Forty-five states allow their state legislatures to impeach judges in a manner similar to the federal judge **impeachment** process. The second method is **legislative address**, which is used in cases of incompetency, senility, and partiality. The third method is a **recall** where a petition has to be signed by a certain percentage of the voters, such as 5 or 10 percent, and then an election is held to determine whether the judge will be removed from office. The fourth method is a **judicial conduct board/commission**, which has the power to censure, suspend, or recommend impeachment proceedings against a judge (Ball 1987).

3.4 LAWSUITS

This section reviews the process of how lawsuits are initiated and tried in the U.S. common law legal system.

Initiating Lawsuits

A lawsuit commences when a party files the legal paperwork required, which is called a **claim**. The claim indicates the purpose of the lawsuit, the parties involved in the lawsuit, and other information pertaining to the lawsuit. The court then hires a **process server** who has to locate the person who is being sued and personally hand them, or send them a registered letter with a return receipt required, a summons stating that they are required by law to appear in court on a certain date. Summons also include information on the reasons lawsuits are being filed against the defendants.

The person who initiates a lawsuit is referred to as a **plaintiff**; the person they are suing is referred to as a **defendant**. When a defendant receives a summons, it is called a **service of process**. The defendant is required to reply to it within a specified length of time, which is normally thirty days. When the defendant responds to the service of process, he or she could indicate that the plaintiff did not correctly represent the facts of the case or state that there is a defense. The defendant could file a **demurrer**, which is used to demonstrate that the plaintiff does not really have a legal claim.

Defendants have the legal right to file a **counterclaim** against the plaintiff, or against someone else who was involved in whatever action resulted in the lawsuit being originally filed by the plaintiff; this is called a **cross-complaint**. The parties named in the cross-complaint are called **cross-defendants**. Plaintiffs have to provide a response to a counterclaim within the stated period of time. Sometimes this process helps prevent any further progress of a lawsuit when it uncovers information that demonstrates that there really is no legitimate legal basis for the disagreement between the plaintiff and the defendant.

One example of a lawsuit with a cross-complaint is when a subcontractor sues a contractor who has caused someone to be injured and the contractor files a cross-complaint against the architect/engineer (A/E) or the owner, claiming that the A/E or the owner was negligent and his or her negligence caused the injury. The contractor could also sue the subcontractor based on the **legal theory of indemnification**, which indicates that the subcontractor holds harmless the contractor based on an indemnity clause in the contract between the subcontractor and the contractor (Sweet and Schneier 2005).

If someone were filing a lawsuit in a state court system, the plaintiff would be required to sue a defendant in the state where the defendant resides since the power of the court system does not extend beyond state lines. Sometimes exceptions are made to this legal requirement when a plaintiff is able to demonstrate some type of hardship that prevents him or her from suing in a different state from where he or she currently reside.

Pretrial Activities—Discovery

Before a trial goes to court, the attorneys for each of the respective parties participate in a process called **discovery**, which includes several different pretrial activities. Potential witnesses, the plaintiff, and the defendant might be provided with written **interrogations**, which are questions they are required to provide answers for in writing, or they may be **deposed**, which requires that they answer questions orally rather than in writing while they are under oath. Being under oath is when a person swears to *tell the truth, the whole truth, and nothing but the truth so help me God, Allah, or other appropriate person of worship* while their hand is on a bible, Qur'an (Koran), or other holy book. During discovery, the attorneys also share all information that pertains to the case collected to date including depositions, interrogatories, and evidence.

Discovery is used so that each party to a lawsuit is aware of what the other side has for evidence and what type of **testimony** (courtroom statements) the opposing party will provide during a trial. It is hoped that discovery will lead to settlements if one of the parties realizes that the opposing party has prepared a more effective case.

Another process used by the courts is **pretrial conferences** that occur in the chamber of a judge. Judges try to have the parties focus on one or two specific issues, or emphasize information that has arisen during the discovery process, to encourage settlements.

Statute of Limitations

In order to prevent **open-ended liability**, which is being responsible for an action for an indefinite period of time, each state has passed **statute of limitation** laws that limit the amount of time a plaintiff has to file a lawsuit. A typical time frame for statute of limitations is seven years, but it varies for different offenses. In the engineering and construction realm, there are disagreements about when the statute of limitations commences because of the nature of the difficulty of determining when defects are first discovered and reported and in determining the exact cause of defects (Bockrath 2000).

Legal Representation

Although not required by law, many individuals hire attorneys to represent them in legal proceedings when they become involved in lawsuits. The cost of using an attorney during litigation varies for every lawsuit. Attorneys usually charge by the hour, but they may require a retainer of several thousands of dollars to take a new case. In the beginning of a case, hourly fees are charged against the retainer until it is depleted and then attorneys bill their clients using an hourly rate with a bill sent once a month.

The fee an attorney charges depends on a number of factors including the following:

- How much a client is able to afford to pay
- The difficulty of a case
- The anticipated amount of time that will be required to settle a case
- Where a case is being **litigated** (tried in a court of law)
- The desired outcome of the litigation process

The fees charged for attorneys range from $30 to $100 per hour for **legal aid** attorneys, attorneys providing services to citizens who are not able to afford regular attorneys fees due to their economic circumstances, to thousands of dollars per hour for high-profile lawyers. Where an attorney graduates from law school also influences what they charge for legal fees, with Ivy League and prestigious private university graduates demanding higher fees than public university graduates.

In legal cases involving personal injuries or death, and the potential for a lucrative settlement, attorneys might agree to take cases on a **contingent fee** basis. For contingent fee cases, the plaintiff is still required to pay any expenses the attorney incurs, but the attorney does not receive any profit on a case unless they are able to recover damages. The profit an attorney will realize on a case depends on whether a case is settled before it goes to trial or a trial is required to obtain a settlement. If a settlement is negotiated by an attorney before trial, the attorney usually receives 25 to 35 percent of the settlement. If a trial is necessary to receive a settlement, then an attorney receives 35 to 50 percent or possibly more of the settlement.

Prejudgment Remedies, Preliminary Injunctions, and Equitable Decrees

This section discusses how **prejudgment remedies**, **preliminary injunctions**, and **equitable decrees** are used in legal cases.

Prejudgment Remedies

When lawsuits commence, there is always a risk that the assets desired as part of a settlement could be sold or disposed of before a settlement is reached in the case. In order to prevent this from happening, the legal system allows a plaintiff to request a prejudgment remedy, which requires the courts to either seize the assets of the defendant or restrict access to the assets (Sweet and Schneier 2005).

Preliminary Injunctions and Equitable Decrees

In situations where the actions of a defendant are alleged to be harming a plaintiff, the plaintiff could request a preliminary injunction or an equitable decree requiring that a defendant cease and desist, which means they must no longer continue doing what the plaintiff claims is responsible for harming the plaintiff (Sweet and Schneier 2005).

Jury Trials

The U.S. Constitution guarantees the right to a trial by a jury in criminal cases. In most jurisdictions, there are twelve members on a jury, but in locations with small populations juries may have only six members. Not all court cases are heard by juries. Plaintiffs have the option of waiving their right to a jury trial; if they do, their case is then decided by a judge.

In most states, criminal cases require a unanimous decision or a 5/6 decision of juries with only six members. Personal injury and criminal trials normally are tried with juries, but juries are not commonly used in commercial disputes including contract disputes.

Lawsuit Judgments

After a case has been heard by a court, a decision will be made by a judge or a jury about whether the plaintiff or possibly the defendant is entitled to a settlement, which is called a **judgment**, in the form of monetary compensation or another form of remedy. Judgments are not supposed to require someone to relinquish what they need for their basic existence such as clothing, whatever is required to perform their job, and other essential items.

If a state has a **Homestead Act**, and a homeowner has applied and received acceptance of protection for their home under the Homestead Act, then their home, or a portion of the proceeds of the sale of their home, could be protected during a lawsuit settlement, demands of creditors, and protecting surviving spouses from losing their home when one spouse passes away. Examples of how Homestead Laws protect property are (1) in Texas, Florida, Iowa, Kansas, and Oklahoma—a home is completely protected and it cannot be seized in a lawsuit, although there may be limitations on the number of acres protected in each of these states; (2) in California, $50,000 of equity per immediate family member in the household is protected; and (3) in North Dakota, $85,000 of equity is exempt no matter how much a home is worth. State Homestead Acts could have exemptions that are set by each state including such items as mortgage foreclosures, property taxes, and mechanic's liens and for these the property is not protected by Homestead Acts.

If a case involves nonpayment of property taxes and a defendant refuses to voluntarily relinquish the judgment amount, then the judgment is sent to the sheriff. The sheriff would then seize the property and sell it at auction with the proceeds from the sale of the property awarded to the plaintiff. Once the property is seized, the defendant is provided with one last opportunity to pay the judgment and to have their assets released before they are auctioned off to the highest bidder.

The Appeals Process

The party that loses a court case is allowed to submit a brief to the appellate court attempting to convince the court that the trial judge in the lower court allowed an error during the trial. The party requesting an appeal is referred to as an **appellant** and the party that prevailed during the lower court case is called a **respondent** or an **appellee**.

After the appellant court hears a case, the appellate judge is required to write an *opinion* providing an explanation of why the case was decided in the manner that it was settled by the court. Once an opinion is written, the case becomes a **precedent** case, which could be referred to during other legal cases. Normally, precedent cases could be cited in legal cases in other states, but the state of California allows only precedent cases from the state of California to be used in its court cases.

When trying cases attorneys review previous cases to locate similar precedent cases and then they introduce these cases during trials. If a precedent case is introduced with facts that are not similar to the case being tried, then a judge will not allow the case to be introduced during a trial. The opposing attorney may also object to the introduction of a precedent case if it is irrelevant to the current legal case.

Chapter Summary

This chapter covered the United States federal and state court systems, and the different levels of courts. Two figures were provided that illustrated the progression of the federal and state judicial systems. The second half of the chapter covered how lawsuits are initiated and explained, the activities included in pretrial activities, statute of limitations, legal representation, prejudgment remedies, preliminary injunctions, jury trials, lawsuit judgments, and the appeals process. The Homestead Act was discussed in the context of how a Homestead exemption protects a home during lawsuits.

Key Terms

Appeal as a Right
appellant
appellee
Certification
circuit courts
claim
class action lawsuits
contingent fee
counterclaim
Court of Claims
Court of Customs and Patent Appeals
cross-complaint
cross-defendants
defendant
demurrer
deposed
discovery
district courts
diversity of citizenship
due process of law
equitable decrees
felony
Homestead Act
impeachment
interrogations
judgment
judicial conduct board/commission
legal aid
legal theory of indemnification
legislative address
litigated
misdemeanor
open-ended liability
plaintiff
plea bargaining
precedent
prejudgment remedy
preliminary injunction
pretrial conferences
process server
recall
respondent
service of process
small claims court
statute of limitation
superior courts
testimony
trial courts
U.S. Constitution
U.S. Court of Appeals
U.S. District Courts
U.S. Military Appeals Court
U.S. Supreme Court
U.S. Supreme Court judges
Writ of Certiorari

Discussion Questions

3.1 What are prejudgment remedies and how are they used in the legal system?

3.2 Which level of the court system reviews criminal cases? How are criminal cases different from other types of legal proceedings?

3.3 Explain why discovery is used during trials.

3.4 Explain how the federal court system is different from state court systems.

3.5 How are lawsuits initiated in the court system?

3.6 Explain counterclaims and how they are used during the initiation of lawsuits.

3.7 Explain statute of limitations and how it affects lawsuits.

3.8 How are preliminary injunctions used in the legal system?

3.9 How do Homestead laws protect the home of a defendant during a lawsuit?

3.10 Explain the difference between a trial court and an appellate court.

3.11 What are legal aid attorneys?

3.12 Explain how contingent fee cases are billed for lawsuits.

3.13 What is the legal theory of indemnification?

3.14 Explain how pretrial conferences are used by judges prior to the court system hearing a case.

3.15 What types of cases are heard by the U.S. Court of Appeals?

3.16 Explain the process for selecting cases to be reviewed by the U.S. Supreme Court.

3.17 How are federal judges removed from office?

3.18 What types of cases are heard by U.S. district courts?

3.19 Explain diversity of citizenship cases.

3.20 What types of cases are under the jurisdiction of the federal criminal court system?

References

Ball, H. 1987. *Courts and Politics: The Federal Judicial System.* Englewood Cliffs, NJ: Prentice Hall.

Bockrath, J. T. 2000. *Contracts and the Legal Environment for Engineers and Architects.* New York: McGraw Hill Publishers.

Calvi, J., and S. Coleman. 1989. *American Law and Legal Systems.* Upper Saddle River, NJ: Prentice Hall.

Congressional Quarterly, 3rd ed. 1993. United States Congress. Washington, DC: Government Printing Office, p. 27.

Sweet, J., and M. Schneier. 2005. *Legal Aspects of Architecture, engineering, and the Construction Process.* Toronto: West Publishing Company.

U.S. Government. United States Constitution. 1776. Washington, DC: Government Printing Office (accessed March 10, 2009).

CHAPTER **FOUR**

UNITED STATES LABOR LAWS THAT AFFECT ENGINEERS AND CONSTRUCTORS

4.1 INTRODUCTION

Managers of engineering and construction firms are normally concerned with four aspects of labor: (1) labor laws, (2) labor organizations, (3) labor costs, and (4) productivity. This chapter focuses on the first two aspects of labor—labor laws and labor organizations.

In order to function effectively in their industry, engineers and construction professionals should have an understanding of which labor laws affect the industry and how they influence it. This chapter provides a summary of the labor laws with the greatest impact on the engineering and construction industry and explains why engineers and constructors need to know about each of the laws discussed in this chapter.

This chapter also covers the formation of labor organizations and how they affect the engineering and construction industry. The following labor organization topics are discussed:

- National labor unions
- Union locals
- Business agents
- Shop stewards
- Union hiring halls
- Jurisdictional disputes
- Secondary boycotts
- Open and closed shop
- Double breasting

4.2 FEDERAL LABOR LAWS

At the federal level, thousands of laws are related to labor, but not all of them directly or indirectly impact engineering and construction operations; therefore, this section will review only the major labor laws that affect the engineering and construction industry. In order to also provide a historical perspective, labor laws are presented in chronological order with a brief synopsis of each law. More extensive coverage of each law discussed in this chapter is available in the ***Federal Register***, which is the official federal record of all the laws passed in the United States (*Federal Register* 2009; *Congressional Digest* 1993).

The Sherman Antitrust Act of 1890

One of the earliest laws that had a major impact on construction operations was section 1 of the **Sherman Antitrust Act** passed in 1890. The original intent of this law was to prevent the formation of large corporate trusts or **cartels** during an era when there were large railroad companies and steel conglomerates. This law is also viewed as an antimonopoly law often used as the grounds for lawsuits by firms that are trying to keep one firm from buying another similar type of firm and thereby eliminating competition. It is also used if there is only one major firm in a market segment. One example of this law being invoked in a court case was when one of the major software companies was purchasing competing firms, thus eliminating those firms from being competitors.

The Sherman Antitrust Act was also used to keep workers from organizing to negotiate together with management at their places of employment. In 1908, the Supreme Court heard a case concerning workers organizing and upheld the legality of this law.

The Clayton Antitrust Act of 1914

In 1914, as industrial, mining, and construction workers were becoming more dissatisfied with their working conditions, they championed for a new labor law, which resulted in section 17 of the **Clayton Antitrust Act** authorizing employees of one firm to form organizations. Being able to legally organize provided workers with additional strength in their negotiations with their employers. Prior to the passing of this act, workers risked losing their job, or bodily harm, when they tried to organize to demand better wages or improved working conditions. The situations in mines and industrial plants were such that employees worked seven days a week ten hours a day. It was risky for workers to complain about their working conditions for fear of reprisal or dismissal from their jobs. Many workers were foreign immigrants who did not speak English; therefore, they were not able to communicate their dissatisfaction with the management for fear of being deported back to their native countries.

The Davis-Bacon Act of 1931

In 1931, in the middle of the Great Depression, the **Davis-Bacon Act** was passed; it requires that prevailing wages and fringe benefits be paid to contractors and subcontractors on all federal and federally funded construction projects. In order to enforce the provisions of the Davis-Bacon Act, the rates used on construction projects are published in contract documents. In union states, the latest union wage rates might be used as local prevailing wage rates on government projects. Another safeguard enforced resulting from this act is that contractors are required to submit copies of their payroll documents to the federal agency that is funding the project. Unfortunately, in some instances, contractors keep two sets of account books and submit one to the funding agency while tracking what is really being paid in the other set of accounts.

The Davis-Bacon Act eliminated the possibility that the government would pay more than the prevailing wage, thus attracting the best workers from other projects to work on government projects. The requirements of this act also keeps the government from depressing wages in rural areas.

The Norris-LaGuardia Act of 1932

The **Norris-LaGuardia Act** of 1932 made it legal for labor unions to organize and strike. This act also outlawed **yellow dog contracts**, which were contracts that required workers to not join unions or to stop being members of unions. This law prohibits federal courts from issuing restraining orders to prevent activities such as joining a union, holding union meetings, striking, discussing labor disputes, and providing legal assistance to individuals involved in labor disputes (*Federal Register* 2009; *Congressional Digest* 1993).

The Wagner Act (The National Labor Relations Act) of 1935

The **Wagner Act (The National Labor Relations Act)** of 1935 allowed for the creation of **closed shop** or **union shop**, which requires workers to belong to unions in order to work on construction projects. This law also has provisions that legally protect union-organizing activities and that protect employees from being discriminated against by their employer when they are involved in union activities. The Wagner Act also created the **National Labor Relations Board (NLRB)**, which hears cases that involve grievances by workers against management. The NLRB later became involved in helping to settle worker strikes against management.

The Fair Labor Standards Act of 1938

The **Fair Labor Standards Act** of 1938 established minimum wages; this law is still in effect today with the legislature voting periodic increases in the minimum wage. The last increase was approved in 2006 and increases in the minimum wage were phased in over several years. The minimum wage effective as of July 24, 2009, was $7.25. Individual states or municipalities may also set minimum wages, but employees are entitled to the higher of the two minimum wages. This law also set the maximum number of hours workers could work at eight hours per day and requires that overtime to be paid one and a half times base wage rates to workers who work more than forty hours per week, excluding management personnel.

The Smith-Connelly Act (The War Disputes Act) of 1943

The **Smith-Connelly Act (The War Disputes Act)** of 1943 was passed to prevent strikes during wars since laborers could stop work on critical wartime operations by striking when they know they are in a good bargaining position due to the necessity of the products that they are creating to the war effort.

During wars, the federal government is allowed to suspend the requirement for awarding construction bids through the competitive bidding process in order to expedite the delivery of services or operations; this was done during the war with Iraq starting in 2003. During peacetime, federal and state governments are required to use competitive bids for all construction projects. In the twenty-first century, some states now allow projects to be built using the design/build approach to help expedite the completion of projects.

The Copeland (Anti-kickback) Act of 1948

Section 874 of the **Copeland (Anti-kickback) Act** addresses the penalties for coercing someone into providing kickbacks for procuring work. Section 874 states the following:

> Whoever, by force, intimidation, or threat of procuring dismissal from employment, or by any other manner whatsoever induces any person employed in the construction, prosecution, completion or repair of any public building, public work, or building or work financed in whole or in part by loans or grants from the United States, to give up any part of the compensation to which he is entitled under his contract of employment, shall be fined under this title or imprisoned for not more than five years, or both. (Title 18, U.S.C. 874, June 25, 1948, Ch. 645, Section 1, 62 Stat. 740 and 862)

In addition, Title 40, U.S.C. 3145 Public Law 107-217, August 21, 2002, amended the original act to include a statement that contractors and subcontractors performing government work have to provide a statement of the wages paid to their employees each week.

The Hobbs Act (Antiracketeering) Act of 1951

The **Hobbs Act**, also known as the **Antiracketeering Act** of 1951, was originally passed to protect employers from threats of violence, the use of force, or general threats by union officials that were being used to **extort** (secure) money from employers. Currently, this law is being used to prosecute firms that bribe owners or government officials to secure work.

The Taft-Hartley Act (The Labor Management Relations) Act of 1947

The **Taft-Hartley Act** of 1947 created federal *mediation* and *conciliation* services to provide an uninvolved third party to help settle grievances between union members and management. The Taft-Hartley Act also includes a section to permit states to determine whether they will enact **right-to-work** laws preventing projects from being completely union shop, which means that every worker has to be a member of the union to work on a construction project. When this law was enacted, union shop was declared to be legal if workers are not prevented from being hired if they are not union members by allowing them thirty days to join the union.

The Taft-Hartley Act gave the president of the United States the right to **enjoin** (force) workers on strike to go back to work for ninety days, which is referred to as a **cooling-off period**, while negotiations are underway to reach new contract agreements if a strike creates a national emergency.

Government workers are not permitted to strike but there have been several incidents where government employees have gone on strike such as postal workers. On August 3, 1981, President Ronald Reagan fired 11,300 of the 17,000 air traffic controllers after they went on strike for shorter workweeks and higher wages. Twelve hundred of the workers returned to work within a week but the other positions were filled by military personnel and newly trained air traffic controllers.

The Taft-Hartley Act added a list of unfair labor practices that unions were no longer allowed to use:

- Restricting or coercing workers that are exercising their right to bargain through representatives of their choosing
- Coercion of an employer in his or her choice of person to represent him or her in discussions with unions
- Refusal of unions to bargain collectively
- Barring a worker from employment because he or she has been denied union membership for any reason except nonpayment of dues
- Striking to force an employer or self-employed person to join a union
- Secondary boycotts
- Various types of strikes or boycotts involving interunion conflict or jurisdictional agreements
- Levying of excessive union initiation fees
- Certain forms of **featherbedding** (payment for work not actually performed) (*Federal Register* 2009, 1)

Other provisions in the Taft-Hartley Act include:

- Authorization of lawsuits against unions for violations of their economic contracts
- Authorization of damage lawsuits for economic losses caused by secondary boycotts and certain strikes
- Relaxation of the Norris-LaGuardia Act to permit injunctions against specified categories of unfair labor practice
- Establishment of a sixty-day no strike and no lockout notice period for any party seeking to cancel an existing collective bargaining agreement
- A requirement that unions desiring status under the law, and recourse to NLRB protection, file specified financial reports and documents with the U.S. Department of Labor
- The abolition of the U.S. Conciliation Service and establishment of the Federal Mediation and Conciliation Service
- A prohibition against corporate or union contributions or expenditures with respect to elections to any federal office
- A reorganization of the NLRB and a limitation on its power
- A prohibition on strikes against the government
- The banning of various types of employer payments to union officials (*Federal Register* 2009, 1)

The Landrum-Grifflin Act (The Labor-Management Reporting Disclosure) Act of 1959

The **Landrum-Grifflin Act** of 1959, also known as the **Labor-Management Reporting and Disclosure Act**, banned organizational and recognition picketing and allowed state labor relations agencies and courts to have jurisdiction over labor disputes not heard by the NLRB.

Title VII of the Civil Rights Act of 1964

Title VII of the Civil Rights Act of 1964 "prohibits discrimination by covered employers on the basis of race, color, religion, sex or national origin. It also prohibits discrimination against an individual because of his or her association with another individual of a particular race, color, religion, sex or national origin. An employer is not allowed to discriminate against a person because of his or her interracial association with another, such as by an interracial marriage" (Title VII of the Civil Rights Act of 1964, codified as Subchapter VI of Chapter 21 of 42 U.S.C. 2000e). The act allows for exceptions for **Bona Fide Occupational Qualifications** where a job requires someone to be a male in order to perform the work but an employer has to be able to prove a direct relationship between sex and the ability to perform the duties of the job, and there is no less-restrictive or reasonable alternative.

Title VII practices are enforced by the **Equal Employment Opportunity Commission (EEOC)** at the federal level and by **Fair Employment Practices Agencies (FEPAs)** at the state level. These organizations "investigate, mediate, and may file lawsuits on behalf of employees" (Title VII of the Civil Rights Act of 1964, codified as Subchapter VI of Chapter 21 of 42 U.S.C. 2000e). Complaints have to be filed within 180 days of learning about the discriminatory act. The act applies only to firms with more than fifteen employees that they hire for more than nineteen weeks a year.

Other forms of discrimination are covered by other acts such as the Age Discrimination Act of 1967, the Pregnancy Discrimination Act of 1978, and the Americans with Disabilities Act of 1990.

The Age Discrimination in Employment Act of 1967

According to the U.S. Equal Employment Opportunity Commission (June 1, 2009a):

> The **Age Discrimination in Employment Act** of 1967 (ADEA) protects individuals who are 40 years of age or older from employment discrimination based on age. The ADEA's protections apply to both employees and job applicants. Under the ADEA, it is unlawful to discriminate against a person because of his/her age with respect to any term, condition, or privilege of employment, including hiring, firing, promotion, layoff, compensation, benefits, job assignments, and training. The ADEA permits employers to favor older workers based on age even when doing so adversely affects a younger worker who is 40 or older.

The Pregnancy Discrimination Act of 1978

The **Pregnancy Discrimination Act** of 1978 requires that employers treat pregnant employees in the same manner as they would a worker who is experiencing health issues or a disability. The act also protects pregnant employees from being discharged from their jobs because of their pregnancy and from not being allowed to return to their jobs after they have been on medical leave while having their babies (U.S. Equal Employment Opportunity Commissions, June 1, 2009b).

The Americans with Disabilities Act of 1990

The **Americans with Disabilities Act (ADA)** of 1990 states that "no covered entity shall discriminate against a qualified individual on the basis of disability in regard to job application procedures, the hiring, advancement, or discharge of employees, employee compensation, job training, and other terms, and privileges of employment" (Title 42, Chapter 126 of the United States Code, 1990, Section 12101 and Title 47, Chapter 5). In addition, this act states that discrimination includes "not making reasonable accommodations to the known physical or mental limitations of an otherwise qualified individual with a disability who is an applicant or employee, unless such covered entity is able to demonstrate that the accommodation would impose an undue hardship on the operation and business of such covered entity" (Title 42, Chapter 126 of the United States Code, Section 12101 and Title 47, Chapter 5).

The ADA Act also requires that both new facilities and alterations to existing facilities have to "be readily accessible to and usable by individuals with disabilities including individuals in wheel chairs" (Title 42, Chapter 126 of the United States Code, 1990, Section 12101 and Title 47, Chapter 5). In order to be in compliance with this requirement, the **Code of Federal Regulations**, published by the U.S. Department of Justice, includes ***ADA Standards for Accessible Design*** (28 CFR Part 36, 1990)—*Nondiscrimination on the Basis of Disability by Public Accommodations and in Commercial Facilities.* This document provides technical specifications in sections 4.2 to 4.35 for public accommodations that are the same as the specifications in *Document A117.1-1980* of the American National Standards Institute (ANSI). The other ADA technical specifications differ from ANSI A117. Engineers and constructors should review the ADA publication before designing or constructing structures to ensure that their designs are in compliance with the ADA Act.

The Civil Rights Act of 1991

The **Civil Rights Act** of 1991 modified the original Civil Rights Act of 1964 to allow for damages in cases of intentional discrimination in order to strengthen the original Civil Rights Act. With the modification of the original act complainants may now recover punitive and compensatory damages if they are able to prove that the defendant (other than a government, government agency, or political subdivision) allowed or "participated in discriminatory practices with malice or with reckless indifference to the federally protected rights of an individual" (The Civil Rights Act of 1991, November 21, 1991, Public Law 102-166).

Exclusions to compensatory damages include back pay, interest on back pay, or any other type of relief authorized under section 706(g) of the Civil Rights Act of 1964. The limits for compensatory damages include: future pecuniary (relating to money) losses, emotional pain, suffering, inconvenience, mental anguish, loss of enjoyment of life, and other nonpecuniary losses. The amount of punitive damages awarded shall not exceed $50,000 for firms with 14 to 99 employees; $100,000 for firms with 100 to 199 employees; $200,000 for firms with 200 to 499 employees; and $300,000 for firms with over 500 employees.

The Family and Medical Leave Act of 1993

Under the Family and Medical Leave Act of 1993 employers have to allow eligible employees twelve "workweeks of unpaid leave during any twelve month period" for one or more of the following reasons:

- For the birth and care of the newborn child of the employee
- For placement with the employee of a son or daughter for adoption or foster care
- To care for an immediate family member (spouse, child, or parent) with a serious health condition
- To take medical leave when the employee is unable to work because of a serious health condition (United States Department of Labor 1993, 1)

The Sarbanes-Oxley Act of 2002

Although not a labor law, the passing of the Sarbanes-Oxley Act of 2002 has affected engineering and construction industry members by requiring firms to employ additional workers in order to comply with the requirements of the Sarbanes-Oxley Act. Violations of this particular act could result in imprisonment; therefore, members of the engineering and construction industry should be aware of its requirements, as follows:

The Sarbanes-Oxley Act:

- Established a five-member Public Company Accounting Oversight Board that oversees audits of the financial statements of public companies through registration, standard setting, inspection, and disciplinary programs
- Requires quarterly and annual certifications by CEOs (chief executive officers) and CFOs (chief financial officers) regarding the annual reports and internal controls of their companies
- Requires board audit committees to include at least one member who is a financial expert
- Prohibits external auditors from performing certain additional services for audit clients such as financial system designs and implementations (Zegarowski 2007, 53)

The Sarbanes-Oxley Act was passed in response to incidents that occurred in 2001 and 2002, when several major firms were not accurately reporting their financial status. When the firms went bankrupt, many employees lost their jobs and their retirement funds and the stockholders lost their investments. When these major firms failed, it demonstrated that self-governance in the area of accounting without oversight could lead to illegal accounting practices and the demise of firms.

The National Defense Authorization Act of 2008

Section 585 of the National Defense Authorization Act of 2008 amends the Family and Medical Leave Act to include the following:

> to permit a spouse, son, daughter, parent or next of kin to take up to twenty-six workweeks of leave to care for a member of the Armed Forces, including a member of the National Guard or Reserves, who is undergoing medical treatment, recuperation, or therapy, is otherwise in outpatient status, or is otherwise on the temporary disability retired list, for a serious injury or illness. Leave may also be taken for any qualifying exigency (urgent need) (as the Secretary of Labor shall by regulation determine) arising out of the fact that the spouse, or a son, daughter, or parent of the employee is on active duty (or has been notified of an impending call or order to active duty) in the Armed Forces in support of a contingency operation. (H.R. 4986 Public Law 110-181 Section 585)

4.3 THE OCCUPATIONAL SAFETY AND HEALTH ACT OF 1970

The Occupational Safety and Health Administration (OSHA) was established by federal law; it has jurisdiction over all industries except for mining and the railroad industry. The Occupational Safety and Health Act of 1970 states that every employer involved in commerce is required to provide a workplace that is free from health and safety hazards. According to the OSHA Public Law 91-596, 84 STAT. 1590, 91st Congress, S. 2193, December 29, 1970, as amended through January 1, 2004:

> An act to assure safe and healthful working conditions for working men and women; by authorizing enforcement of the standards developed under the Act; by assisting and encouraging the States in their efforts to assure safe and healthful working conditions; by providing for research, information, education, and training in the field of occupational safety and health; and for other purposes.

Employers are required to comply with all of the OSHA standards and to keep records of occupational illnesses and injuries. In the construction industry, the OSHA safety inspections require either a voluntary commitment for an inspection or a search warrant. When an inspection is being conducted, the employer and a representative of the employees have the right to accompany the inspector.

When an OSHA violation is discovered, a citation is issued that describes the violation and penalty and that provides a time frame in which the violation needs to be corrected by the employer. All OSHA citations have to be posted close to where the violations have occurred at job sites. Employers have fifteen days in which to contest a violation and have a review commission hold a formal hearing. The OSHA citations are enforced through the U.S. Court of Appeals.

One case involving an OSHA violation resulted in half a million dollar fine:

> The U.S. Labor Department plans to fine two key contractors that worked on the demolition of a bank building at Ground Zero almost half a million dollars for violations discovered after two firefighters died fighting a blaze in the structure last summer. In an announcement made February 19, 2008, the Occupational Health and Safety Administration cited the general contractor, and a defunct demolition subcontractor for 44 alleged violations. The contractor fired the subcontractor within days of the August 18 fire, apparently started by cigarettes that killed two firefighters. "Employees must adhere to safety and health standards, and prepare completely and effectively for workplace emergencies, says the OSHA's area director. Failure to do so can, and in this case, did cost lives." (ENR.com, February 25, 2008)

Another lawsuit involving an engineering firm that was issued an OSHA citation was overturned on appeal as the following case synopsis states:

> A construction firm used improper electrical equipment during tunnel construction resulting in a methane gas explosion and the death of three employees. The engineering services contract between the owner and the engineering firm had the usual clause imposing no responsibility on the engineers for the safety practices of the contractors and permitting only the owner to stop work. The engineer was required to redesign the work, when requested by the owner, which was underway at the time of the explosion. OHSA issued a citation to the engineering firm, finding that is was actually involved in construction activities at the time of the explosion under OSHA construction law. The Court of Appeals vacated the citation, finding that while the engineering services and other contracts do not per se exclude design and other professionals from liability for OSHA construction standards, in this case the engineering firm was not engaged in construction work and thus was not subject to OSHA construction standards, under OSHA construction law. [*CH2M Hill, Inc. v. Alexis Herman*, Secretary of Labor, et. al., 192 F. 3d 711 (7th Cir. 1999)]

Employees are allowed to request inspections by filing a complaint with the U.S. Labor Department. If an employee files a complaint, they cannot be discriminated against or discharged for doing so. The U.S. Supreme Court has ruled that employees have the right to refuse to perform any task they think might result in serious injury or death. If there is imminent danger to the safety or health of employees, then an inspector notifies the U.S. Secretary of Labor. If the danger is not eliminated, then the Secretary of Labor could seek a temporary injunction to close down part of the job site where the danger exists.

The OSHA requires that companies report the following:

- Occupational deaths
- Injuries
- Loss of consciousness
- Restriction of work or motion
- Transfer to another job
- Work-related illnesses
- Medical treatments other than first aid (OSHA 2009, 1)

Incidence rates have to be calculated and reported to the OSHA; incident rates are the number of cases or days lost per year times 200,000 work hours per year divided by the total employee hours per year. The 200,000 work hours are 100 employees times 2,000 hours per year. Recordable cases (Clough, Sears, and Sears 2005, 428) are classified as the following:

1. Total recordable cases—The sum of all recordable occupational injuries and illnesses, including deaths, lost-workday cases, and cases without lost workdays
2. Deaths
3. Total lost-workday cases—The sum of cases involving days away from work and/or days of restricted activity
4. Nonfatal cases without lost workdays—The sum of cases that are recordable injuries or illnesses not resulting in death or lost workdays, either days away from work, or days of restricted activity
5. Total lost workdays—The sum of days away from work and days of restricted work activity

4.4 UNIONS AND UNION LABOR

In the United States, there are many different types of unions. The unions that govern the construction industry are (1) the **American Federation of Labor (AFL)** established in 1886 to organize skilled and craft workers, (2) the **Building Trades Department** formed in 1908 that governs craft unions, and (3) the **Congress of Industrial Organizations (CIO)** created in 1945. Union members that work in the construction industry are governed by the AFL-CIO. These two organizations merged in the twentieth century, split in the later part of the twentieth century, and merged once again in the first decade of the twenty-first century. Unions were created to increase the bargaining power of workers when they negotiate with management for wages and fringe benefits. In the beginning of the twentieth century, during each decade the balance of power between unions and management fluctuated. The laws relating to unions discussed in previous sections in this chapter and passed during each of the early decades reflect whether management or labor was more powerful during particular decades.

Union membership has been declining in the twenty-first century with only 12 percent of workers belonging to unions in 2009. The peak union membership took place during the 1950s when 50 percent of workers were union members.

Agency Shop

If workers choose not to join a union involved in negotiations that affect the salary and benefits of the worker, then the worker would still be subject to **agency shop** dues. Agency shop employees are required to pay dues at a slightly lower rate than union dues and they do not have voting rights when it comes to union contracts. They are required to pay agency dues because they benefit from the negotiating power of unions.

4.5 UNION STRUCTURES

This section explains how unions (closed shop) are organized and how they operate in the construction industry.

National Unions

National unions are distinguished from other unions in that they have **collective bargaining agreements** (union contracts) with employers in more than one state. National unions have elevated levels of bargaining power because of their size and national prominence. National union representatives are involved in negotiating contracts for their members and ensuring that contract provisions are followed by employers. National unions are also the umbrella organizations for all of the union divisions.

Union Locals

The lowest level of the union structure is **union locals**. Most engineering and construction professionals only have contact with officials from local unions unless a dispute is not settled at the local level and union officials from higher levels of the organization structure become involved in trying to settle disputes. Union locals are organized by the different crafts, by job sites, or by plants. The union locals are administered by **business agents**.

Business Agents

Business agents are full-time employees of the union local; they provide advice to members of their local union and to the elected union officials in each union local. Business agents canvas the job sites where there are union workers in their area to ensure that union contracts are being followed by employers; they also help settle grievances that union members have with their employers. Business agents are directly involved in negotiating union agreements and are also responsible for informing employers about any violations of trade agreements.

Shop Stewards

Shop stewards are not union officials but they are union representatives that have close contacts with union members. Shop stewards are elected by members of the local union at their plant or job site; they help ensure that union requirements are met by employers. Shop stewards become directly involved in grievances of the workers against employers and help negotiate grievance settlements.

Union Hiring Halls

In addition to their responsibilities related to union locals, business agents are also responsible for operating local union **hiring halls**. Whenever an employer needs to hire a union worker, they have to contact the local union hiring hall. At the hiring hall, union workers sign up on a list to indicate that they are eligible to work. When an employer needs workers, they are assigned workers in the order they appear on the hiring hall lists. Employers are not able to choose which workers they desire because they are required to accept the workers at the top of the hiring hall list. The skill level of individual workers could be unpredictable since workers have varying degrees of experience. The unions indicate that all of the workers are capable of performing their required job functions since they have completed union **apprenticeship programs**. Apprenticeship programs are used to train union workers; these programs are paid for by employers when they are included in union contracts.

If union workers report to a job site and they are not needed that day, their employer is required to pay them for four hours of work—the minimum number of hours that they have to be paid according to union contracts. For every seven union workers hired from one craft to perform a task, a foreman is required; the foreman is paid according to the wages set forth in union contracts.

4.6 JURISDICTIONAL DISPUTES

Jurisdictional disputes occur when more than one union claims jurisdiction (the right to perform a task) over a particular work item. If a construction project is a union project, then each member of the craft trades is required to perform only work that is specific to their trade. But in some circumstances it may be difficult to determine which trade has jurisdiction over a particular work item, this could result in a jurisdictional dispute where two, or possibly more, trades claim that they should be performing the work (Sweet and Schneier 2005). When a jurisdictional dispute arises and it is not settled expeditiously (quickly), there is the risk that the workers will all stop working until the dispute is settled in favor of one craft performing the work over another craft.

The document that contains the jurisdictions of each craft is published by the **AFL-CIO Building Trades Department** and is called the **Green Book** (AFL-CIO 2009). As soon as a jurisdictional dispute arises at a construction job site, a member of the construction management team will try to determine which craft is entitled to perform the task being disputed by the workers. The Associated General Contractors publish a document, called the **Gray Book**, that is commonly referred to during jurisdictional disputes, because it contains information on the different construction trades and the types of work that they are allowed to perform while working on construction projects (Associated General Contractors 2009).

If management and business agents are not able to determine who has jurisdiction over a particular work item, then the matter could be referred to the **National Joint Board (NJB)**, which hears jurisdictional dispute cases and tries to settle them before workers shut down job sites by going on strike.

Strikes

If jurisdictional disputes are not settled quickly, and to the satisfaction of the two crafts involved in the dispute, it could lead to workers going on **strike**. When a strike occurs, the crafts involved in the dispute stop working. The workers that are on strike may picket the job site by walking back and forth in front of the job site carrying signs that refer to their grievance. The strike will last until the dispute is settled to the satisfaction of the union members.

Strikes may also occur when labor and management are not able to reach an agreement on union contracts. If a union contract expires in the middle of a project, there is always the risk that the workers will strike if a union agreement is not accepted by the union members before their old union contract expires. If contract negotiations stall, an experienced mediator may be brought in to help labor and management reach an agreement before old union contracts expire so that workers will not walk off the job while an agreement is being negotiated for a new union contract.

During strikes management might attempt to hire nonunion workers, referred to by union workers as **scabs**, to perform the work of the union workers who are on strike. In order for the scabs to enter the job site, they might have to cross the picket line, which could lead to acts of violence perpetrated by the union workers on the picket line. For a strike to continue, at least one worker has to be on the picket line during normal business hours.

Secondary Boycotts

Sometimes at construction job sites when there is a strike by one union, the union workers will convince different union workers, or many other union workers, to honor their strike and boycott their work and join them on the picket line. The Taft-Hartley Act outlawed **secondary boycotts** but if management personnel are not aware of this, some of their workers might attempt secondary boycotts. Another term for secondary boycotts is **common situs picketing** (Bockrath 2009).

Separate Gates

If a prime contractor does not want to involve all of his subcontractors in a dispute, they could establish a second entrance to the job site. Having a **separate gate** also provides an alternative location where union members are able to picket the project without disrupting the other workers as they enter and leave the job site.

4.7 OPEN SHOP (NONUNION) ENVIRONMENTS

Open shop (nonunion shop) occurs when there are no union agreements and the workers do not have an organization that performs collective bargaining for them. In the United States, twenty-two states have **right to work** laws that allow employees to decide whether to join a union or not (U.S. Department of Labor 2009):

- Alabama
- Arizona
- Arkansas
- Florida
- Georgia
- Idaho
- Iowa
- Kansas
- Louisiana
- Mississippi
- Nebraska
- Nevada
- North Carolina
- North Dakota
- Oklahoma
- South Carolina
- South Dakota
- Tennessee
- Texas
- Utah
- Virginia
- Wyoming

The rest of the states are either union (closed) shop states or states in which there are both union and nonunion operations. The autonomy of states allows state legislatures to choose to be either a union or a nonunion state.

4.8 DOUBLE BREASTING

If members of a firm decide that it would be advantageous to bid on both union and nonunion projects, they could create a subsidiary of the firm that bids on only nonunion contracts while the main firm bids on union contracts. By operating in this manner, the firm is able to secure work in states that are nonunion without violating nonunion laws. The term for this is **double breasting**.

Chapter Summary

This chapter provided information on the labor laws that affect engineering and construction professionals and provided a historical perspective on their evolution. The laws that were discussed were only those laws directly affecting the operations of engineers and constructors. The laws presented in this chapter were the Sherman Antitrust Act, the Clayton Antitrust Act, the Davis-Bacon Act, the Norris-LaGuardia Act, the Wagner Act, the Fair Labor Standards Act, the Smith-Connelly Act, the Copeland "Anti-kickback" Act, the Hobbs Act, the Taft-Hartley Act, the Landrum-Griffin Act, Title VII of the Civil Rights Act of 1964, the Age Discrimination in Employment Act, the Pregnancy Discrimination Act, the Americans with Disabilities Act, the Civil Rights Act of 1991, the Family and Medical Leave Act, the Sarbanes-Oxley Act, and the National Defense Authorization Act. For additional information on labor laws, see the *Federal Register*. This chapter also provided information on the Occupational Safety and Health Administration and the role it fulfills in the engineering and construction industry.

The second half of the chapter explained unions, the union organizational structure, agency shop, union locals, business agents, shop stewards, jurisdictional disputes, strikes, secondary boycotts, separate gates, and double breasting. All of these topics are related to the day-to-day operations of unions at construction job sites.

Key Terms

ADA Standards for Accessible Design
AFL-CIO Building Trades Department
Age Discrimination in Employment Act
agency shop
American Federation of Labor (AFL)
Americans with Disabilities Act (ADA)
Antiracketeering Act
apprenticeship programs
Bona Fide Occupational Qualifications
Building Trades Department
business agents
cartels
Civil Rights Act
Clayton Antitrust Act
closed shop
Code of Federal Regulations
collective bargaining agreements
common situs picketing
Congress of Industrial Organizations (CIO)
cooling-off period
Copeland (Anti-kickbacks) Act
Davis-Bacon Act
double breasting
enjoin
Equal Employment Opportunity Commission (EEOC)
extort
Fair Employment Practices Agencies, (FEPAs)
Fair Labor Standards Act
Family and Medical Leave Act
featherbedding
Federal Register
Gray Book
Green Book
hiring halls
Hobbs Act
jurisdictional disputes
Labor-Management Reporting and Disclosure Act
Landrum-Griffin Act
National Defense Authorization Act
National Joint Board (NJB)
National Labor Relations Act
National Labor Relations Board (NLRB)
national unions
nonunion shop
Norris-LaGuardia Act
Occupational Safety and Health Act
Occupational Safety and Health Administration (OSHA)
open shop
Pregnancy Discrimination Act
right to work
Sarbanes-Oxley Act
scabs
secondary boycotts
separate gate
Sherman Antitrust Act
shop stewards
Smith-Connelly Act (The War Disputes Act)
strike
Taft-Hartley Act
Title VII of the Civil Rights Act
union locals
union shop
Wagner Act
yellow dog contracts

Discussion Questions

4.1 How is the Sherman Antitrust Act being applied to large corporations in the United States that are trying to expand the size of their firm by purchasing other firms?

4.2 Who is the Sherman Antitrust Act applicable to when trying to prevent workers from organizing at work to negotiate with management?

4.3 Which labor law provides workers with the legal right to form labor organizations?

4.4 What is the requirement related to wages and benefits that resulted from the Davis-Bacon Act?

4.5 Which labor law gives workers the legal right to go on strike without the government interceding on behalf of management?

4.6 In what year did workers obtain the legal right to create closed-shop projects?

4.7 Which labor law protects workers from retaliation from management if they become involved in union-organizing activities?

4.8 What is the purpose of the National Labor Relations Board?

4.9 How does the Labor Standards Act affect the engineering and construction industry?

4.10 How is the Hobbs Act currently used in the construction industry?

4.11 Why should the U.S. government be allowed to forbid strikes during wars?

4.12 Which labor law created the right-to-work states? What is a right-to-work state?

4.13 What rights did the Landrum-Griffith Act provide to state labor relations agencies and courts?

4.14 When a firm advertises a position and indicates that they are "an equal opportunity employer," what does this mean?

4.15 How does the Occupational Safety and Health Administration (OSHA) have the right to shut down construction job sites? If they do, what are the reasons?

4.16 How does the Occupational Safety and Health Administration enforce safety violations at construction job sites?

4.17 Is it legal to prevent a worker from working on a union construction project? Why or why not?

4.18 What is meant by a cooling-off period and how is it used during strikes?

4.19 Which union has jurisdiction over construction job sites?

4.20 Explain the purpose of union locals.

4.21 Who is responsible for ensuring that union contracts are followed at construction job sites? How do they do this?

4.22 If an employer needs union construction workers, what does the employer have to do to locate them for a job?

4.23 What are two types of activities that would cause construction workers to go on strike?

4.24 Explain what scabs are and how they are used on construction projects.

4.25 Why would an employer try to prevent secondary boycotts?

4.26 What advantage is gained by having secondary gates during a strike at a construction job site?

4.27 Explain the difference between open shop and closed shop.

4.28 How is double breasting used in relation to unions and union work?

4.29 What usually happens when a labor agreement expires before a new labor agreement is voted on by union members?

4.30 Which labor law directly affects every construction project? How does it affect them?

References

ADA Standards for Accessible Design (28 CFR Part 36, 1990)—*Nondiscrimination on the Basis of Disability by Public Accommodations and in Commercial Facilities.* Washington, DC: Government Printing Office.

AFL-CIO. 2009. *Green Book.* Washington, DC: American Federation of Labor—Congress of International Organizations.

Associated General Contractors. 2009. *Gray Book.* Washington, DC: Associated General Contractors.

Bockrath, J. T. 2000. *Contracts and the Legal Environment for Engineers and Architects.* New York: McGraw Hill Publishers.

CH2M Hill, Inc. v. Alexis Herman, Secretary of Labor, et al., 192 F. 3d 711 (7th Cir. 1999).

Clough, Jr., J. and R. Nash Jr. 1981. Administration of Government Contracts. Washington, DC: George Washington University.

Congressional Digest (June–July 1993). Washington, DC: Government Printing Office.

ENR.com. February 25, 2008. OSHA Proposes Fine of $486,000 Afrex Bank Fire. New York: McGraw Hill Publishing.

Federal Register. 2009. *United States Federal Labor Laws.* Washington, DC: Government Printing Office.

H.R. 4986 Public Law 110-181 Section 585. 2008. National Defense Authorization Act. Washington, DC: Government Printing Office.

OSHA. 2009. *Occupational Safety and Health Administration.* www.osha.gov.

OSHA Public Law 91-596, 84 STAT. 1590, 91st Congress, S. 2193 1970 as amended through January 1, 2004. Washington, DC: Government Printing Office.

Sweet, J., and M. Schneier. 2005. *Legal Aspects of Architecture, Engineering, and the Construction Process.* Toronto: West Publishing Company.

The Civil Rights Act of 1991. November 21, 1991. Public Law 102-166. Washington, DC: Government Printing Office.

Title VII of the Civil Rights Act of 1964, codified as Subchapter VI of Chapter 21 of 42 U.S.C. 2000e. Washington, DC: Government Printing Office.

Title 18, U.S.C. 874. June 25, 1948. Ch. 645, Section 1, 62 Stat. 740 and 862. Washington, DC: Government Printing Office.

Title 40, U.S.C. 3145, Public Law 107-217. August 21, 2002. Washington, DC: Government Printing Office.

Title 42, Chapter 126 of the United States Code. 1990. Section 12101 and Title 47, Chapter 5. Washington, DC: Government Printing Office.

U.S. Department of Labor. 1993. *Family and Medical Leave Act.* Washington, DC: Government Printing Office.

U.S. Department of Labor. January 1, 2009. *State Right-to-Work Laws and Constitutional Amendment in Effect as of January 1, 2009.* Employment Standards Division, Wage and Hour Division. Washington, DC: Government Printing Office. www.dol.gov/esa/whd/state/righttowork.htm.empl.stds.adm (accessed June 1, 2009).

U.S. Equal Employment Opportunity Commission. June 1, 2009a. *The Age Discrimination in Employment Act of 1967—Facts About Age Discrimination.* Washington, DC: Government Printing Office.

U.S. Equal Employment Opportunity Commission. June 1, 2009b. *Facts About Pregnancy Discrimination.* Washington, DC: Government Printing Office.

Zegarowski, G. 2007. Corporate sustainability after sarbanes-oxley linking social-political initiatives and small and medium-sized enterprise resources. *International Journal of Disclosure and Governance*, 4 (1): 52–58.

ENVIRONMENTAL AND SUSTAINABILITY LAWS THAT AFFECT ENGINEERS AND CONSTRUCTORS

5.1 INTRODUCTION

When engineers create designs and contractors build structures, they should be aware of the major environmental and sustainability laws that could affect their work. In the United States, environmental laws did not significantly impact society until the 1950s when environmental legislation started moving to the forefront of citizen concerns. The 1970s ushered in a period when the most far-reaching federal environmental legislation was passed and implemented by administrative agencies. In the first decade of the twenty-first century there was a resurgence in concern for the environment.

This chapter introduces major federal environmental and sustainability legislation and explains how different environmental and sustainability laws affect the engineering and construction (E&C) industry. This chapter also discusses environmental impact statements and how they are used for construction projects and the Council on Environmental Quality and its responsibilities. The last section of this chapter introduces the concept of sustainable design and construction and how sustainability issues affect the E&C industry.

5.2 THE COUNCIL ON ENVIRONMENTAL QUALITY (CEQ)

In 1969, the **National Environmental Policy Act (NEPA)** was passed in the United States, which created the **Council on Environmental Quality (CEQ)**. This act also established a requirement that all federal projects, all federally funded projects, and every proposed legislative act that affects the environment has to include an **environmental impact statement (EIS)**. The CEQ was formed in order to advise and provide studies to the president of the United States on environmental matters and to produce the environmental quality report that is required each year, as per the National Environmental Policy Act (Ortolono 1997). The yearly CEQ environmental report (Public Law 91-190, 1969) provides:

1. the status and condition of the major natural, man-made, or altered environmental classes of the Nation including, but not limited to, the air; the aquatic including marine, estuarine, and fresh water; and the terrestrial environment, including but not limited to, the forest, dry land, wetland, range, urban, suburban and rural environment;
2. current and foreseeable trends in the quality, management and utilization of such environments and the effects of those trends on the social, economic, and other requirements of the Nation;
3. the adequacy of available natural resources for fulfilling human and economic requirements of the Nation in the light of expected population pressures;
4. a review of the programs and activities (including regulatory activities) of the Federal Government, the State and local governments, and nongovernmental entities or individuals with particular reference to their effect on the environment and on the conservation, development, and utilization of natural resources; and
5. a program for remedying the deficiencies of existing programs and activities, together with recommendations for legislation.

Members of the CEQ also analyze and interpret environmental information for the president and his or her staff members and review environmental programs and activities proposed by the federal government. Along with their advisory role, members of the CEQ develop policies that help to improve environmental quality and document changes to the environment. Members of the CEQ also develop the guidelines used for preparing environmental impact statements.

5.3 THE ENVIRONMENTAL PROTECTION AGENCY AND ENVIRONMENTAL IMPACT STATEMENTS

The Environmental Protection Agency

The **Environmental Protection Agency (EPA)** was created in September 1970 when the president at that time, Richard Nixon, presented to Congress proposed changes to the organization of the U.S. government. The programs transferred to the EPA were in the areas of water quality management, air quality and solid waste management, pesticides, radiological health, and water hygiene. The EPA would be responsible for implementing environmental laws, "developing policies and regulations, conducting research and monitoring activities, imposing sanctions, and engaging in numerous other activities. The EPA can influence legislation by proposing new programs to Congress and by informing Congress of actions that could be taken to avoid future environmental problems" (Ortolano 1997, 46).

Oversight for the EPA is provided by the U.S. Congress:

> Congress is able to monitor EPA's implementation of environmental statutes by calling for reports on progress and requesting appraisals of performance from the General Accounting Office. Moreover, congressional committees and subcommittees frequently hold hearings that allow Congress to monitor EPA's implementation of a statute or to amend a statute. These hearings give Congress a chance to hear from all interested parties, including those regulated by the laws. (Ortolano 1997, 50)

Members of the engineering and construction industry have to follow all the rules developed by the EPA when designing and constructing structures especially in the areas of air quality, water quality management, solid waste management, and hazardous waste mitigation.

Environmental Impact Statements

Environmental impact statements are required on all federal and federally funded projects; they may also be required by state governments. Environmental impact statements are used by Congress, federal agencies, and the public when they are required to make decisions that affect the environment. They are used on federal and federally funded projects during licensing and permitting procedures and for funding considerations (Ortolono 1997).

Environmental impact statements provide an analysis of the environmental costs and benefits of projects and explain the primary consequences of proposed projects along with potential secondary consequences. Environmental impact statements provide a description of the potential environmental risks of all proposed alternative projects to allow decision makers to make more informed decisions. Members of engineering firms are hired to investigate the environmental consequences of projects and to write environmental impact statements.

5.4 FEDERAL ENVIRONMENTAL LAWS OF CONCERN TO ENGINEERS AND CONSTRUCTORS

This section reviews some of the environmental laws that are pertinent to engineering and construction professionals, because these laws could affect how they design and construct their projects. Additional information on these and other environmental laws is available in the *Federal Register* or in the book *Environmental Regulation and Impact Assessment* by L. Ortolono, published by John Wiley and Sons (1997).

The Air Pollution Control Act of 1955

In 1953, U.S. Army General Dwight D. Eisenhower became president of the United States. General Eisenhower had been the General of the Army during World War II, Military Governor of the American Occupation Zone in Germany in 1945, and Supreme Allied Commander in Europe from April 1951 to May 1952. While Eisenhower was directing military operations in Europe, he realized the importance of transportation systems in the success of military strategy and maneuvers. As a result of this experience, President Eisenhower implemented plans for a nationwide interstate highway system in the United States. President Eisenhower envisioned a federal highway system that would provide both east–west and north–south major highways through each state in the union.

Although it took decades to complete his vision, one of the early effects of increased automobile travel, because of the availability of highways, was an increase in the level of air pollution. In 1955, the **Air Pollution Control Act** indicated for the first time that air pollution is a danger to public health, but it left the regulation of air pollution to the states. This act allowed the federal government to conduct research to investigate the effects of air pollution. This act was replaced by the Clean Air Act of 1963.

The Clean Air Acts of 1963, 1970, and 1990

In 1963, the **Clean Air Act** was developed and passed to help abate interstate air pollution. Prior to the passing of this act, individual states did not have any recourse when adjacent states had facilities that were polluting the local environment, including the environment across state boundaries. This act established the **U.S. Public Health Service** to conduct research into developing techniques for monitoring and controlling air pollution.

The Clean Air Act was amended in 1970 to establish **National Ambient Air Quality Standards** and again in 1990 to authorize a program for **Acid Deposition Control**, to control 189 toxic pollutants, and to establish permit program requirements. The Clean Air Act and the Air Quality Act provide power to the **U.S. Secretary of Health** to establish air quality standards for different pollutants and combinations of pollutants. These acts also provide the Environmental

Protection Agency (EPA) with the power to require the abatement of pollutants.

The Clean Air Act of 1970 set updated emissions standards for new vehicles and engines and authorized emissions testing of vehicles. Not all of the fifty states have adopted vehicle emissions tests and some states, such as California and Colorado, require emissions standards that are more rigid than the emissions standards in other states.

The Motor Vehicle Air Pollution Control Act of 1965

The **Motor Vehicle Air Pollution Control Act** of 1965 established auto emissions standards that set the maximum emissions that automobiles are allowed to eject through exhaust systems.

The Air Quality Act of 1967

It was not until 1967, when the **Air Quality Act** was passed, that a regional framework was created that allowed the enforcement of federal and state air quality standards. The Air Quality Act is also used to regulate a variety of toxic emissions. One situation that illustrates this occurred in Florida in 2009. A major class-action lawsuit was filed in March 2009 that involved toxic gases being emitted from drywall that had been installed in homes in Florida and other states:

> A group of Florida homeowners filed a class-action lawsuit on Monday against a German drywall maker, its Chinese subsidiaries and several U.S. homebuilders, alleging they put toxic drywall in thousands of U.S. homes.
>
> The lawsuit alleges defective Chinese drywall, that emits sulfur gases, was used during a building materials shortage at the height of the construction boom and installed in thousands of homes, where it was corroding the wiring, wrecking air conditioners, and making residents sick.
>
> The lawsuit, which could represent the owners of up to 30,000 Florida homes, named three Chinese manufacturers of plasterboard and three homebuilding companies as defendants.
>
> At least 550 million pounds of Chinese drywall was brought into the United States from 2004 to 2006, the peak of the U.S. housing boom, and up to 60,000 homes may be affected.
>
> The only way to fix the problem is to move the homeowners out, gut the houses and rebuild the interior, as well as replacing drapes, furniture, and other property that may have been contaminated by the gases. (ENR.com, March 3, 2009)

The National Environmental Policy Act (NEPA) of 1969

In addition to detailing the requirements for the Council on Environmental Quality, the National Environmental Policy Act of 1969 includes Section 102, which requires that:

> every recommendation or report on proposals for legislation and other major Federal actions significantly affecting the quality of the human environment requires a detailed statement by the responsible official on—
>
> (i) the environmental impact of the proposed action,
>
> (ii) any adverse environmental affects which cannot be avoided should the proposal be implemented,
>
> (iii) the relationships between local short-term uses of man's environment and the maintenance and enhancement of long-term productivity, and
>
> (iv) any irreversible and irretrievable commitments of resources which would be involved in the proposed action should it be implemented. (Public Law 91-190, 1969)

All of the above listed requirements are addressed in environmental impact statements.

The National Environmental Policy Act of 1970

The **Environmental Protection Agency** was established by executive order by President Nixon in 1970. The EPA was charged with the enforcement of air pollution laws and for establishing criteria to help create a cleaner environment. The National Environmental Policy Act followed the executive order and established that the EPA would be responsible for implementing the requirements included in the Clean Air Act of 1970.

The Noise Pollution Act of 1972

The **Noise Pollution Act** of 1972 provides the EPA with the legal right to control noise levels of products used for commerce and also to regulate the noise levels of railroads and freight carriers. Noise pollution includes unwanted and disturbing sounds. Chronic exposure to high levels of noise could lead to health-related illnesses such as "high blood pressure, speech interference, hearing loss, sleep disruption, and lost productivity" (EPA, June 10, 2009, 3).

The types of noise pollution that the EPA regulates include "low noise emission products, construction equipment, transport equipment, trucks, motorcycles, and the labeling of hearing protection devices" (EPA, June 10, 2009, 2). Noise caused by airplanes is regulated by the **Federal Aviation Administration (FAA)**.

In 1981, the responsibility for addressing noise issues, other than the ones cited previously, was transferred to state governments. Members of the federal EPA will still investigate noise issues, provide information on noise pollution, and evaluate existing regulations to determine whether they are protecting public welfare (EPA, June 10, 2009). Many states regulate the hours in which construction activities may be conducted such as not starting before 8 AM and not continuing after 6 PM.

The EPA regulates hearing protection devices through the **Labeling of Hearing Protection Devices** (HPD) **Regulation** (40CFR, Part 211 Subpart B). The devices that are regulated by the EPA include earplugs, earmuffs, and communication headsets, all of which are used routinely at construction job sites. Hearing protection devices are rated by the maximum decibel level that they protect the user from and their effectiveness in reducing unwanted noise.

The Federal Water Pollution Act of 1948, 1972, and 1977

The **Federal Water Pollution Act** of 1948 was amended in 1972 and became known as the **Clean Water Act**. This act was amended again in 1977 to establish environmental standards for water and waterways and to create a system for the issuance of permits for discharging pollutants from point sources such as pipelines, drainage ditches, ships, floating facilities, and other point sources. The act also made it against the law to discharge pollutants from a point source into a navigable waterway without first obtaining a permit from the EPA through the **National Pollutant Discharge Elimination System** (NPDES).

The EPA also levies penalties for oil spills into waterways and requires firms to pay to remediate the surrounding areas after oil spills, as the following case illustrates:

> One of the most environmentally damaging oil spills in the history of the U.S. occurred on March 23, in 1989 when an oil tanker ran aground and eight of its eleven cargo tanks were compromised releasing eleven million gallons of oil into Prudhoe Bay in the Prince William Sound in Valdez, Alaska. For three days, the oil was not skimmed off the water surface and when a storm hit the area the oil was spread onto the coastline. The oil company that owned the tanker was fined and required to pay for cleaning up the oil and the adjacent environment. (*National Geographic* 2004, 1).

Another landmark case involving water pollution was settled in 2009:

> Several major oil companies, including ConocoPhillips, Shell, BP, Amoco and Chevron, agreed to pay $422 million for the cleanup of wells and their surrounding areas owned by more than 153 water providers in 17 states and contaminated by the gasoline additive methyl tertiary butyl ether, known as MTBE. Under the terms of the settlement announced on May 7, pending approval by the U.S. District Court for the Southern District of New York, the oil companies will not only help pay for current remediation costs but will also pay 70% of future costs over the next 30 years. If approved, the settlement will be the largest of its kind to date. (*Engineering News Record Insider*, May 15, 2009)

The EPA also has the authority to require equipment to help prevent oil spills at oil-handling facilities and when it is being transported by oil tankers. One example of an oil spill prevention requirement is for ships to be built with double hulls in case the exterior hull is compromised in any way.

The Federal Insecticide, Fungicide, and Rodenticide Act of 1972 and 1996

The **Federal Insecticide, Fungicide, and Rodenticide Act** was passed in 1972 and amended in 1996. This act requires federal regulation of pesticide distribution, sale, and use. It also requires that all pesticides have to be registered (licensed) by the EPA. In order to obtain a license from the EPA the applicant has to be able to demonstrate that the pesticide will not adversely affect the environment.

The Toxic Substance Control Act of 1976

The **Toxic Substance Control Act** (TSCA) of 1976 "provided the EPA with the authority to require reporting, record-keeping and testing requirements, and restrictions relating to chemical substances and/or mixtures" and "addresses the production, importation, use, and disposal of specific chemicals including polychlorinated biphenyls (PCBs), radon, and lead based paint" (15 U.S.C. 2601 et seq.). The EPA maintains a list of toxic chemicals that included 83,000 chemicals in 2009.

The Solid Waste Disposal Act of 1965, the Resource Conservation and Recovery Act of 1976, and the Hazardous and Solid Waste Act of 1984

The **Solid Waste Disposal Act** of 1965 was amended in 1976 and became the **Resource Conservation and Recovery Act** (RCRA), and another amendment in 1984 changed the act into the **Hazardous and Solid Waste Act**. This act provides the federal government with the power to regulate hazardous wastes and has a provision for providing assistance to states in developing solid waste management plans and in implementing new technology to reduce or abate hazardous wastes.

The 1976 amendment provided the EPA with the legal authority to control hazardous wastes from "cradle to grave" including generation, transportation, treatment, storage, and disposal (42 U.S.C. 6901 et seq.). The 1984 amendment requires federal facilities to pay fines and penalties for violations of hazardous and solid waste requirements and addresses problems that result from underground petroleum storage tanks. Also included in this amendment are more stringent hazardous waste management standards and the phasing out of land disposal of hazardous wastes. Permits are required for treatment, storage, and disposal facilities for hazardous wastes. The facilities allowed for hazardous wastes include "container storage areas, tanks, surface impoundments, waste piles, land treatment facilities, landfills, incinerators, containment buildings, and/or drip pads" (42 U.S.C. 6901 et seq.).

The RCR Act also provides for whistleblower protection for employees in the United States who are fired or suffer other adverse actions, because of their involvement in the enforcement of the RCR law. Employees have thirty days to file a complaint with the Occupational Safety and Health Administration. For additional information, contact the National Whistleblower Center (www.whistleblowers.org) or the **U.S. Department of Labor Whistleblower Program** (www.osha.gov.oia.whistleblower/index/html).

The first time the government took drastic measures to mitigate hazardous wastes was at Love Canal, in Niagara Falls, New York, in 1979:

> An incident occurred during the 1970s that affected the manner in which the federal government dealt with hazardous waste sites. In Niagara Falls, New York there was an incident where citizens in a neighborhood were falling ill with cancer and leukemia and there was a high level of miscarriages. The *Federal Emergency Management Administration* (FEMA) was called in to investigate why there were so many cases of cancer and miscarriages occurring in one neighborhood. Between 1974 and 1978 fifty-six percent of the children born to parents that lived close to a canal, called Love Canal, had birth defects. After investigating the area it was discovered that a grade school had been built over a dump site that was used for decades and no one knew for sure what had been dumped in the area but there were indications of different toxic chemicals being in the dumpsite.
>
> The regional director of the Federal Emergency Management Administration (FEMA), Dr. Rita Meyninger, an environmental engineer, recommended to President Carter that the federal government purchase the homes in the surrounding area and move all of the residents out of the area. This was the first time in the history of the United States that the federal government intervened in a situation involving hazardous wastes and allocated funds to move residents out of an infected area. Since Love Canal other incidences have been discovered of citizens being affected by the toxic wastes in dump sites and the federal government has dealt with these incidences in a variety of ways. (Meyninger 1994)

For additional information on the Health History of Environmental Remediation that occurred at Love Canal, see Appendix B.

Comprehensive Environmental Response Compensation and Liability Act (CERLA) of 1980

The **Comprehensive Environmental Response Compensation and Liability Act** of 1990 created the **Superfund Program** that documents the location of the identified hazardous waste sites throughout the United States and tries to determine which organizations are responsible for the dumping of hazardous wastes into the identified sites. As of December 12, 2008, there were 1,288 Superfund sites on the *Superfund National Priority List*, 322 sites had been removed from the list after having been remediated, and 63 new sites were proposed to be added to the list. Figure 5.1 provides a map of the United States showing the Superfund sites as of January 1, 2009.

The CERLA Act requires that the organizations responsible for creating the hazardous waste sites pay for remediation costs, as the following case illustrates:

> Columbia, Maryland based W.R. Grace has agreed to pay $250 million, the highest sum in the history of the Superfund program, to help pay the cleanup costs of asbestos contamination in Libby, Montana the U.S. Justice Department and Environmental Protection Agency announced March 11.
>
> The settlement by the federal bankruptcy court overseeing W.R. Grace's reorganization, would settle the federal government's bankruptcy claim against the company. But the settlement does not resolve an on-going federal criminal case alleging that senior company officials covered up the extent of the contamination in Libby.
>
> W.R. Grace, a supplier of specialty chemicals, owned and operated a vermiculite mine and vermiculite processing facilities in Libby from 1963 to 1990. The vermiculite ore was contaminated with asbestos, and vermiculite and asbestos have been found in various locations in and around Libby since then. Hundreds of people in Libby have gotten sick or died from asbestos-related illnesses such as lung cancer and mesothelioma.
>
> The EPA has been removing asbestos-contaminated soil and other materials in Libby since May 2000. The federal government filed suit against W.R. Grace in the March of 2001 to recover its cleanup costs through the Superfund program. W.R. Grace filed for bankruptcy the same year. Although the federal district court in Montana awarded the EPA more than $54 million for cleanup costs in 2003, that award has not been paid because of Grace's bankruptcy.
>
> The March 11 settlement resolves the 2003 judgment and covers future cleanup costs that W. R. Grace might incur. The EPA will place the settlement funds into a special account within the Superfund program that will be used to pay for future cleanup work at the site.
>
> (On May 11, 2009 a jury in Libby, Montana acquitted W.R. Grace, and three of its former executives, of intentional exposure of mineworkers and Libby residents to asbestos.) (ENR.com, May 12, 2008; Asbestos.com, June 25, 2009)

Asbestos is a natural substance that is mined and used in the manufacture of insulation boards, sheetrock, wall boards, ceiling tiles, floor tiles, and other construction materials. As asbestos breaks down, or it is disturbed by drilling or other means of penetration, it releases a fine dust that is toxic to humans, but the affects of asbestos exposure may not be apparent for decades. Asbestos poisoning manifests itself as silicosis, which is a fatal lung disease; therefore, asbestos was banned in Great Britain in 1985 by the Health and Safety Code of Practice and by the Environmental Protection Agency in the United States in 1970. Other countries throughout the world still use asbestos in the manufacture of wall board for construction. Construction workers should be aware that they might be exposed to asbestos dust while installing or demolishing wall board (sheetrock) if it was originally installed prior to 1970 or if it was manufactured in a country that has not banned the use of asbestos.

The **Superfund Amendments and Reauthorization Act** (SARA) of 1986 reauthorized the CERLA to continue cleaning up hazardous waste sites in the United States. It also included

- Real Estate Developers
- Building Owners
- Financiers
- Contractors
- Utility Providers
- Federal, State, and Local Code and Regulatory Officials

In addition to the New Construction rating system (LEED-NC), LEED standards are currently available for the following:

- Commercial Interiors (LEED-CI)
- Existing Building Operations and Maintenance (LEED-EB)
- LEED Core and Shell
- LEED for Homes
- LEED for Retail
- LEED for Healthcare
- LEED for Schools
- LEED for Neighborhood Development (pilot) (USGBC 2008)

LEED was created to promote integrated building design, construction, and operation and maintenance practices that minimize the environmental impact of buildings. The LEED certification program promotes this by doing the following:

- Defining "green building" by establishing a common standard of measurement
- Recognizing environmental leadership in the building industry and preventing false claims
- Stimulating green competition
- Raising consumer awareness of green building benefits
- Transforming the building market (USGBC 2008)

LEED emphasizes utilizing integrated technologies that promote the following:

- Water savings
- Energy efficiency
- Sustainable materials selection
- Improved indoor environmental quality

The LEED certification process is used to recognize structures that have met the requirements in the area of green building, as set forth by the USGBC. The USGBC LEED certification process recognizes achievements and promotes expertise in green building through a comprehensive system offering project certification, professional accreditation, training programs, and other resources.

LEED includes information that could be utilized during the design phase by design team members to help them create sustainable projects. This information is also used as an evaluation system that assesses sustainability achievements using industry standards. The LEED checklist for projects includes different sustainability categories and a scoring system. The total number of points achieved is used to determine an overall LEED rating of certified, silver, gold, or platinum. Examples of LEED categories (USGBC 2008), with related sustainable strategies, are as follows:

1. Sustainable Sites
   - Site selection
   - Urban or brownfield redevelopment
   - Alternative transportation
2. Water Efficiency
   - Landscaping
   - Wastewater technology
3. Energy and Atmosphere
   - Energy optimization
   - CFC reduction
   - Green power
4. Materials and Resources
   - Building reuse
   - Construction waste management
   - Certified wood
5. Indoor Environmental Quality
   - Low emitting materials
   - Ventilation effectiveness
6. Innovation and Design Process
   - Having a LEED Accredited Professional on design teams
   - Development of a sustainable education program
   - Achievement of sustainability goals far in excess of requirements

The LEED certification process is evaluated and updated by the U.S. Green Building Council committee members. Samples of new initiatives in 2008 include a LEED-NC 3.0 that will include regionally weighted credits, more seamless online registration and certification processes, and planned integration with 3-Dimensional Computer-Aided Drafting software to help monitor the viability of various sustainability strategies and technologies.

Many government agencies at all levels have specified LEED, or an equivalent process, for use on their projects. Government projects represent more than 40 percent of all LEED-NC projects. Since the introduction of the LEED certification process in 2002, there have been more than 47 million square feet of LEED-EB registered projects. Over 3,000 projects are registered to LEED, at a value of more than $10 billion, and over 1,000 projects are certified under LEED for New Construction (USGBC 2008).

There are three ways to help avoid damaging the environment or to help mitigate the negative impacts on the global environment:

1. Reduce energy use.
2. Minimize pollution in order to reduce environmental damage and minimize health risks.

3. Minimize the amount of resources used in order to reduce embodied energy and resource depletion in:
   - the direct and indirect methods used to extract materials in manufacturing
   - the reduction of transportation energy
   - the reduction of waste of materials on site
   - using renewable and sustainable green materials

The LEED certification process is being applied to a wide range of building types but it does not apply to industrial facilities.

Sustainable Industrial Construction

A research project that was funded by the construction industry institute investigated sustainable development practices that are currently being used on industrial construction projects and the results of the research are published in the research report *Design and Construction for Sustainable Industrial Construction* (Yates 2008). The research project published several implementation resources including a Checklist for Sustainable Industrial Construction Job Sites, a Quick Start Guide, and a Maturity Matrix, which are included in the publication *Sustainable Industrial Construction: Implementation Resource* (Yates 2010a). A primer was also written that summarizes industrial construction sustainability practices; it is called *Sustainable Industrial Construction: Primer* (Yates 2010b).

5.6 ENVIRONMENTAL PROTECTION AGENCY SUSTAINABILITY (EPA) LAWS

In the Resource Conservation and Recovery Act (RCRA) of the U.S. Environmental Protection Agency there is a section called the Affirmative Procurement Program (APP) that was established to encourage the procurement of recycled products or products that contain some recycled components (EPA 2007). In addition, Executive Order 13101 was implemented to mandate that federal government agencies use recycled and environmentally friendly products. Executive Order 13423 replaced 13101 in January 2007 and emphasizes energy and environmental issues (EPA 2007).

The EPA also issued Comprehensive Procurement Guidelines (CPG) that specify environmentally friendly products and the minimum recycled content requirements for products. Federal agencies are required to purchase these products. Fifty-one items are listed in the CPG as compliant items and new products are added each year. The product categories (EPA 2007, 7) that are listed in the CPG are the following:

- Paper and Paper Products—including sanitary tissue, printing and writing paper, newsprint, paperboard and packaging, and paper office supplies (e.g., file folders and hanging files).
- Non-Paper Office Products—including binders, recycling and trash containers, plastic desktop accessories, plastic envelopes, trash bags, printer ribbons and toner cartridges, report covers, plastic file folders, and plastic clipboards.
- Construction Products—including insulation, carpet, cement and concrete, latex paint, floor tiles, patio blocks, shower and restroom dividers, structural fiberboard, and laminated paperboard.
- Transportation Products—including channelizers, delineators, parking stops, barricades, and cones.
- Landscaping Products—including garden and soaker hoses, mulch, edging, and compost.
- Miscellaneous Products—including pallets, mats, awards, and plaques.

In March 2008, the EPA passed new ozone requirements that reduced the allowable ozone emissions from eighty parts per billion to seventy-five parts per billion. The original ozone standard was passed in 1997 and eighty-five counties have not been able to meet the requirements of the original standard. The enforcement mechanism will be penalties, and the withholding of federal transportation funding, until counties are able to improve their air quality. If counties are evaluated at the new ozone emissions level then 345 counties in the United States will not meet the new requirements (*Engineering News Record*, March 24, 2008).

5.7 UNITED STATES GOVERNMENT SUSTAINABILITY LAWS

This section discusses some of the U.S. government sustainability laws, and potential laws, that affect members of the engineering and construction industry. This section also provides a discussion of the ramifications of these laws.

America's Climate Security Act of 2007 (S. 2191)

America's Climate Security Act of 2007 was introduced on October 18, 2007, and was still under consideration by the U.S. Congress in 2010. The act is described as a "bill that sets a mid-term goal of reducing emissions from the power, industrial, and transportation fuel sectors by 15% in 2020 and 70% by 2050, compared to 2005 emissions levels. These sectors account for about 75% of U.S. greenhouse gas emissions (GHG). By ratcheting down emissions by nearly 2% per year, the bill should reduce total U.S. greenhouse gas emissions about 51–63% by 2050 from 2008 levels, taking an important step toward the 80% emissions reduction goal the international scientific community says is necessary to limit global warming" (Pew Charitable Trusts 2008, 1).

In order to meet the requirements of the act, the bill provides methods for achieving emissions reductions or offsetting emissions with credits. A summary of the methods being explored in this act (U.S. Congress, October 18, 2007, 16) include the following:

- The firms affected by the act include those that emit more than 10,000 carbon dioxide equivalents of greenhouse gases per year.
- The act sets up a **cap-and-trade** system that sets mandatory limits on CO_2 emissions. Firms could borrow reductions from future years at 10 percent.
- The act creates a **Carbon Market Efficiency Board** (CMEB) that is responsible for implementing cost-relief measures if the cost of reducing emissions is higher than the original estimates.
- Fifteen percent of allowances could be met by offsets that come from sources not covered by the bill or they could be satisfied by international trading.
- After eight years the president is allowed to enact requirements that importers of products that create CO_2 emissions have to submit emissions credits to sell their products in the United States.
- Pollution allowances will be auctioned off at a rate of 23 percent by the year 2012 and 73 percent by the year 2036. The proceeds from the auctions will be used to help workers and states transition to climate-friendly energy sources, help the poor with energy bills, and invest in clean technology.
- The act proposes to cut energy requirements of new buildings and homes by 50 percent by the year 2020 and to adopt new building codes to help meet this requirement.
- Firms will be required to disclose to the **Security and Exchange Commission** their global warming–related financial risks to shareholders.
- Phantom reductions will be hard to verify.

The Climate Security Act is still being debated by the U.S. Congress and the act had not been approved when this book was submitted for publication.

Energy Independence and Security Act of 2007

The **U.S. Energy Independence and Security Act** of 2007 was passed to promote energy efficiency and the development of renewable energy sources. The four major provisions (U.S. Congress 2007) of the act are:

- The **Corporate Average Fuel Economy** (CAFÉ)—Sets a requirement of thirty-five miles per gallon as the combined fleet average for automobiles and light trucks to be achieved by the year 2020.
- **Renewable Fuels Standard (RFS)**—Requires that 9 billion gallons of fuel be from renewable sources by the year 2008 and 36 billion gallons by the year 2022.
- **Energy Efficiency Equipment Standards**—Sets new standards for lighting and residential and commercial appliances and equipment.
- **Repeal of Oil and Gas Tax Incentives**—Repeals two previously implemented tax subsidies to help fund the cost of the CAFÉ.

Individual states are also passing new environmental legislation that relates to global warming such as the *Global Warming Solutions Act* (AB 32) that was passed in 2006 in the state of California.

Climate Change Legislation Design—U.S. Government White Paper of 2007

The U.S. Committee on Energy and Commerce and the **Subcommittee on Energy and Air Quality** issued a white paper on **Climate Change Legislation Design**—Scope of the Cap and Trade Program in 2007 (Lieberman et al. 2007). The white paper discusses legislation that if enacted would require reductions in green house gas emissions by 60 to 80 percent by the year 2050. It seeks to stabilize atmospheric green house gas concentrations of CO_2 equivalents to a level between 450 and 550 parts per million. The gases that are covered include: (Lieberman et al. 2007):

- Carbon dioxide CO_2
- Methane CH_4
- Nitric oxide N_2O
- Fluorinated gases
- Hydrocarbons HFCs
- Perflourocarbons PFCs
- Sulfur hexaflouride SF_6

The toxic affects of some of these greenhouse gases are:

> Nitrogen oxides, and sulfur oxides, can be grouped into pollutants, which causes acidification when mixed with water in the air. Acid rain is one effect of acidification, which leads to damage to agriculture, public health, buildings and materials. These pollutants together with suspended particulate matter cause detrimental effects to the human health or human toxicity. Moreover, NOx also causes eutrophication, a phenomenon that depletes the nutrients of the soil, thereby decreasing agricultural productivity. (Gerilla, Teknomo, and Hokao 2007, 2782)

In the year 2005, the United States produced 7,280 million metric tons of carbon dioxide equivalents. The percentage of emissions in 2005 (EPA 2008) for each of the gas types were the following:

- Hydrocarbons, perflourocarbons, and sulfur hexaflouride—22 percent
- Nitric oxide—5 percent
- Methane—7.4 percent
- Carbon dioxide—83.9 percent

The sources for GHG emissions (EPA 2008) were from the following industries:

- Electric Generation—34 percent
- Transportation—28 percent
- Industrial—10 percent

- Commercial—6 percent
- Residential—5 percent
- Agriculture—8 percent

In the Industrial Sector and the Electrical Generation Sector, the GHG emissions (EPA 2008) were from the following areas:

- Petroleum Refining—3 percent
- Fossil Fuel Exploration and Production—3 percent
- Chemical Manufacturing—5 percent
- Others—8 percent
- Coal—27 percent
- Natural Gas—4 percent
- Petroleum—2 percent
- Others—1 percent

In the Electrical Generation Sector, the power generated (EPA 2008) is divided into the following areas:

- Coal—49.7 percent
- Petroleum—3 percent
- Natural Gas—18.7 percent
- Hydroelectric—6.5 percent
- Renewables—2.3 percent
- Other gases—0.4 percent

Any legislation resulting from the white papers will attempt to reduce greenhouse gas emissions in the future by regulating the firms that produce them.

5.8 GLOBAL ENVIRONMENTAL MANAGEMENT STANDARDS

The **International Organization for Standardization (ISO)** publishes 350 international environmental standards that could be incorporated into the designs of projects. For more information on the specific standards that are available, see the ISO website (www.iso.org). The ISO website also includes information on environmental management standards, called the ISO 14,000, addressing environmental management systems and describing the ISO 14,000 certification process.

5.9 GLOBAL SUSTAINABILITY—THE UNITED NATIONS FRAMEWORK CONVENTION ON CLIMATE CHANGE (UNFCCC)

Environmental concerns related to global climate change led to the formation of the **United Nations Framework Convention on Climate Change (UNFCCC)** and development of the **Kyoto Protocol**, which introduced measures for controlling global climate change caused by the emission of **greenhouse gases (GHGs)** in industrialized and developing countries. The Kyoto Protocol established baseline principles and commitments that countries that are a party to the convention have to follow in order to help reduce GHGs.

Table 5.1 Sources of Greenhouse Gases

Fuel combustion	Industrial processes	Rice cultivation
Energy industries	Mineral products	Agricultural soils
Manufacturing industries	Chemical industry	Prescribed burning of savannas
Construction	Metal production	Field burning of agricultural residues
Transportation	Other production	Solid waste disposal on land
Fugitive emissions from fuels	Solvents	Wastewater handling
Solid fuels	Agriculture	Waste incineration
Oil and natural gas	Enteric fermentation	
	Manure management	

UNFCCC, 2005a, *Essential Background*, United Nations Framework Convention on Climate Change, http://unfcc.int/essential_background/items/2877.php

Table 5.1 contains a list of the processes that emit greenhouse gases. Scientists have indicated that GHGs are responsible for depleting the **ozone layer** surrounding the earth. The ozone layer protects the surface of the earth from the damaging ultraviolet light rays of the sun, therefore, if the ozone layer is compromised, it could cause climate changes throughout the world such as increasing temperatures and melting of the polar ice caps. Greenhouse gases include the following:

- Carbon monoxide (CO_2)
- Nitrous oxide (N_2O)
- Hydrofluorocarbons (HFC)
- Perfluorocarbons (PFC)
- Sulphur hexafluoride (SF_6) (UNFCCC 2005a)

The Kyoto Protocol

The Kyoto Protocol is an amendment to the United Nations Framework Convention on Climate Change; it formalizes the intentions of the UNFCCC, which is an international agreement that sets binding targets for the reduction of GHG emissions by industrialized countries by 2012. It is available in Arabic, Chinese, English, French, Russian, and Spanish. The goals of the Kyoto Protocol include the following:

- Changing consumer patterns
- Protecting and promoting human health conditions
- Promoting sustainable human settlement development
- Protecting the environment, air, water, and ecosystems and combating deforestation and managing wastes (UNFCCC, 2005b website)

The Kyoto Protocol sets specific targets for reducing GHG emissions that each country that has ratified the Protocol has to meet by 2012. The **emission targets** for each country represent a percentage of greenhouse gases that ranges from 5 to 8 percent below 1990 emission levels (Ho-won Jeong 2001). Emission targets vary by country with targets of 8 percent reduction in emissions being set for the European Union, Switzerland, and most Central and Eastern European states, 6 percent for Canada, 7 percent for the United States, and 6 percent for Hungary, Japan, Poland, New Zealand, Russia, and the Ukraine. Because Norway, Australia, and Iceland produce low levels of greenhouse gases, they are able to increase their emissions by up to 1 percent in Norway, 8 percent in Australia, and 10 percent in Iceland.

The European Union is able to balance its emission targets between countries by allowing countries with low emissions to increase their emissions and reduce emissions in countries with high levels of GHG emissions. Developed countries increased their GHG emissions from 1990 to 2000 by 8.2 percent except for Eastern Europe and the Former Soviet Union, which saw an overall drop in GHG emissions during that time period due to the decline of their economies.

The premise behind the target emissions in the Kyoto Protocol is to not restrict growth in economies that are in transition (which include the former Soviet Union and Central and Eastern European nations) and developing countries, but to limit emissions in developed countries. Economies in transition (EITs) are also allowed to choose a different base year from 1990, because they may not have measurements for 1990. Any country may choose a base year of either 1990, or 1995, for the emissions of hydroflourocarbons (HFCs), perflourocarbons (PFCs), and sulfur hexafluoride (SF_6). Sustainable development is part of the goal of the Kyoto Protocol.

The Kyoto Protocol became effective in February 2005 when fifty-five countries throughout the world ratified it. By the end of 2005, eighty-four countries had ratified the Kyoto Protocol; they are listed in Table 5.2 (UNFCCC 2005b). The United States and Australia have not ratified the Kyoto Protocol, although the United States produces the highest level of greenhouse gases of any nation in the world followed by China, Russia, India, Japan, Germany, Brazil, Canada, the United Kingdom, Italy, Korea, the Ukraine, France, Mexico, and Australia. Fifteen countries produce 70 percent of the greenhouse gases in the world.

Table 5.3 contains a list of the target emission reductions, or increases, that each country has to meet by 2012, which are a percentage of 1990 emissions that was developed from information from the UNFCCC website (www.UNFCCC.org).

There is a **Clean Development Mechanism** in the Kyoto Protocol that permits industrialized countries to partly meet their emission targets through **credits** earned by sponsoring GHG-reducing projects in developing countries (Elliott 1998). **Joint implementation** allows developed countries to invest in clean technology that reduces greenhouse gases in other developed countries and both countries are awarded emission credits. Some countries are able to meet their emission targets by **emissions trading** whereby countries are allowed to sell their emission credits, or debt, to other countries.

Countries are allowed to counterbalance GHG emissions by removing greenhouse gases from the atmosphere by **carbon sinks** such as reforestation, which is the process of planting trees that absorb carbon monoxide and other pollutants from the air. Countries are also able to bank their emission credits for use in other years or sell them to other countries in subsequent years.

The techniques and methods for estimating GHG emissions by sources and the removal of emissions by sinks have to be approved by the **Intergovernmental Panel on Climate Change** and also be agreed upon by a **Conference of the Parties**, which are the countries that have ratified the Kyoto Protocol. The Conference of the Parties determines the consequences of a country not meeting their emission targets by the year 2012. If the Kyoto Protocol is not extended beyond 2012 or there is no other treaty in place by 2012, then countries will be able to continue to produce GHGs at toxic levels (UNFCCC 2005c).

The Basel Convention

The UNFCCC also developed and implemented the **Basel Convention** that controls the trans-boundary movement of hazardous wastes and their disposal. One hundred and sixty-four countries have agreed to minimize the generation of hazardous wastes and ninety-five countries have banned the export of hazardous waste materials from developed to developing countries including toxic, poisonous, explosive, flammable, ecotoxic (toxic to the environment), and infectious wastes. Some countries require prior notification before any other country is able to export a hazardous waste to their country. Prior approval also has to be obtained from nations where hazardous wastes will be in transit (UNFCCC 2005a).

The Rio Declaration

The UNFCCC facilitated the development and implementation of the **Rio Declaration** that requires countries to enact environmental legislation that facilitates the exchange of environmental information including environmental impact statements (EIS) or environmental impact assessments (EIA), the results of decisions related to the environment, and the results of judicial and administrative proceedings between countries that share natural resources (UNFCCC 2005a). International environmental impact assessments are used during decision-making processes to evaluate the potential physical, biological, cultural, and socioeconomic effects of a project and its alternatives.

Table 5.2 Countries that Have Ratified the Kyoto Protocol

Albania
Algeria
Angola
Antigua and Barbuda
Argentina
Armenia
Australia
Austria
Azerbaijan
Bahamas
Bahrain
Bangladesh
Barbados
Belarus
Belgium
Belize
Benin
Bhutan
Bolivia
Bosnia and Herzegovina
Botswana
Brazil
Bulgaria
Burkina Faso
Burundi
Cambodia
Cameroon
Canada
Cape Verde
Central African Republic
Chile
China
Columbia
Cook Islands
Costa Rica
Croatia
Cuba
Cyprus
Czech Republic
Cote d'Ivoire
Democratic People's Republic of Korea
Democratic Republic of the Congo
Denmark
Djibouti
Dominica
Dominican Republic
Ecuador
Egypt
El Salvador
Equatorial Guinea
Eritrea
Estonia
Ethiopia
European Community
Fiji
Finland
the former Yugoslav Republic of Macedonia
France
Gabon
Gambia
Georgia
Germany
Ghana
Greece
Grenada
Guatemala
Guinea
Guinea-Bissau
Guyana
Haiti
Honduras
Hungary
Ireland
India
Indonesia
Iran (Islamic Republic of)
Ireland
Israel
Italy
Jamaica
Japan
Jordan
Kenya
Kiribati
Kuwait
Kyrgyzstan
Lao People's Democratic Republic
Latvia
Lebanon
Lesotho
Liberia
Libyan Arab Jamahiriya
Liechtenstein
Lithuania
Luxembourg
Madagascar
Mali
Malta
Marshall Islands
Mauritania
Mauritius
Mexico
Micronesia (Federated States of)
Moldova
Monaco
Mozambique
Myanmar
Namibia
Nauru
Nepal
Netherlands
New Zealand
Nicaragua
Niger
Nigeria
Niue
Norway
Oman
Pakistan
Panama
Papua New Guinea
Paraguay
Peru
Philippines
Poland
Portugal
Qatar
Republic of Korea
Romania
Russian Federation
Rwanda
Saint Lucia
St. Vincent and the Grenadines
Saudi Arabia
Senegal
Serbia
Seychelles
Sierra Leone
Singapore
Slovakia
Slovenia
Solomon Islands
South Africa
Spain
Sri Lanka
Sudan
Suriname
Swaziland
Sweden
Switzerland
Syrian Arab Republic
Tajikistan
Thailand
Timor-Leste
Togo
Tonga
Trinidad/Tobago
Tunisia
Turkmenistan
Tuvalu
Uganda
Ukraine
United Arab Emirates
United Kingdom of Great Britain and Northern Ireland
United Republic of Tanzania
Uruguay
Uzbekistan
Vanuatu
Venezuela (Bolivarian Republic of)
Vietnam
Yemen
Zambia

Source: *United Nations Framework Convention on Climate Change*, 2009, maindb.unfcc.int/public/country.pl?group=kyoto

Table 5.3 Percentage of 1990 Greenhouse Gas Emission Reductions, or Increases, Required by 2012

Country Name	Percentage of 1990 GHG Emissions	Country Name	Percentage of 1990 GHG Emissions
Australia	108	Monaco	92
Austria	92	Netherlands	92
Belgium	92	NewZealand	100
Bulgaria*	92	Norway	101
Canada	94	Poland*	94
Croatia*	95	Portugal	92
CzechRepublic*	92	Romania*	92
Denmark	92	Russian Federation*	100
Estonia*	92	Slovakia*	92
European Com.	92	Slovenia*	92
Finland	92	Spain	92
France	92	Sweden	92
Germany	92	Switzerland	92
Greece	92	Ukraine*	100
Hungary*	94	United Kingdom of Great Britain and Northern Ireland	92
Iceland	110	United States of America	93
Ireland	92		
Italy	92		
Japan	94		
Latvia*	92		
Liechtenstein	92		
Lithuania*	92		
Luxembourg	92		

*Economies inTransition

Source: UNFCCC, 2005a, *Essential Background*, United Nations Framework Convention on Climate Change, http://unfcc.int/essential_background/items/2877.php.

The Stockholm Convention

The **Stockholm Convention** is a treaty that was developed by the UNFCCC to reduce global production, use, and release of twelve of the most harmful chemicals called **persistent organic pollutants** (POPs). Table 5.4 provides a list of the persistent organic pollutants.

Table 5.4 Persistent Organic Pollutants

Pesticide	Industrial Chemical	By-product
Aldrin	Hexachlorobenzene	Hexaclorobenzene
Chlordane	Mirex	Mirex
DDT	Toxaphene	Toxaphene
Dieldrin	Polychlorinated biphenyls	(PCBs)
Endrin	Polychlorinated dibenzo-p-dioxins	Dioxins
Heptachlor	Polychlorinated dibenzo-p-furans	Furans

Source: *Regulating Pesticides*, March 2008, U.S. Environmental Protection Agency, www.epa.gov/oppfead/international/pops.htm.

Environmental Compliance

The UNFCCC is developing international compliance methods and penalties for noncompliance for the Kyoto Protocol. In the international arena, the enforcement of international treaties is difficult unless countries submit to voluntary compliance. If a treaty is an **international customary law**, it is a law that is defined by the International Court of Justice (a general practice that is accepted as law) and governments enforce it because they see themselves as legally obligated to do so. Treaties and customs, such as the Kyoto Protocol Treaty, are called **hard laws** and these types of laws could be enforced through economic sanctions set by international legal systems. **Soft laws** are nonbinding laws that are based on international diplomacy and customs; countries enforce them because they fear retribution by other countries if they do not enforce them.

The **International Network for Environmental Compliance and Enforcement** is a network of governmental and nongovernmental practitioners from over one hundred countries that are trying to raise awareness on complying with environmental standards and regulations. They are developing methods for the enforcement of standards and trying to increase cooperation between nations to strengthen the capacity, implementation, and enforcement of environmental regulations (OIA-EPA 2005).

Chapter Summary

This chapter included information on some of the environmental laws that affect engineering designs and construction. The environmental laws that were covered were related to:

- Environmental quality
- Environmental impact statements
- Air pollution
- Water quality
- Noise pollution
- Hazardous wastes
- Solid waste disposal
- Insecticides

Sustainability issues were also discussed including life-cycle environmental cost analysis, the Leadership in Energy and Environmental Design (LEED) certification process, current and pending sustainability legislation, and global environmental management standards. Global sustainability issues were covered including the United Nations Framework Convention on Climate Change, the Kyoto Protocol, the Basel Convention, the Rio Declaration, and the Stockholm Convention.

Key Terms

Acid Deposition Control
Affirmative Procurement Program
Air Pollution Control Act
Air Quality Act
America's Climate Security Act
asbestos
Basel Convention
cap-and-trade
carbon dioxide
Carbon Market Efficiency Board
carbon sinks
carcinogen
Clean Air Act
Clean Development Mechanism
Clean Water Act
Climate Change Legislation Design
Comprehensive Environmental Response Compensation and Liability Act (CERLA)
Comprehensive Procurement Guidelines
Conference of the Parties
Corporate Average Fuel Economy
Corporate Social Responsibility
corporate sustainability
Council on Environmental Quality (CEQ)
credits
Dow Jones Global Sustainability Index
Emergency Planning and Community Right-to-know Act
emission targets
emissions trading
Energy Efficiency Equipment Standards
environmental impact statement (EIS)
Environmental Protection Agency (EPA)
Environmental Protection Agency Office of Research and Development Strategic Plan
Environmental Protection Agency Strategic Goals
Federal Aviation Administration (FAA)
Federal Insecticide, Fungicide, and Rodenticide Act
Federal Water Pollution Act
Green Building Rating System
greenhouse gases (GHGs)
hard laws
Hazardous and Solid Waste Act
Intergovernmental Panel on Climate Change
international customary law
International Network for Environmental Compliance and Enforcement
International Organization for Standardization (ISO)
joint implementation
Key Performance Indicators
Kyoto Protocol
Labeling of Hearing Protection Regulation
Leadership in Energy and Environmental Design (LEED)
life-cycle environmental and cost analysis
Material Safety Data Sheets
Motor Vehicle Air Pollution Control Act
National Ambient Air Quality Standards
National Environmental Policy Act (NEPA)
National Pollution Discharge Elimination System
Noise Pollution Act
Occupational Safety and Health Communication Standard
ozone layer
persistent organic pollutants
Renewable Fuel Standards
Repeal of Oil and Gas Tax Incentives
Resource Conservation and Recovery Act
Rio Declaration
Security and Exchange Commission
Social Responsibility Investment
soft laws
Solid Waste Disposal Act
Stockholm Convention
Subcommittee on Energy and Air Quality
Superfund Amendments and Reauthorization Act
Superfund Program
sustainable development
Toxic Substance Control Act
United National Framework Convention on Climate Change (UNFCCC)
U.S. Department of Labor Whistleblower Program
U.S. Energy Independence and Security Act
U.S. Green Building Council (USGBC)
U.S. Public Health Service
U.S. Secretary of Health

Discussion Questions

5.1 How was the interstate highway system established in the United States that led to the proliferation of air pollution problems?

5.2 Why are vehicle emissions tests required for vehicles in the United States?

5.3 Explain how the National Environmental Policy Act of 1969 affects engineers when they are designing projects.

5.4 Discuss the responsibilities of the Council on Environmental Quality (CEQ).

5.5 How would engineers be involved in Environmental Impact Statements?

5.6 What federal law impacts the generation of hazardous wastes and how does the law impact it?

5.7 How are pesticides regulated in the United States? What law requires this form of regulation?

5.8 What agency regulates noise pollution? What agency regulates noise pollution generated by airplanes?

5.9 How does the EPA try to regulate oil spills? What types of sanctions does the agency have available when there are oil spills?

5.10 Which law sets emissions standards for vehicles in the United States? Discuss whether states are allowed to enact more stringent requirements than the federal standards and why they would do it.

5.11 Which law helps protect the citizens of one state from air pollution generated in another state?

5.12 How are Environmental Impact Statements used by the federal government?

5.13 If the president of the United States needed to obtain information about the potential hazards of a proposed project, which agency would assist the president? How would the members of this agency assist the president?

5.14 Why was the incident at Love Canal, which was related to the toxicity of hazardous wastes, an important case in the history of U.S. environmental policies?

5.15 Why is it necessary for engineers and constructors to know about sustainability practices? How are they affected by sustainability?

5.16 What affect does the Kyoto Protocol have on the U. S. construction industry?

5.17 Which industries generate the highest amount of greenhouse gas emissions?

5.18 How does the Superfund program affect the remediation of hazardous waste dump sites that are included in it?

5.19 If a country does not meet its target emissions reductions as set by the Kyoto Protocol, what sanctions are available to help enforce the target emissions?

5.20 Explain emissions trading and carbon sinks and how they affect climate change.

References

15 U.S.C. 2601 et seq. 1976. Toxic Substance Control Act. Washington, DC: Government Printing Office.

42 U.S.C. 6901 et seq. 1976. Resource Conservation and Recovery Act. Washington, DC: Government Printing Office.

Asbestos.com. June 25, 2009. Definition of Asbestos. Washington, DC: American Asbestos Association.

Clough, R., G. Sears, and K. Sears. 2005. *A Practical Guide to Contract Management*. New York: John Wiley and Sons.

Elliott, L. M. 1998. *The Global Politics of the Environment*. New York: New York University Press.

Engineering News Record. March 24, 2008. New EPA ozone standard sharply criticized. *Engineering News Record*, McGraw Hill Publishers, 260 (10): 9.

Engineering News Record Insider. May 15, 2009. Oil companies agree to $422 million deal. New York: McGraw Hill Publishers, www.enr.com.

ENR.com. March 3, 2009. Florida lawsuit alleges defective drywall in homes. *Engineering News Record.* New York: McGraw Hill Publishers, www.enr.com.

ENR.com. May 12, 2009. W. R. Grace to pay huge Montana site cleanup bill. *Engineering News Record.* New York: McGraw Hill Publishers.

EPA. 2007. U.S. Environmental Protection Agency (EPA), p. 1. http://www.miamidade.gov/derm/tips_business_waste.asp and http://www.epa.state.oh.us/opp/tanbook/fppgch1.txt.

EPA. 2008. *Inventory of the United States Greenhouse Gas Emissions and Sinks 1990–2005.* U.S. Environmental Protection Agency. Washington, DC: Government Printing Office.

EPA. *Regulating Pesticides.* March 2008. U.S. Environmental Protection Agency. www.epa.gov/oppfead/international/pops.htm.

EPA. January 1, 2009. *Superfund Hazardous Waste Sites in the United States.* Environmental Protection Agency. www.epa.org.

EPA. June 10, 2009. Allowable Noise Pollution Levels. Washington, DC: Government Printing Office.

Gerilla, G., K. Teknomo, and K. Hokao. 2007. An environmental assessment of wood and steel reinforced concrete housing construction. *Building and Environment*, 42 (7): 2778–2784.

Global Warming Solutions Act (AB 32). 2006. California Assembly Bill 32. Sacremento, CA: California Legislative.

Ho-Won Jeong. 2001. *Global Environmental Policy Making: Institutions and Procedures.* New York: Palgrave Publications.

Lieberman, J., et al. 2007. *S. 2191, America's Climate Security Act of 2007.* Washington, DC: U.S. Library of Congress, pp. 1–10.

Meyninger, R. 1994. *The Affects of Toxic Wastes on Humans.* Ph.D. Dissertation. Polytechnic University, Brooklyn, New York, 294 pages.

National Geographic. 2004. Exxon Valdez spill, 15 years later: Damage lingers. www.news.nationalgeographic.com/news/2004/03/0318_04038_exxonvaldez.html.

OIA-EPA. 2005. Office of International Affairs—United States, Environmental Protection Agency. www.epa.org.

ORD Strategic Plan. January 2001. Office of Research and Development Strategic Plan. Washington, DC: The Environmental Protection Agency.

Ortolano, L. 1997. *Environmental Regulation and Impact Assessment.* New York: John Wiley and Sons.

Pew Charitable Trusts. 2008. *National Environmental Trust.* www.pewtrust.org.

Public Law 91–190. 1969. Section 201, 42 U.S.C. 4341. Washington, DC: Government Printing Office.

Samaras, C. 2004. *Sustainable Development and the Construction Industry—Status and Implementation.* Pittsburg, PA: Carnegie Mellon, pp. 1–7.

UNFCCC. 2005a. *Essential Background.* United Nations Framework Convention on Climate Change, http://unfcc.int/essential_background /items/2877.php.

UNFCCC. 2005b. *Kyoto Mechanisms.* United Nations Framework Convention of Climate Change, http://unfcc.int/kyoto_mechanisms /items/1673.php.

UNFCCC. 2005c. *Parties and Observers.* United Nations Framework Convention of Climate Change, http://unfcc.int/parties_and_observers/items /2704.php.

United States Congress. October 2007. *United States Energy Independence and Security Act of 2007.* Washington, DC: Government Printing Office.

USGBC. 2008. *Leadership in Energy and Environmental Design.* United States Green Building Council, http://www.usgbc.org.

Yates, J. 2008. *Design and Construction for Sustainable Industrial Construction.* Austin, TX: Construction Industry Institute.

Yates, J. 2010a. *Sustainable Industrial Construction: Implementation Resource.* Austin, TX: Construction Industry Institute.

Yates, J. 2010b. *Sustainable Industrial Construction: Primer.* Austin, TX: Construction Industry Institute.

ENGINEERING PROFESSIONAL ETHICS

6.1 INTRODUCTION

This chapter covers one of the more difficult aspects of working as an engineer, which is the area of professional ethics. The other chapters in this book provide information on legal and contractual issues, but there are many day-to-day issues that engineers have to address that are not legal issues but that affect the safety and well-being of their fellow citizens. Engineers should be aware of their obligations as guardians of public safety; this chapter attempts to quantify an area that is difficult to assess because it is also influenced by the personal ethics of the individuals practicing in the engineering profession.

The ethical responsibilities of engineers have been prescribed by professional societies dating back almost two centuries. The first chartered professional society, the **Institute of Civil Engineers**, was formed in Great Britain in 1818 and was the first civil engineering professional organization. A decade and a half later in 1838 the **Royal Institute of British Architects** was formed; the **Institution of Mechanical Engineers** was created in 1847.

In the United States, civil engineers were once again the first engineering profession to form a professional society. The **American Society of Civil Engineers** was formed in 1853, followed in 1857 by the **American Institute of Architects**. In 1880, the **American Society of Mechanical Engineers** was created, and the **Institute of Electrical and Electronic Engineers** was chartered in 1884. The **American Society of Agricultural and Biological Engineers** was founded in 1907. The **American Institute of Chemical Engineers** was organized in 1908 and half a century later the Institute of Industrial Engineers was formed in 1948. Although each of the engineering professional societies evolved separately, they all attempt to foster similar professional behavior and to pursue similar goals, especially in the area of engineering ethics.

In order to help engineers cope with the ethical dilemmas they will encounter during their career, this chapter discusses morals, ethics, the role of engineering societies in ethics, the engineering codes of ethics, how the codes of ethics are enforced, whistle-blowing, and methods for whistle-blowing.

The Importance of Engineering Professional Ethics

Of all areas that engineers and constructors have to be aware of from a legal perspective, **engineering professional ethics** is one of the most important areas because of the potential for loss of life if mistakes are made by engineers. In the medical profession, if a physician makes a mistake, normally only one patient dies, or maybe several patients if the records of patients are inaccurate or become mixed up. However, in the engineering profession if an engineer makes a mistake, hundreds or possibly thousands of people could perish. Architects rely on structural engineers to ensure that their structures are safe, so it becomes the responsibility of the structural engineers, and not the architects, to be concerned about the safety of the occupants of the buildings and structures that they design.

Traditionally, when structural engineering courses are being taught, diagrams, such as the ones shown in Figure 6.1, are shown to university students. Live loads on the structural diagrams are shown as arrows that point in the direction that the load is exerting its force, such as the arrows shown in Figure 6.1. From an ethical perspective, it might be more beneficial to draw live loads as humans, such as the ones shown in Figure 6.2, since this would be a more accurate depiction of reality and instill in engineers the concept of live loads being human beings that are relying on engineers to design safe structures for them to inhabit.

An extreme example that demonstrates the importance of engineers and contractors not being careless when designing or constructing structures, or unethical in performing their required duties, occurred in 2003—it resulted in a nightclub fire that occurred in West Warwick, Rhode Island. One hundred people died in this fire and two hundred people were injured when fireworks being used on the stage hit

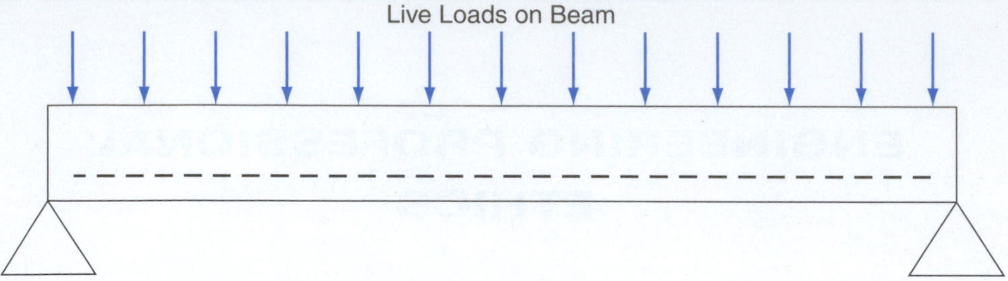

FIGURE 6.1 Traditional Structural Analysis Designs.

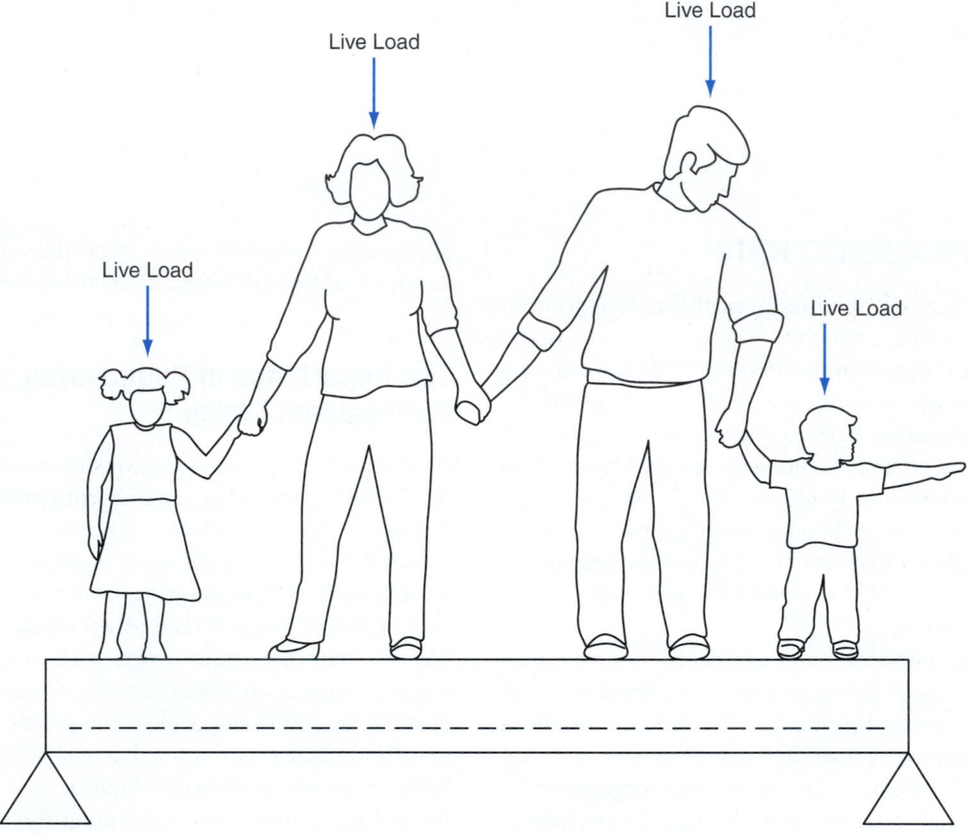

FIGURE 6.2 Revised Structural Analysis Designs.

the walls and ignited the wall insulation, which was highly flammable. Also contributing to so many deaths and injuries was the fact that the exit doors opened in rather than out, as required by most building codes. When the patrons in the night club rushed to exit the building, they pulled the doors in and the crowd surging through the doors blocked the doors and the patrons who fell down were crushed under the wave of people trying to get out of the building.

In the ensuing forensic engineering investigation, it was determined that improper insulation was used in the building. Criminal charges were filed against the manager of the band that was performing for using fireworks during the performance and against the nightclub owner. The charges leveled against the building owner by the families of the victims in the negligence case included cutting corners, allowing too many patrons into the nightclub, using cheap foam that was flammable as soundproofing, and not carrying any worker's compensation insurance for the building. The building owner was also fined $1 million by the state. Blame was also apportioned to the engineering inspectors who allowed the use of flammable foam as insulation in the building (Cannon 2004). Criminal charges were filed against the owner and the band manager and resulted in both of them serving jail sentences.

One of the patrons who survived the nightclub fire has undergone seventy different surgeries including operations to close his eyes and his head. The fire survivor lost "his fingers, part of his nose, and both ears. He lost his left eye, and his right is damaged, too. His legs are a patchwork where doctors have removed healthy skin to replace dead skin" (Cannon 2004, 35). This survivor of the nightclub fire will probably have to undergo seventy more surgeries. The hundred people who perished in the fire, and the survivor who

FIGURE 6.3 Construction Crane Accident March 21, 2005, in California.

will continue to undergo surgery for many more years of his life, are a reminder of why engineers and constructors should always be aware that everything they do at work could impact the lives of members of society and they need to always carefully analyze and review their work to try and ensure that no one will be harmed as a result.

Another example of what could happen if an engineer or constructor is careless in their daily work responsibilities is shown in Figure 6.3. This figure shows a crane accident that occurred on March 21, 2005, in California. The accident interrupted construction of the adjacent building and delayed the construction project for over a month while a forensic engineering investigation was conducted to determine the cause of the accident.

Both of the previously mentioned incidents demonstrate the important responsibilities engineers and constructors have in relation to the work they perform on a daily basis and why ethics are so crucial in the engineering and construction industry. Engineers who have tried to prevent accidents from happening, such as the engineer who attempted to convince management not to launch the ill-fated Challenger space shuttle, which blew up a few minutes after being launched, when he determined that the "o" rings would fail at the predicted launch temperature, have to live the rest of their lives with the guilt they carry because they did not do more to prevent a horrible accident from occurring when they were involved with a project. If every engineer and contractor is reminded of the results of these types of occurrences on a daily basis, it might help to reduce the number of accidents, injuries, and deaths that occur in the engineering and construction industry.

6.2 THE DEFINITION OF ETHICS AND MORALS

In order for societies to continue evolving, their economic base is intertwined with the pursuit of free enterprise, which thrives on competition. In any open society, the entire way of life of its citizens is held together by mutual trust and respect. Without this, the activities of commerce have to be formalized in written agreements or contracts. This section defines **ethics** and **morals** as they relate to their use in society and in professional pursuit of excellence.

Ethics

In a society where there are many different influences, a person's ethics are affected by their exposure to the people and activities surrounding their life. People develop their ethics based on their interactions with their parents; ministers, priests, rabbis, or imans; teachers and professors; family members; peers; and also through the mass media.

When the following three definitions of ethics were presented to college students in six regions of the United States including the West Coast, the South, the East, the Midwest, the Rockies, and the Great Plains, there were regional differences in their collective responses. In a nationwide survey of the general population, each of the definitions received the percentages listed next to each of the questions.

1. What my feelings tell me is right—50 percent
2. In accordance with my religious beliefs—25 percent
3. Conforms to the Golden Rule (do unto others as you would have them do unto you)—25 percent

When the college students were asked to select one of the definitions, the responses to "in accordance to my religious beliefs" responses were higher in the South than other regions of the United States. In the Midwest and the Great Plains, "conforms to the Golden Rule" received higher responses than in other regions of the United States.

According to *Webster's Dictionary*, the definition of ethics is: (McKechinie 2008, 454):

1. The study of standards of conduct, moral judgment, or moral philosophy.
2. Tie together the system of morals of a particular philosopher, religion, group or profession.

Morals

The definition of morals is "relating to, dealing with, or capable of making the distinction between right and wrong in conduct" (McKechinie 2008, 687).

Ethics versus Morals

Ethics allude to people conforming to codes of moral conduct such as those set forth by professional societies. Morals are related to people behaving in a manner that is acceptable to society.

6.3 THE ROLE OF ENGINEERING PROFESSIONAL SOCEITIES IN ETHICS

In order to help protect the welfare of the public engineering professional societies, each have a code of ethics. In the United States, there are over 150 engineering associations and societies. In addition to the discipline specific codes of ethics, there are other organizations such as the **National Society of Professional Engineers** (NSPE) and the **Project Management Institute** that also have codes of ethics. Professional societies monitor and try to enforce their particular code of ethics; they review cases and discipline their members for violations of the code of ethics.

6.4 THE PURPOSE OF PROFESSIONAL CODES OF ETHICS

The original intent of most engineering **codes of ethics** was to ensure that while engineers are trying to excel in their profession that they do it with moral purpose, rather than merely for personal gain. Another reason for codes of ethics is for engineering professional societies to be able to regulate themselves rather than being regulated by members of government agencies. The reasoning behind self-regulation is that only individuals who are proficient in a particular engineering discipline are qualified to evaluate the professional conduct of those who are practicing that engineering profession.

In the engineering professions, a code of ethics defines the standards of ethical behavior and of professional performance. A code of ethics outlines the duties and responsibilities required of those performing a particular engineering profession and provides a standard to which members of the profession should aspire while they are practicing the profession.

Original American Society of Civil Engineers Codes of Ethics

The American Society of Civil Engineers (ASCE) wrote the first U.S. engineering code of ethics in 1914. The original ASCE code of ethics (American Society of Civil Engineers 1914) was the following:

> It shall be considered unprofessional and inconsistent with honorable and dignified bearing for any member of the American Society of Civil Engineers:
>
> 1. To act for his clients in professional matters otherwise than as a faithful agent or trustee, or to accept any renumeration other than his stated charges for services rendered his clients.
> 2. To attempt to injure falsely or maliciously, directly or indirectly, the professional reputation, prospects, or business, of another Engineer.
> 3. To attempt to supplant (displace) another Engineer after definite steps have been taken toward his employment.
> 4. To compete with another Engineer for employment on the basis of professional charges, by reducing his usual charges and in this manner attempting to underbid after being informed of the charges named by another.
> 5. To review the work of another Engineer for the same client, except with the knowledge or consent of such Engineer, or unless the connection of such Engineer with the work has been terminated.
> 6. To advertise in self-laudatory language, or in any other manner derogatory to the dignity of the Profession.

The latest version of the ASCE code of ethics is more extensive than the original version and a copy of the 2008 version is provided in Appendix C.

Part of the ASCE code of ethics was written to help engineers understand how they should deal with one another: Do not compete with each other on the basis of price, do not advertise, and do not review the work of another engineer without their permission. Number one is still being practiced today in that engineers are not allowed to competitively bid for work, because lowering the price of design services to secure work could potentially compromise the safety of the structures being designed by a firm if the engineers designing them do not spend enough time on design details. Civil engineers are still restricted in how they advertise their services. It is acceptable to place a name, or the name of a firm, in an advertisement, such as in the yellow pages of a phone book or on the Internet, but it should be done in a professional manner that does not embarrass members of the profession. The medical and legal code of ethics removed this restriction in the later part of the twentieth century, which resulted in full-page advertisements and freeway billboards promising all types of positive results.

The third restriction of not reviewing the work of another engineer without their permission helps to prevent engineers from attempting to take work away from another engineer who is already working on a design for a client. It is acceptable to review the work of an engineer if the engineer has given their permission to do so.

Current Engineering Professional Society and Contractor Codes of Ethics

Appendix D provides samples of engineering codes of ethics for the American Society of Agricultural and Biological Engineers, the American Society of Civil Engineers, the American Society of Mechanical Engineers, the American Society of Chemical Engineers, the Institute of Electrical and Electronic Engineers, and the Institute of Industrial Engineers.

Current codes of ethics have expanded upon their original intentions to include other areas such as sustainability. Engineering codes of ethics were also formed to help lay people recognize legitimate engineers and to give citizens security and confidence that they will be provided with the engineering services they contract for to obtain their designs. The codes of ethics also help to maintain the integrity of the engineering profession and to provide certain privileges and an elite status. Unlike some of the other professions, such as law and medicine, there are many different codes of ethics in engineering because of the specific nature of engineering disciplines and subdisciplines.

Only five professions are registered by State Professional Boards of Registration and they are the following:

1. Accounting
2. Architecture
3. Engineering
4. Law
5. Medicine

The registration process for engineers is discussed in detail in Chapter 7.

Even though one of the main purposes of having engineering codes of ethics is self-regulation, it is not an entirely effective process since only 20 percent of U.S. engineers are professionally registered in a specific engineering discipline and 50 percent are members of professional societies (2009 data). Professional societies are only able to censure their members; therefore, nonmembers are not subject to the code of ethics of a professional society. Another regulatory obstacle is that the professional registration process is administered by individual states; therefore, someone who is banned from practicing in one state could conceivably obtain a license and practice in another state.

Contractors are represented by professional organizations such as the Associated General Contractors, the Associated Building Contractors, or other state-based professional organizations. Currently there is no all encompassing code of ethics for contractors; however, the U.S. government requires contractors to follow its code of ethics for contractors when they are working on federal government projects. The U.S. government's **Federal Acquisition Regulations** contain a section on **Contractor Business Ethics Compliance Programs and Disclosure Requirements**, the **Contractor Code of Business Ethics and Conduct**, which has to be followed by contractors performing for work the U.S. government. The U.S. government's Contractor Code of Business Ethics and Conduct is the following:

CONTRACTOR CODE OF BUSINESS ETHICS AND CONDUCT

Summary:

The Federal Acquisition Regulation (FAR) Council today issued the final rule on its "Contractor Business Ethics Compliance Program and Disclosure Requirements" on November 12, 2008 and it becomes effective on December 12, 2008. The new rulemaking amends the requirements for a contractor code of business ethics and conduct, an internal control system, and disclosure to the Government of certain violations of criminal law, violations of the civil False Claims Act, or significant overpayments.

In summary, the final rule requires Federal contractors to:

- Establish and maintain specific internal controls to detect and prevent improper conduct in connection with the award or performance of any Government contract or subcontract; and
- Timely disclosure to the agency Office of the Inspector General, with a copy to the contracting officer, whenever, in connection with the award, performance, or closeout of a Government contract performed by the contractor or a subcontract awarded there under, when the contractor has credible evidence of a violation of Federal criminal law involving fraud, conflict of interest, bribery, or gratuity violations found in Title 18 of the United States Code; or a violation of the civil False Claims Act (31 U.S.C. 3729 3733).

The rule also provides as cause for suspension or debarment, knowing failure by a principal, until 3 years after final payment on any Government contract awarded to the contractor, to timely disclose to the Government, in connection with the

(*continued*)

award, performance, or closeout of the contract or a subcontract there under, credible evidence of:

- Violation of Federal criminal law involving fraud, conflict of interest, bribery, or gratuity violations found in Title 18 of the United States Code;
- Violation of the Civil False Claims Act; or
- Significant overpayment(s) on the contract, other than overpayments resulting from contract financing payments as defined in FAR 32.001, Definitions. (Federal Acquisition Regulations, December 12, 2008)

6.5 ENFORCING ENGINEERING CODES OF ETHICS

Engineering professional societies strive to protect the integrity of the profession and to set standards of professional conduct and service, they oversee professional education, and they help to advance the profession. In order to be able to accomplish these objectives, members of each professional engineering society are responsible for policing their profession. Members of engineering professional societies have to notify the licensing boards or their professional society if they become aware of another engineer violating their code of ethics.

The license of an engineer could either be suspended or revoked for submitting inaccurate or misleading information, improper conduct, gross incompetence, or gross misconduct. In addition to license revocation, a member could also be suspended from their professional society. If the conduct of an engineer goes beyond mere ethical violations and they violate the law, then they would be tried in the court system; this is discussed in Chapter 9 of Section 9.

One violation that should always be reported by engineers is if they encounter an engineer who is using a professional title such as civil engineer, mechanical engineer, or electrical engineer if he or she is not a professionally licensed engineer. Unregistered engineers are able to call themselves other types of engineers such as a design engineer, a project engineer, a field engineer, or a traffic engineer, as long as they do not call themselves one of the licensed engineering titles.

6.6 CONFLICTS CREATED BY ENGINEERING PROFESSIONAL CODES OF ETHICS

Ninety percent of engineers work for a business or the government rather than being self-employed; therefore, their work is directed by their employers. Engineers have to be able to justify their actions to their employers and perform under the restrictions of their professional code of ethics. The conflict created by this duality is the protection of public welfare versus employer confidentiality. Should engineers do whatever they are being paid to do or should they refuse to perform work when the work would endanger members of society? This dilemma haunts engineers throughout their careers.

Confidentiality Statements

Engineering and construction firms may require that all of their employees sign a **confidentiality statement** when they are hired to work for the firm. A confidentiality statement indicates that the employee will not share information about their work with anyone outside of the firm unless requested to do so by the firm. The main objective of using confidentiality statements is to protect proprietary information that is used within a firm; however, this causes a conflict when an engineer discovers that members of a firm are doing something that could potentially harm the public and the engineer is not able to legally share that information outside of the firm. Companies will sue employees if they violate their confidentiality statements that they have signed at their firm. A firm could also apply for an injunction against an employee to stop them from divulging company information.

Dr. Jeffrey Wygant chose to violate his confidentiality statement and provide company information to the news media and the courts demonstrating that the firm he was working for knew that cigarettes were harmful to the health of people due to tests conducted inside of the firm, but the firm continued to lie about the harmful effects of cigarettes. Members of the firm lied under oath to members of the U.S. government when they were asked if they knew whether cigarettes were harmful.

Dr. Wygant was dismissed from the firm without his pension or medical benefits and was hounded by members of the firm for disclosing company information to the press. For a while he required protection by the FBI since he was receiving threats through the mail and e-mail. In order to provide testimony in a court case, he had to defy a court order to suppress information. He also appeared on the television show ***60 Minutes*** to discuss the information he had divulged concerning his former employer. Dr. Wygant was not able to find employment in his profession after he blew the whistle but he was able to find employment as a high school chemistry teacher. He continues to speak in locations throughout the United States about his experience and the importance of whistle-blowing to stop unethical behavior from happening within firms.

Examples of Engineering Ethical Dilemmas

Some examples that help to illustrate the conflicts that engineers might face while practicing their profession are the following:

1. A structural engineer is working on the design of a building and his or her employer asks him or her to make the structural steel members "slightly" smaller than the design calculations indicate that they need to be, thus saving the owner money.
2. Rather than include the amount of rebar indicated on the drawings and in the specifications for a concrete slab, a contractor provides money to an inspector and the inspector approves a reduced amount of rebar in the slab.

3. An engineer is provided with gifts by a vendor that is competing with other vendors during the process of deciding where to order long lead-time items or a vendor offers the engineer a monetary incentive to use his or her firm.
4. An engineer is hired to work on a project that is being paid for by the owner on a cost-plus-a-percentage payment system. His or her employer is applying a multiplier of three times their salary and charging the owner that amount. The engineer is not given any work to do, and when he or she asks about being idle, the boss tells him or her not to worry and to just keep showing up for work and do nothing since the employer is making money based on the number of people he or she says are working on the project.
5. An engineer has been working on designs for client "A" and client "B." After a few months when he or she is filling out his or her time sheets, he or she is told by his or her employer to charge his or her time to client "B," even though he or she is actually working on designs for client "A." When questioned about why he or she is doing this, the employee is told they have expended all of the money from client "A," but they have not completed the work for client "A" yet.
6. An engineer is hired by a firm that is working under contract to the government. Toward the end of the fiscal year (which is October for the federal government), the engineer notices that his or her firm is spending money on equipment that is not necessary for the job. When the engineer asks their employer why the equipment is being ordered, he or she is told that if they do not spend out their budget by the end of the fiscal year, they will not receive the same amount for the next year but rather a lower amount of funds.
7. An engineer is working for a firm that is required to x-ray steel welds on a pipeline. During the course of their employment they discover that some of the X-rays of the welds are merely duplicates of one weld with different location numbers written on each one of the welds.
8. An engineer is ordering steel from a manufacturer and he or she visits the manufacturer to watch tests being performed on the steel to determine its strength. After the tests are completed, the engineer approves the steel and an order for the steel is placed. Eventually it is shipped to the project job site. The engineer is present as the steel is loaded onto a barge that will take it to the project job site. During transit the steel is replaced by inferior steel and the inferior steel is delivered to the project job site instead of the steel that was approved by the engineer.
9. An electrical engineer is working for a firm that builds airplanes. During one of his projects he or she discovers that one of the circuit boards used in the instrument panel has a defect in it. The engineer informs his or her boss about the defect and is told by the boss to not worry about it and to continue working on the design. A few years later one of the firm's airplanes crashes and kills 364 people. During the inquiry into the cause of the accident and why the airplane crashed, it is suggested that there was a defect in the instrument panel and that could have caused the plane to crash.
10. The suspended country building inspector, who claims he and other inspectors routinely accepted gifts from contractors and took materials from construction sites, will get a chance to fight for his job. The inspector will meet with Manatee County officials Friday for a pre-termination meeting, during which he can respond to misconduct charges or present information that may affect Manatee County's decision on whether to fire him, according to a January 4 letter to him from the interim director of the building department. The inspector had worked for Manatee County for five years. On December 3, the inspector signed a sworn affidavit attesting that he had taken goods and materials from contractors while employed as a building inspector. He also admitted, as alleged in the pre-termination letter, he knew of other country building inspectors who engaged in similar, unauthorized practices and alleged that the practice was widespread in the department. Under country ethics policy, building department employees are prohibited from accepting gifts (ENR.com, January 9, 2008, p. 1).

Unfortunately, all of the examples cited in this section actually happened to engineers when they were working for firms in the industry. Some of them left their jobs when they discovered what was happening, some decided to stay and keep quiet about what was happening, and others blew the whistle and reported what was happening to people outside of the firm. The next section discusses whistle-blowing and how it could be accomplished by engineers when they discover that something members of their firm are doing could be potentially harmful to society.

6.7 WHISTLE-BLOWING IN THE ENGINEERING PROFESSION

The examples listed in Section 6.6 illustrate the types of ethical conflicts that engineers could face during their career. When they have to address these types of ethical issues, they have to decide whether they will report them or not report them. If they report the ethical violation to their superior, and to upper level managers, and nothing is done to stop it from happening, then their next decision is whether to blow the whistle, or **whistle-blowing**, which is a term for going public or to the authorities with the information they possess from working for the firm.

It is at this point that an engineer has to determine whether their allegiance is to the firm they are working for or to protecting public safety. The engineering codes of ethics state that the primary responsibility of an engineer is public safety. However, when engineers make high salaries, they worry about losing their job if they blow the whistle, so it is hard for them to decide whether to blow the whistle. The code of ethics of the U.S. Government Service states, "put loyalty to the highest moral principles and to the country above loyalty to person, party, or government department" and no less should be expected of engineers (U.S. Government Service 2008, 1).

6.8 DECIDING WHEN TO BLOW THE WHISTLE

Unfortunately, ethical issues may not be obvious and may require that judgment be exercised by an individual in determining the risk they pose to the public. Some examples of ethical decisions that occur in many industries are: (Nadar 1973, 4):

1. Defective parts, vehicles, or equipment in the process of being marketed to unsuspecting consumers.
2. Vast waste of government funds by private contractors or government agencies.
3. The industrial dumping of mercury or other toxic substances in waterways.
4. The connection between companies and campaign contributions.
5. A pattern of discrimination by age, race, or sex in a labor union or a company.
6. Mishandling the operations of a worker's pension fund.
7. Willful deception in advertising a worthless or harmful product.
8. The use of government power for private, corporate, or industry gain.
9. The knowing nonenforcement of laws being seriously violated such as pesticide laws.
10. Corruption in an agency or company.

According to Ralph Nadar, who has been an advocate for safe products in the marketplace, in his article "An Anatomy of Whistle Blowing" before someone decides to blow the whistle they should ask themselves the following questions: (Nadar 1973, 5):

1. Is my knowledge of the matter complete and accurate?
2. What are the objectionable practices and what public interest do they harm?
3. How far should I go, and can I go, inside of the organization with my concern or objection?
4. Will I be violating any rules by contacting outside parties and, if so, is whistle-blowing nevertheless justified?
5. Will I be violating any laws or ethical duties by not contacting external parties?
6. Once I have decided to act, what is the best way to blow the whistle—anonymously, overtly, by resigning prior to speaking out, or in some other way?
7. What will be the likely responses from various sources—inside and outside the organization—to the whistle-blowing activities?
8. What is expected to be achieved by whistle-blowing in the particular situation?

The next section discusses potential methods that could be used when blowing the whistle on unethical or illegal behavior.

6.9 METHODS FOR WHISTLE-BLOWING

Whistle-blowing is one of the more difficult aspects of engineering ethics because there are no right answers and engineers are required to make a determination of whether a situation warrants someone going outside of a firm to report an ethical violation. Suggestions for how to blow the whistle are provided by Lester A. Myers, Ph.D., J.D., C.P.A., director of advisory services at KMPG Forensic in Washington, D.C. (PM Network, March 2008, 26–27):

1. **Look outside of the inner circle.** Try to find a "separate apparatus," such as an ombudsman's office, where you are able to report concerns. An internal group could qualify if it is independent from the company's own legal counsel, but he suggests avoiding any corporate attorney. "Attorneys that work for the organization, work for the organization."
2. **Respect the chain of command.** If possible, follow the established process for making a complaint. Going around or over someone could brand you as reactionary or rebellious and diminish your credibility, which is why it is important for an employee to get as many facts as possible, to deliberate carefully, and to act in good faith when filing a report alleging fraud or misconduct. "Sometimes, alleged violations may turn out to be benign or innocuous and employees and other covered persons must respect the process to preserve the integrity of their conscientious concern and their credibility of their message."
3. **Use an internal standard.** Adopt a company code of conduct as the basis for your report. Although an outside standard such as the ***PMI* Code of Ethics and Professional Conduct** is helpful, using an internal one could show how the company is "at variance with everything the organization says about itself."

The most difficult aspect of whistle-blowing is suffering the consequences when supervisors within the organization try to discredit the whistleblower in an attempt to protect whoever is guilty of the ethical violation. This is especially true if it is the supervisor who is the one who either was involved in the ethics violation or was covering for someone else who was unethical.

Some whistleblowers wait until they have secured another job before blowing the whistle on their former firm. One example of a successful whistleblower that occurred in the engineering industry involved an employee who discovered that her boss, and a few other employees, were taking kickbacks from contractors when they awarded contracts to contractors through their government agency. When she confronted her boss, she was also offered money and told she would receive a portion of each of the kickbacks if she would keep quiet about what was happening in the department. The engineer had no interest in being a part of this scheme and she sought advice from a former professor. She was advised by the professor to copy everything on her computer, to make copies of all of the evidence that verified that the kickbacks were happening, and to secure the computer copies and documentation in a safety deposit box.

The engineer started to be harassed by her boss and other employees and it escalated into a hostile work environment. Her home was broken into, and her computer was stolen out of her house. After a month, she negotiated a severance package with the company of six months pay and left the company to seek employment elsewhere. After she found another position as an engineer, she presented her evidence to the district attorney. The district attorney pressed charges and after a brief trial three of the employees of the firm received jail sentences.

In some instances, unethical behavior crosses way over the line and becomes criminal, as the above case illustrates. Another much higher stakes bribery case follows:

A Houston-based construction firm pleaded guilty to federal criminal charges alleging it paid millions of dollars in bribes to Nigerian officials to win contracts to build a massive natural gas project in the country.

The engineering giant and military contractor will pay $402 million in fines and spend three years under watch of a court-appointed monitor, according to the plea agreement entered before the U.S. District Judge. The former parent company of the construction firm will pay most of the fine under terms of the two companies' separation agreement of almost two years ago. A guilty plea was entered on behalf of the company.

The five charges—one count of conspiring to violate the Foreign Corrupt Practices Act and four counts of violating the act—stem from allegations of a decade-long scheme in which the construction firm, through intermediaries, paid more than $180 million in bribes to Nigerian officials to land contracts to build a $6 billion gas liquefaction plant on Nigeria's Bonny Island.

Gas liquefaction is the process of turning natural gas into a liquid so it can be transported on tankers. (ENR.com, February 23, 2009)

6.10 WHISTLEBLOWER PROTECTION LAWS

Some states now have **Whistleblower Protection Laws** to provide individuals who report unethical or illegal behavior of government employees to the state with protection if they are also state employees. The Whistleblower Protection laws normally state that an employee will not be terminated if they report unethical or illegal activities and that others will be censured if they harass the employee in any manner. If an employee chooses to report an unethical or illegal incident anonymously, then the Whistleblower Protection laws would not apply to them. For additional information on individual state Whistleblower Protection laws, see the individual state government websites.

Chapter Summary

In this chapter, professional ethics were covered including a discussion on the difference between morals and ethics. Engineering professional societies play an important role in developing and enforcing engineering codes of ethics; this influence on the ethics of the engineering profession was explained in this chapter along with how they enforce codes of ethics. The reasons for having engineering codes of ethics were presented along with a discussion on the conflicts that engineers experience, as a result of being independent autonomous agents that have to also protect the public welfare.

The concept of whistle-blowing, as it relates to exposing potentially dangerous or harmful consequences of the actions of members of firms, was explained. Information was provided on how to decide whether to blow the whistle and procedures for blowing the whistle. Examples of engineering ethical dilemmas were also included in this chapter that have occurred in the engineering and construction industry. Whistleblower protection laws were mentioned in terms of the type of protection that they provide to state employees who are whistleblowers.

Key Terms

60 Minutes
American Institute of Architects
American Institute of Chemical Engineers
American Society of Agricultural and Biological Engineers
American Society of Civil Engineers
American Society of Mechanical Engineers
code of ethics
confidentiality statements
Contractor Business Ethics Compliance Program and Disclosure Requirements
Contractor Code of Business Ethics and Conduct
engineering professional ethics
ethics
Federal Acquisition Regulations
Institute of Civil Engineers
Institute of Electrical and Electronic Engineers
Institution of Mechanical Engineers
morals
National Society of Professional Engineers
Project Management Institute
Project Management Institute Code of Ethics and Professional Conduct
Royal Institute of British Architects
Whistleblower Protection Laws
whistle-blowing

Discussion Questions

6.1 What is whistle-blowing and why is it necessary in the engineering profession?

6.2 When should someone blow the whistle when they are working in the engineering profession?

6.3 Explain the conflict that engineers face when they are working for employers but they are also individual autonomous agents.

6.4 What types of ethical violations should engineers report to their engineering professional society when they discover other engineers doing them?

6.5 How does the issue of confidentiality affect engineers when they are working for employers that require confidentiality statements to be signed in order to work for them?

6.6 How should an engineer blow the whistle when they discover unethical behavior within their organization?

6.7 In what manner are engineers allowed to advertise their services and what is the reason they are required to do it in this manner?

6.8 How does a layperson know when hiring someone whether this person is legally qualified to perform engineering services or not?

6.9 Are engineers allowed to review the work of another engineer and if so under what circumstances?

6.10 Are engineers allowed to compete for work on the basis of price? Why or why not?

6.11 Why do engineering professional societies have engineering codes of ethics?

6.12 Explain the difference between morals and ethics.

6.13 Why would it be hard for engineering professions to be regulated by government officials rather than by members of their own profession?

6.14 Should engineers be members of their engineering professional society? Why or why not?

6.15 Why is it important for engineers to understand the ramifications of signing confidentiality statements?

References

American Society of Civil Engineers. 1914. *Civil Engineering Code of Ethics*. Reston, VA: American Society of Civil Engineers.

ENR.com. January 9, 2008. Inspector gets shot to fight for his job. *Engineering News Record*. New York: McGraw Hill Publishing Company.

Federal Acquisition Regulations. December 12, 2008. *Contractor Business Ethics Compliance Program and Disclosure Requirements*. Washington, DC: Government Printing Office.

Cannon, A. February 16, 2004. Looking for the answer in the ashes. *U.S. News and World Report*, 33–37.

Nadar, R. 1973. Anatomy of Whistle Blowing. *Whistleblowing: The Report of the Conference of Special Responsibility*. Grossman, NY.

McKechinie, J. L., ed. 2008. *Webster's New Universal Unabridged Dictionary*. New York: Simon and Shuster.

Meyers, L. March 2008. Steps for whistleblowing. Newtown Square, PA: PMI Network.

CHAPTER **SEVEN**

ENGINEERING PROFESSIONAL REGISTRATION AND THE CONTRACTOR LICENSING PROCESS

7.1 INTRODUCTION

Obtaining professional engineering registration or contractor licenses is a major milestone in the career of engineers and constructors. Not only does a license allow an engineer to design projects that require professional registration, but it also provides a measure of their professional ability to potential clients. Engineering professional registration also allows engineers to legally use the title of their profession. Licensing allow contractors to operate legally in their state of residence if licenses are required in their state. State agencies issue professional engineering licenses and licenses for contractors through **State Boards of Professional Registration**. Engineers and contractors have to renew their license every year, and some states require that an engineer show evidence of continuing education credits to renew a license.

This chapter provides information on licensing procedures and the importance of obtaining professional engineering registration or a license as a contractor. This chapter also covers out-of-state practice, violations of engineering licensing laws, moonlighting, expert witness testimony, and depositions.

7.2 JUSTIFICATION FOR ENGINEERING PROFESSIONAL REGISTRATION

One of the primary reasons that states require professional registration is because it is difficult for nonengineers to judge for themselves whether an engineer is performing work properly or not. State licensing requirements provide a method that demonstrates to potential clients whether an engineer meets minimum competency and integrity requirements. Professional registration also ensures that an engineer has the required education and experience to be practicing engineering.

7.3 THE PROFESSIONAL ENGINEERING LICENSING PROCESS

Each state determines the requirements that have to be meet in order for the state to issue a professional engineering license to an individual. The State Board of Professional Registration in the state where someone will be practicing engineering should be contacted to obtain specific licensing requirements. Some states require that an engineer be a U.S. citizen, or that they meet minimum residency requirements, before applying to take the licensing examination. States also set the minimum age at which an engineer is eligible to take their first examination; it ranges from between twenty-one and twenty-four years of age.

Fundamentals of Engineering Examination (FE)

In order to be eligible to take the **Fundamentals of Engineering (FE) Examination**, an engineer has to obtain their engineering degree from a university that is accredited by the **Accreditation Board for Engineering and Technology** (ABET), as an engineering program not a technology program. Universities are reviewed on six-year cycles to ensure that they meet the minimum requirements as set forth by the ABET on courses, course content, faculty qualifications, facilities, educational procedures, and other criteria. If an engineering program falls below the minimum requirements in any area, they receive probationary status and they are provided with a set time frame to alleviate program deficiencies.

Two eight-hour examinations are required for licensing and the first examination, which is called the Fundamentals of Engineering Examination (FE), formerly known as the **Engineering-in-Training Examination** (EIT), is normally taken during the last semester or quarter of engineering programs or in the fall after an engineer graduates from college.

There are two sessions during the FE examination: (1) a morning session and (2) an afternoon session. The morning session has one hundred and twenty questions in twelve topic areas:

- Mathematics—15 percent
- Engineering Probability and Statistics—7 percent
- Chemistry—9 percent
- Computers—7 percent
- Ethics and Business Practices—7 percent
- Engineering Economics—8 percent
- Engineering Mechanics (Statics and Dynamics)—10 percent
- Strength of Materials—7 percent
- Material Properties—7 percent
- Fluid Mechanics—7 percent
- Electricity and Magnetism—9 percent
- Thermodynamics—7 percent

The afternoon session has sixty questions in seven topic areas as discipline-specific multiple-choice questions that cover chemical, civil, electrical, environmental, industrial, mechanical, or other/general engineering. For detailed information on the afternoon examination subjects for each engineering discipline, see the **National Council of Examiners for Engineering and Surveying** website at www.ncees.org/exams/professional/fe.

The FE examination is scored electronically with several months passing before the scores are released to the individuals who have taken the examination. Although each individual state administers the examinations in their state, they could use common examinations. The FE examination occurs twice a year in October and April. Engineers take the FE examination in a location assigned by the state based on zip codes. Application deadlines for the FE examination are several months prior to the examination and the specific deadlines are listed on the State Board of Professional Registration websites. The State Board of Registration issues a reference handbook that could be used during the examination. There are restrictions on the types of calculators that may be used during the examinations and information on which calculators are allowed should be verified prior to the examination by reviewing the guidelines on the State Board of Registration websites or the National Council of Examiners for Engineering and Surveying website.

The Professional Engineers Principals and Practice Examination

After an engineer has passed the FE examination, they have to meet the minimum requirement for work experience before applying to become a **Professional Engineer** (PE) by taking the **Principals and Practice Examinations**. States require between two and four years of professional engineering experience prior to taking the principals and practice examinations. In some states, master's degrees and Ph.D. degrees in specific engineering disciplines are accepted as work experience. According to the **National Council of Examiners for Engineers and Surveying Model Law** Section 230.40, if an applicant has a Ph.D. from an institution that offers ABET-accredited undergraduate engineering degrees and if his or her undergraduate degree is from an ABET-accredited program, then he or she is not required to take the Fundamentals of Engineering Examination before taking the **Professional Engineering Examination** (NCEES 2009)

When an application to take the principals and practice examination is submitted, it should include letters of reference and other letters might be required that address the character, honesty, and integrity of the applicant from coworkers, supervisors who have a PE license, or other individuals who have worked with the engineer.

The principals and practice examinations for civil and mechanical engineering require a discipline-specific examination in the morning, and a breadth examination in the afternoon that is subdiscipline specific. The afternoon examination requires civil engineering examinees to complete one of five breath areas: (1) construction, (2) geotechnical, (3) structural, (4) transportation, (5) water resources and environmental. Mechanical engineers have to complete one of three areas: (1) heating, ventilation, and air conditioning; (2) mechanical systems and materials; (3) thermal and fluids systems. The electrical engineering examination has subdiscipline areas of computers, electrical and electronics, and power. The industrial engineering examination covers facilities engineering and planning, systems analysis and design, logistics, work design, ergonomics and safety, and quality engineering. The chemical engineering examination covers mass/energy balance and thermodynamics, fluids, heat transfer, mass transfer, kinetics, and plant design and operation. The environmental examination covers water; air; solid, hazardous, and special waste; and environmental assessment, remediation, and emergency response. The structural I examination covers analysis of structures and design and details of structures; the structural II examination covers buildings and bridges.

The principals and practice exams have eighty to one hundred multiple-choice questions except for the structural II examination, which includes eight essay questions, four of which have to be answered during the examination. The principals and practice examinations that are administered by states are listed in Table 7.1.

Some engineering disciplines have multiple professional engineering examinations that are required for obtaining additional licenses. One example is the discipline of civil engineering where there are two separate structural eight-hour examinations. A civil engineer could be both a registered professional engineer and a registered structural engineer.

Once an engineer has passed the principals and practice examination, they are said to be **admitted to practice** and may now legally refer to themselves as a civil engineer, a mechanical engineer, a chemical engineer, an electrical engineer, an industrial engineer, or other type of professional engineer. If an unlicensed engineer refers to himself or herself as a particular type of engineer such as civil engineer, mechanical engineer, or industrial engineer, this practice is referred to as **holding out**

Table 7.1 Principals and Practice Examinations

Agricultural	Environmental	Naval Architecture and Marine Engineering
Architectural	Fire Protection	Nuclear
Chemical	Industrial	Petroleum
Civil	Mechanical	Structural I
Control Systems	Metallurgical and Materials	Structural II
Electrical and Computer	Mining and Mineral Processing	

Source: National Council of Examiners for Engineering and Surveying, May 13, 2009, *Principles and Practice Exams*, www.ncees.org/exams/professional/pe.

since they are not licensed by the state. They may still legally practice engineering under certain circumstances that are described in Section 7.5. Unlicensed engineers may not design structures that require a Professional Engineer to design them, such as institutional structures. Some projects under a certain size (in square feet) or a certain dollar volume are exempt from having to be designed by a Professional Engineer.

When an engineer passes the principals and practice examination, they receive a stamp with their unique license number on it, which is used to stamp drawings after they have been reviewed, analyzed, and signed by the Professional Engineer. The engineer who signs off on a drawing is the one who is legally liable for the accuracy of the drawings. This is one of the reasons that some engineering firms prefer to pay someone outside of their firm to review, sign, and stamp drawings since it reduces their legal liability.

7.4 OUT-OF-STATE ENGINEERING PRACTICE

In order to practice engineering in another state, an engineer needs to ascertain what the requirements are for that particular state. Many states allow **reciprocity**, or **comity**, which is a process whereby one state accepts the registration of another state. Some states accept a license with the provision that an engineer take any sections of the PE exam required in the new state that were not included in the state where the engineer passed his or her examination. One example of this occurs in the sate of California where Professional Engineers are required to take an earthquake examination in order to practice in California with an out of-state-PE license.

States have the right to not allow someone an architecture or engineering license if they have been censured in other states, as the following lawsuit indicates [*Hughes v. Board of Architectural Examiners*, 68 Cal. App. 4th 685, 80 Cal. Rptr.2d 317 (Cal. App. 3rd dist. 1998)]:

> An architect's license in California was properly revoked for conduct in other states before he was admitted to California, even though no violation occurred in California and no California clients were harmed.

In some states, an engineer may obtain a temporary license if they will only be practicing engineering in that state for a limited amount of time. Foreign countries do not have the same professional registration process as the United States; therefore, citizens of foreign countries might take the FE and the PE examinations in the United States in order to demonstrate their professional credibility. Some of the engineering departments at foreign universities have ABET accreditation so that their graduates are eligible to take the FE and PE examinations.

7.5 CRIMINAL VIOLATIONS OF ENGINEERING LICENSING LAWS

In Chapter 6, the enforcement options for violations of engineering codes of ethics were discussed in relation to the jurisdiction of professional societies. If an engineer violates criminal statutes, then a case is heard in the court system not reviewed by their engineering professional society. If an infraction was minor, an engineer would probably be granted probation or possibly a suspended sentence. For major criminal offenses engineers would receive jail sentences. In addition to **criminal sanctions**, there are **quasi-criminal sanctions** that are used to stop unlicensed design professionals from practicing their profession by issuing a court order. If an engineer continues to practice after a court order has been issued, they could be held in **contempt of court** and be subject to paying a fine or serving time in jail.

If an engineer works on a design where a license is required by law, and the engineer is not a licensed engineer, then he or she would have no legal recourse if not paid for his or her services by a client. This is probably the most important reason for an engineer to obtain a Professional Engineers license, although this person could be working for a firm where someone has a license and he or she would not have to be licensed for that firm to collect their fees since someone in the firm has an engineering license.

7.6 MOONLIGHTING AS AN ENGINEER—UNREGISTERED DESIGN PROFESSIONALS

As mentioned in Section 7.4, if an engineer performs work for a client that requires a license, the engineer may not be able to recover the fee for services rendered to the client, but may be allowed to work on designs for projects that are not regulated by licensing requirements. Engineers should notify their clients of their licensing status, especially if a project involves issues of public safety. Clients have the option of waiving the liability of the engineer for their performance, since the engineer would not be eligible for liability insurance without having a PE license.

The term **moonlighting** is used to describe engineers who have the educational requirements to be practicing

Expert witnesses provide four types of services:

1. They investigate the particular scene, or event; research everything written on the subject; run tests; and then analyze and evaluate their findings.
2. They evaluate the merits of a potential claim and document their work with a written report on their findings. They express their opinion about the case or the problem and the merits of the claim.
3. They recommend certain aspects of litigation strategy. The client-attorney may not be suing the right party. The expert may know more about the law in this specific area because experts tend to collect articles and citations on matters of interest to them. They will suggest other areas to be investigated or tested. They probably know the opposing expert and can project the arguments that he or she may use. These recommendations make the expert a valuable assistant to the client-attorney.
4. They testify in depositions and at trial to explain and then defend the technical conclusions they have reached. The Federal Trade Commission even allows expert witnesses to cross-examine opposing witnesses before an administrative law judge. This is permitted because they know their subjects better than the lawyers. (Poynter 1987, 25)

One example of when expert witnesses are used in law cases is when a lawyer needs an evaluation of additional information on studies surrounding a site investigation such as:

- soils reports
- surveys
- water flow analysis
- water quality analysis
- soil stability
- concrete strength evaluations

Expert witnesses are also used in traffic accident cases to evaluate roadway conditions, or design elements of sections of roads where an accident has occurred along a highway or freeway. Expert witnesses provide testimony on the integrity of a structure at the time an accident occurred, or a person was injured on a job site in cases where there are injuries or deaths due to a construction or structural failure.

Expert witnesses are paid on an hourly basis to review the case documents prior to their testimony, to provide input on cases to lawyers, to provide depositions, or to testify in court. They may also review the depositions of other witnesses or expert witnesses to provide feedback on the statements provided by other individuals involved in the case.

In order to be an expert witness, an engineer must not have had any type of affiliation with the case or the firms involved in the case. Expert witnesses shall not have worked for either of the firms involved in the case or their subsidiaries at any time in the past. There should not be any evidence of conflict of interest on the part of expert witnesses. The lawyers involved with cases need unbiased, expert opinions; therefore, they carefully screen potential expert witnesses to ensure that there is nothing in their past that would taint their testimony during a case.

Chapter Summary

This chapter presented information on engineering professional registration and contractor licensing processes. It also covered the responsibilities that are associated with being admitted to practice as an engineer or a licensed contractor, and the processes required for being able to practice in other states as a professional engineer including reciprocity (comity). This chapter also discussed the types of sanctions that are used when engineers violate their licensing agreement.

The second half of this chapter discussed moonlighting as an unregistered design professional. Information was also provided on being deposed for legal cases and how engineers are involved in court cases as expert witnesses.

Key Terms

Accreditation Board for Engineering and Technology
admitted to practice
comity
contempt of court
criminal sanction
depositions
Engineer-in-Training Examination
expert witness
Fundamentals of Engineering Examination
holding out
moonlighting
National Council of Examiners for Engineering and Surveying
National Council of Examiners for Engineering and Surveying Model Law
Principals and Practice examinations
Professional Engineer
Professional Engineering Examination
quasi-criminal sanction
reciprocity
State Boards of Professional Registration

Discussion Questions

7.1 If a student plans to take the Fundamentals of Engineering Examination, what types of questions should they be prepared to answer?

7.2 How do the requirements for being eligible for the Professional Engineers Examination differ from the requirements for the Fundamentals of Engineering Examination?

7.3 Investigate the requirements for taking the Professional Engineering and Contractor Licensing Examination in the state where this course is being taught and list them.

7.4 Why should engineers obtain a Professional Engineer's license?

7.5 Why are letters of reference required when an individual applies to take the Professional Engineers Examination?

7.6 Investigate the Accreditation Board for Engineering and Technology and explain the purpose of this organization and how this organization is related to the Professional Engineering licensing process.

7.7 If registered professional engineers plan to perform design work in a state other than the state in which they obtained their registration, what do they need to do?

7.8 What happens to professional engineers when their actions are a criminal offense? What types of penalties are they likely to receive if they are tried in a criminal court?

7.9 Explain what moonlighting is in relation to engineers. Why it is allowed to happen?

7.10 Discuss why there are problems in enforcing contractor licensing requirements.

References

Clough, R. H., G. A. Sears, and S. K. Sears. 2005. *Construction Contracting*. Hoboken, NJ: John Wiley and Sons.

ENR.com. September 24, 2008. Contractors held liable for independents; Court extends responsibilities. *Engineering News Record*. New York: McGraw Hill Publisher.

Lewis & Queen v. N. M. Ball Sons (1957) 49 Cal. 2d 141, 308 P.2d 713.

NCEES. 2009. *National Council of Examiners for Engineering and Surveying Model Law*. Washington DC: National Council of Examiners for Engineering and Surveying.

Poynter, D. 1987. *The Expert Witness Handbook: Tips and Techniques for the Litigation Consultant*. Santa Barbara, CA: Para Publishing.

The State Board of Architects v. Clark, 689 A.2d 1247 (Md. Appl 1997).

CHAPTER EIGHT

AGENCY RELATIONSHIPS

8.1 INTRODUCTION

When engineers represent clients and perform work on the behalf of their clients, a relationship has to be legally established between them. An agency relationship exists when one party is given the legal right to act on the behalf of another party. The term **agency** is used to describe a **fiduciary** relationship between a principal, such as a client, and the person who is representing the principal and their interests. A fiduciary relationship is a relationship that includes mutual trust and confidence (Sweet and Schneier 2005).

This chapter explains agency relationships and how they are created in the U.S. legal system. Issues of authority are also included in this chapter along with a discussion on agency disputes and how to terminate agency relationships.

8.2 DEFINITION OF AGENCY

Agency is a legally binding relationship that creates a situation where one person is trusted to perform the duties of another person. Agency agreements allow one person to represent someone else and their actions create an obligation for the person they are representing as an agent. "As a rule, the relationship of principal and agent is created by a contract expressed in words or manifested by acts. The agent may be a regular employee of the principal or an independent person hired for a specific purpose and not controlled as to the details of the requested activities" (Sweet and Schneier 2005, 26).

When an agent is legally representing a principal, this provides third parties that are working with the agent with assurance that they would be able to hold the principal responsible for the actions of the agent. Unfortunately, an agent may inadvertently bind a principal to a commitment that the principal does not desire to be a party to, but the principal is obligated for it, due to an agency agreement. The legal system attempts to provide some measure of protection to principals when an agent binds the principal but the legal system also has to try to protect the rights of third parties that are relying on the commitments made by agents.

8.3 CREATING AGENCY RELATIONSHIPS

Agency relationships are created either with written contracts stating that an agent may act on the behalf of a principal or by the principal acting in a manner that is construed by third parties to indicate that a particular person is an agent of the principal. Agency relationships must be in writing in the states in which it is a legal requirement, but they could be verbal agreements in states that do not require written agency agreements.

When an engineer or a constructor is hired to work for an employer, they may automatically become an agent of their employer since any obligations they undertake while working bind their employer. Agents could also be hired by a principal to perform only one obligation, as specified in an agency agreement.

If an agency agreement is not definitive about the duties of an agent, then the legal system might assume that the terms were meant to be the ones that the legal system recognizes as being part of agency relationships. In legal terms, agency could also be created by **estoppel**. Estoppel exists when a third party relies on the actions of someone because they appear to be an agent of the principal (Sweet and Scheiener 2005). The principal does have to contribute in some manner to the third party being able to assume that someone is acting as an agent of the principal. The inaction of a principal could also cause a third party to think that someone is acting as the agent of the principal.

Actual Authority in Agency

When a written agency contract is formed, an agent is provided with legal authority to perform work, and other acts, on the behalf of the principal (Bockrath 2010). One difficulty that agents face is knowing what the limit is on their authority. In some circumstances, agents are able to make all different kinds of decisions and enforce them on the behalf of a principal; in other situations the authority of agents is limited to specific areas or situations. Therefore, it is crucial that an agency contract carefully delineates where and when an agent has the authority to act on the behalf of the principal; this is called **actual authority**.

When agents are trying to determine how far their authority extends, they should analyze the intent of the principal when they were establishing the agency relationship. A principal might desire to be actively involved in decision-making processes, and if this is the case, the authority of the agent may be of a limited nature. If a principal is not knowledgeable on the type of work being performed by the agent, the principal may relinquish their authority and expect the agent to make all of the decisions related to their assigned work.

When a principal forms a relationship with an engineer, unless he or she is an engineer, the principal will rely heavily on the expertise of the engineer and invest them with the authority to perform their work with little or no interference from the principal. In some instances, even though a principal is not knowledgeable about the work they have commissioned, they may be actively involved in decision-making processes, thus diminishing the authority of the engineer who is their agent.

When the legal system is involved in determining whether an agent had actual authority to perform certain duties, it will analyze the circumstances of each of the particular situations where the authority of an agent is in question, it will review the normal procedures and practices of the business owned by the principal, and it will also take into consideration the objective that the principal was attempting to accomplish that required the principal to establish an agency relationship to assist the principal in obtaining his or her objective.

Apparent Authority in Agency

From a legal perspective, **apparent authority** is a situation that is not clearly defined with an agency contract, but third parties rely on the actions of someone who appears to be an agent acting on the behalf of a principal. Sometimes people act as if they have the authority to make decisions or perform functions, when in fact they do not have the legal authority to be performing these functions. To a third party it is almost impossible to distinguish whether someone has the legal authority to act on the behalf of the principal; therefore, a principal is only held liable if they contribute in any way to the appearance of an agent performing duties on the behalf of the principal (Bockrath 2010).

One example of how a principal might inadvertently contribute to the misimpression that someone is their agent could occur at construction job sites. An owner shows up at a job site with another individual and introduces the individual to the workers and he or she is seen showing the individual around the job site. The next day the individual returns to the job site alone and starts issuing orders to the workers. The workers would assume that the individual has the authority to be instructing them on what to do because the owner caused the workers to view the individual as someone working directly for the owner.

When owners introduce individuals at job sites, they should include job titles in introductions and explain the function that will be performed by the person they are introducing and how it relates to the construction project. Providing job titles and job functions helps to eliminate any confusion related to authority issues that potentially could arise when new people are assigned to work at job sites.

8.4 TERMINATING AGENCY CONTRACTS

The most legally unambiguous manner in which to terminate an agency contract is to include an end date in an agency contract. If a specific date for completion is not known when an agency contract is formed, the contract could state that it will end when the objective of the agency contract is completed, or when a specific event is finished by the agent. An agency contract should state that if a project that the agent is working on is inadvertently destroyed, this also terminates the contract. Another method for terminating an agency contract is for the principal and the agent to agree to terminate the contract. This usually occurs when the working relationship is not satisfactory to an owner or the agent. Rather than suffer through a difficult situation they mutually agree to end the contract. If the principal or the agent pass away, then this terminates an agency contract.

If an agency contract were written for a specified period of time, and either the principal or the agent terminates it earlier than the specified time period, then he or she would be in breach of contract and be liable for damages. If an agency contract does not provide any information about when the contract ends, the legal system would view it as terminating in a reasonable amount of time. In order to determine what is considered to be a reasonable amount of time, the courts would review similar types of agency contracts and rely on the testimony of expert witnesses that would be required to testify about what is considered to be a reasonable term for an agency contract in their particular profession.

8.5 DISPUTES BETWEEN THIRD PARTIES AND AGENTS

Agents should inform third parties that they are an agent of the principal and provide the name of the principal in order to limit their personal liability during transactions with third parties. If a third party sues a principal, they could be waiving their right to sue the agent, although they could also have a legitimate lawsuit against an agent if the agent performed any acts that falsely represented the principal.

Chapter Summary

This chapter defined agency relationships as they pertain to engineers working for principals or clients and employees working for any type of a firm. How agency relationships are formed was explained along with the difference between the actual and apparent authority of agents. Methods for settling disputes between principals and agents and between agents and third parties were presented along with a discussion on possible methods for terminating agency relationships.

Key Terms

actual authority
agency
apparent authority
estoppel
fiduciary

Discussion Questions

8.1 Explain the different ways in which an agency contract could be terminated between a principal and an agent.

8.2 What is the best method for indicating when an agency contract terminates? Why is this the best method?

8.3 How might an agent protect himself or herself from liability if the project he or she is working on for a principal is destroyed by fire, an earthquake, or flooding?

8.4 Describe how apparent authority might be conferred on someone who is working at a construction job site.

8.5 Explain the difference between apparent authority and actual authority.

8.6 Explain what the best method is for limiting the power of an agent.

8.7 Discuss why a principal needs to rely on an agent and why he or she should hire an agent.

8.8 Explain how the court system would determine whether an agent had actual authority to perform on the behalf of a principal.

8.9 What is the most efficient method for forming an agency relationship?

8.10 Explain what agency is and how agents are used in the engineering and construction industry.

References

Bockrath, J.T. 2000. *Contracts and the Legal Environment for Engineers and Architects*. New York: McGraw Hill Publishers.

Sweet, J., and M. Schneier. 2005. *Legal Aspects of Architecture, Engineering, and the Construction Process*. Toronto: West Publishing Company.

CHAPTER NINE

TORT LAW

9.1 INTRODUCTION—DEFINITION OF TORT LAW

The focus of engineering and construction legal issues is mainly on contract law, but **negligence (tort)** lawsuits are becoming more prevalent in the engineering and construction industry. In Latin, the word *tort* means twisted and it refers to behavior that is crooked or that is wrong, but not necessarily against the law, such as negligence (Sweet and Scheier 2005). There are **personal torts,** where an individual is injured or killed, and there are **property torts**, where the property of someone is damaged or destroyed by another person or persons.

Tort law is the area of the legal system that addresses civil wrongs by providing a means for negligence victims to file a lawsuit in order to establish whether they are entitled to compensation for their injuries, property damage, or the death of a family member. Tort law is usually a separate area of the legal system due to the volume of cases that the courts hear each year. Tort law proceedings may only be used by individuals who are suing other individuals or corporations. Tort courts are separate from criminal courts because criminal cases involve legal actions that are initiated by states, since criminal acts are crimes against the state (the public) not individuals. Criminal cases result in defendants receiving jail sentences rather than monetary compensation being paid to the victims or their families.

In some instances, victims of crimes will also sue in tort courts for monetary compensation in addition to the case being prosecuted in the criminal court system. One high-profile criminal case where this occurred was in the O. J. Simpson trial at which the defendant was tried for allegedly having murdered his wife and another man. O. J. Simpson was found not guilty in the criminal case, but he was found guilty in the tort lawsuit; the families of the victims, his wife and another individual, were awarded monetary compensation in the millions of dollars for the loss of their family members.

Another concept related to tort law is **tortuous interference**, which happens when one party interferes with the work of another and prevents the other party from performing their work (Sweet and Schneier 2005). Tort liability results from negligence but it could also be assessed when someone intentionally interferes and it causes harm, or in situations where someone is not following public policy and it results in someone being injured because they did not follow public policy. The difficult part of proving tortiuous interference is being able to determine whether the party being accused had the right to be doing whatever interfered with the other party. In one legal case a tortiuous interference claim was made by a construction manager/contractor against the architect. The court ruled in favor of the architect, as described in the following legal case [Fleischer-Seeger Construction Co. v. Hellmuth, Obata, and Kassabaum, 870 S.W. 2d 832 (Mo. App 1993)]:

> The construction manager/contractor on the Admiral reconstruction project did not have a cause of action for tortious interference with contract rights against the architect. The contracts between the architect and the owner and between the owner and construction manager/contractor required the architect to perform the activities, which the construction manager/contractor claimed to be tortious interference with its rights under its contract with the owner.

9.2 NEGLIGENCE

Negligence is an area of the law that is different from the other legal venues because the person who sues in tort court has to try to prove that the person they are suing caused them to be injured because they did not adhere to what is considered to be the legally required **standard of conduct**. This means that the plaintiff in a tort lawsuit is responsible for proving that the defendant did not perform as a reasonable person would have under the same circumstances and that the injury sustained by the plaintiff resulted from the negligent act of the defendant. One example of a tort case involving professional engineers resulting in their being found liable for negligence in inspecting a home is the following [*Moransais v. Heathman et al.*, 744 So.2d 973 (Fla. 1999)]:

> Individual professional engineers may be held liable to homeowners for negligently inspecting a house, even though the inspection services contract is between the homeowners and the professional engineering corporation, which employs the individual engineers. The statutes regulating professional engineering services do not permit the individual engineers to escape responsibility for their conduct, and they may not contractually limit their common law and statutory duties. The economic loss rule also does not preclude the homeowners from receiving purely economic losses (there were no personal injuries or property damage) in tort. The economic loss rule is limited to products liability contexts and similar circumstances.

Even though contractors do not have contracts with engineering firms that design the projects they build, engineering firms still have a legal obligation to not be negligent when designing structures, as the following case illustrates [*Tommy L. Griggin Plumbing and Heating Co. v. Jordan, Jones and Goulding, Inc.* 463 S.E.2d 85 (S.C. 1995)]:

> A general contractor could recover purely economic damages from the engineer that designed and administered the project under negligent design, negligent administration, and breach of implied warranty theories notwithstanding the absence of a contract between them. The engineer owed a duty to the contractor to refrain from negligently designing or administering the project. The economic loss rule did not bar the claim.

When individuals file lawsuits in the tort court system against corporations, the corporations have some measure of protection because of the **fault doctrine** that is slanted toward industrial firms to protect them from open-ended liability to help ensure that the firms do not go out of business because of all of the negligence lawsuits filed against large companies. Without the fault doctrine, members of large firms would be hesitant to conduct business and this could negatively affect consumers.

The following sections discuss some of the legal issues that are examined during tort liability cases.

Assumption of Risk

In some instances, plaintiffs knowingly put themselves in situations that they are aware are dangerous. If a defendant is able to prove that the plaintiff was informed about the danger of an activity before they performed it, and the plaintiff chose to continue to be exposed to the danger, then it would be difficult for the plaintiff to be compensated for the negligent actions of the defendants if they are injured while performing that activity. The courts will review the circumstances that led to the plaintiff being injured to determine whether the plaintiff understood the danger associated with the activity they undertook before being injured by the negligence of the defendant.

One example that illustrates the concept of **assumption of risk** is when someone signs up to perform extreme sports and they are required to sign an assumption of risk form, or a **waiver of liability** form, which states that they waive all liability against the firm that is operating the extreme sports venue. The signed waiver of liability may or may not protect the firm. If the plaintiff assumed the risk during the normal functioning of equipment used for the sport, and because of a negligent act by one of the operators of the equipment they were injured, then the firm might not be legally protected by the waiver of liability form that was signed by the plaintiff. Some courts do not permit the contracting away of liability; therefore, they would not consider the signed waiver of liability as valid.

Standard of Care

In tort liability lawsuits, it is the responsibility of the plaintiff to demonstrate that the defendant was required to perform some type of action that would help ensure that individuals would not be harmed while on the property owned by the defendant or when working with the defendant; this is called demonstrating that the **standard of care** was being followed by the defendant (Bockrath 2010). The plaintiff has to prove that the injury or loss they sustained was caused by some type of an act on the part of the defendant and demonstrate that they suffered some type of loss as a result of the act of the defendant.

If a plaintiff would have sustained an injury even if the defendant had performed everything required of him or her, then the plaintiff would have a difficult time proving that the defendant caused them to be injured or to sustain a loss. In all tort liability cases, the burden of proof for proving the alleged negligence of defendants is on plaintiffs.

The legal system considers construction job sites to be attractive nuisances. It is the responsibility of contractors to protect third parties, especially children, from being harmed while they are on construction job sites, even after normal work hours. The typical methods for protecting construction job sites are providing twenty-four hour security and fencing the area. When lawsuits are filed for injuries or deaths that occur at construction job sites, it is the contractors who are liable for damages in the tort system, as the following legal cases demonstrate [*Potts v. Halsted Financial Corporation* (1983) 142 Cal. App. 3d 727, 191 Cal. Rptr. 60]:

> The plaintiff and his friends, along with a 14 year old sister Bambi, drove down to Malibu beach to celebrate the 4th of July. They found two buildings under construction, and went to the roof. Two loose boards connected the roofs of the two buildings. Bambi reached the roof of the second building,
>
> *(continued)*

> and screamed for help. Plaintiff guided her back across the boards, but as she stepped to safety the boards slipped apart and the plaintiff fell two stories. Nonsuit granted based on Cal Civil Code 846 was **REVERSED**. The exemption from liability for injuries flowing from public recreation on privately owned property did not apply here. The legislature did not intend to encourage landowners to allow public access to places as unsuitable for recreation as the rafters or roofs of new homes.

[*Crain v. Sestak* (1968) 262 Cal. App. 2d 478, 68 Cal Rptr. 849]:

> The builder was held liable when a 12 year old, playing on the scaffolding at a residence, stopped on the end of a board and it tilted like a teeter-totter, and the 12 year old was injured. Attractive nuisance doctrine applied.

Proximate Cause

During tort liability legal proceedings the courts require that the plaintiff prove that there was some type of connection between the actions of the defendants and the injuries sustained by the plaintiff. The courts review the **proximate cause** of the injuries and evaluate whether the conduct of a defendant caused the injuries that were sustained by the plaintiff.

Contributory Negligence

One defense that could be used by defendants in tort liability cases is **contributory negligence**. A defendant would attempt to demonstrate that some act or actions performed by the plaintiff contributed to the plaintiff being injured or killed while on the property of the defendant or while working with the defendant. Even if a defendant proves that the plaintiff in some manner contributed to their being injured, the courts could still award the plaintiff compensation but the award might be less than if the plaintiff had not contributed to their own injury (Bockrath 2010).

Comparative Negligence

Comparative negligence would be the result of a tort liability case if a defendant were able to prove that the acts or actions of the plaintiff also contributed to the plaintiff sustaining their injuries. The courts would not prevent the plaintiff from recovering compensation for their injuries but the court would apportion the responsibility for causing the injuries between the plaintiff and the defendant and award damages in relation to the portion of responsibility apportioned to the defendant (Bockrath 2010). The following legal case provides an example of how the courts support the right to apportion negligence in tort lawsuits [*American Motorcycle Association v. Superior Court of Los Angeles County* (1978) 20 Cal. 3d 578, 146 Cal. Rptr. 182, 578 P.2d 899]:

> 1. First, we conclude that our adoption of comparative negligence to ameliorate the inequitable consequences of the contributory negligence rule does not warrant the abolition or concentration of the established 'joint and several liability' doctrine . . . [which] continues to play an important and legitimate role in protecting the ability of a negligently injured person to obtain adequate compensation." 578 p2d at 901, 146 Cal. Rptr. at 184.
> 2. The doctrine of equitable indemnity 'should be modified to permit, in appropriate cases, a right of partial indemnity, under which liability among multiple tortfeasors may be apportioned on a comparative negligence basis." 2d at 902, 146 Cal. Rptr. at 185.
> 3. A named defendant may file a cross-complaint for equitable indemnity against any person to seek total or partial indemnity, but a tortfeasor who enters into a good faith settlement is discharged from any liability for further contribution to any other tortfeasor.

Respondent Superior and Vicarious Liability

When an agent is acting on the behalf of a principal, and a third party is injured due to the negligence of the agent, the principal could be held liable for the injury under the legal concept of **respondent superior** (Sweet and Schneier 2005). In many instances, a plaintiff will sue a principal rather than an agent because of the financial position of the principal versus the agent; this is referred to as suing the party with **deep pockets**. Respondent superior would also affect the liability of joint venture partners, since they may be held liable for the acts of each other under the legal concept of respondent superior. **Vicarious liability** is another legal concept that governs the liability of principals for the actions of their agents. Vicarious liability means that employers are responsible for the acts and actions of their employees.

9.3 NEGLIGENCE OF ENGINEERS AND CONTRACTORS

Engineers have a **duty of care** to not only their clients but also to third parties and for the welfare of the public. The duty of care of engineers to owners is discussed in the next section, followed by a discussion of the duty of care engineers and contractors have to third parties.

Engineers Duty of Care to Owners

When engineers are working for owners, if they commit any type of fraud, they could be legally liable for any damage that occurs as a result of their fraudulent actions. Engineers are liable to their employer for any negligent acts where they do not exercise due diligence in relation to their expertise. Some examples of the areas where engineers could be found to be negligent are the following (Sweet and Schneier 2005):

- Preparing defective engineering designs or drawings that are defective
- Not completing drawings when they are needed during construction
- Providing cost estimates to owners that are significantly lower than the resulting actual costs

In order for an owner to be awarded damages for negligence on the part of an engineer, the owner has to be able to prove that the engineer did not perform up to the appropriate standard for the industry. Owners normally want the work corrected rather than monetary damages for negligence, but if there is no way to correct the work, then an owner is entitled to the difference between the current value of the defective structure and the value of the structure if it had been correctly designed and built. Owners may also be entitled to damages for any time that the completion of a structure was delayed or for any loss of business due to the negligent acts of the engineer.

Responsibilities of Engineers and Constructors to Third Parties

When a third party is at a construction job site and they are injured or killed, as the result of an engineer or constructor being negligent in their responsibility to carefully plan and inspect the work, and it is determined by the court system that the harm done to the third party was foreseeable, and there was a direct connection between the negligence and the injury, then the engineer or constructor could be liable for the injuries or death to the third party, as the following case illustrates (ENR.com, January 23, 2008, 1):

> The joint venture program manager of the problem plagued Boston Central Artery project, and the project designers, have agreed to pay $458 million to settle major claims that include tunnel leaks, and a fatal tunnel plenum collapse. The deal with the state of Massachusetts and the U.S. Justice Department enables the firms to avoid possible prosecution and debarment from federal and state contracts.
>
> . . . Under the terms, the joint venture partners would pay a total of $407 million. Prosecutors say the settlement covers "defects in slurry-wall construction, use of out-of-specification concrete by certain contractors, failure to disclose financial information, and various cost recovery matters for deficient work. Under the deal one of the joint venture partners will pay $352 million and the other will pay $47.2 million. About 24 project design firms, whose names were not disclosed, will pay $51 million. But the settlement notes that the joint venture partners did not engage in criminal conduct or commit a fraud, and the firms did not admit liability. Of the total to be paid, $415 million will be held by authorities in a special interest bearing account that will fund future Central Artery "non routine" repairs. The agreement does not release the joint venture partners from liability in the event of a future "catastrophic failure" of artery infrastructure that causes more than $50 million but it caps its financial responsibility at $100 million. The state opted for the settlement because manslaughter charges against corporations carry penalties of only $1,000.

Negligence lawsuits are also filed against engineers when structures are defective, or if the designs for structures do not follow required building codes, as the following legal case illustrates [*Huang v. Garner* (1984) 157 Cal. App. 3d 404, 203 Cal. Rptr. 800]:

> Purchaser v. the developer, the building designer (licensed, or not, unclear), and the project engineer. Theories of strict liability, negligence, and implied warranty. Nonsuit was granted on all theories to the building designer and the project engineer. Nonsuit was granted of the breach of warranty theory against the developer and the construction company. Partial nonsuit was granted on the other theories limiting recovery to physical damages only. Verdict for $40,300 was **REVERSED**. An apartment building was built in 1965 on the developer's land by its construction company. The developer hired the building designer, who in turn hired the project engineer. The buildings were defective due to: 1) insufficient fire retardation and shear walls; 2) inadequate structure; and 3) deviation from building plans, as well as other deviations. Many of these defects were in violation of the 1961 Uniform Building Code (UBD). In 1970, the developer sold the building, which after two additional transactions was sold to the purchaser in 1974. When the purchaser attempted to convert the building to condominiums it discovered the defects. **HELD** against building designer and project engineer. It was error to nonsuit. Violations of the UBC were sufficient to send the case to the jury under negligence *per se*.

The liability of engineers and constructors to third parties varies depending on the circumstances of each individual case and on the jurisdiction where particular cases are heard by courts of law. As with so many other laws and statutes, liability laws vary between states and within the different jurisdictions within individual states. The most useful source of information related to the liability of engineers and constructors to third parties is precedent law cases from the state where the legal case is being prosecuted and these are available in the individual state registers of legal cases.

9.4 DEFENSES TO THE LIABILITY OF ENGINEERS OR CONTRACTORS

Two defenses to negligence that could be used by engineers or constructors are contributory negligence or assumption of risk. If a third party was in a location at a job site where they were not supposed to be when they were injured that could be construed as contributory negligence. If the third party was advised of the dangers associated with being at a job site, and they willfully chose to be present at the job site, that could be categorized as assumption of risk.

Statute of Limitation

To prevent engineers from having **open-ended liability** for their work, each state has a **statute of limitation** that ranges from seven to ten years. A statute of limitation requires that lawsuits be filed during the time frame of the limits set by the state. Unfortunately, there have been some legal cases that were allowed to be filed beyond the statute of limitation due to considerations related to latent defects that are not discovered prior to the end of the statute of limitation and defects that are discovered by new owners of structures that purchase structures after the statute of limitation expires. Whether the courts will allow these types of cases to be heard is up to the individual courts. An example of a court case on a statute of limitation is the following [*Businessmen's Assurance Company of American v. Graham, representative for Skidmore, Owings, and Merrill, et al.*, 984 S.W.2d 501 (Mo. Banc 1999)]:

> A judgment against the designing architect for more than $5,800,000 (with interest) under design professional law, was affirmed by the Missouri Supreme Court, which found that the damage occurring when marble panels fell from the side of a building were not capable of ascertainment until then. As a result the five year statute of limitations for negligence actions began from the date the marble panels fell, not the date on which the design work was performed.

In a different statute of limitation case the judge ruled that the statute of limitation was applicable to an engineering design that resulted in a similar type of incident [*Harbor Court Associates v. Leo A. Daly Co.*, 179 F.3d 147 (4th Cir. 1999)]:

> A clause in an A.I.A. form contract, overriding statutes of limitations and repose, providing that all statutes of limitations run from the date of substantial and final completion was enforced. An owner's claim for negligent design for damages occurring eight years after final completion, when a fifteen foot area of brick veneer exploded off the face of the complex, was barred by the contract language. The "date of discovery" rule, that otherwise would have applied, was overridden by the contract.

9.5 TORT LIABILITY COMPENSATION

The purpose of tort liability judgments is to try to compensate individuals who have been injured, or the family members of individuals who have been killed, but the original intention of tort law was to help injured parties to obtain a settlement that returns them to the position they were in before they were injured by the negligent act of someone else. Since it is impossible to undo the damage caused by an injury or death, the only means for compensating victims of negligence is to provide monetary awards to plaintiffs that are paid by defendants or the insurance company of the defendants.

The amounts awarded for compensation in negligence lawsuits varies depending on the circumstances surrounding the case and the injuries sustained due to the negligence of the defendant. Plaintiffs are entitled to receive compensation for several types of losses such as the following:

- Medical expenses
- Pain and suffering
- Disfigurement
- Decreased life expectancy
- Loss of wages, salary increases, and promotions

If the medical bills of a plaintiff are paid by their medical insurance provider, the plaintiff is required to reimburse their medical insurer for all of their medical expenses if the plaintiff receives a settlement from the defendant. Insurance companies have actuary tables that provide a dollar amount for the loss or disfigurement of body parts. When an insurance company is estimating the amount they will offer as a settlement in a negligence lawsuit, the actuary tables take into account the importance of particular body parts and a higher compensation rate is used for more essential body parts, such as eyes or thumbs.

In order to estimate an appropriate settlement for loss of wages, salary increases, and promotions, an economic analysis is performed utilizing time value of money formulas to estimate the present worth of lost future earnings with periodic increases for potential promotions or raises. In negligence settlements, the most unpredictable part of the awards is the monetary compensation for pain and suffering. If a negligence case is being presented in front of a jury, then the jury may determine the monetary compensation for pain and suffering. Awards could be influenced by emotional factors such as severe disfigurement or by persuasive lawyers who work for plaintiffs.

Individual states may pass legislation to limit the amount of awards for noneconomic or economic damages in tort lawsuits. One example is the state of California that limits noneconomic damage awards in medical malpractice cases to $250,000 per case to cover "pain and suffering, inconvenience, physical impairment, disfigurement, and other pecuniary (relating to money) injury" (California Civil Code 3333.2). California does not limit the awards for economic damage to cover medical costs, loss of earnings, and other economic losses.

9.6 INDEMNIFICATION CLAUSES

Construction contracts will typically contain some type of **indemnification clause** that attempts to shift liability to contractors if there is any type of lawsuit related to construction projects. Indemnity clauses are also referred to as **hold harmless clauses**. Liability cannot be shifted to contractors for design errors or for engineers not properly transmitting instructions to contractors. The majority of construction contracts will contain indemnity clauses, and if they were always upheld by the courts, contractors would never agree to sign contracts that contain this clause. The clause is included in case a situation arises that would support the invocation of this clause to protect the owner or the engineer of record. Unfortunately, some contractors will increase their bid proposal estimate price if a contract contains an indemnity clause to compensate for agreeing to absorb additional liability during construction.

Chapter Summary

Negligence is becoming one of more high-profile areas of the law because of the proliferation of negligence lawsuits. This chapter explained negligence and defined tort law, which is the area of law that processes negligence lawsuits. Negligence was discussed in terms of what constitutes an act of negligence, how assumption of risk affects negligence lawsuits, how proximate cause and foreseeability influences negligence, and how contributory and comparative negligence affect tort settlements.

This chapter also provided information on the basis for tort liability, and the duty of engineers to owners and third parties. Potential forms of negligence by engineers were covered along with a discussion on the types of defenses that engineers and constructors could use when they are accused of negligence. Indemnification clauses were mentioned in the context of how they could increase the cost of construction if they are included in construction contracts.

Key Terms

assumption of risk
comparative negligence
contributory negligence
deep pockets
duty of care
fault doctrine
hold harmless clause
indemnification clause
negligence
open-ended liability
personal torts
property torts
proximate cause
respondent superior
standard of care
standard of conduct
statute of limitation
tort
tortuous interference
vicarious liability
waiver of liability

Discussion Questions

9.1 What types of losses are victims of negligence entitled to receive from defendants when a lawsuit is decided in favor of a plaintiff?

9.2 Explain why lawsuits that involve engineering defects would be allowed to be filed after a statute of limitation has expired in a state.

9.3 Explain what is meant by deep pockets and how this term applies to tort liability lawsuits.

9.4 How might a lawyer for a defendant demonstrate that their client performed according to the standard of care for their particular engineering profession?

9.5 What would be considered to be negligence on the part of design engineers?

9.6 Explain contributory negligence and how it could be used as a defense by a defendant in a negligence lawsuit.

9.7 How is the term *proximate cause* applied to negligence lawsuits?

9.8 Under what circumstances might an assumption of risk agreement not be admissible during a negligence lawsuit?

9.9 Why are negligence lawsuits referred to as tort cases?

9.10 How do tort lawsuits differ from criminal lawsuits?

References

American Motorcycle Association v. Superior Court of Los Angeles County (1978) 20 Cal. 3d 578, 146 Cal. Rptr. 182, 578 P.2d 899.

Bockrath, J.T. 2000. *Contracts and the Legal Environment for Engineers and Architects*. New York: McGraw Hill Publishers.

Businessmen's Assurance Company of American v. Graham, representative for Skidmore, Owings, and Merrill, et al. (Mo. Banc 1999) 984 S.W.2d 501.

Crain v. Sestak (1968) 262 Cal. App. 2d 478, 68 Cal. Rptr. 849.

ENR.com. January 23, 2008. Bechtel, PB, and Subs sign $458-million big dig settlement. *Engineering News Record*. New York: McGraw Hill Publishers.

Fleischer-Seeger Construction Co. v. Hellmuth, Obata, and Kassabaum (Mo. App 1993) 870 S.W. 2d 832.

Harbor Court Associates v. Leo A. Daly Co. (4th Cir. 1999) 179 F.3d 147.

Huang v. Garner (1984) 157 Cal. App. 3d 404, 203 Cal. Rptr. 800.

Moransais v. Heathman et al. (Fla. 1999) 744 So.2d 973.

Potts v. Halsted Financial Corporation (1983) 142 Cal. App. 3d 727, 191 Cal. Rptr. 60.

Sweet, J., and M. Schneier. 2005. *Legal Aspects of Architecture, Engineering, and the Construction Process*. Toronto: West Publishing Company.

Tommy L. Griggin Plumbing and Heating Co. v. Jordan, Jones and Goulding, Inc. (S.C. 1995) 463 S.E.2d 85.

LEGAL ISSUES RELATED TO REAL PROPERTY

10.1 INTRODUCTION

This chapter explains how some of the real property laws could affect the design and construction of projects including: options for ownership of real property, deeds, zoning, servitudes, covenants, eminent domain, condemnation, right-of-way, tax liens, mechanic's liens, and contracting for the sale of property.

10.2 DEFINITION OF REAL PROPERTY

When engineers design or construct projects, a major component of their work revolves around the land used for projects. In the engineering and construction industry, land is referred to as **real property** or **real estate** and these terms are used to describe land and anything that is attached to the land such as structures, trees, and scrubs. It could also include water and mineral rights beneath the property. In the United States, there are specific laws that define the rights associated with real property. The engineers who design projects, and the constructors who built them, should be aware of the real property laws that could influence the design and construction of their projects.

A **cadastral survey** is conducted for the purpose of locating and establishing the boundaries of real property in order to legally define and officially register "the quantity, value, and ownership of real estate in apportioning taxes . . . is confined to the location and subdivision of the public domain" (Robillard, Wilson, and Brown 2006, 4). A boundary survey is "a survey that is conducted for the location and establishment of lines between legal estates, or it may be a physical feature erected to mark limits of a parcel or political unit" (Robillard, Wilson, and Brown 2006, 4).

The **Model Registration law** by the National Council for Examiners and Engineering and Surveying (NCEES) states the following (NCEES 1960):

> The term Land Surveying used in this act shall mean and shall include assuming responsible charge for and/or executing: the surveying of areas for their correct determination and description and for conveyancing; the establishment of corners, lines, boundaries and monuments; the platting of land and subdivisions therefore including as required, the functions of topography, grading, street design, drainage and minor structures, and extensions of sewer and water lines; the defining and location of corners, lines, boundaries and monuments of land after they have been established; and preparing the maps and accurate records and descriptions thereof.

Surveyors provide legal descriptions of property and its boundaries, called **metes and bounds**, and these descriptions are used to (Robillard, Wilson, and Brown 2006, 422):

1. Identify unequivocally and definitively one and only one unique parcel of land;
2. Identify for legal purposes;
3. Identify for the purposes of describing, recovering, and retracing land boundaries;
4. Identify for cadastral purposes (taxation and valuation);
5. Locate encumbrances positively (e.g., easements); and
6. Identify for addresses and indexing.

"A person could gain or lose land rights or modify boundary lines by use of the following" (Robillard, Wilson, and Brown 2006, 359):

1. Agreement, either expressed or implied, including agreement on practical location, silent recognition and acquiescence, and the doctrine of estoppel, where both agreement and dishonesty enter;

2. Adverse relationships (hostile) either expressed or implied (**adverse possession**);
3. Statutory (legal) proceedings; and
4. Prescription ripening into easements or other rights.

In order for a real property owner to prove ownership, the property must have a **deed** recorded in a public place, such as the country clerk or county recorder's office; the deed and registration of the property is **evidence of land ownership**. In addition to this process, in some states **Torrens registration** is also required for land and it "includes both registration of title and registration of ownership" and new transfer of land must be accompanied by a change in **registration of ownership**. The difference between the two systems is that one registers title, whereas the other registers ownership (Robillard, Wilson, and Brown 2006, 397 and 407). The following states require Torrens registration:

- Illinois
- Massachusetts
- Minnesota
- Georgia
- Tennessee
- North Dakota
- South Dakota
- Washington
- Utah
- Virginia
- Colorado
- Hawaii

The term deed was legally defined by the courts in Arizona in 1907 (*Test Oil Company v. La Tourette,* 19 Ok. 214, 91P. 1025):

> Formally, the word "deed" was synonymous with the word "act" and, as the law recognized a peculiar sanctity in the use of a seal, and act under seal became something finished, sacred and worthy of great respect. Hence a man's act under seal became his deed, and likewise any instrument under seal became his deed, irrespective of the nature of the transaction involved. But the term "deed" is used now in a more limited sense and as the meaning a written instrument, duly acknowledged by a competent party, conveying title to land.

In order for real property to be sold, the following conditions have to be met (Robillard, Wilson, and Brown 2006, 413):

1. Be in writing,
2. Identify grantors and grantees,
3. Identify the interests being conveyed,
4. Express an intent to convey the interest identified,
5. Identify the location of the land or the interests that are being conveyed with legal certainty, and
6. Be signed by the grantor.

10.3 OPTIONS FOR OWNERSHIP OF REAL PROPERTY

When an owner purchases land that will be used for a construction project, he or she has to decide on the form of ownership that will be recorded during the purchase of the property. The ownership options available are:

- Sole ownership
- Joint tenancy
- Tenancy in common
- Fee simple
- A life estate

The following sections describe each of the ownership options and some of the ways in which they could affect the use or sale of property:

Sole Ownership

The most elementary form of real property ownership is **sole ownership** where only one person is recorded on the deed as being the owner of the property. Sole ownership allows an owner to be the only person responsible for decisions concerning the disposition of the real property, how the property will be used, and what could be constructed on the property. For engineers and constructors it is easier to work with a sole owner because decisions could be made quicker, there are less legal ramifications, and access to the client is less restrictive.

Joint Tenancy

Joint tenancy occurs when two or more individuals own real property. Joint tenancy is a common form of ownership among married couples. The major drawback of joint tenancy occurs when one of the owners would like to sell the property. If the other owner does not want to sell the property, then the only option for selling the property is by **partitioning** it. Partitioning is a legal process whereby one of the owners files with the court to have the land partitioned by the court. During the partitioning process the court appoints a referee who determines a fair market value for the property, hires a real estate agent to sell it, and then divides the proceeds from the sale between the joint tenants minus court costs, the fee charged by the real estate agent, and the fee charged by the court-appointed referee. Difficulties could also arise when joint tenants are constructing a structure on their property. When decisions are made, all of the joint tenants have to agree on decisions, which could be a long and possibly contentious process.

Tenancy in Common

Tenancy in common is a form of ownership of real property whereby multiple parties own property and they have a designated share of the property but it is an undivided interest in the property.

Fee Simple

Someone could inherit real property rather than purchase it and the person who originally owned the property determines the form of ownership that will be transferred to the person or persons who inherit the property upon the death of the original owner. If the original owner does not want any restrictions on the real property when it is inherited, they could set up their estate in a **fee simple** legal structure. Once the property is inherited, the person, or persons, who inherit the property are able to manage it in any way that they would like to manage it or they may sell the property if they desire to sell it.

Life Estate

In some instances, the owner of the property does not want the property to be transferred to his or her beneficiaries yet they want the beneficiaries to be able to use the property during their lifetime. In this case, a **life estate** could be formed that permits a beneficiary or beneficiaries to use the property during their lifetime or the life estate may be set up to allow the beneficiaries use of the property during the lifetime of another individual named in the life estate. Upon the death of the person named in the life estate, the property would revert to whoever is designated in the life estate or it may pass to an organization or a charity if they are designated in the life estate. A life estate prevents the beneficiaries from ever dividing or selling the real property.

10.4 MORTGAGES

Mortgages are loans that are provided by banks, savings and loans, credit unions, or other lending institutions, for the purchase of land or structures that are secured by the property that is pledged as **collateral** for the loans. Collateral is real property, or other property, that is pledged in order to secure a loan. If there is a default on the loan, the bank or lending institution will take possession of the collateral. When an individual secures a mortgage, it creates a lien on the property; if the property is sold, the proceeds of the sale are used to retire the loan and remove the lien. If the mortgage is not repaid according to the monthly schedule set up at the time the mortgage is secured, then the lending institution will foreclose on the loan, repossess the property, and sell the property at an auction, and then the proceeds of the sale are applied toward repayment of the loan. In addition, if the property is sold for less than the mortgage value, the lending institution has the legal right to sue the mortgage holder for the difference between the selling price of the property and the value of the mortgage plus the cost of legal fees, court costs, and any interest accrued on the loan during the lawsuit.

If a lending institution has foreclosed on many loans, and they have not been able to sell the properties seized in the foreclosure proceedings for the value of the mortgages, the lending institution may hire a law firm and have the law firm try to collect any amount possible from the mortgage holders. The law firm is paid a percentage, such as 30 percent, of whatever they are able to collect from the mortgage holders who defaulted on their mortgages. Each of the mortgage holders will be sued by the law firm, but members of the law firm are willing to negotiate a lower settlement in order to prevent the cases from going to trial.

10.5 ENCUMBRANCES

Encumbrances are amounts of money owed that are attached to property through legal channels such as back taxes owed, city or country assessments that are in arrears, and mechanic's liens against the property for work performed by contractors, subcontractors, or suppliers. All of the encumbrances attached to real property have to be paid before the property could be sold to another party or they will be paid out of the proceeds of the sale of the property.

10.6 SERVITUDES

Servitudes are a legal instrument that allows someone other than the owner of real property to use the property or portions of it. The three different types of servitudes are (1) easements, (2) licenses, and (3) profit a prendre. Each of these types of servitudes is described in the following sections.

Easements

Easements allow someone to use the land, or a portion of the land, of another for a specific purpose. Common types of easements are utility lines such as electrical transmission, gas, telephone, cable, fiber optics, high-speed Internet, water, and sewer. Before purchasing property it is important to know where all of the easements are located on the property, because if the owner builds anything on an easement, it could be removed when access to the easement is required. The person using the easement is not under any obligation to return the easement to its former state after they have excavated a utility line that is buried on the easement property. An engineer should know where all of the easements are located before they design a structure for a piece of property to avoid locating the structure on, or close to, any easements.

Easements could also be formed for reasons other than for utility access. They could allow the public to travel across property to reach public property such as beaches or parks and allow private citizens access to their property if their property is **land locked** (surrounded by property owned by other individuals). Easements could be inherited, they could be assigned to other parties, and they could also be transferred separately from the rest of the property. Easements could be created by eminent domain, where the government takes possession for public use, or by adverse possession, which takes place when someone has used the property of someone else for a long period of time and the owner of the property has been aware of it and they have not stopped it from happening (Bockrath 2000).

Licenses

Anyone who owns real property may provide others with a license to use their property for a specific purpose, as stated in the license.

Profit a Prendre

Profit a prendre is an expression that when translated from French means "profit to take." It refers to a process whereby a party is legally entitled to utilize something that is on or beneath land and sell it for a profit. Examples of items that could be sold for profit are soil, trees, minerals, oil, or natural gas (Bockrath 2000).

10.7 DOCTRINE OF SUPPORT

An owner of property could enter into an agreement called a **doctrine of support**, which allows an adjacent land owner to use their property for a specified period of time with the provision that they return the property back to its original status once the doctrine of support time frame expires (Bockrath 2000).

10.8 RENTAL PROPERTY

An owner of real property may form a contract with another individual that allows them to use the property for a specified amount of money per month and for a time frame designated in a rental contract, which is called a **lease**. The person who owns the property is called the **landlord** and those who rent the property are referred to as **tenants**.

Subleases

If a rental agreement permits it, a tenant may sign a lease with a third party that allows the third party to occupy the rental property for whatever time that is remaining on the lease between the landlord and the tenant. The legal document that is signed by the tenant and the person that will occupy the property is called a **sublease**.

10.9 LIENS

Liens are a legal process that allows mortgage holders, the Internal Revenue Service, state tax authorities, contractors, subcontractors, vendors, and suppliers to put an attachment on a piece of property for an amount that they are owed by the owner for unpaid contractual obligations or services rendered by the party applying for the lien. The owner is not able to sell their property until all of the liens have been satisfied by repaying all of the debts. If the liens are not satisfied when the property is sold, the funds from the sale of the property will be used to pay the lien holders.

Contractors, subcontractors, and suppliers file **mechanic's liens** at the county courthouse by providing proof of their services, and nonpayment of the amount they are owed by the owner that they are filing the lien against for nonpayment. In the case of mechanic's liens, the workers and suppliers are paid before contractors and subcontractors. Mechanic's liens are liens that are filed by contractors, subcontractors, workers, and suppliers against owners usually during the construction of a structure when they are not paid after they have performed services for the owner. The government does not allow liens to be attached to public property but subcontractors are able to file **stop notices**. After a stop notice is filed, an owner will not pay the contractor until the subcontractors and supplier are paid from the funds that have already been paid to the contractor. Owners are able to withhold the amount owed to a contractor until the court decides who gets paid after stop notices have been filed against a contractor. Stop notices are covered in detail in Chapter 16. The government requires payment bonds to be secured by contractors; payments bonds provide payments to subcontractors and suppliers if they have not been paid by the contractor. Payment bonds are also discussed in Chapter 16.

Tax Liens

Tax liens are filed for the amount owed in back taxes on property. The government has the legal right to take possession of the property. If the lien is not discharged during a designated time period, the property and its contents will be sold at auction and the proceeds will be used to pay the back taxes that are owed on the property.

10.10 EMINENT DOMAIN

Eminent domain is a legal process that allows federal or state governments to use **due process of law** to try to take possession of private property and use it for a purpose that benefits the public, such as highways, roadways, parks, schools, or government facilities. Due process of law requires that the government provide property owners with adequate notice about their intention to take possession of property owned by private citizens. Private citizens are allowed to present their case against the intention of the government to seize their property. If the government prevails, they have to provide adequate compensation for the property, as determined during the eminent domain legal proceedings in the court system, which is normally fair market value, but a higher amount might be negotiated by property owners.

During the early part of the twenty-first century there were a few prominent legal cases related to eminent domain procedures that altered how property could be used after it has been seized by the government. In one case in Norwood, Ohio, the government seized homes under condemnation proceedings to allow a developer to build a $125 million project that included condominiums, office buildings, and stores. "From 1998 through 2002, state and local governments seized or threatened to seize more than 10,000 homes, businesses, churches, and pieces of land, not for 'public use' but to enrich private interests, some of whose enhanced riches can be siphoned off in taxes" (Will, April 24, 2008, 94). The original intention of eminent domain laws was to allow land to be seized for government facilities not for use by private developers but in 2005 in the landmark case of *Kelo v. New London* the court ruled that land could be seized to cure urban blight (Will, April 24, 2008).

Condemnation

Condemnation is a part of the process of eminent domain where the government takes possession of property that it is attempting to seize for public use.

10.11 RIGHT-OF-WAY

Right-of-ways are portions of private land that are designated for public use for purposes such as highways or public utilities. Right-of-ways are legal agreements that state the intended purpose of the portion of a piece of property that will be used as a right-of-way and they include:

1. The term that the property may be used for such as ninety-nine years or other time periods
2. The use of the property during the designated time period
3. A statement about who may use the property during the designated time period

10.12 ZONING

Zoning is a process whereby cities designate where certain types of structures could be built within city limits. Typical zoning designations are R1, R2, R3; C1, C2, C3; and M1, M2, M3. Residential zones are referred to as R1, R2, and R3; the numbers refer to whether structures may be single family, duplexes, or multiple family residential structures. Commercial structures are C designations and manufacturing structures are M designations. Other designations refer to rural and recreational facilities. The C and M numbers refer to the densities of the structures in a particular area. Additional restrictions within the zoning categories include:

- Designating the minimum lot sizes
- Setbacks—how close structures on the property could be located to the street or adjacent property
- The type of structures allowed
- The height of the structures
- The number of stories
- What percentage of a lot could be covered by a structure
- Any other restrictions imposed by the city

Zoning requirements could be changed or waived by a majority vote of city council members.

Chapter Summary

This chapter introduced some of the legal concepts that influence the ability of engineers and constructors to perform their occupations when they are working on projects that involve the acquisition of land, determining the use of land, designing projects to fit the land available, and adapting designs to meet zoning requirements. The legal topics covered in this chapter included real estate, liens, eminent domain, condemnation, right-of-ways, and zoning. Information was also provided on legal forms for the ownership of land, and mortgages to purchase real estate.

Key Terms

adverse possession
cadastral survey
collateral
condemnation
deed
doctrine of support
due process of law
easements
eminent domain
encumbrances
evidence of land ownership
fee simple
joint tenancy
land locked
landlord
lease
liens
life estate
mechanic's liens
metes and bounds
Model Registration law
mortgages
partitioning
profit a prendre
real estate
real property
registration of ownership
right-of-ways
servitudes
sole ownership
stop notices
sublease
tax liens
tenancy in common
tenants
Torrens registration
zoning

Discussion Questions

10.1 Why would someone want to include the real property they own in a life estate?

10.2 When someone secures a mortgage to be used to purchase real property, what type of a legal obligation are they entering into and how does it obligate them financially?

10.3 Explain how zoning is used in cities.

10.4 If an engineer will be working with a client on designing a structure for the client on property that is owned by the client, which form of ownership of the property would expedite the decision-making process?

10.5 Explain why joint tenancy would be a contentious form of ownership if a married couple is divorcing and one of them would like to sell their home and the other person does not want to sell the house.

10.6 Explain why it is important for an engineer to know whether there are easements on a piece of property that they are designing a structure for and what they should do if there are easements on the property.

10.7 Provide examples of situations where a landowner would provide a doctrine of support to someone else for using their property.

10.8 What are mechanic's liens? How are they used by subcontractors and suppliers during the construction of a project?

10.9 Discuss whether the government should or should not be allowed to use eminent domain proceedings to secure land to build stores or shopping malls.

10.10 Provide an explanation that could be given to a client on why he or she should not build a housing development next to an area that is zoned as C1 or M1.

References

Bockrath, J. T. 2000. *Contracts and the Legal Environmental for Engineers and Architects*. New York: McGraw Hill Publishers.

NCEES. 1960. *Model Registration Law*. Washington, DC: National Council of Examiners for Engineering and Surveying.

Robillard, W., D. Wilson, and C. Brown. 2006. *Evidence and Procedures for Boundary Location*. New York: John Wiley and Sons.

Test Oil Company v. La Tourette. 1907. 19 Ok. 214, 91P. 1025.

Will, G. April 24, 2008. Legal theft in Norwood. *Newsweek* CL (17): 94.

CHAPTER ELEVEN

FORMS OF OWNERSHIP OF FIRMS

11.1 INTRODUCTION

When a company is being formed, a decision has to be made by the individuals creating the company about the type of ownership of the firm. In the United States, the legal system allows several different options on the form of ownership for firms. Each form of ownership offers advantages and disadvantages and this chapter discusses these for each potential form of ownership including: sole proprietorship, partnerships, limited partnerships, corporations, joint ventures, and foreign corporations.

11.2 SOLE PROPRIETORSHIP

The most elementary form of ownership is when one person owns a firm; this is called a **sole proprietorship**. When an individual owns a firm, it allows them the freedom to manage the firm in their own way. They do not have to confer with anyone when there are decisions that need to be made concerning how to operate the firm. The ability to expand the firm is limited by the ability of the owner to obtain financing for any proposed ventures of the firm. The life of the organization is limited to the life of the owner, unless provisions are included in the will of the owner for continuation of the firm upon his or her death. Two major disadvantages of sole proprietorships are that the profits of the firm are taxed at the individual tax rate of the owner. In addition, the owner is personally liable for the debts of the firm and also personally liable if the firm is sued by a third party for any reason.

11.3 PARTNERSHIPS

A **partnership** exists when two or more individuals sign a contract agreeing to jointly operate a firm. Partnerships form a contractual relationship not a legal entity, but when a firm is being sued, a partnership is considered to be a legal entity. Partnerships are formed for a variety of reasons including providing assistance in managing a firm, injecting additional assets into a firm, securing the talents of individuals, and combining resources such as facilities or equipment.

When a partnership is formed, whatever proportion of ownership that each partner has determines the proportion they will receive as profits and also the proportion of losses that they are each liable for if the firm is not profitable for any reason. When a partnership makes a profit, the partners are taxed on their proportion of the profits at their individual tax rates and their salaries are also taxed at their individual tax rates.

In a partnership, the partners are liable for all obligations that are undertaken by any of the partners. The term for this liability is **joint and severable** liability for the entire obligation. This means that any of the partners could be sued for the entire obligation. Partnerships create **mutual agency**, which means that every partner is an agent and a principal of the partnership; they could enter into obligations that bind the other partners without the express authority of the other partners (Sweet and Schneier 2005).

When the partners have to make a decision, it could be blocked by one of the other partners. If a partner no longer desires to be in the partnership, and they would like to transfer their share of the partnership to someone else, this has to be approved by all of the other partners. Unless there is a provision in the partnership agreement that states how the partnership could be dissolved, then it is only dissolved upon death, bankruptcy, one partner withdrawing from the partnership, mutual consent of the partners, or expiration of the time limit for the partnership contract.

Limited Partnerships

Limited partnerships require that there be at least one general partner and one or multiple limited partners. The limited partners mainly contribute cash or property to the partnership; they usually do not have any influence on how a firm is managed by the general partner. Limited partnerships protect the limited partners from liability but allow them to share in the profits of a firm. In some jurisdictions, a certificate has to be filed with the state that outlines the provisions of the limited partnership agreement (Sweet and Schneier 2005).

11.4 CORPORATIONS

Unlike sole proprietorships and partnerships, **corporations** are separate legal entities. Corporations are created by filing for incorporation with the state where the corporation will perform its primary operations. When a corporation is formed, as part of the incorporation application, it files a charter with the state that indicates what type of operations the firm will be performing after its charter is approved by the state. Corporations are either:

- Public
- Quasi public
- Private
- For profit
- Nonprofit
- Domestic
- Foreign

When a firm is formed as a corporation, it restricts the control of the managers in decision making, since the firm will be controlled by a board of directors and stockholders. In corporations, capital is raised by issuing stock and selling it to stockholders. Decisions are made by the board of directors who are elected by the stockholders. The officers of corporations are agents of the corporation. In some corporations, all of the stock is owned by family members, which means it is not traded on the **New York Stock Exchange**; therefore, the firm does not have to file statements with the **Security and Exchange Commission**. Privately held stock ensures the right to secrecy of firms, since the value of the shares of the firm, and ultimately the value of the firm, are not known by the public.

The existence of corporations continues even if any of the principals in the firm pass away. In order to transfer ownership in corporations, the stock is either given or sold to other individuals.

Corporations pay taxes on profits at federal and state business tax rates after deducting ordinary business expenses and depreciation on its capital assets. In addition to corporate taxes, any dividends paid to stockholders or profits from the sale of stock are taxed at individual tax rates. If the owners of a corporation pay themselves a salary, then their salaries are taxed at individual tax rates. If profits are reinvested in the corporation, then the profits are not taxed by either federal or state governments.

A major advantage that corporations have over other forms of ownership of firms is that their liability is limited to corporate assets not the personal assets of its owners. If there is a lawsuit against a corporation, and the corporation does not win the lawsuit, only the corporate assets could be attached to pay the judgment against the firm.

Domestic Corporations

Domestic corporations are firms that are operating in the state where they are incorporated to conduct business.

Foreign Corporations

A corporation is considered to be a **foreign corporation** in all states except for the state where the firm is incorporated and they are required to be certified in each state where they will be conducting business.

Individual Corporations—Living Trusts

An individual may be incorporated in order to allow their personal assets to be transferred to the designated secondary trustees of the corporation upon their death; this is called a **living trust**. The person who writes the living trust is the **primary trustee** of the corporation and they designate the **secondary trustees**. The personal assets that are to be included in the trust are listed in the incorporation documents. The primary trustee also designates who will receive each of the assets stipulated in the living trust when the corporation is transferred to the secondary trustees upon his or her death. A major advantage of forming a living trust is that the assets in the living trust are not subject to probate or probate taxes when they are transferred upon the death of the primary trustee; therefore, there is no delay in the transfer of assets. Living trusts have to be registered with the county where the primary trustee resides. Another advantage of a living trust is that the heirs are not allowed to contest the living trust, which could happen if someone has a will rather than a living trust.

The cost to establish a living trust corporation varies depending on whether someone writes the incorporation document or hires a lawyer to write it and register it with the county. References are available at bookstores and online that provide instructions on how to write a living trust. Some of the references include computer disks that contain programs with the living trust forms.

11.5 JOINT VENTURES

Another form of a legal entity for conducting business is **joint ventures**. They are created to increase resources and to spread the risks associated with a particular project. Joint ventures are formed when two or more firms combine to form a separate firm for the purpose of completing a particular project.

All of the joint venture partners are legally bound to perform their contract obligations and for all of their construction costs. Each joint venture member is treated as a general partner in the enterprise. The liability of each joint venture partner is limited to the assets involved in the joint venture, not the assets of the parent companies. The legal liability of the joint venture partners is determined by the type of contract that is signed by the joint venture partners, whether it is a joint and severable or an entire and severable agreement (see Chapter 12). Joint ventures are formed by all of the parties signing a contract and all of the parties have an equal share in controlling performance on a project unless the contract states otherwise. Joint ventures pay taxes in the same manner as the firms involved in the joint venture.

According to Bockrath in *Contracts and the Legal Environment for Engineering and Architects* (2000, 240):

> Persons engaged in a true joint enterprise are, without any wrongful conduct of their own, each responsible to injured outsiders for the fault of other participants in the undertaking as long as the tortuous behavior occurred within the purview of the joint enterprise. To constitute such an enterprise, wherein the negligence of one participant is imputable to the others, there must be a community of interest in the purpose of the undertaking and an equal right for each affiliated individual to govern the conduct of thereof.

In the construction industry, one primary reason for forming joint ventures is to increase performance bonding capacity for government construction projects. If one firm is not able to secure a bond from a surety company in the amount required for a government project, they could form a joint venture with one or more other firms. The bonding capacity of each firm would be added together to provide enough bonding capacity to bid on the government project. When the Hoover Dam was built in the 1930s, there were no construction companies large enough to be able to secure the bond being required by the government; therefore, seven of some of the largest construction firms at the time formed a joint venture to construct Hoover Dam.

Chapter Summary

The forms of ownership that were explained in this chapter were sole proprietorships, partnerships, corporations, and joint ventures. This chapter also provided a discussion on the advantages and disadvantages of each form of ownership and on how the manner in which a firm is organized, and its legal status, determines how the firm is taxed and the extent to which its principals are legally liable when the firm is sued by third parties.

Key Terms

- corporations
- domestic corporations
- foreign corporation
- joint and severable
- joint ventures
- limited partnerships
- living trust
- mutual agency
- New York Stock Exchange
- partnership
- primary trustee
- secondary trustee
- Security and Exchange Commission
- sole proprietorship

Discussion Questions

11.1 Explain the advantages of being in a limited partnership rather than a general partnership.

11.2 What is the major advantage of forming a firm as a sole proprietorship?

11.3 What is the major risk involved in being in a partnership? Why is it such a major risk?

11.4 Why is it harder to make decisions in a partnership than in a sole proprietorship?

11.5 What is the liability of partners in relation to debt obligations and lawsuits?

11.6 What is the major advantage of forming a corporation rather than operating as a sole proprietorship?

11.7 Which form of ownership offers the best tax advantage and why?

11.8 Why is it advantageous to form joint ventures in relation to the financial obligations of the joint venture partnership?

11.9 What types of government projects would be so large as to require firms to form a joint venture in order to bid on them?

11.10 What form of ownership should be recommended to someone thinking of forming their own firm and why?

References

Bockrath, J.T. 2000. *Contracts and the Legal Environment for Engineers and Architects*. New York: McGraw Hill Publishers.

Sweet, J., and M. Schneier. 2005. *Legal Aspects of Architecture, Engineering, and the Construction Process*. Toronto: West Publishing Company.

CHAPTER **TWELVE**

FORMING ENGINEERING AND CONSTRUCTION CONTRACTS

12.1 INTRODUCTION

This chapter provides information on the processes required to form engineering and construction contracts. It explains what is required for legally binding contracts, the formation principles for contracts, how contracts are formed, and it discusses the different types of contracts that are used in the engineering and construction industry.

12.2 THE DEFINITION OF CONTRACTS

According to the ***Construction Specification Institute (CSI) Manual of Practice***, a **contract** is "a promise or a set of promises for the breach of which the law gives a remedy of the performance of which the law recognizes a duty" (CSI 2008, 1). What this definition is indicating is that once they are formed, contracts are a legally binding obligation, and if the terms of a contract are not completed, the contract is in default and the party that has been defaulted upon has the right to seek a remedy for the default through the legal system.

The Character of Contracts

The *CSI Manual of Practice* also provides a definition for the **character of a contract**—"the character of a contract is determined by the nature of the obligation imposed by the provisions of the contract as a whole. The specifications are part of the contract and along with all other parts of the contract determine its character. Each part must be read and construed together (in context)" (CSI 2008, 2).

When someone enters into a contract, they become legally obligated for the entire contract, as set forth in all of the contract documents. If any type of a dispute arises over the performance of a contract, then the contract documents are used to determine the nature and extent of the legal obligation. The majority of the components contained in construction contracts are not referred to unless there is a contract default. This is why commercially produced standard contract forms such as the ones published by the **American Institute of Architects**, the **Associated General Contractors**, the **Construction Management Association of America**, and the **Engineers Joint Construction Document Committee** are sometimes used in construction contracts.

The importance of contract clauses is only fully realized when the performance of a contract is interrupted and the parties to the contract need to locate a clause that could be used as a defense against being accused of defaulting on their legal obligation. This is the reason that some contracts appear to have repetitious clauses that are expressed slightly differently. When someone involved in a contract is defending himself or herself against a breach of contract claim, the outcome of their defense might depend upon which clause they use to justify their defense. This concept is elaborated upon in Chapter 14.

Scope of Work

Contacts are used to outline the **scope of work** for projects. The scope of work "defines the obligations and financial risks and describes the undertaking prior to bidding. The requirements of contracts are constantly changing" (CSI 2008, 2). Contracts are considered to be imperfect expressions of what owners think they want for their projects. Contracts have to be written with the flexibility to change as the scope of work for projects is refined by owners.

Project Baselines

The original scope of work, as outlined in the contract documents, is used as the **project baseline** against which progress and changes are measured during the monitoring of a project. Project baselines should include enough detailed information in order to be able to monitor cost expenditures, equipment usage, manpower, materials, and the project schedule. If computer software is being used to help monitor work progress, then the original project baselines are input into project control software and monitored on a regular basis.

The original scope of work should be clearly defined because it is used to determine whether changes to the scope are occurring that would entitle contractors to change orders and additional compensation if the scope of work is altered to reflect a change that increases or decreases the value of a contract. If the party performing the contract is not aware of their legal rights, owners might try to increase the scope of work without providing additional compensation.

If an owner desires additional work, above and beyond the original scope of work, that would result in the value of the contract increasing by more than 10 percent, then their request is considered a **cardinal change** since it substantially alters the original intent of the contract agreement. Rather than unilaterally having to perform the additionally required work, an engineer or a contractor has the right to request that the contract be voided and a new contract be written that incorporates the changes that will be performed and that includes a newly agreed upon cost. Contractors also have the option of performing additional work without voiding the original contract but for additional compensation that is calculated based on the unit prices of the bid items in the original bid proposal estimate.

12.3 CONSTRUCTION INDUSTRY PROJECT TEAM MEMBERS

This section describes the personnel who are directly involved in construction contracts and some of their responsibilities when they are assigned to work on construction projects.

Owners

Owners are responsible for financing projects, the initial planning of projects, and providing information to designers on project requirements and the operation and maintenance requirements of their facilities. During the design and construction phases, owners request modifications to projects that could lead to change orders. Owners could also request that **value engineering** be part of their projects. Value engineering allows designers or contractors to propose alternative designs or construction processes that would save owners money; it allows designers and contractors to share in any cost savings generated by their alternative schemes.

Designers

Designers are selected by owners through negotiations, not competitive bidding. The code of ethics of professional engineering societies prohibits engineers from competing for commissions on the basis of cost, because it might compromise the quality of projects. Architects and engineers are responsible for converting design concepts into plans and specifications that are used for the construction and operation of projects. In addition to designing projects, designers perform engineering analysis to verify that their designs conform to legal requirements and standards, such as building codes, and to ensure that their designs meet the required factors of safety for each particular type of structure.

Designers have to effectively coordinate multiple disciplines and many different agencies, and act as a representative of the owner as their agent during construction. Designers work with owners, contractors, vendors, suppliers, lawyers, and employees of government agencies. It is the responsibility of designers to verify conformance to standards, codes, and contract plans and specifications. Designers could also be responsible for verifying that quality control procedures are properly implemented and that quality assurance checks are completed in a timely manner.

Contractors

Contactors interpret contract documents, drawings, and specifications and use them to create physical facilities. Contractors monitor the progress of construction projects and provide leadership and management expertise during construction. Contactors assemble the workforce, issue instructions, provide work plans and methods, order and expedite materials, and provide or rent equipment and tools. The main concern of contractors is profitability. If construction costs exceed productivity gains, profits decrease. Contractors work under time constraints and their work is negatively impacted by procurement delays, unskilled workers, shortages of workers, the job site environment, site accessibility issues, equipment mismanagement, and poor planning. If contractors are able to improve their productivity at job sites, it improves profits.

Construction Managers

Construction managers have a contract with owners to manage construction on the behalf of owners. In many instances, construction managers do not have legal liability to ensure that a project is completed on time and within the projected budget. Construction managers do not have contractual relationships with any of the other parties involved in the construction process except for owners.

Sureties

Sureties are involved in construction project teams on government projects because of the requirement for sureties to approve change orders before they provide additional bonding capacity to cover any additional amounts included in change orders. Sureties are discussed in detail in Chapter 16.

The Labor Force

The **labor force** is involved in construction processes through foreman and skilled craft workers. Laborers either work individually or in teams and are responsible for transforming plans and specifications into completed facilities. Construction workers use methods that they have developed or methods stipulated by project management team members to perform their jobs. Construction workers combine the essential ingredients of plans, specifications, materials, equipment, and tools to create projects. Construction workers that do not demonstrate that they possess the proper skills and attitude, or that are not adequately trained to perform their tasks, have a difficult time adjusting to construction job site work requirements. Workers and materials have to be delivered to job sites in a timely manner. In order for construction workers to perform their tasks, their supervisors should provide proper directions and leadership and make any necessary corrections to the work, approve the work as it is being completed, and provide training to the workers.

12.4 CONSTRUCTION PROJECT DEVELOPMENT

When initiating construction projects, clients first identify a particular need and then **feasibility studies** are conducted to provide financial information that is used to determine whether the projected rate of return meets or exceeds the **minimum attractive rate of return** (MARR) for a firm. If the projected rate of return exceeds the MARR, clients will then use the feasibility study to help them obtain financing for projects and to solicit public or political support for projects.

Figure 12.1 shows the project development phases that are required in order to design and construct construction projects. A more detailed list of the project development phases includes:

1. Preliminary planning and feasibility studies
2. Preliminary engineering and design
3. Engineering
4. Procurement
5. Construction
6. Operation
7. Demolition

12.5 CONTRACTS USED DURING THE PHASES OF CONSTRUCTION PROJECTS

Contracts are used during all of the project phases of construction projects. In the first project phase, which is the **feasibility phase**, contracts are formed between an owner and a consultant, an engineering firm, or an architectural firm to perform a feasibility and site selection study. This first contract is a negotiated contact that describes the services that will be rendered by the firm performing the feasibility and site selection study.

During the next project phase, the **preliminary engineering design phase**, a contract is negotiated with a design firm to develop a preliminary project design that includes basic design considerations. The third and fourth project phases, **engineering** and **procurement**, may also be performed by members of the firm that developed the preliminary project design or by a different firm. The third project phase requires the production of detailed drawings, specifications, and contract documents. Contracts could be signed during this phase for major items of equipment, or other long lead-time items, including materials or equipment that require long periods of time to be manufactured or fabricated such as structural steel or turbines.

The fourth phase is the **bidding and project award phase**. Preparing a project to be competitively bid, and the evaluation of bids, is normally performed by the firm that designed a project as part of the original design contract or as a separate contract for services as a representative of the owner.

The fifth phase is the **construction phase**. During this phase, there may be one contract between an owner and a general contracting firm who is the prime contractor or several contracts between the owner and multiple prime contracts. If a single prime contractor is used as the general contractor on a project, there will be subcontracts between the prime contractor and the subcontractors.

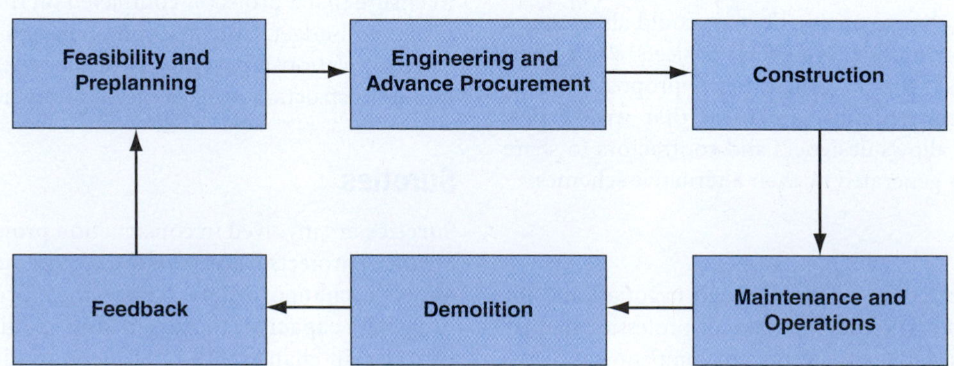

FIGURE 12.1 Project Development Phases.

During the construction phase there will also be contracts between the prime contractor and material and equipment suppliers. If union labor is used on a construction project, then there will be labor union contracts in force during construction.

The sixth phase involves the **testing and operation** of facilities. General contractors could contract to perform these services for owners or other firms might be hired under new contracts to test and operate facilities.

At the completion of the useful life of a facility, the structure is normally demolished by specialty **demolition** contractors who are familiar with proper demolition techniques and the environmental requirements for dismantling existing structures.

The following sections describe in detail each of the project phases.

The Feasibility Study and Preplanning Phase

If an owner is having a difficult time articulating what he or she desires in a structure, he or she may hire an architect or an engineer to define the initial scope of the project and to help them perform a feasibility study. Feasibility studies require owners and architect/engineers to work together to evaluate potential project alternatives and project sites. For a privately funded project, the main thrust of feasibility studies is to determine whether a potential project will be profitable in terms of rate of return (ROR), which is the amount that an owner would make in potential profit including the time value of money. Projections are developed for all of the revenue and income streams and they are analyzed using either **present worth analysis** (PW) or **equivalent uniform annual cost** (EUAC). After the engineering analysis equations are developed, the rate of return is calculated as the point at which the PW is equal to zero or the EUAC is equal to zero.

Owners will normally compare the estimated rate of return to the minimum attractive rate of return (MARR) of their firm. Each firm uses a different MARR, which might be based on the amount of money that an owner would be able to obtain if they were to invest their money in other more secure investments such as certificates of deposits, money market accounts, savings accounts, or other secure investments. If several potential projects meet the MARR, then further analyses are performed in order to evaluate the different parameters of the potential projects, such as:

- Location including desirability of an area
- Accessibility to mass transit systems
- Accessibility to schools or shopping areas
- Accessibility to major arterials such as freeways or feeder roads
- Value of adjacent property
- Availability of resources
- Availability of labor
- Availability of utilities
- Right-of-way availability
- Potential risks analysis
- Environmental impacts report/study
- Traffic impacts in construction
- Potential business/resident relocations
- Alignment feasibility studies
- Project management plan
- Contract implementation plan
- Real estate acquisition management plan
- Relocation plans

In order for engineers or architects to perform the above mentioned evaluations, the following are first completed:

- Developing the project scope
- Gathering supporting information and data
- Analyzing the data
- Preparing a plan
- Disseminating the appropriate information to all affected parties
- Evaluating the results of the planning effort

During the preliminary planning and feasibility study phase environmental studies are conducted along with site evaluations and economic studies. Other studies that could be generated during this phase include:

- Initial transportation alignment studies
- Facility sizing
- Project scope development
- Aerial and preliminary surveys
- Justification plans for the preferred alternatives

The Design Phase

The engineering design phase starts after funding is obtained for the initial design effort. All throughout this phase owners promote their projects to try to secure the support of the local community. The financial plan is refined as the design evolves during this phase. During this phase engineers develop the contract, the plans, and the specifications and develop an estimate and other supporting documentation. Engineers also develop plans for construction methods and processes and a preliminary project schedule.

Once the planning effort is sufficiently far along, the owner either hires the architect/engineers who performed the feasibility study, or another firm, to develop the contract, the specifications, the plans, and supporting documentation. While a project is being designed, studies that might have to be conducted include:

- Soil/geotechnical reports
- Traffic studies
- Vibration studies
- Noise studies
- Utility impacts and utility relocation studies
- Surveys
- Amendments to environmental studies

The Procurement Phase

Usually some materials will be needed during construction that are long-lead time items and these should to be ordered during the engineering design phase; therefore, procurement starts during the engineering phase and continues throughout the construction phase as changes are initiated during construction. The procurement phase not only includes ordering materials but it also includes expediting materials to make sure that they arrive at job sites when they are needed for activities. A procurement schedule is developed that indicates when materials are ordered, their expected delivery dates, and when they are needed for activities.

While a project is being designed, the engineers working on the design have to inform the owner of any materials that require a long period of time to be fabricated and shipped to the job site. An example of a long-lead time item is steel, which requires time for it to be fabricated in the sizes and shapes required for each project. Normally the structural frames for buildings will be designed in the early design stage and then the other elements will be designed around the structural frame. If any changes have to be made to the design of steel frames once the steel has been ordered, it will be more expensive than if the change is incorporated prior to the steel being ordered from steel fabricators. Major steel elements such as columns and beams or prefabricated sections containing columns and beams welded together at steel fabrication plants could take a year or more to reach job sites once they are ordered by engineers.

Other long-lead time items include major pieces of equipment such as turbines or equipment that will be built from unique specifications for a particular project. Turbines for nuclear power plants or hydroelectric power plants may have to be ordered years in advance.

The Construction Phase

The construction phase requires contractors to interpret the contract documents and to construct facilities according to the contract. In addition, contractors perform field modifications and manage construction using project control, quality assurance, and quality control procedures. Construction projects are constructed according to the schedules submitted with bid proposals that are developed by contractors, not owners or their representative. If owners include project schedules in the contract documents, then subcontractors would have claims against owners every time there is a delay on a project.

The Operation and Maintenance Phase

During the operation and maintenance phase owners operate and maintain their projects unless they have made prior arrangements for a contractor to operate the facility. Before a facility is commissioned for operation, it goes through a series of tests to ensure that all of the systems are operating properly. The testing phase could require a few days for a simple project or a few years, as is the case with nuclear power plants.

The Demolition Phase

Once a facility has served its purpose or has been operated for its useful life, it is demolished and another facility is constructed in its place; this is done during the demolition phase of a project. Facilities are demolished using heavy construction equipment, explosives, wrecking balls, or other methods to dismantle structures. The materials from demolished structures either have to be recycled for use in other facilities or transported to landfills. If any of the materials contain hazardous wastes, they have to be disposed of in special waste disposal sites that usually contain liners to prevent the hazardous materials from leaching into the soil or water systems.

12.6 PROJECT CONTRACT ADMINSTRATION MODELS

Several different types of project contract administration models are used on construction projects. Owners select different contractual relationships in order to provide checks and balances while they are attempting to achieve their desired level of quality and timely completion of projects without exceeding project budgets.

Typical contractual relationships sometimes result in **adversarial relationships** developing between the various parties that are involved in construction contracts. The reason adversarial relationships develop on projects is that each of the major parties is trying to achieve different objectives. Owners want the lowest cost project at the highest quality. Contractors want to complete projects quickly and make the most profit possible. Engineers try to make sure that projects are built according to the plans, specifications, and the contract and that they will be safe for the occupants of the structures.

There are several reasons for carefully selecting which type of contracting procedure to use on a construction project. Some of the reasons that owners select a particular type of contractual relationship are:

- Obtaining representation for their interests
- Fast-tracking to expedite the project construction phase
- Reducing processing costs associated with design and construction
- Having assistance in maintaining the schedule and with project control processes
- Outsourcing to obtain management expertise that is not available in-house
- Reducing the liability of owners

Single Prime Contracts

The traditional form of a construction contractual relationship is a **single prime contract** by which the owner hires an architect/engineer (A/E) to design a project and then hires a general contractor to construct the project. This contractual relationship is shown in Figure 12.2. The contractual agreements in force in this contracting arrangement are the owner has a contract with the architect/engineer and the owner has a contract with the general contractor. There is no contractual agreement between the architect/engineer and the general contractor. Although in a single prime contract, the architect/engineer does not have a contractual relationship with the contractor he or she may be called upon to administer the contract for an owner and act as an agent for an owner.

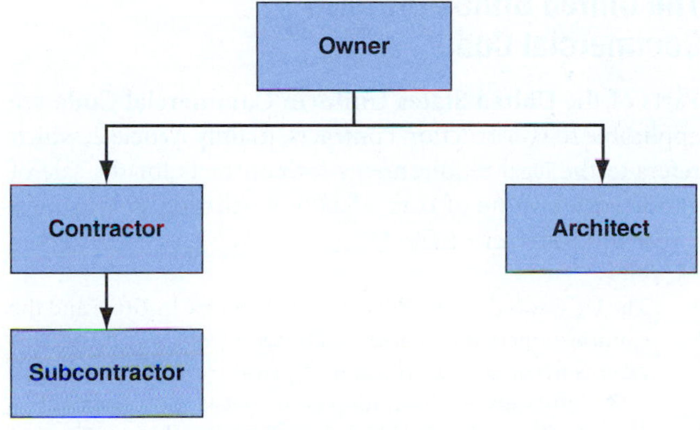

FIGURE 12.2 Single Prime Contracts.

Separate Prime Contracts

If an owner has the expertise to manage a project, to perform the duties of a general contractor, and to coordinate **separate prime contracts**, then they will use separate prime contractual agreements, which is the contractual relationship shown in Figure 12.3.

Construction Management Contracts

If an owner needs additional assistance in managing a project, they may hire a construction manager to oversee construction; this contractual relationship is shown in Figure 12.4. A construction manager has a contract with an owner, but the construction manager does not have a contract with either the architect/engineer or the general contractor, which dilutes their authority with these other parties unless the owner firmly emphasizes the authority of the construction manager. The owner has contractual agreements with the architect/engineer, the general contractor, and the construction manager.

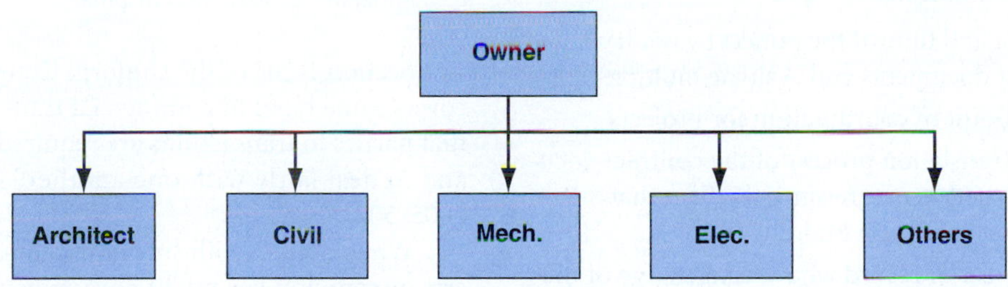

FIGURE 12.3 Separate Prime Contracts.

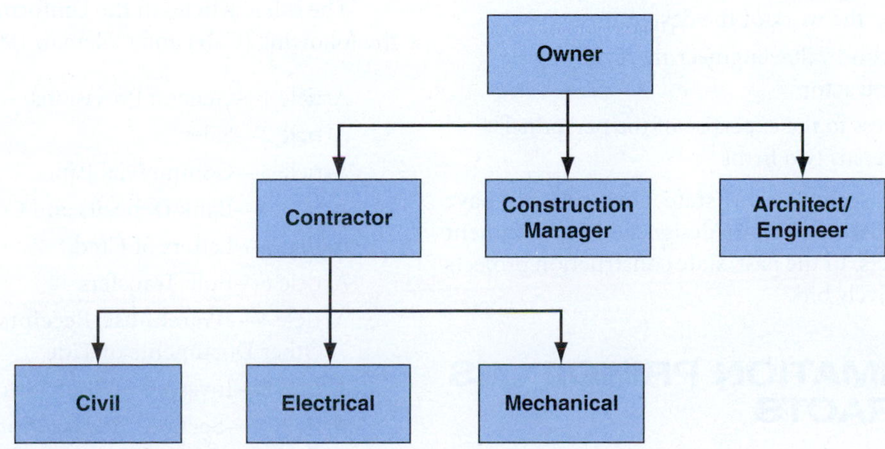

FIGURE 12.4 Construction Management Contract.

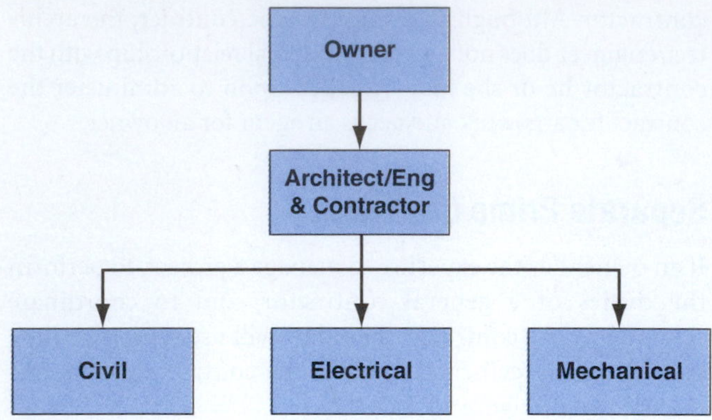

FIGURE 12.5 Design/Build Contract.

Design/Build Contracts

Another form of contractual relationships that places more responsibility and liability with one firm is the **design/build** contracting method that is shown in Figure 12.5. In addition to the traditional form of design/build contractual relationship shown in Figure 12.5, there could also be alternative forms, including (1) when a contractor subcontracts to perform design services, (2) when a construction management firm subcontracts to perform both design and construction, or (3) when a designer subcontracts to perform construction.

In this contracting agreement one firm both designs and builds a structure. Advantages to using this contracting method include the following:

- Reducing the cost and time of the project by not having to issue the contract documents and evaluate multiple bids
- Having a single point of coordination for projects
- Simplifying the translation process of the contract documents into a finished structure since the firm that designed the structure is also building it
- Having the engineer of record who was in charge of the design process available during construction

Disadvantages include the following:

- No unbiased oversight since there is not a separate contractor reviewing the work of the design firm
- Not having a second value engineering review by an independent contractor
- Only having access to the expertise of the personnel from one firm versus two firms

In the public sector, some state governments have passed legislation that now allows design/build government construction projects. In the past, state construction projects had to be competitively bid.

12.7 FORMATION PRINCIPLES OF CONTRACTS

In order for parties to avoid unknowingly entering into contracts, there are specific procedures that have to be followed during the formation of valid contracts. The following sections explain contract formation principles and the legal requirements for entering into contracts.

In order for contracts to be legal, they have to adhere to all of the requirements, as set forth in contract laws. As was mentioned in Chapter 3, in some areas of the law there are exclusive jurisdiction of states; therefore, parties that enter into contracts are subject to following state contract laws except for situations in which the Uniform Commercial Code takes precedence over common law commerce transactions.

The United States Uniform Commercial Code

Parts of the **United States Uniform Commercial Code** are applicable to construction contracts, mainly Article 2, which refers to the legal requirements for contracts for the sale of goods with a value of over $5,000. According to attorneys Sweet and Schneier (2005, 3):

> The UCC was developed by the America Law Institute and the Commissioners on Uniform State Laws, both private organizations devoted to unification of private laws. At present, all states but Louisiana have adopted the UCC (Louisiana, influenced by its civil law tradition, has adopted only some parts of the UCC). Article 2 regulates the sale of materials and supplies used in a project. Although it does not regulate construction itself (inasmuch as it does not govern services), it can be influential in a construction dispute.

Section 1-203 of the Uniform Commercial Code (UCC) covers some types of commercial transactions and indicates that parties to transactions are required to "act in good faith and to deal fairly with one another" (Sweet and Schneier 2005, 507).

When dealing with subcontractors, the UCC could govern, or common law might govern, depending on whether a subcontract is mainly for goods or services. If a subcontract is mostly for services, it is governed by common law; if it is for materials, it is governed by the UCC.

The other articles in the Uniform Commercial Code are the following (Calvi and Coleman 1997, 252):

Article 1—General Provisions
Article 2—Sales
Article 3—Commercial Paper
Article 4—Bank Deposits and Collections
Article 5—Letters of Credit
Article 6—Bulk Transfers
Article 7—Warehouse Receipts, Bills of Lading, and Other Documents of Title
Article 8—Investment Securities
Article 9—Secured Transactions, Sales of Accounts, Contract Rights, and Chattel (title of tangible property other than real estate) Paper
Article 10—Effective Date and Repealer

Contract Agreements

Contract agreements are used to legally bind parties to the terms of contracts. Agreements are part of engineering and construction contracts and they are usually located at the beginning of contracts. When the parties to a contract sign the agreement, they are signifying that they have not only read the contract but that they also understood the terms of a contract. The agreements in engineering contracts are signed by owners and engineers. Construction contracts are signed by owners and contractors. If they are government contracts, sureties also sign the contracts. The reason sureties also sign government contracts is discussed in Chapter 16 in Section 16.3.

Agreements are only legal if they are signed by competent parties. In order to be considered to be a competent party, the person signing an agreement must not be under eighteen years of age and they need to be mentally competent. The parties to an agreement have to signify their participation in the contract in whatever form is legally required, such as their signature and the date on an agreement form or their signature, date, and the stamp and signature of a notary republic. Figure 12.6 shows a sample agreement signature form that is used in engineering contracts. Figure 12.7 contains an example of an agreement signature form that is used in construction contracts. Appendix E provides a sample *Suggested Form of Agreement Between Owner and Contractor for Construction Contract (Stipulated Price)* that is published by the Engineers Joint Contract Documents Committee.

There are also agreements to agree "that are documents that indicate that two or more parties will be entering into an agreement with certain terms in the future" (Construction Specification Institute 2008, 23). Agreements to agree are valid as long as they are not formed while the parties are still negotiating the terms of the agreements.

Meeting of the Minds

All of the parties to a contract have to understand the terms of a contract, and what they will be obligated to perform, before there is an actual contract; this is called **meeting of the minds**. If a contract is being disputed, one party might attempt to say that what they thought was in the contract was different from what the other party to the contract thought had to be performed. But if the terms of the contract are clearly written, then this defense would probably not be successful in a court of law. If the terms of a contract are written in a convoluted manner, then a defense of there not being a meeting of the minds might provide relief to the party using this defense.

Proper Subject Matter

In order for a contract to be legally valid, the subject matter of the contract has to be legal; this is referred to as **proper subject matter**. Contracts are not legal if they contract for services that violate public policy or laws. An extreme example is if one party hires another person to either harm or kill a third party. Another example is if a party tries to form a contract to illegally dump hazardous wastes. Some states have laws that invalidate a contract if it is formed on a Sunday.

Duress and Undue Influence

When parties enter into a contract, they have to do it of their own free will and there should not be any **undue influence** used to force a party to sign a contract. If **economic duress** is used to force someone to sign a contract, it could invalidate the contract. Economic duress exists when one party

Standard Form of Agreement Between Owner and Engineer

with Standard Form of Architect's Services

BETWEEN the Engineer's client identified as the Owner: ______________________

(Name, address, and other information)

and the Engineer: ______________________

(Name, address, and other information)

For the Following Project: ______________________

(Include detailed description of Project)

The Owner and the Engineer agree as follows: ______________________

FIGURE 12.6 Sample of an Engineering Contract Agreement Form.

Standard Form of Agreement Between Owner and Contractor
where the basis of payment is a STIPULATED SUM

AGREEMENT made as of the ______________________ day of ___________

in the year ______________

(in words indicate day, month, and year)

BETWEEN the Owner: ____

(Name, address, and other information)

and the Contractor: ____

(Name, address, and other information)

The Project is: ____

(Name and location)

The Architect is: ____

(Name, address, and other information)

The Owner and Contractor agree as follows: _____________________________

FIGURE 12.7 Sample of a Construction Contract Agreement Form.

exerts excessive pressure on the other party and due to their financial circumstances the other party agrees to the contract terms, even though the contract terms are not favorable to him or her. One example of economic duress is when a contractor has not been employed for awhile and an owner knows that the contractor has bills that are past due and the owner convinces the contractor to perform work for him or her far below the fair market value of the work.

Fraud

Fraud crosses the boundaries of contract law into the realm of federal law. If fraud is used to secure money, then it is considered to be a federal crime; anyone found guilty of fraud could serve time in federal prison. Fraud is the intentional deception of someone. When fraud is involved in the formation of a contract, then the defrauded party is allowed to void the contract and also may be able to recover some of the funds that were paid to the party that committed fraud based on **restitution** principles, which were designed to return the defrauded party back to where they were before the fraud was committed by the other party.

Unfortunately, many examples of fraud have occurred in the construction industry in the United States. The following is an example of a legal case that occurred in the construction industry related to minority-owned businesses and fraud (ENR.com, October 10, 2007):

> A smalltime contractor, who is at the center of a federal investigation into fraud nationally in the granting of highway construction work to minority owned firms, pleaded not guilty yesterday to fraud charges at the U.S. District Court of Central Islip, according to officials. The owner of the contracting firm was released on $100,000 bail by the U.S. Magistrate, pending future hearings.
>
> The contractor, who is of Brazilian background, is eligible for a federal program that allocates up to 15 percent of highway contraction contracts to businesses owned by minorities or the disabled, according to an investigation by the Criminal Investigation Division of the Internal Revenue Service.
>
> But while the contractor received contracts on road repair in Freeport and landscaping on the West Side Highway in Manhattan, the actual work was performed by a number of other companies. In return for, in effect, lending his business' name to the work, the contractor allegedly received $10 million in kickbacks, according to the federal prosecutor.

Each state has a requirement that a certain percentage of government construction contracts be awarded to **minority business entities** (MBE), **women business entities** (WBE), and **disadvantaged business entities** (DBE). The percentage of minority contracts that have to be awarded is based on the minority population in each state. In addition, the federal government requires that 15 percent of construction contracts be awarded to MBE or DBE contractors.

Another example of a legal case involving fraud that occurred at the Pentagon is the following (ENR.com, February 25, 2008):

> A former Pittsburgh area contractor pleaded guilty to participating in a scheme to defraud the federal government in the reconstruction of the Pentagon after the September 11, 2001 terrorist attacks. The defendant pleaded guilty to mall fraud, major fraud, and conspiracy. The three defendants helped to overcharge the government $850,000 for reconstruction work at the Pentagon, prosecutors said. The false bills padded labor hours and cost of materials used at the Pentagon, when some of the labor and materials were actually used on a personal business owned by the contractor. The contractor billed $13.9 million for its work on the Pentagon between September 2001 and May 2002. The project manager pleaded guilty in October 2008 to conspiracy for accepting kickbacks from the contractor.

A third example of fraud during construction occurred on the Big Dig project in Boston, Massachusetts (ENR.com, March 3, 2008):

> One of the Big Dig's largest contractors admits that it bilked taxpayers out of more than $300,000 by charging journeyman-labor rates for work that lower-paid apprentices actually did. The U.S. attorney announced a plea bargain under which the contractor admitted to conspiracy to defraud the federal government. The project manager also pleaded guilty to making false statements regarding a federal highway project. Prosecutors agreed to recommend that the contractor receive three years' probation, a $500,000 fine, and unspecified restitution payments. The project manager faces up to five years in prison and a $250,000 fine, but the court will seek a sentence and fine at the low end of the applicable guideline range. The project manager will probably not get any jail time. The case involved overcharging on a $245 million contract that the contractor had for finish work on the Tip O'Neill tunnel—one of the Big Dig deals the firm had. The contractor admitted that between 2002 and 2005, it faked nearly 2,000 records to claim journeyman trades people did work that apprentices actually performed. That jacked up the government's bill by $314,000.

Another incidence of fraud related to asphalt construction occurred in Plymouth, Massachusetts (ENR.com, December 1, 2008):

> A Hanover, Massachusetts contractor has agreed to pay the U.S. Government $900,000 to settle a 2005 civil claim that firm officials submitted fake tickets inflating load weights for asphalt delivered to federal paving jobs from 1995 to 2003. The U.S. attorney in Boston says the firm created fake paperwork by manually overriding computer controls on its Plymouth, Massachusetts asphalt plant. A state criminal jury in 2007 found the firm's CEO and a manager guilty of similar charges. The CEO was sentenced to 42 months in prison and fined $15,000.

Concealing information creates a situation where an individual could file a claim against the party that concealed the information. **Misrepresentation** is different than fraud or concealing information if it is not done intentionally. If inaccurate information is supplied, but the party providing it does not know it is not accurate, and if the other party relies on the information when they form a contract, then the party who received the inaccurate information has the basis for a performance claim that could encompass claims for both costs and time.

Mutual Mistake and Unilateral Mistakes

When contracts are being formed, one or both of the parties could make an unintentional mistake, which differs from fraud because fraud is intentional. When there is a **mutual mistake** during the formation of a contract, both parties could agree to void the contract. When a **unilateral mistake** is made by a contractor in a bid estimate, the contractor might have to still perform the work. If contractors were excused from performance for unilateral mistakes, then any time a contractor submitted a bid that was substantially lower than the other bids they would claim that a unilateral mistake had occurred and attempt to withdraw their bids. Rather than create this type of a situation, owners are legally able to require that a contractor perform the work for their bid price even if it contains an error. Some owners prefer not to obligate a contractor to perform the work for an underbid amount because it could compromise the quality of the project. The following case illustrates how one court ruled in a case when a unilateral mistake occurred on a project [*Architects and Constructors Estimating Service, Inc. v. Smith* (1985) Cal. App. 3d 1001, 211 Cal. Rptr. 45]:

> Plaintiff Noble contracted with the Navy to remodel an engine maintenance shop at El Toro Marine Corps air station for $647,500. He used a $90,000 "budgetary bid" to cover HVAC. The bid was worked up by Russ. Another subcontractor bid $94,000 excluding the chilling water system, which would cost $16,000. Coast, the competing subcontractor, warned Noble, the general contractor, that $86,500 was an unreasonable bid. Russ' employer refused to sign the contract at $86,500. The first bidder then told Smith that the $86,500 bid had a 20 percent profit margin, and Smith, who had never done a job for more than $10,000, and had not thoroughly examined the plans and specifications, signed a contract with Noble. Smith then checked with Ruff, who told Smith the cost would far exceed $86,500. Smith refused to do the work, and Noble responded that it would contract with Coast and sue Smith for the difference. **HELD:** There was a unilateral mistake of fact by Smith,
>
> *(continued)*

and Noble knew it. Under Cal Civil Code 1577, Smith must show its mistake was not caused by "neglect of legal duty." "Smith's business acumen was something less than sharp and business was slow. He was ambitious." 211 Cal. Rptr. at 48. Smith's conduct was not preposterous, irrational, or grossly negligent; therefore, the conduct was not neglect of duty. Alternatively, Noble obtained Smith's consent by fraud. Noble suppressed the fact that the bid did not include the cost of the chilling water system.

Statute of Frauds

Each state sets the requirements for contracting within the state. Part of the requirements is related to when a contract is required and what is the legal format for contracts. **Statutes of fraud** (laws) require that contracts have to be in writing if they are for the purchase of products that are above a certain amount of money that is set within each state. Typical minimum amounts are $500 to $5,000 in states with large economies. This practice is required in order to protect consumers from fraud or unscrupulous merchants.

Another situation where contracts are required by law is if the content of a contract will not be performed within one year. The impetus for this legal requirement is the difficulty that parties to contracts have in remembering the terms of contracts after a substantial amount of time has passed after contracts are signed by all of the parties to the contract.

The third type of contracts that have to be in writing are contracts for the sale of land or interest in land. This requirement stems from the high incidence of fraudulent schemes for selling land to unsuspecting multiple parties when the parties purchasing the land think they are the only ones purchasing the land. An example of this is schemes to sell beachfront property in the state of Florida. The parties attempting to sell the property sell it to more than one party in different states. The requirement for written contracts and for contracts to be registered with the county where the property is located helps to prevent sellers from selling property to more than one person.

Contracts also have to be in writing if they are a prenuptial agreement that is written to provide a legal means of apportioning property if the marriage is terminated by a divorce. The prenuptial agreement must be signed and notarized by a notary republic prior to the marriage not after it has occurred.

If a surety is involved in a contract, then the contract has to be in writing and signed by all of the concerned parties and the surety.

12.8 FORMING CONTRACTS FOR ENGINEERING AND CONSTRUCTION SERVICES

This section discusses the legal requirements for forming contracts for engineering and construction services.

Contract Offers

Before a legally binding contract occurs, someone has to make an **offer** to form a contract for goods or services. Offers have to be made in such a manner that they will not be misunderstood. If an offer is not clear as to the intention of the offer, the judicial system may not recognize the obligation, as set forth in the contract. Offers to contract also need to be of a definite nature not merely **invitations to negotiate an offer**, such as the advertisements in store windows.

Offers to form a contract have to be for obligations that are legally permissible. This means that contracts must not include provisions for obligations that violate state or federal laws and regulations. One example of an illegal offer would be offering to dump hazardous wastes in a location that is not a government-approved hazardous waste dumpsite.

Offers should also be serious since the parties to which an offer is being made is **justifiably relying** on it being a legitimate offer when they are contemplating entering into a contract based on an offer.

If someone makes a **gratuitous offer**, it is not considered to be an offer, because it may not be based on valid information. When a contractor calls a subcontractor and obtains a bid over the telephone, this is not considered to be a valid offer. The subcontractor has to first review the construction documents related to the work they are providing a bid for and then they need to submit a written bid to the contractor that is soliciting bids.

Contract Acceptance

Several requirements have to be met in order to legally accept an offer for forming a contract. **Contract acceptance** has to occur when the offer is made unless arrangements are made for an alterative location or deadline at the time of the offer. If an offer indicates that acceptance has to be made through the mail, then that is the only acceptable method for accepting the offer; acceptance occurs when it is mailed, not when it is received by the party extending the offer.

An offer should not be made in such a manner that no reply is considered to indicate acceptance of the offer. Laws were passed to prohibit this practice after there were solicitations through the mail indicating that if the recipient of the letter did not respond within a certain time frame, then they would start receiving merchandise in the mail and be billed for it at a later date. The most common form of this practice was solicitations for records, cassettes, or CDs through the mail.

Adequate Consideration for Contracts

Due to the volume of cases being heard by court systems, they are not able to enforce each and every promise that is made by two or more parties. In order to determine whether a contract is enforceable, the legal system evaluates agreements in the context of whether **adequate consideration** was offered as payment for services that would be rendered by one or more parties. Payments for services or goods are considered to be adequate if they represent a fair exchange, or a reasonable payment for services or goods. Fair market value would be adequate payment for services or goods.

Courts would also enforce promises when parties are able to prove that there was **justifiable reliance** on the actions of another party such as when a contractor relies on a bid provided by a subcontractor and uses the bid of the subcontractor when preparing a bid proposal for a construction project.

Parties to contracts could also help prove that a contract exists for adequate consideration by claiming that their actions were motivated by having received benefits in the past from a party that they have performed services for or provided products to, such as vendors or suppliers that have worked with contractors in the past. Since they were able to rely on being paid for their services or goods in the past, they continue to expect payments in the future from the same contractor.

Contracts Under Seal

Some state governments require that the signatures on all contracts be witnessed by a licensed **notary republic** in order for the contract to be legally enforceable. Sometimes the use of a notary republic and fingerprinting is required for certain types of contracts such as for the sale of real property in the state of California. This requirement is enforced because of instances when someone would state that a person was their spouse and the person who was signing as the spouse was not really their spouse. These circumstances arise when there are contested divorce proceedings and one spouse would not agree to sell a home. Rather than having to use the court system to partition the home, which could take from several months to possibly years, one spouse tries to sell the house without the knowledge of the other spouse.

If a contract states that it has to be signed, sealed, and delivered at a set date and time, this means that there is a requirement for validation of the signature by a notary republic. The terminology of sealed refers to the former method of verifying a signature by a wax seal being affixed below or next to the signature. Each person had their own personal stamp and wax would be heated and the seal stamped into the wax to indicate the validity of the signature so that someone could not forge signatures. Currently, sealed means that a notary republic has to verify the signature of the person signing the contract against their driver's license, or another form of photo identification, and signifying their verification by singing and affixing their notary stamp on the document.

Promissory Estoppel

Promissory estoppel is an exception to the requirement that contracts require adequate consideration to be involved in the contractual exchange. In promissory estoppel, a situation may arise where one party relies upon the promise of another party and they utilize that reliance to perform some act or action, or they suffer some type of consequences because of their reliance on the promise. The courts could view the promise as binding in these situations even if there is no transfer of adequate consideration. The aspect that the courts will examine is whether one party relied on what they were promised by the other party to a degree that they invested their funds in an action based upon their reliance on the promise, they contemplated actions or inaction based on their reliance on the promise, or they were financially or materially harmed by their reliance on the promise.

The following court case illustrates how promissory estoppel could be used during a legal dispute over a contract [*C and K Engineering Contractors v. Amber Steel Company, Inc.* (1978) 23 Cal. 3d 1, 151 Cal. Rptr. 323, 587 P. 2d 1136]:

> The plaintiff, the prime contractor, used defendant's bid in its master bid. The master bid was ultimately accepted (installation of reinforcing steel in a wastewater treatment plant). The written bid was $129,511. The defendant refused to sign the contract. The plaintiff employed another contractor. The work actually cost $242,171. The judgment was for the plaintiff $102,660. A jury trial was demanded by the defendant, but there is no right to a jury trial in a promissory estoppel action, which is equitable in nature. Proffered testimony that it is customary for the general contractor to disclose to subcontractors the appropriate disparity between bids was properly rejected. The evidence showed that the plaintiff called the defendant to confirm the bid and told the defendant its bid was "a hell of a lot" lower than 20 percent below the other bids and asked the defendant to recheck, which the defendant did, and confirmed the bid. Thus, the plaintiff did not dispute the existence of the custom to warn subcontractors of unusually low bids, but the plaintiff complied with the custom. Therefore, the offered testimony was cumulative and of doubtful relevance.

Contract Privity

Privity refers to who is a party to a contract; it determines who is able to sue for breach of contract when there is a default related to a contract. In a construction contract between an owner and a prime contractor only the owner and the prime contractor have privity. Subcontractors working for the contractor do not have privity with the contract with the owner, but they do have privity with the prime contractor through their subcontract. Without privity between the subcontractor and the owner, if an owner defaults on a contract, the only recourse for subcontractors is to sue the prime contractor for breach of contact if the contractor stops work on the contract since they have no legal right to sue the owner for breach of contract.

Two examples of court cases that addressed the issue of privity are the following:

> A subcontractor who has been fully paid by the owner for performance of the construction contract may, despite the lack of contractual privity, recover in tort from the construction manager for economic losses caused by negligent misrepresentations by the construction manager and its on-site superintendent. The negligent misrepresentations may be in the form of either directions or supervision. [*John Martin Co. v. Morse/Diesel, Inc.*, 819 S.W.2d 428 (Tenn. in banc 1991)]

In Wisconsin a commercial user (owner, lessee) of a construction product, may recover for diminution of value of the building, resulting from the defective product, in tort. The user may also recover the cost of repairing the damaged building. Privity of contract between the user-plaintiff and the product manufacturer is not required. [*Northridge Co. v. W. R. Grace and Co.*, N.W.2d 179 (Wisc. 1991)]

Assignment

When contracts are formed, the parties to the contract could arrange for the **assignment** of the rights of one party to another party. Assignment would be used by owners for paying subcontractors directly rather than paying a prime contractor and then having the prime contractor pay the subcontractors. Assignment would not be used in this manner by an owner unless the prime contractor is delinquent in paying subcontractors and the nonpayment is increasing the risk of the project not being completed on time.

12.9 ENGINEERING CONTRACTS

Engineering contracts are always negotiated contracts because it is against the codes of ethics of engineering professional societies to competitively bid for engineering work. Owners negotiate with engineers for the price of the engineering contract; this price includes a fee for engineering services. Engineering contracts include a clause that indicates at what stage each progress payment will be made to the engineer or the engineering firm. Typical payment percentages for engineering services are the following:

- Preliminary investigation—10 percent
- Preliminary design—35 percent
- Final design but unchecked—65 percent
- Final design and checked—100 percent

In addition to providing the design documents to owners, engineers also perform other services during projects. Engineers conduct prebid meetings with contractors to answer their questions and to clarify the bid documents. They also issue **addenda** when there are changes to the contract documents. Addenda are required to legally change or revise contract documents prior to bids being submitted. They are normally issued using a standard format that provides detailed information on the changes or revisions being made by the owner.

Engineers might also be responsible for representing the owner during construction and are referred to as the owner's representative if they are acting in this capacity. They are an agent of the owner and as such they are required to monitor the progress of a project for the owner and to report to the owner. Agency is discussed in detail in Chapter 8.

During construction the owner's representative performs additional duties on the behalf of the owner such as the following:

- Participating in preconstruction meetings with the contractor
- Reviewing and signing off on change orders
- Reviewing shop drawings
- Reviewing submittals
- Inspecting the job
- Providing written and visual documentation of the progress of projects
- Interpreting the contract documents and drawings

Engineers are also involved in permitting processes on the behalf of the owner. They arrange for geotechnical engineers to perform soil tests, they arrange for site surveys, they apply for right-of-way certifications, and they have to submit the plans to the city or county for their review and approval.

The fees charged by engineers should reflect the cost of performing all of these services in addition to developing the design drawings and the contract documents. In some cases, an engineering firm might only be hired to provide the design drawings and the fee charged for this service would be less than if the firm would be acting as the representative of the owner during construction. A sample of a *Standard Form of Agreement Between Owner and Engineer for Professional Services* contract is shown in Appendix F and Appendix G shows a sample of a *Short Form of Agreement Between Owner and Engineer for Professional Services* contract.

12.10 CONSTRUCTION CONTRACTS

Construction contracts are not limited to being negotiated between owners and contractors because they may also be competitively bid. Federal and state government contracts are required by law to be competitively bid and awarded to the lowest responsible, responsive bidder. In order to be a **responsible bidder**, a contractor has to be able to meet the minimum requirements for performing the work. Examples of the reasons to reject bids from the lowest bidder as not being responsible are the following (Bockrath 2008, 140):

1. The lowest bidder may not have sufficient finances to handle the project.
2. The lowest bidder may not have sufficient experience in the particular kinds of work involved.
3. The lowest bidder may have an unsatisfactory reputation. The contractor in question may have been guilty in the past of careless work, of always maneuvering for extras, of being uncooperative, or of making it necessary for the engineer to inspect the construction work with extreme care. Although rejection on the rather nebulous

ground of "reputation" could lead to legal ramifications, it is a proper course of action when circumstances require it.

4. The staff and equipment of the lowest bidder may be inadequate. The engineer may ask the bidder to submit an itemized list of his or her proposed staff and of the equipment he or she intends using on the proposed project.

A **responsive bidder** is a contractor who submits all of the required bid documents with their bid proposal. If an owner provides bid forms, these have to be used and every form has to be filled out completely by the contractor. Every bid requirement has to be submitted by the bid deadline. Bid proposals are not accepted after the bid deadline that is stated in the contract documents. Bids have to be submitted in the format required by owners such as electronically or hard copy. If bids are to be submitted by mail, then they have to be mailed and postmarked by the bid deadline. A bid is considered to be accepted when it is postmarked, not when it is received by an owner or the owner's representative.

Owners may restrict the number of bidders by listing **prequalification** requirements that have to be met before a bid is accepted by an owner. Examples of prequalification requirements include having performed similar types of work, having completed projects of comparable size or dollar volume, or having a particular type of expertise.

Once all of the competitive bids are submitted for a potential project, there is either a public or a private **bid opening**. Government construction bids require public bid openings. During public bid openings, each of the bids are opened and announced to the audience. At the conclusion of the bid opening, the **apparent low bidder** is indicated, but it could require up to ninety days to confirm that the apparent low bidder is responsible and responsive to the bid. A team of engineers **canvases** the bids to determine which bids are responsive and which bids are not responsive bids. After the bid canvassing process is complete, the bid is awarded to the lowest responsible, responsive bidder.

Unit Price Contracts

In order to prepare a bid, contractors have to be able to determine a price that they will charge an owner. Some owners require that contractors submit a price for each major unit of work in the construction contract. Examples of major units of work are mobilization, earthwork, steel, and concrete. The price for each major unit would include materials, labor, equipment, overhead, and profit for all of the activities that are required to complete the major unit of work. Once a price is calculated for a major unit, it is called a **unit price**. **Unit price bids** are a compilation of the unit prices for each major item of work required for a construction project.

Unit price bids are used on projects where it is hard to determine the overall extent of the work or actual quantities prior to the commencement of work. An owner who uses the unit price bidding process will not know what the total price of the bid contract will be prior to the completion of the project. Since it is difficult to exactly determine what is included in each particular unit, an owner has to trust that the design engineers will be able to develop a viable system for quantifying each of the units. When bids are submitted, contractors will not know exactly how many units there will be; therefore, the quantities could vary by up to 10 percent before a contractor is entitled to an adjustment to the contract.

Lump Sum Contracts

Lump sum contracts require one price to be listed for the proposed construction project. Lump sum contracts are required for government contracts, they are also used on commercial projects where there are too many different types of units to quantify, or they are used on projects where there are numerous subcontractors. A lump sum contract could also include unit prices for major units of work in order to expedite the negotiation of change orders. When a change in the quantity occurs, the contractor is compensated based on the unit price that they included in their bid for the work item that changes.

Lump sum contracts could be risky for contractors because they prepare their bids based on imperfect construction documents, but lump sum contracts benefit owners since an owner will know up front what the total contract price will be before any contract changes. Owners who need to secure financing or government agencies that have to issue and sell bonds to the public to finance projects need to have a total price for their projects before construction starts.

Unfortunately, owners could be taken advantage of by contractors who suspect that the estimated quantities for a particular unit will end up being much higher than the engineer's estimated quantity. This will increase the bid amount for that unit over the actual price, thus increasing their profit for each additional unit. In order to increase their unit price for one item, a contractor would also need to decrease the unit cost of another item; therefore, this technique creates an **unbalanced bid**. By unbalancing the bid a contractor is able to still keep his or her lump sum bid price the same. If the quantities in the original bid documents were correct, then a contractor would have to make up the difference in his or her actual costs for the underbid item. If the bid item that was overbid has increased quantities, then the contractor would be able to make additional money through change orders that use the overbid price for each additional unit of the item. Using this technique a contractor would be paid more than the lump sum bid price for a project. After bids have been submitted and they are being canvassed, if a unit price for one contractor is obviously higher than it is in the other bids, it might be noted; therefore, contractors attempt to be subtle about unbalancing their bids.

Another form of unbalancing bids takes place when a contractor uses **front end loading** in a bid. Rather than having to use their own funds, or borrow funds at the beginning of a construction project, a contractor might increase their bid cost for mobilization and decrease the cost of a bid item that will occur later in the project. This provides additional funds in the first progress payment beyond the realistic mobilization costs that could be used by a contractor rather than having to use their own or borrowed funds.

Cost Plus a Percentage of Cost Contracts

In **cost plus a percentage of the cost contracts** an owner and a contractor negotiate a set percentage of the direct costs of materials, labor, and equipment for overhead and profit. This type of contract could be used (1) when the contract documents are not well defined enough to allow for competitive bids, (2) when a project needs to be completed quickly, or (3) when a contractor defaults and a new contractor is brought in to complete a job.

Cost plus a percentage of the cost contracts could be abused by contractors because the more they spend on direct costs the more they will be paid in overhead and profit and there is no incentive for a contractor to keep costs low. One way to prevent this from happening is to set a maximum cost for a project or to provide an **incentive clause** that rewards a contractor for keeping the costs to a minimum.

Cost Plus a Fixed Fee and Cost Plus a Fixed Sum Contracts

Cost plus a fixed fee or **fixed sum contracts** are similar to cost plus a percentage of the cost contracts in that contractors are paid for actual direct costs but their overhead and profit are paid to them in a set fee or sum that is negotiated with the owner. This type of contract could include monetary incentives for early completion or a **guaranteed maximum cost**.

Broker Contracts

In addition to the traditional form of contracts, some contractors enter into **broker contracts** where they subcontract 80 to 90 percent or more of the work. This contract form is used because the overhead costs are lower, but the broker contractors could increase their markup on costs. In order for owners to prevent their contracts from being performed by a broker contractor, a construction contract would have to state that the general contractor has to perform a certain percentage of the work that is typically between 15 and 40 percent.

Unilateral and Bilateral Contracts

Contracts are formed as either **unilateral** or **bilateral contracts**. Bilateral contracts require that one promise be exchanged for another promise of performing future activities. This type of contract requires something to be done by both parties to the contract such as is the case with marriage contracts. Unilateral contracts require that only one of the parties to a contract make a promise to immediately perform an act in response to an offer that is legally enforceable, such as construction contracts. If a contract is written in such a manner that the court system is not able to decipher whether it was intended to be a unilateral or a bilateral contract, then the contract will be treated as if it is a bilateral contract.

Express and Implied Contracts

Express contracts are contracts where the terms of the contract promise are explicit and summarized in either a written or an oral agreement. An **implied contract** occurs when the court system concludes that a contract exists based on an interpretation of the activities surrounding the circumstances that occurred indicating the parties involved intended for there to be a contract.

Joint and Several and Entire and Severable

Joint and several and **entire and severable** refer to the liability associated with the parties to a contract. If a contact is a **joint contract**, or it has a joint clause, then all of the parties that sign the contract agreement have to be sued together; if one of the signatories reaches a settlement, it releases all of the other signatories from their legal obligation.

In a **several contract**, or a contract that contains a several clause, each of the signatories to the contract is only legally liable for their portion of the contract, as prearranged prior to the signing of the contract. If one or more of the signatories to the contract does not perform, they are each sued individually for nonperformance of their pro rata share of the contract.

In **joint and severable contracts**, or a contract containing a joint and severable clause, every one of the signatories to the contract is legally liable for the entire contract obligation, not just their pro rata share of the contract. The parties who have signed the contract could all be sued together or one individual or firm could be sued for the entire contract obligation.

Void and Voidable

A contract may be **void** from its inception if it includes agreements that are not legal or if the parties are trying to form a contract that is against public policy. The term **voidable** refers to contracts that are viable contracts until they become voidable when something in the contract comes to light during the execution of the contract that voids them such as discovering that one of the parties to the contract is a minor, incompetent, signed the contract based on fraudulent information, or signed the contract under duress.

Subcontracts

Subcontracts are contracts that are formed by someone who has a primary contract obligation and they hire other parties to perform part of the work that is in the original contract. A standard contract agreement form may be used to form a contract obligation between the party forming the primary contract and the parties hired to perform portions of the contract or standard subcontract agreements are also available.

Chapter Summary

This chapter provided a definition for contracts that included a description of the character of contracts, scope of work, and project baselines. A discussion was provided on each of the project team members including owners, designers, contractors, construction managers, sureties, and the labor force. Different types of contracts were covered, as they apply to each of the phases of construction. A detailed discussion was provided on each of the construction phases and on the different types of project contract administration models including:

- Single prime
- Separate prime
- Construction management
- Design/build

The second half of the chapter covered the formation principles of contractors including the U.S. Uniform Commercial Code, contract agreements, forming contracts for engineering and construction services, and the legal requirements involved in executing contrasts. The last part of the chapter discussed different types of contracts.

Key Terms

addenda
adequate consideration
adversarial relationships
American Institute of Architects
apparent low bidder
assignment
Associated General Contractors
bid opening
bidding and project award phase
bilateral contract
broker contract
canvas
cardinal change
character of a contract
concealing information
construction contracts
Construction Management Association of America
construction managers
construction phase
Construction Specification Institute (CSI) Manual of Practice
contract
contract acceptance
contract agreements
contractors
cost plus a fixed fee or fixed sum contract
cost plus a percentage of the cost contract
demolition phase
design/build
designers
disadvantaged business entity
economic duress
engineering contracts
engineering phase
Engineers Joint Construction Document Committee
entire and severable contract
equivalent uniform annual cost
express contract
feasibility phase
feasibility studies
fraud
front end loading
gratuitous offer
guaranteed maximum cost
implied contract
incentive clause
invitation to negotiate an offer
joint and several contract
joint contract
justifiable reliance
labor force
lump sum contracts
meeting of the minds
minimum attractive rate of return
minority business entity
misrepresentation
mutual mistake
National Society of Professional Engineers
notary republic
offer
owners
preliminary engineering design phase
prequalification
present worth analysis
privity
procurement phase
project baseline
promissory estoppel
proper subject matter
responsible bidder
responsive bidder
restitution
scope of work
separate prime contract
several contract
single prime contract
statute of fraud
subcontract
sureties
testing and operation phase
unbalanced bids
undue influence
unilateral contract
unilateral mistake
unit price
unit price bids
United States Uniform Commercial Code
value engineering
void
voidable
women business entity

Discussion Questions

12.1 What are some of the reasons that a contractor would be disqualified from a bid for not being a responsible bidder?

12.2 What would cause a contractor to be disqualified for not being a responsive bidder?

12.3 What is promissory estoppel? How does it affect construction contracting practices?

12.4 What are the requirements for the acceptance of contracts?

12.5 When and how does the U.S. Uniform Commercial Code apply to construction?

12.6 Explain what a contract agreement is and why it is used in construction contracts.

12.7 What are the project contract administration models that could be used on construction projects?

12.8 What are the project phases of construction projects?

12.9 What is the first phase in the development of a construction project? What takes place during this phase?

12.10 What are the responsibilities of designers related to construction projects?

12.11 Why is it important to have accurate baselines before a project starts? What establishes the baselines?

12.12 Explain the difference between unit price and lump sum contracts.

12.13 Explain the difference between a contract that is void and voidable.

12.14 Explain why an owner might not want to be involved in a broker construction contract.

12.15 Explain why a joint and severable contract would increase the liability of one of the signatories.

12.16 Explain why cost plus a percentage of the cost contracts could be a disadvantageous contracting type for an owner.

12.17 Explain the difference between express and implied contracts.

12.18 In engineering contracts, what is a typical format for payments to the engineering firm for design services?

12.19 How does privity affect the rights of parties that do not have privity to a contract?

12.20 Explain why the legal system investigates whether there is adequate consideration in order for a contract to be valid.

12.21 Explain statute of limitations and how they affect the work of engineers.

12.22 Why is fraud considered to be a federal rather than a state offense?

12.23 What are the responsibilities of owners during construction projects?

12.24 What is required of the labor force during construction projects?

12.25 Explain why for construction contracts a notary would be required to validate the signatures on the agreement forms.

References

Architects and Constructors Estimating Service, Inc. v. Smith (1985) Cal. App. 3d 1001, 211 Cal. Rptr. 45.

Bockrath, J. T. 2000. *Contracts and the Legal Environment for Engineers and Architects*. New York: McGraw Hill Publishers.

C and K Engineering Contractors v. Amber Steel Company, Inc. (1978) 23 Cal. 3d 1, 151 Cal. Rptr. 323, 587 P. 2d 1136.

Calvi, J., and S. Coleman. 1997. *American Law and Legal Systems*. New Jersey: Prentice Hall Publishers.

Construction Specification Institute. 2008. *Manual of Practice*. Washington, DC: Construction Specification Institute.

ENR.com. October 10, 2007. Minority contractor pleads guilty to fraud. *Engineering News Record*. New York: McGraw Hill Publishers.

ENR.com. February 25, 2008. Pa. Contractor pleads guilty to fraud. *Engineering News Record*. New York: McGraw Hill Publishers.

ENR.com. March 3, 2008. Big Dig cheat caught; McCourt Co. to pay $500G-plus. *Engineering News Record*. New York: McGraw Hill Publishers.

ENR.com. December 1, 2008. Contractor to pay feds $900,000 in road inflated-billing case. *Engineering News Record*. New York: McGraw Hill Publishers.

John Martin Co., v. Morse/Diesel, Inc. (Tenn. in banc 1991) 819 S.W. 2d 428.

Northridge Co. v. W. R. Grace and Co. (Wisc. 1991) N. W. 2d 179.

Sweet, J., and M. Schneier. 2005. *Legal Aspects of Architecture, Engineering, and the Construction Process*. Toronto: West Publishing Company.

CONTRACTS FOR ENGINEERING AND CONSTRUCTION SERVICES

13.1 INTRODUCTION

This chapter covers the main components of construction contracts, the requirements for construction bid advertisements, and the procedures for submitting construction bid proposals. In addition to bid proposals, this chapter explains engineering contracts and the main components of construction contracts including general conditions, supplementary conditions, specifications, and plans. Standard engineering and construction contracts that are available through professional organizations and societies are also mentioned in the context of how they are used for engineering and construction projects.

13.2 CONSTRUCTION BID PROPOSALS

After a contractor has carefully reviewed the contract documents issued by owners, or the engineers representing owners, he or she prepares a **bid proposal** that contains the estimated cost to build the structure. Construction bid proposals provide either a lump sum price that is an estimate of what it would cost a contractor to build a project or unit prices for each of the major items of work in a project. Construction contracts normally stipulate the format for bid proposals and they might include **bid forms** that have to be filled out and submitted with the bid estimate. The bid forms that are supplied by the owner in the contract documents have to be used in order for a bid to be **responsive**. After a bid proposal is submitted by a contractor, it is evaluated by the owner, or a representative of the owner. If anything that is required in the contract documents is missing, then the bid proposal is rejected by the owner as being nonresponsive.

Federal and state government agencies are required by law to competitively bid construction projects to ensure that projects will be constructed at the lowest possible cost to tax payers. Private owners could competitively bid their projects or they also have the option of negotiating with different contractors to arrive at a bid price. If private owners prefer to restrict the number of contractors who will be bidding on their project, they could invite only a certain number of contractors to bid on the project or they could require that contracts be able to prequalify to bid on a project by meeting a certain set of criteria specified by the owner.

Instructions to Bidders

In addition to the main construction contract components of the general conditions, the supplementary conditions, the specifications, and the plans construction contract documents also include a section called the **instructions to bidders** that provides the procedures that have to be followed by whoever is bidding on a construction project. The instructions to bidders provide the following information:

- The nature of the work.
- How and where bids are to be submitted including the time, the date, the location of the bid opening, and whether the bid opening is public or private.
- The location of where the contract, drawings, and specifications could be viewed such as the plan rooms of the Associated General Contractors (AGC) or the Associated Building Contractors (ABC).
- The agreements that will be used on the project.
- Whether **bonds** are required on the project and if they are a list of the bonds that are required such as bid, performance, or payment bonds.
- When the project will commence and when it has to be completed (in calendar days not working days).
- Instructions on the preparation of the bids, the number of copies that are required, and who should sign the bid.
- Whether a list of subcontractors has to be submitted with the bid to prevent bid shopping (see a later section for a description of bid shopping).

- Information to help identify bids after they have been submitted, such as:
  - The project name
  - The project number
  - The bid number
  - The agency
  - The bid date and time
  - The name of the bidder
- If bonds are required, then a statement will be included that indicates that if a bid is withdrawn after it has been submitted, the bid bond will be forfeited by the bidder.
- Information on the disqualification of bidders.
- Applicable laws and regulations, such as:
  - Contractor licensing requirements
  - Permits required for the project
  - Wage rate requirements such as Davis-Bacon Act requirements
- Prebid conference information, including:
  - The date and time
  - The location
  - Who needs to attend
  - Whether attendance is required
- Whether there is a liquidated damage clause.
- Information on how the bids will be evaluated by the owner.
- How the contract will be executed and signed by all parties.
- Other bid requirements.

If a contractor does not understand any of the requirements, as stated in the instructions to bidders, he or she should seek clarification from the owner or the owner's representative.

The Bid Submission Process

The **bid submission process** begins with an owner issuing the construction documents and advertising for bids. The amount of time allowed for the development and submission of bids varies depending on the type of project; it could be from a few weeks to several months. The amount of time designated for the development of the bid proposals needs to be long enough for contractors to take off quantities, determine the cost of bid items, analyze equipment and labor requirements, request approvals for **bid substitutions** if they plan on having any in their bid proposals, and evaluate subcontractor bids. If an owner needs their facility built quickly, the bid cycle time could be short in order to expedite the construction process.

In addition to the contract documents issued by owners, other information might be available from the architectural or engineering firm that designed the project including items such as geotechnical reports, existing site data, and property surveys. It is the responsibility of contractors to contact the architect or engineer to determine what types of additional information are available to the contractors bidding on the project.

Bids have to be submitted at the location indicated in the instructions and in the manner stipulated in the contract documents. Bids are not accepted late under any circumstances. It is possible to withdraw a bid before the time indicated in the instructions to bidders, but bids cannot be withdrawn after the time that the bids are due without forfeiting the bid bond. To ensure confidentiality and security, owners could stipulate that the bids have to be submitted to a **bid depository** such as a local construction association.

For privately funded projects, the owner and the A/E have the option of opening the bid proposals in private and for public bid openings the bids are opened one at a time and the amount of each bid is announced to the public. After all of the bids have been opened, the bid with the lowest bid price is announced as the **apparent low bidder**. After bid openings, the bids are evaluated by the A/E before the apparent low bidder is confirmed as being responsive and responsible; this is called **canvassing the bids**. On government projects, the bid evaluations usually require one to three months and for private projects the bid evaluation period varies depending on the number of bidders.

All of the required bid forms have to be completed by each of the bidders including the basis of the bid and the bid prices. In addition, the bids might require bidder qualifications, statements of **noncollusion**, and financial statements. Statements of noncollusion indicate that a contractor has not worked with other contractors to arrive at his or her bid price. By signing a noncollusion statement contractors are verifying that they did not share their bid price with other contractors and they did not form any type of agreement with other contractors stating that one of them would bid lower on this job and the other contractors would each take turns bidding the lowest on other jobs to ensure that all of the contractors would eventually be awarded contracts.

When subcontractors are submitting bids to a prime contractor, they might not submit their bids until right before the bids are due; therefore, general contractors should calculate a rough estimate of the subcontractor bid estimates that they could substitute at the last minute if subcontractor bids are not received in time to be included in the bid estimate.

Bid Advertisements

Owners publicize potential projects in two ways: (1) through invitations to bid and/or (2) by advertising for bids. Government public works projects are published as legal notices indicating that there will be a bid opening. At a minimum **bid advertisements** will include the following information:

1. Project identification information including the name of the owner, the name of the A/E, the address of the A/E, the title of project, and the date of issue

2. A description of the work including the characteristics of the project, the type of project, and the size of the project
3. The basis of the bids indicating whether the project will be bid as lump sum, unit price, cost plus a percentage of cost, cost plus a fixed fee, or some other type of bid
4. The time of completion in the number of calendar days allowed for the completion of the project
5. Bid opening information such as who, when, where, and whether it will be a public or a private bid opening
6. Examination and procurement of documents including the addresses of plan rooms, the charges for plans, the amount of the deposits required for plans, and refunds
7. Whether a bid security is required and the percentage of the bid amount required or a designated amount
8. Bidder prequalification requirements that designate the type of information that has to be submitted with the bid
9. A statement on the right to reject any and all bids
10. References to government laws and regulations such as nondiscrimination laws and equal opportunity employment laws

Bid Shopping

Bid shopping is illegal and in order to prevent it from occurring contractors might have to submit a list of the subcontractors they plan to hire if they are awarded a project. Bid shopping occurs after a contractor is awarded a contract. If the prime contractor did not provide a list of their subcontractors when the bid was submitted, then he or she could contact different subcontractors from the ones they originally received sub-bids from and determine whether they would be willing to perform the work for a lower price and if the subcontractor is willing to reduce their price, then they are awarded the subcontract. If prime contractors bid shop, the contractors are able to increase their profit margin by the difference between the originally quoted subcontractor bid prices and the lower prices quoted by the subcontractors after they are awarded the bid.

13.3 THE MAIN COMPONENTS OF CONSTRUCTION CONTRACTS

The main components of construction contracts are reviewed in this section including general conditions, supplementary conditions, specifications, plans, and schedules. When all of the main components are combined, they are referred to as the construction contract documents. Construction contract documents are issued by the owner's representative, or by an owner, and the contract documents are used by contractors to develop bid proposals or to negotiate a contract price. Construction contract documents also include the invitation to bid; instructions to bidders; the agreement; any addenda that are issued by the engineer of record; bid proposal forms; bid, performance, and payment bond requirements; and the bid forms. Contract documents are used to illustrate design requirements and provide information on how contracts are to be administered during construction. After construction contracts are signed by all of the parties to the contract, they become legally enforceable.

The General Conditions

The **general conditions** of construction contracts are referred to as **boilerplate**, a term that originated in the nineteenth century in the steel and boiler industry. The general conditions are the section of construction contracts that include common construction practices and information that are used to interpret the contract. If there are conflicts during a construction project that are not resolved by the parties to the contract, and the conflicts lead to arbitration or litigation, the general conditions provide arbitrators or the court system with information that helps in the interpretation of the intent of the contract. If there are two or more clauses in a construction contract that conflict, the clause(s) that is more specific has precedence over the other clause(s) unless the contract states the order of precedence of the clauses.

Documents containing standard general conditions could be purchased from organizations such as the Associated General Contractors, the Construction Management Association of America, the American Institute of Architects, and the Engineers Joint Contracts Documents Committee (EJCDC). The articles in the general conditions normally include definitions of terms used in the contract and descriptions of administrative procedures. A sample (Associated General Contractors 2008a) of the types of specific headings for the articles usually included in the general conditions is the following:

1. Definitions
2. Contract documents
3. Site investigations
4. Differing site conditions
5. Site access and rights of way
6. Prosecution of the work
7. Materials, equipment, and appliances
8. Labor and supervision
9. Royalties
10. Permits, licenses, and regulations
11. Inspection of work
12. Warranty
13. Payments
14. Changes
15. Rights and remedies
16. Termination
17. Suspension of work
18. Completion and acceptance

19. Surety bonds
20. Protection of the public and of work and property
21. Insurance
22. Assignment
23. Work by owner or by separate contractors
24. Subcontracts
25. Architect/engineer's authority
26. Arbitration
27. Governing law
28. Notice
29. Miscellaneous provisions

In each of the articles under the headings listed previously there are individual clauses that describe the contract requirements. Chapter 14 elaborates on contract conditions and terms, and it also provides explanations for some of the clauses that cause confusion that are in the general conditions. Appendix H provides a sample of *Standard General Conditions of the Construction Contract* that was written by the Engineers Joint Contract Documents Committee of the National Society of Professional Engineers (NSPE).

The general conditions might also include a variety of standard contract forms (Associated General Contractors 2008a) such as the following:

1. Prime contract between an owner and a contractor
2. Prime contract change order
3. Long-form standard subcontract
4. Short-form standard subcontract
5. Standard subcontract change order
6. Long-form purchase order
7. Confirmation of offer
8. Preliminary notice
9. Conditional waiver and release upon progress payment
10. Unconditional waiver and release upon progress payment
11. Conditional waiver and release upon final payment
12. Unconditional waiver and release upon final payment
13. Notice of cessation
14. Notice of completion
15. Mechanic's lien (claim of lien)
16. Partial release of mechanic's lien
17. Stop notice (private work)
18. Stop notice (public work)
19. Release of stop notice (public or private work)

Standard subcontracts are also available (Associated General Contractors 2008b) and they usually include the following information:

1. The agreement
2. The scope of work
3. The schedule of work
4. The contract price
5. Payment
6. Changes, claims, and delays
7. Contractor's obligations
8. Subcontract provisions
9. Recourse by contractor
10. Labor relations
11. Indemnification
12. Insurance
13. Arbitration
14. Contract interpretation
15. Special provisions

Information that is specific to a particular project is covered in the supplementary conditions, which are discussed next.

Supplementary Conditions

The **supplementary conditions**, which are also referred to as **special conditions**, are included in contracts to expand upon the information contained in the general conditions. They include clauses that describe the unique features of each particular project. The supplementary conditions are also used to modify the general conditions, especially if standard general conditions are being used for a project. Architects or engineers are responsible for writing the supplementary conditions that describe the unique aspects of construction projects. The Engineers Joint Contract Documents Committee publishes a *Guide to the Preparation of Supplementary Conditions*, which is provided in Appendix I. The EJCDC guide provides information on the clauses that are normally included in the supplementary conditions along with explanations on how the different clauses apply to construction contracts. In addition, there are other publications available that describe how to write supplementary conditions, such as the *Construction Specification Institute Manual of Practice.*

Some examples of the types of articles that might be included in supplementary conditions (Clough, Sears, and Sears 2005, 70) are the following:

- The number of sets of contract documents to be furnished to the contractor
- Limitations on surveys to be provided by the owner
- Special instructions to the contractor when requesting material substitutions
- Changes in insurance requirements
- Special documentation required by the owner as a condition of final payment

Other examples (Clough, et al. 2005, 70) of the types of information provided in the supplementary conditions are:

- Order of procedures
- Times during which the work must proceed
- Owner-provided materials or equipment

- Other contracts
- Unusual contract administration procedures
- Early occupancy by owner
- Time of project completion

In addition to the general conditions and the supplementary conditions, contracts contain specifications that provide detailed descriptions of how construction projects are to be built by contractors and these are referred to in the next section.

Specifications

Contract **specifications** are the technical instructions that describe how a project is to be built, the materials to be used in projects, and the quality of work. In order to properly address the importance of contract specifications, an entire chapter is devoted to a discussion of specifications—Chapter 15.

Contract Drawings (Plans)

Contract **drawings** (**plans**) provide visual representations of the intent of the architect and engineers. The plans show how a structure is to be arranged on a site, they include dimensions for every object and structure, and they provide details and sections. Additional information that is necessary for building structures is elaborated upon in the specifications. Construction drawings, at a minimum, include the following types of drawings:

- Index and title page
- Architectural
- Structural
- Mechanical
- Electrical
- Plumbing
- Fire protection
- Landscape
- Schedules for doors, windows, and floors
- Sections and details

Figures 13.1 to 13.7 demonstrate some of these types of drawings by providing examples of drawings for a recreation center on the moon, and Figures 13.8 to 13.10 include examples of steel details for a girt-to-column connection, a metal decking support, and a wall panel to floor joist connection that illustrate how details are drawn.

For large construction projects the plans will include numerous drawings that are designed by architects and civil,

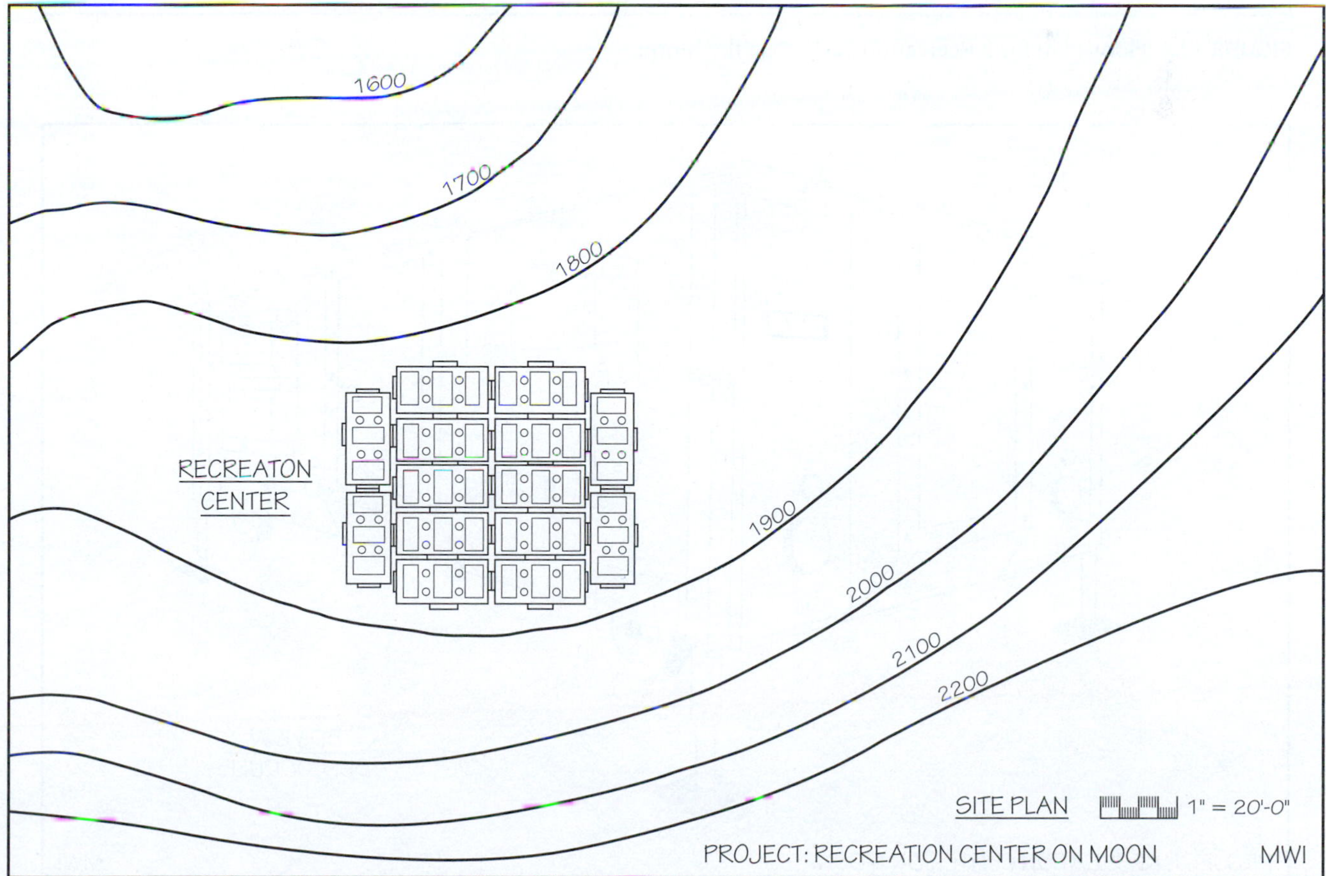

FIGURE 13.1 Site Plan for a Recreation Center on the Moon.

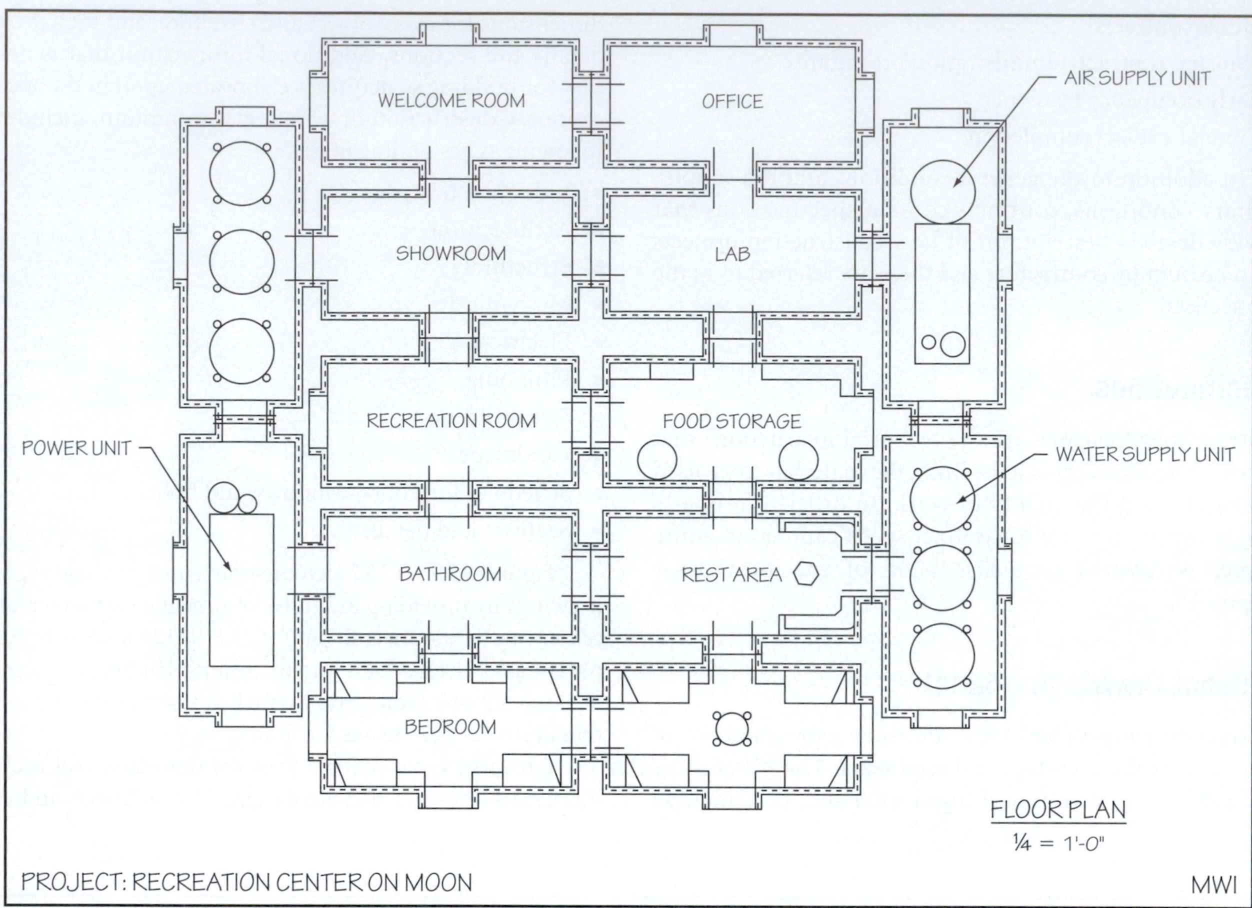

FIGURE 13.2 Floor Plan for a Recreation Center on the Moon.

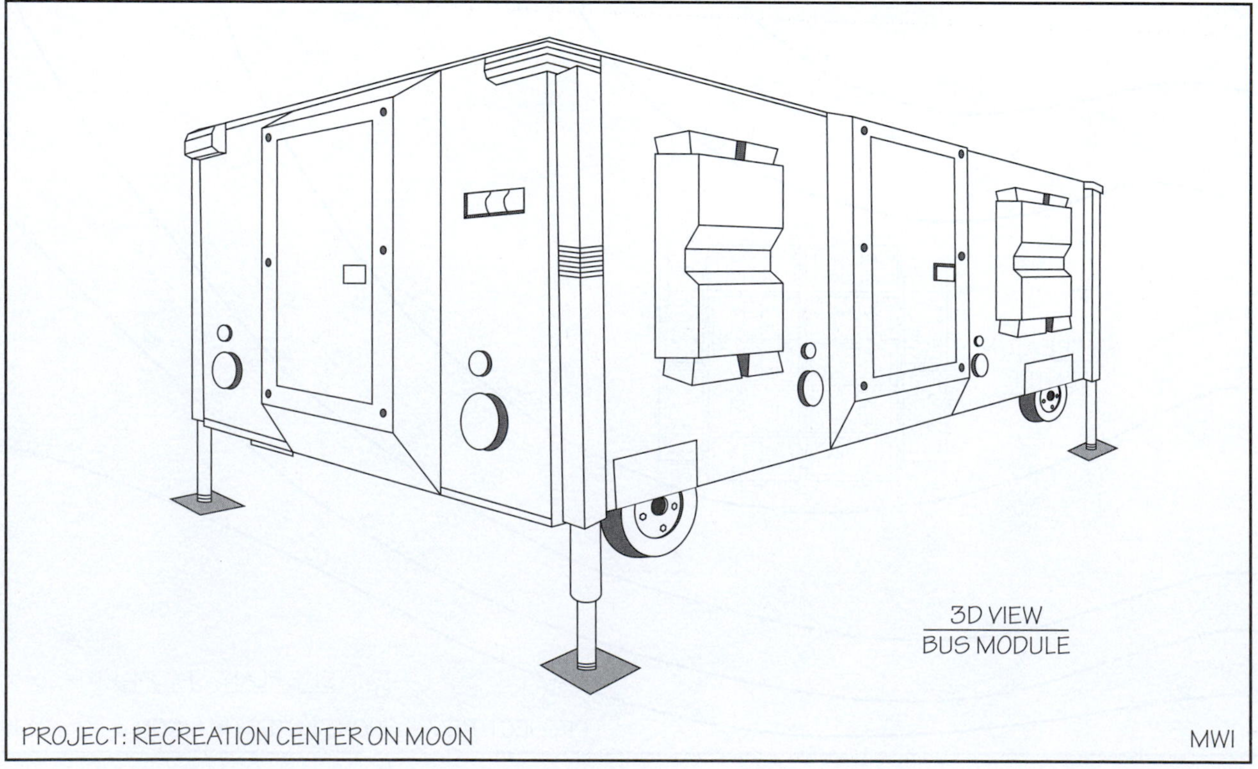

FIGURE 13.3 Three-Dimensional View of the Bus Module.

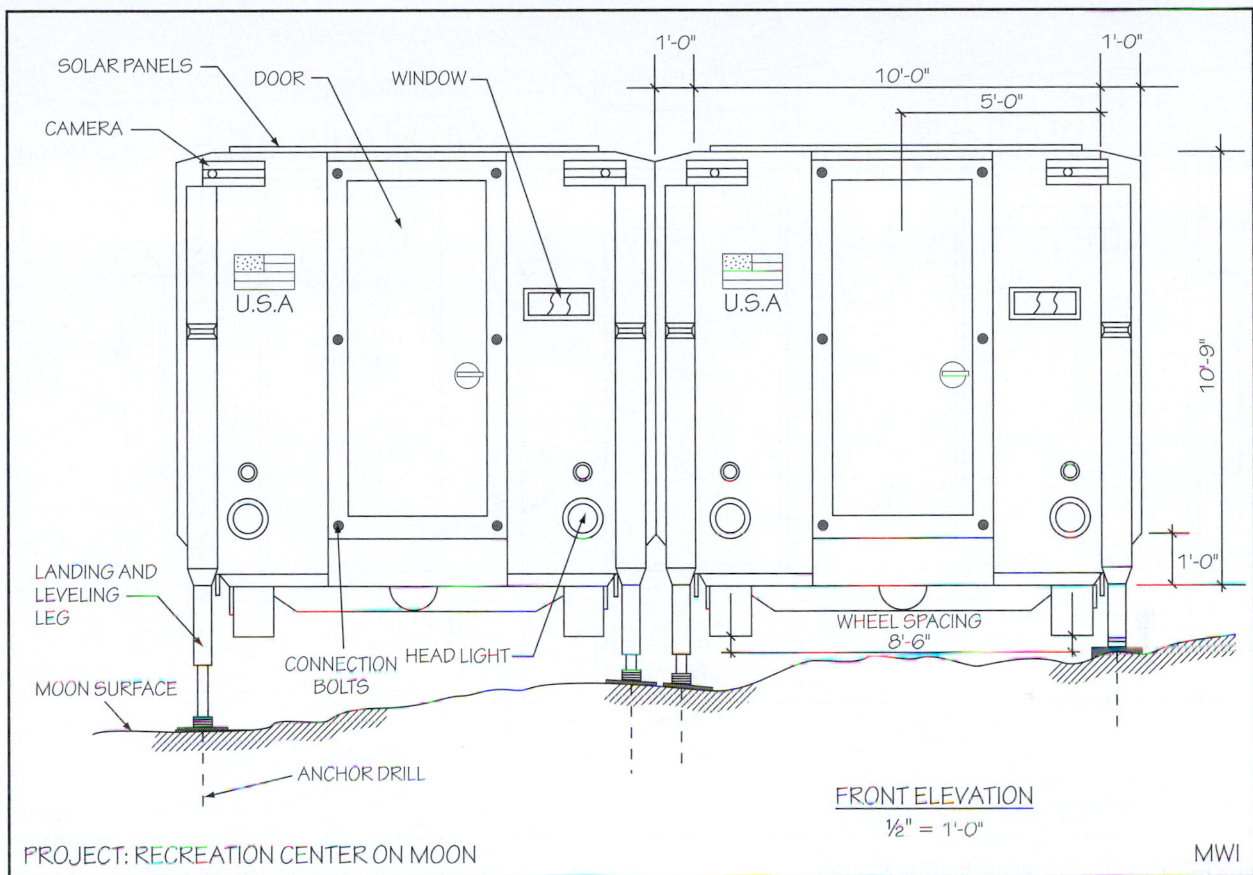

FIGURE 13.4 Front Elevation of the Bus Module.

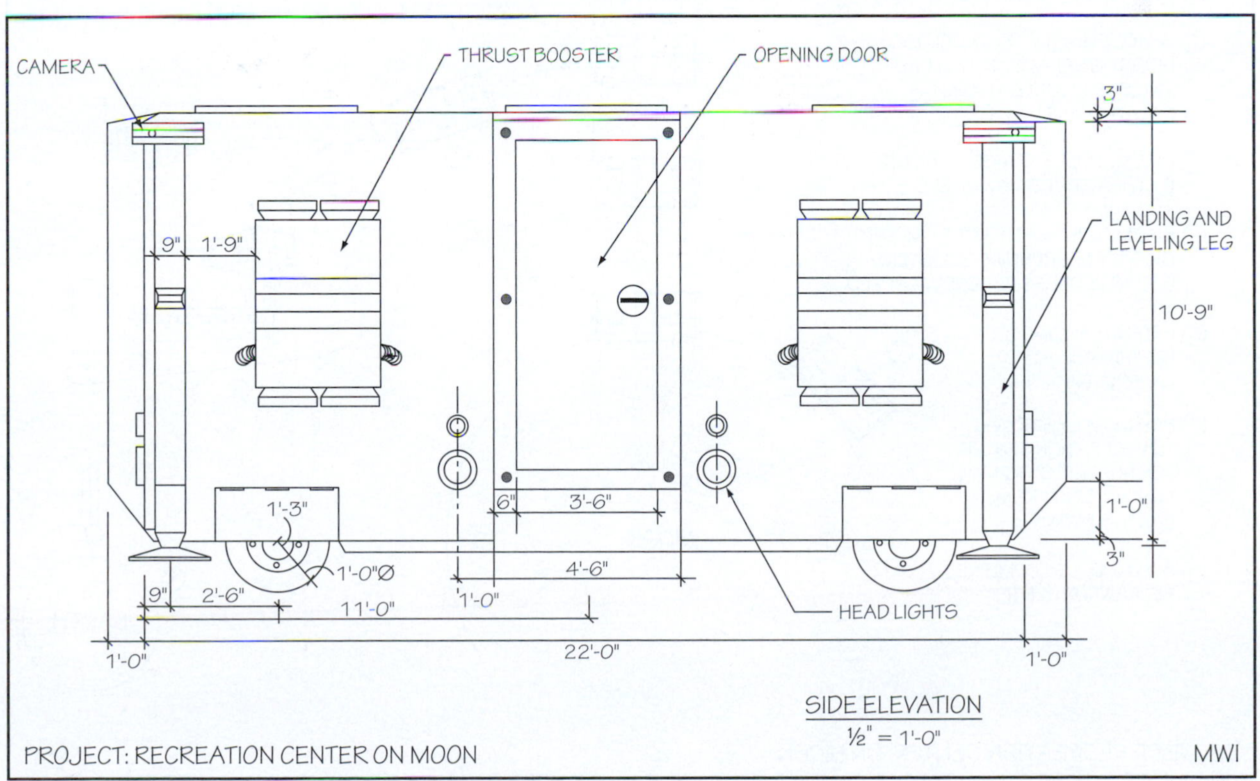

FIGURE 13.5 Side Elevation of the Bus Module.

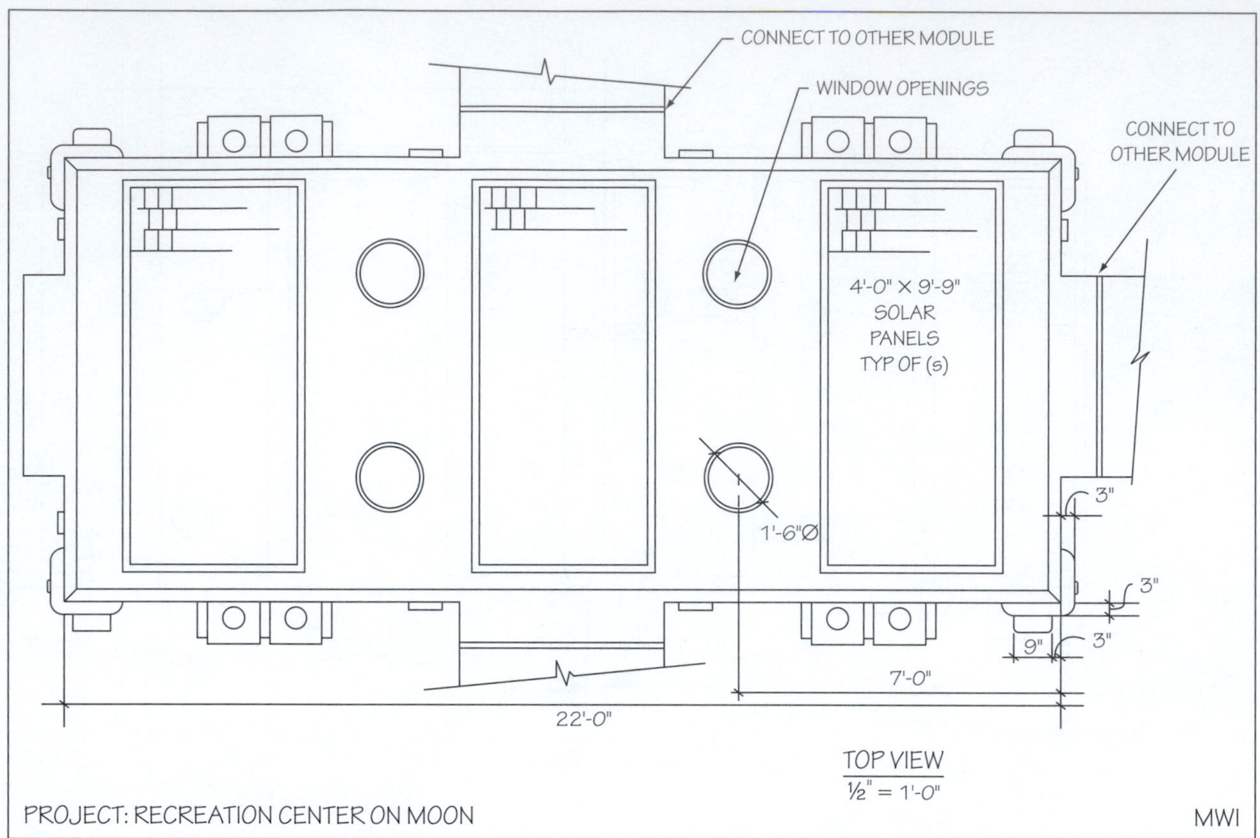

FIGURE 13.6 Top View of the Bus Module.

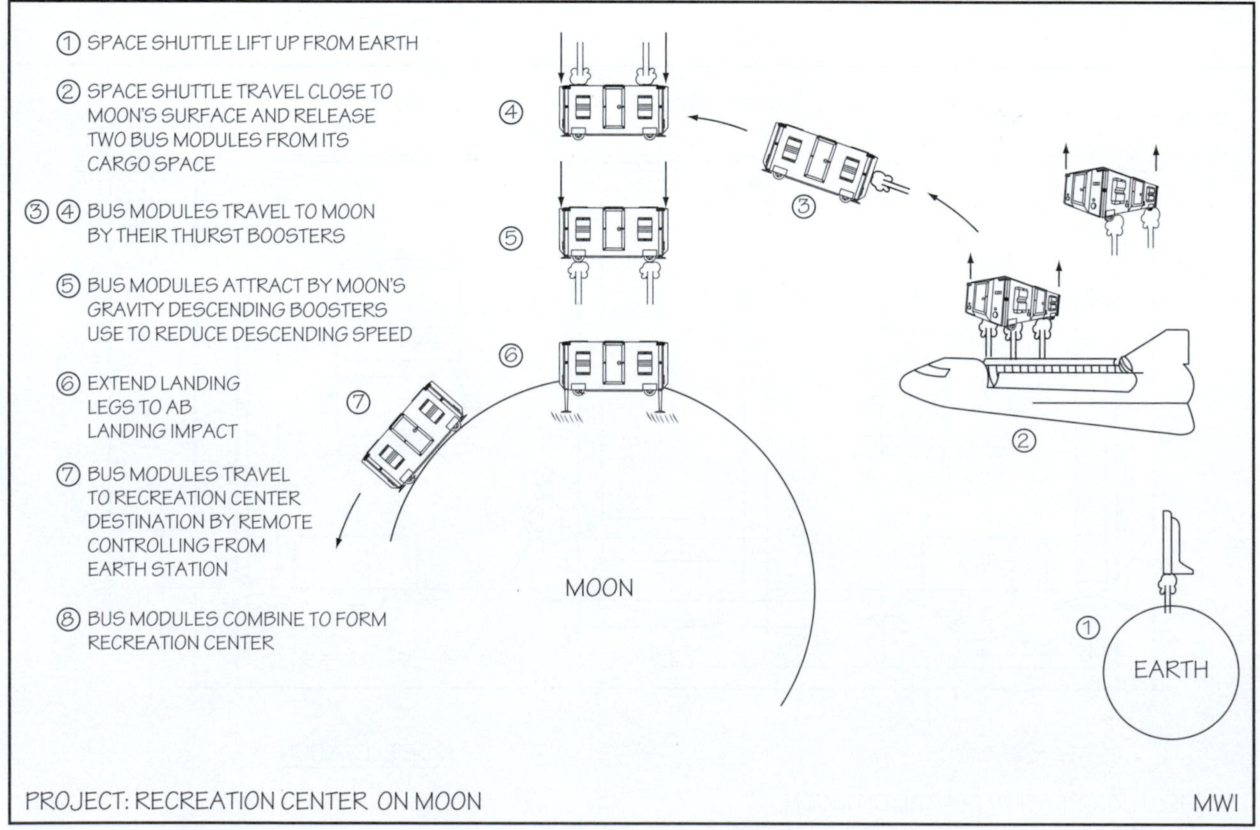

FIGURE 13.7 Bus Module Delivery Details.

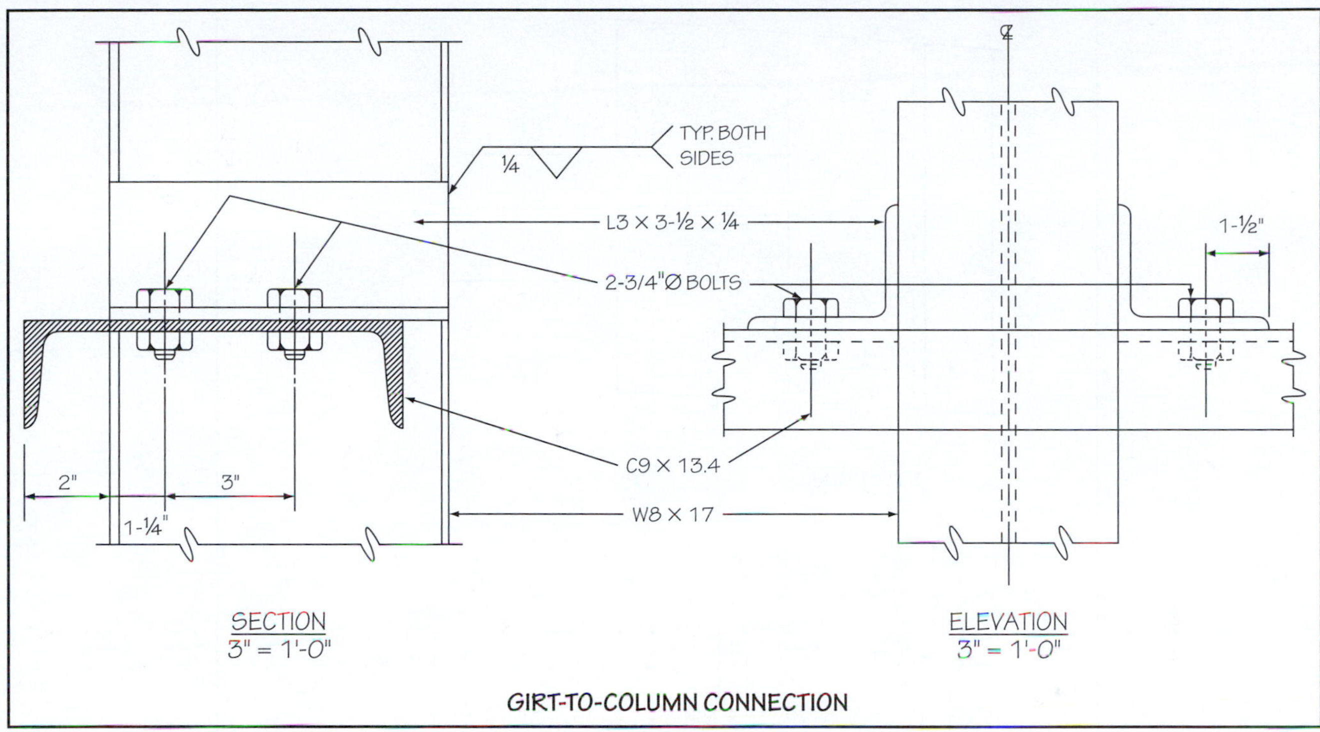

FIGURE 13.8 Girt-to-Column Connection Detail.

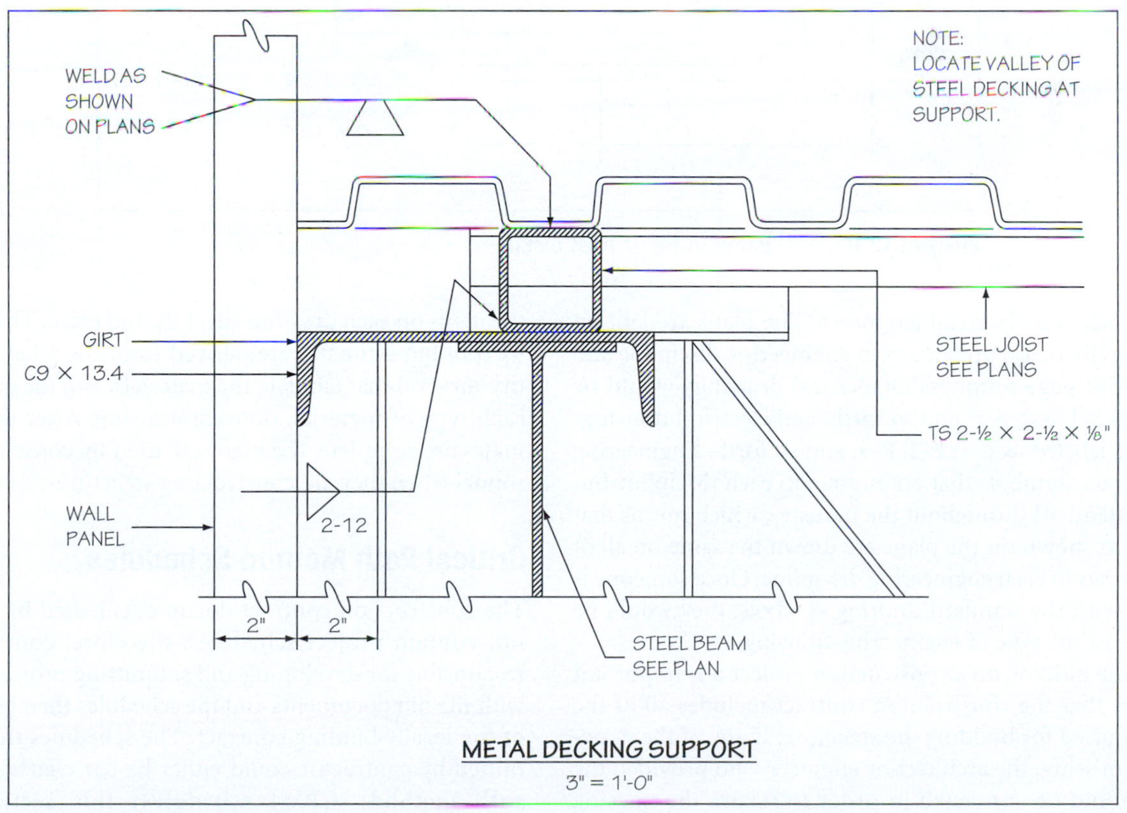

FIGURE 13.9 Metal Decking Support Detail.

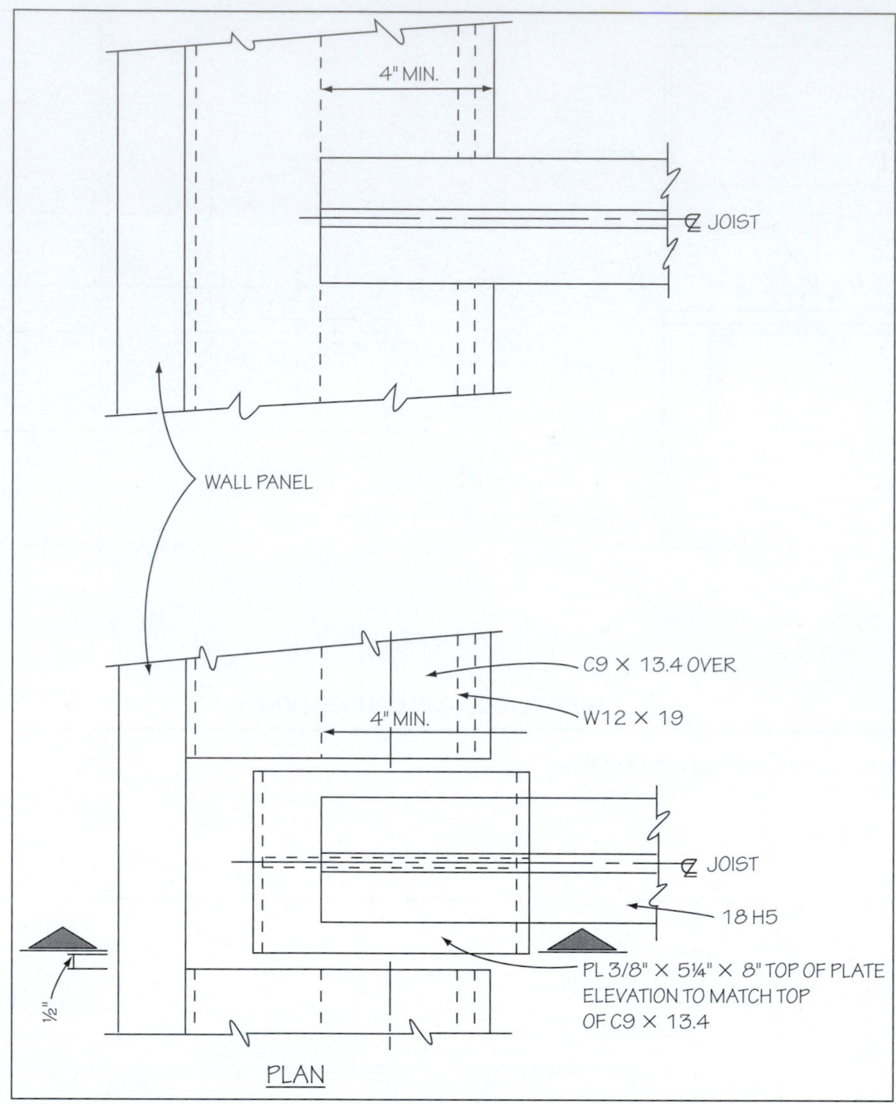

FIGURE 13.10 Wall Panel to Floor Joist Detail.

mechanical, and electrical engineers. The plans are labeled using a prefix that represents each engineering discipline and consecutive page numbers. Structural drawings would be labeled as S-1, S-2, S-3, and so forth, and electrical drawings would be labeled as E-1, E-2, E-3, and so forth. Engineering drawings use symbols that are unique to each discipline but that are standard throughout the industry, which means that the objects shown on the plans are drawn the same on all of the drawings in each engineering discipline. Once someone is familiar with the standard drafting symbols, they would be able to read any type of engineering drawing.

Before bidding on a construction project it is important to ensure that the construction contract includes all of the plans required for building the structure. If any of the drawings are missing, the architect or engineer who provided the plans should be contacted in order to secure the missing plans. When construction bid estimates are being prepared, the drawings should be reviewed individually, and in context to the other drawings provided in the contract in order to determine whether there are any unusual features and how the items on each drawing affect the bid price. The quantities used in bid estimates are derived from the take-offs (quantity surveys) that tabulate the materials, and the quantities of each type of material, on each drawing. After the bid estimates are complete, the plans are used by construction personnel when they are constructing structures.

Critical Path Method Schedules

The construction contract documents issued by owners do not contain project schedules; therefore, contractors are responsible for developing and submitting project schedules with the bid documents and the schedules then become part of the legally binding contract. The schedules that are submitted by contractor could either be bar charts or **Critical path method** (CPM) **schedules**; this section briefly describes both of these types of schedules to illustrate their relevance to construction contracts. Detailed treatments on the subject of scheduling are available in construction scheduling textbooks.

In order to develop project schedules all of the tasks required to perform a project have to be identified. Sometimes a **work breakdown structure** is used to divide projects into smaller and smaller work units until all of the specific activities are identified that are required for a project. Figure 13.11 shows an example of a work breakdown structure for a building.

After all of the project tasks have been identified, a project schedule is developed that includes all of the activities required to complete the project. Critical path method schedules could be used on construction projects to plan and monitor progress during construction. Critical path method schedules provide information on the longest sequential path through a project that determines the minimum project duration. A project could have multiple critical paths and the critical path(s) could change as a project progresses during construction. Every activity in a critical path method schedule has a duration, an early start, a late start, an early finish, a late finish, free float, and total float. Critical path method schedules also show which activities have to be completed before succeeding activities may start; this is called **logical precedence** or **technological dependences**. Critical path method schedules could either be drawn by hand or generated using computer software programs such as Primavera Project Planner (P^6) or Microsoft Project. A sample section of a CPM schedule is shown in Figure 13.11.

The terminology used in Figure 13.12 is arrow on arrow notation. The arrow on arrow notation represents the following:

(10) Activity Name or Number ^{TF}(ES, EF)^{FF} → (20)
Duration (LS, LF)

Where:

ES = Early Start—the latest of the early finishes of all preceding activities

EF = Early Finish—the early start plus the duration

LS = Late Start—the late finish minus duration

LF = Late Finish—the earliest of the late starts of all of the activities that come after the activity

TF = Total Float—the maximum time that an activity may be delayed without delaying the completion of the project. Total float is the early start minus the late start.

FF = Free Float—the maximum time that an activity may be delayed without delaying the start of any of the activities that come after the activity. The free float is the early finish of an activity minus the early start of the lowest early start of all of the activities that come after the activity.

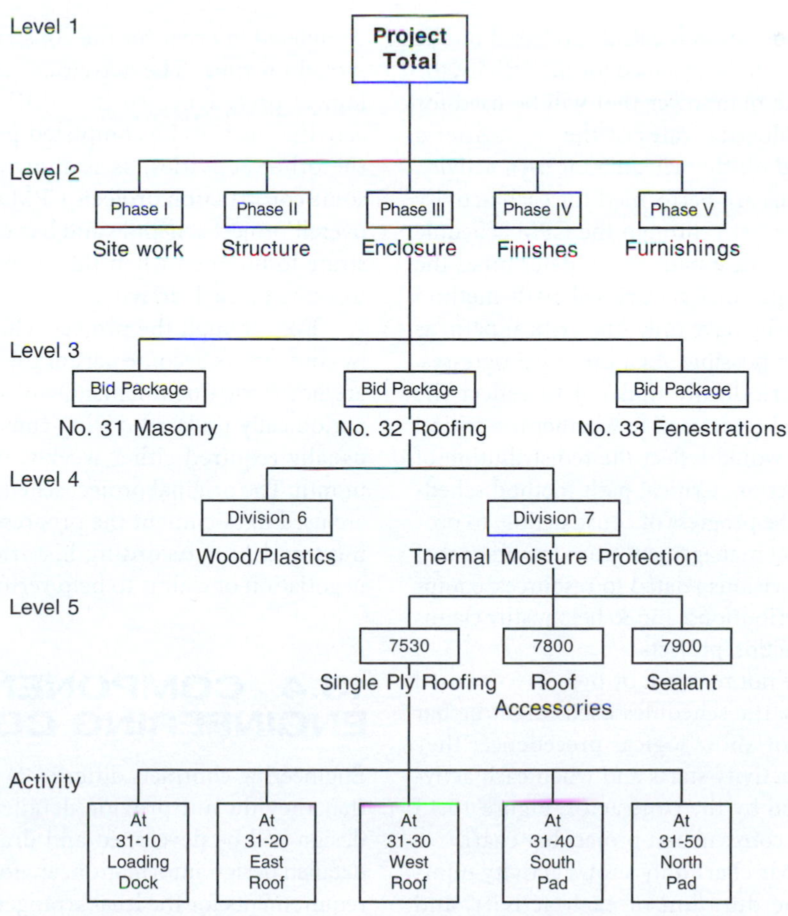

FIGURE 13.11 Example of a Work Breakdown Structure for a Building.

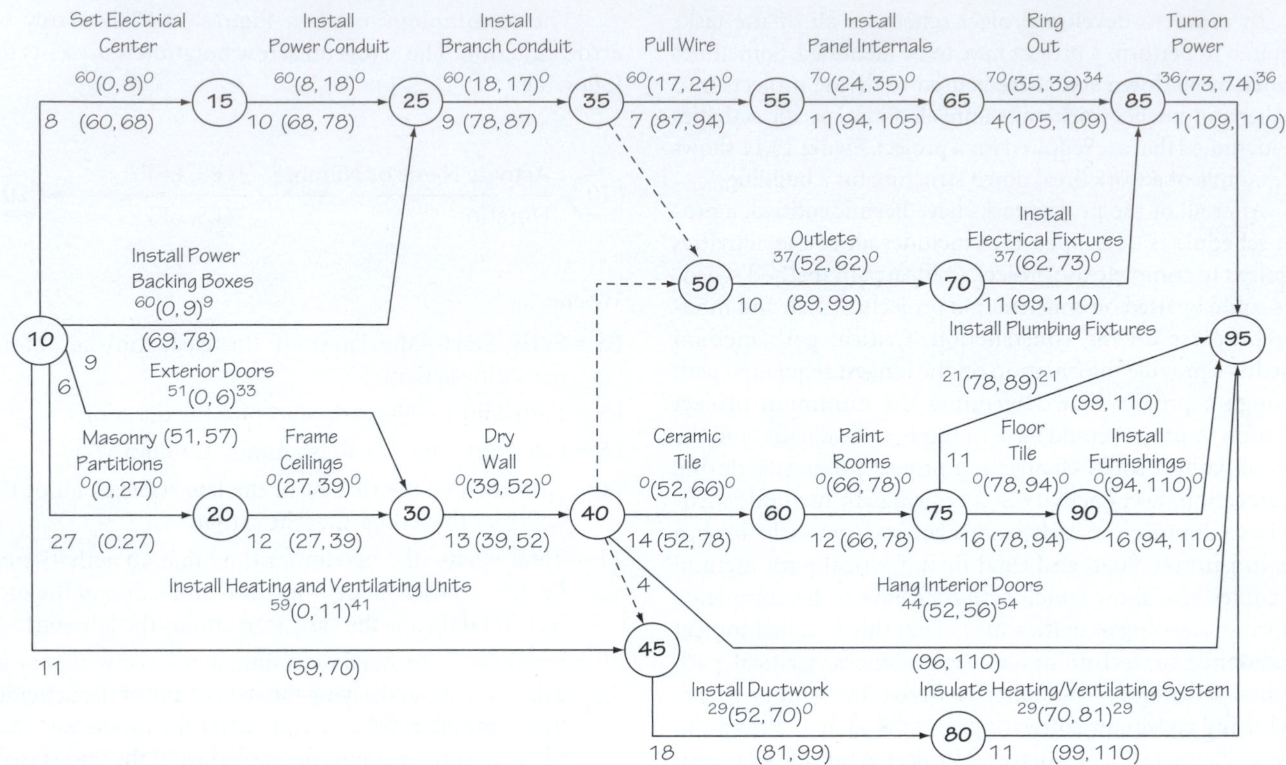

FIGURE 13.12 Sample Critical Path Method Schedule.

For each activity the duration is calculated based on the total quantities of materials to be installed for the activity, the crew configuration of the manpower that will be used for the activity, and the production rates of the crew. After a duration has been entered on the schedule for each activity, then the CPM calculations are performed for each activity and the longest sequential path through the CPM schedule that has zero float is the critical path and it determines the minimum duration of the project. Critical path method schedules do not necessarily have only one critical path, as multiple critical paths are possible. As a project progresses, the CPM schedule is periodically updated to reflect the progress of the project and to determine whether the critical path has changed, which would reflect the redistribution of resources during construction. Critical path method schedules are used to monitor the progress of construction; to provide information to project management team members that assists them in making decisions related to resources, equipment, and manpower distributions; and to help justify claims that occur during construction projects.

If CPM schedules are not required by owners to be used on construction projects, the schedules used might be bar charts. Bar charts do not show logical precedence; they merely show when each activity starts and when each activity needs to be completed by the contractor. Figure 13.13 provides an example of a construction project bar chart.

In Figure 13.13 the bar chart only shows activity numbers, activity names, the duration of each activity, and when each activity needs to start and when it needs to be completed in order for the construction project to be completed on time. The activities are not linked together by logical precedence so it is difficult to determine which activities have to be completed prior to the start of any of the other activities, as is shown in CPM schedules. On some construction projects CPM schedules are used for the overall project schedule and bar charts are used to demonstrate to foremen when they need to have the work crews working on each activity.

Even though the project schedules that are submitted by contractors become part of the contract documents, they are not static documents. Owners require schedules to be periodically updated during construction and updates are usually required either weekly, twice a month, or once a month. The original project schedules and schedule updates are used to document the progress of projects, for forecasting trends, for recording historical data, and during the negotiation of claims to help verify claims.

13.4 COMPONENTS OF ENGINEERING CONTRACTS

Engineering contracts differ from construction contracts in that they do not provide detailed information on how a design will be developed and drawn. Instead of providing detailed design information, engineering contracts cover the requirements for the legal arrangement between owners and engineers and describe the responsibilities of the engineers

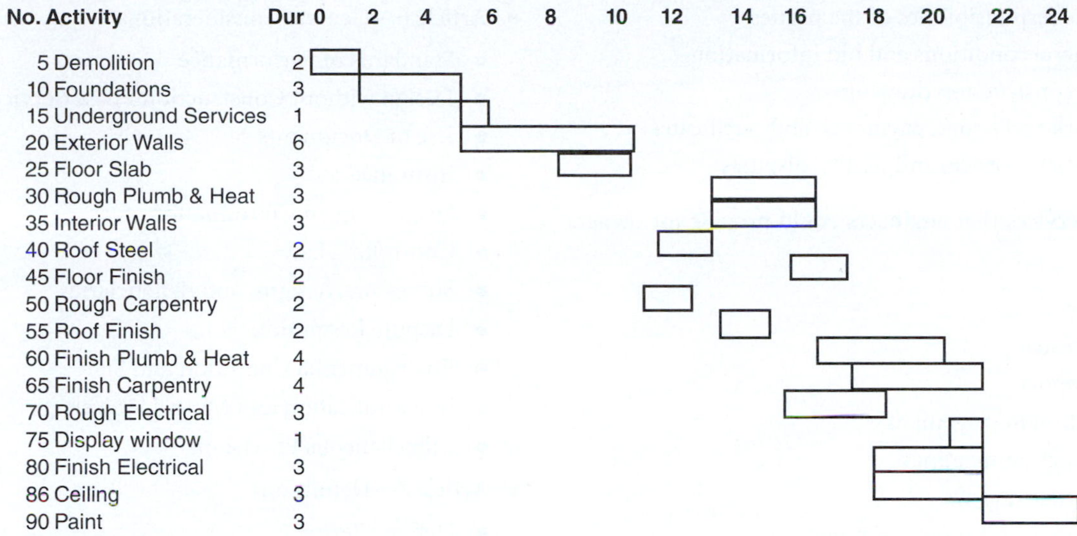

FIGURE 13.13 Sample Bar Chart.

who will be representing clients. Engineering contracts establish the authority of the engineers in an agency agreement. Engineering agreements usually provide information on the authority of the engineer, the limits of the agency agreement, and the terms of the agreement. Engineering contracts also contain an agreement page for the signatures of owners and the engineers or their representatives.

Before an engineering contract is created, an owner should determine whether the person representing the engineering firm has the authority to sign an engineering contract. If the engineering firm is a sole proprietorship, then the owner of the firm is the party who is required to sign the contract. For partnerships, if all of the partners sign the contract, it reduces potential future problems. In corporations, the corporate bylaws, or the articles of incorporation, usually stipulate who has the authority to sign contracts. Contracts for large projects might have to be signed by the board of directors of corporations.

Engineering contracts are not awarded on a competitive basis since it would violate engineering codes of ethics that state that engineers do not compete on the basis of professional charges. The **Brooks Act** that was passed by the federal legislature in 1972 states that "the federal government will negotiate engineering contracts on the basis of demonstrated competence and qualifications for the type of professional services required and at fair and reasonable prices" (Sweet and Schneier 2004, 154). Engineering contracts are awarded through negotiations between owners and the principal, or his or her agent, of engineering firms.

For government projects, engineering firms might be required to submit information that outlines the qualifications and the performance achievements of the firm. Government representatives then rank the top three firms based on their qualifications to perform the work. Negotiations commence with the top-ranked firm and if they are not successful, the second-ranked firm enters into negotiations. If the second-ranked firm does not reach an agreement, the third-ranked firm starts negotiating with the government representative.

Engineers are paid for their services in phases starting with a retainer at the beginning of a project. Typical phases and the percentage of the total price that the firm will have been paid by the end of each phase are shown in Table 13.1. If a design firm will be representing an owner during construction, an additional fee schedule is negotiated with the owner.

Additional services that an engineering firm might provide to owners are the following (Sweet and Schneier 2004, 165):

- Develop schematic design
- Prepare drawings and models of the plan
- Site development
- Stipulate features of construction equipment and appearance
- Outline specifications
- Revise costs
- Prepare working drawings and specifications for general construction, structural and mechanical systems, materials, workmanship, and site development

Table 13.1 Typical Engineering Services Fee Schedule

Phase	Percentage of Total Paid
Preliminary Investigation	10%
Preliminary Design	35%
Final Design—Unchecked	65%
Final Design—Checked	100%

- Define the responsibilities of the parties
- Draft general conditions and bid information
- Interpret construction drawings
- Verify work and issues, payments, and certificates
- Process change orders and resolve disputes

Other services that engineers could provide for owners include:

- Prebid meetings
- Issue addenda
- Client support
- Construction management
- Preconstruction meetings
- Change order reviews
- Shop drawing and submittal reviews
- Inspection services or site visits
- Provide photo, video, or written documentation
- Complete record drawings
- Supplementary survey information
- Right-of-way certifications
- Permit applications
- Construction schedule
- Construction documents
- Resident engineer's file
- Project startup
- Project verification

Contracts for engineering services include a variety of articles that outline the responsibilities of the engineer. Examples of the types of items that are included in engineering contracts are outlined in the table of contents from a *Standard Form of Agreement Between Owner and Engineer for Professional Services*, which is published by the Engineers Joint Contract Documents Committee (EJCDC 2002b):

- Article 1—Services of Engineer
  - Scope
- Article 2—Owner's Responsibilities
  - General
- Article 3—Schedule for Rendering Services
  - Commencement
  - Time for Completion
- Article 4—Invoices and Payments
  - Invoices
  - Payments
- Article 5—Opinions of Cost
  - Opinions of Probable Construction Cost
  - Designing to Construction Cost Limit
  - Opinions of Total Project Costs
- Article 6—General Considerations
  - Standards of Performance
  - Design without Construction Phase Services
  - Use of Documents
  - Insurance
  - Suspension and Termination
  - Controlling Law
  - Successors, Assigns, and Beneficiaries
  - Dispute Resolution
  - Environmental Condition and Site
  - Indemnification and Mutual Waiver
  - Miscellaneous Provisions
- Article 7—Definitions
  - Defined Terms
- Article 8—Exhibits and Special Provisions
  - Exhibits Included
  - Total Agreement
  - Designated Representatives

Other items covered in engineering contracts include the following (EJCDC 2002b):

- Engineer's Services
  - Study and Report Phase
  - Preliminary Design Phase
  - Final Design Phase
  - Bidding and Negotiating Phase
  - Construction Phase
  - Post Construction Phase
- Additional Services
  - Additional Services Requiring Owner's Written Authorization
  - Additional Services Not Requiring Owner's Written Authorization
- Owner's Responsibilities
- Construction Cost Limit
- Dispute Resolution
- Allocation of Risks

Appendix F contains an example of a *Standard Form of Agreement Between Owner and Engineer for Professional Services*, which is published by the Engineers Joint Contract Document Committee (EJCDC), and provides details about what is contained in these types of contracts. Appendix G provides a sample of a *Short Form of Agreement Between Owner and Engineer for Professional Services* from the EJCDC that could be used rather than the long version of engineering services contracts.

In 2008, the American Institute of Architects (AIA) issued several new engineering services contracts to address the growing demand for projects that require **Integrated Project Delivery** (IPD) and **Building Information Modeling** (BIM). The following describes the new IPD documents released by the AIA on May 15, 2008 (ENR.com, May 15, 2008, 1):

> The American Institute of Architects, at its convention in Boston, released new model agreements for integrated project delivery. The standard contract documents follow the concepts in Integrated Project Delivery: A Guide. They provide two levels of design and construction integration, says the Washington D.C.-based AIA. One is a transitional document for those unaccustomed to IPD. The other, called the single purpose entity, offers a fully integrated way to deliver a building.
>
> The transitional agreements use familiar, contracting models B195-2008, an owner agreement, and A295-2008, an owner-contractor agreement with guaranteed maximum price amendment, and A295-2008, a shared general conditions document. The general conditions document also includes the architects design services and the contractor's preconstruction services. The A295 also details how the parties will work together at each phase of the project. Both types of the IPD agreements require use of building information models, says AIA. Also the documents divide a project into different phases than is conventionally done. These are the conceptualization phase, which correlates to an expanded schematic design phase; a detailed design phase, which correlates to an expanded design development phase; an implementation documents phase, which correlates to an expanded construction documents phase; and a construction phase.

13.5 CONSTRUCTION MANAGEMENT CONTRACTS

Standard contracts are also available for construction management contracts. In April 2009, the AIA released a new version of its **Construction Management Contract Documents**. These documents cover both construction managers as advisors and construction managers as constructors. Some of the new features of the construction management contract documents (American Institute of Architects 2008, 1) are:

- Provisions for selecting binding dispute resolution techniques
- Provisions for the owner and contractor to identify an **initial decision maker** (IDM) other than the architect
- Digital data provisions that encourage the use of standard protocols for the transmission and use of digital data
- Selections for contractor payments of stipulated sum, cost of the work plus a fee with a guaranteed maximum price, or cost of the work plus a fee without a guaranteed maximum price

13.6 OTHER CONTRACT DOCUMENTS

In addition to the types of standard contracts mentioned in the previous sections, professional organizations also publish contract documents for a variety of other types of services. Examples of some of the other contract document categories (Engineers Joint Contract Document Committee 2009a) are the following:

- Construction
- Design/Build
- Environmental Remediation
- Contract Document Collections
- Joint Venture, Peer Review, and Other Agreements
- Owner and Engineer
- Engineer and Subconsultant
- Funding Agency Editions
- Procurement

Complete sets of construction documents could also be purchased from professional organizations and they typically include the following documents (Engineers Joint Contract Document Committee 2009a):

- Bid Bond: Damages Form
- Bid Bond: Penal Sum Form
- Certificate of Substantial Completion
- Change Order
- Contractor's Application for Payment
- Payment Bond
- Performance Bond
- Engineer's Letter to Owner Requesting Instructions Concerning Bonds and Insurance
- Field Order
- Suggested Instructions to Bidders for Construction Contracts
- Guide to the Preparation of Supplementary Conditions
- Narrative Guide to Construction Documents
- Notice of Award
- Notice to Proceed
- Bidding Procedures and Construction Contract Documents
- Owner's Instructions to Engineer Concerning Bonds and Insurance
- Suggested Form of Agreement Between Owner and Contractor for Construction Contract (Stipulated Price)
- Suggested Form of Agreement Between Owner and Construction for Construction Contract (Cost-Plus)
- Standard General Conditions of the Construction Contract
- Suggested Bid Form for Construction Contracts
- Work Change Order Directive

Chapter Summary

The first part of this chapter included a discussion on bid proposals, instructions to bidders, the bid submission process, bid advertisements, and bid shopping. This chapter also covered the main components of construction contracts including general conditions, supplementary conditions, specifications, drawings, critical path method schedules, and bar charts. The last part of the chapter provided information on the components of engineering contracts and provided a synopsis of the types of standard contracts available from professional organizations that could be used for construction projects.

Key Terms

advertising for bids
apparent low bidder
bid advertisements
bid depository
bid forms
bid proposals
bid shopping
bid submission process
bid substitutions
boilerplate
bonds
Brooks Act
Building Information Modeling
canvassing the bids
Construction Management Contract Documents
critical path method schedules
Drawings
general conditions
initial decision maker
instructions to bidders
Integrated Project Delivery
logical precedence
noncollusion
plans
responsive
special conditions
specifications
supplementary conditions
technological dependences
work breakdown structure

Discussion Questions

13.1 Which section of a construction contract describes the unique features of projects?

13.2 Explain the main difference between engineering and construction contractors.

13.3 If an engineer does not want to have to write the general conditions of a construction contract, where could standard general conditions be purchased?

13.4 Which of the standard contracts would be used most frequently during construction projects and why?

13.5 What is the purpose of the instructions to bidders? Why are they required?

13.6 Why are state and federal government agencies required to competitively bid all construction contracts?

13.7 What are the main components of construction contracts?

13.8 What information is required in order to be able to identify bids after they are submitted to owners for canvassing?

13.9 What type of information on prebid conferences is typically included in the instructions to bidders?

13.10 Why should contractors perform an estimate of items that will be subcontracted during construction?

13.11 If a contractor does not want to purchase bid documents, where are they able to view copies of the documents?

13.12 Why would a bid depository be used when bids are submitted for projects?

13.13 Describe what is meant by apparent low bidder.

13.14 Explain collusion and how it pertains to construction bids.

13.15 Explain bid shopping and how it could be prevented on construction projects.

References

American Institute of Architects. 2008. *Construction Management Contract Documents*. Washington, DC: American Institute of Architects.

Associated General Contractors. 2008a. *Standard Form Prime Contract Between Owner and Contractor*, Table of Contents, Form AGCC-1. Washington, DC: Associated General Contractors.

Associated General Contractors. 2008b. *Standard Form Subcontract Between Owner and Contractor*, Table of Contents, Form AGCC-1. Washington, DC: Associated General Contractors.

Clough, R., G. Sears, and K., Sears. 2005. *A Practical Guide to Contract Management*. New York: John Wiley and Sons.

ENR.com. May 15, 2008. AIA issues new docs for integrated delivery. *Engineering News Record*. New York: McGraw Hill Publishers.

The Engineers Joint Contract Documents Committee (EJCDC). 2002a. *Standard Form Complete Set of Construction Documents*. National Society of Professional Engineers, American Council of Engineering Companies, and the American Society of Civil Engineers, Alexandria, Virginia; Washington, D.C.; and Reston, Virginia.

The Engineers Joint Contract Documents Committee (EJCDC). 2002b. *Standard Form of Agreement Between Owner and Engineer for Professional Services*. National Society of Professional Engineers, American Council of Engineering Companies, and the American Society of Civil Engineers, Alexandria, Virginia; Washington, D.C.; and Reston, Virginia.

Sweet, J., and M., Schneier. 2004. *Legal Aspects of Architecture, Engineering, and the Construction Process*. Toronto: Thomson Publishers.

CHAPTER **FOURTEEN**

CONTRACT TERMS AND CONDITIONS

14.1 INTRODUCTION

The importance of including numerous clauses in construction contracts is not realized until something unplanned happens during the building of a project and either the owner or the contractor is in need of a defense for their actions or inactions. Construction contract clauses not only outline the scope of work and the responsibilities of each of the parties to the contract, but they also provide detailed descriptions on who is responsible when work on the project does not progress as envisioned by the owner. When problems arise during construction, owners or contractors rely on the construction contract to provide them with guidance about who is responsible for deviations from the contract and whether the deviations constitute a breach of contract claim.

When contractors evaluate contracts in order to determine whether they will prepare and submit a bid estimate, they do so from the perspective of determining whether the contract is written in such a manner as to favor the owner or the contractor. If a contract is slanted toward an owner, a contractor might either decide not to bid on the project or he or she might bid the project at a higher cost than the actual estimated cost to compensate for the contract favoring the owner. One method used to establish who a contract favors—a contractor would read and evaluate each individual clause to determine whether there are more clauses that protect the owner than there are clauses that would allow the contractor to defend their actions if something were to go wrong during construction.

This chapter discusses standard contract clauses along with providing an analysis of some of the more crucial construction contract clauses. Information is also included that explains why some of the construction contract clauses should be carefully reviewed and how some clauses could be used to help defend against breach of contract.

This chapter is divided into five major sections that present information on contract conditions. The first section is a brief description of the duties and the responsibilities of the architect or engineer, as they relate to construction contracts. A detailed description of the overall duties of architects and engineers is included in Chapter 13. The second section of this chapter discusses topics frequently covered or avoided in contracts. The third section provides an analysis of some of the contract clauses that cause confusion but that are essential to include in construction contracts. The fourth section discusses clauses that could affect performance of a contract or that could be used as defenses for nonperformance or breach of contract. The fifth section lists different methods for terminating contracts.

14.2 DUTIES AND RESPONSIBILITES OF THE ARCHITECT/ENGINEER (A/E)

Chapter 13, Section 13.4, includes a detailed description of the duties of architects/engineers (A/E) in relation to the services they provide owners during the design phase and during construction, as the owner's representative. The major duties and responsibilities that the A/E is expected to perform during construction are the following:

1. Act as the owner's representative and visit the site to observe and evaluate progress
2. Validate progress and progress payments
3. Initiate, prepare, and sign change orders
4. Clarify and interpret drawings and specifications
5. Approval and disapproval authority on submittals
6. Reject defective work
7. Determine dates of substantial completion and final completion

In addition to the items listed above, the A/E would be responsible for performing any other duties as described in the contract between the owner and the A/E. If a construction project has a construction manager involved in supervising construction, then the construction manager would have a contract with the owner and they could potentially perform the duties previously described as the duties of the A/E during construction.

14.3 TOPICS FREQUENTLY COVERED OR AVOIDED IN CONSTRUCTION CONTRACTS

A variety of different topics are addressed in the clauses in construction contracts and some traditional clauses are included in most construction contracts. The traditional clauses constitute the minimum requirements in order to provide contractors with an adequate amount of information for them to be able to bid on a construction contract. The topics that are frequently covered in construction contracts are listed in this section. The next section provides more insight into the types of clauses that should be included in construction contracts to provide defenses to breach of contract if there are difficulties during construction or construction is stopped due to unforeseen circumstances.

General Contract Topics

The **general conditions** in construction contracts cover a variety of topics but they traditionally include information (Acret 2002; Acret 1990; Bartholomew, 2001; Bockrath 2000; Clough, Sears, and Sears 2005; Sweet and Schneier 2004; Smith, Currie, and Hancock 2005; White 2002; and Yohara, Turange, and Gatlin 2001) on the following:

1. A description of the project including its location and the nature of the project.
2. The **scope of work** including a statement that the contractor agrees to furnish all materials and perform all work in accordance with the contract documents.
3. A description of the role of the architect/engineer.
4. Information on **access to the work site** including how it will be provided and by whom.
5. A list of what the **contractor is required to furnish** during construction.
6. A description of what the **owner will furnish** during construction.
7. **Adjustments for alternatives** or revisions required after the bid is awarded.
8. A location where the lump sum contract price is entered or a description of how the contract price will be determined.
9. The method that will be used to verify and issue **progress payments**.
10. A discussion on the amount that will be withheld for **retainage** until all items in the contract are complete. The amount is typically 10 percent of the contract price.
11. **Schedule of values**—The contractor provides a statement of the amount it will cost for each division of work (unit prices).
12. **Work in place** and **stored materials**—The work in place is calculated as a percentage of the total work to be completed for each work item. Contractors may be able to receive payment for materials stored on site, or at other locations, if they are able to verify that the materials have been paid for with a bill of sale or other proof.
13. A section on the **interest** that is required to be paid during construction.
14. Statements about the right of a contractor to **suspend work** for nonpayment by the owner.
15. The time at which construction will commence and when it has to be completed to avoid paying liquidated damages if they are to be assessed on a project. This is provided in number of calendar days not workdays.
16. Information on what would allow time extensions for a contractor, including **Force Majeure** (**Acts of God** or circumstances beyond the control of contractors such as floods, lightening strikes, hurricanes, earthquakes, tsunamis, monsoons, and tornadoes if they are not common occurrences in a particular geographical area).
17. Circumstances under which a **suspension of work** will be enforced by the owner.
18. The **project schedule** provided by the contractor. If owners were to provide a schedule in the contract, then whenever there was a delay, the subcontractors would be able to sue the owner not the contractor.
19. **Time is of the essence**—This phrase refers to the importance of completing a project within the designated time frame.
20. **Damages for delays**—In this clause the owner sets forth the amount of damages per day that would have to be paid to the owner by the contractor if a project is not completed on time. Damages could only be assessed if owners are able to prove the amount they would be losing per day if a project were not completed on time.
21. **Incentive clause**—This clause states the amount per day that an owner would pay a contractor for each day that the contractor is able to complete a project ahead of schedule.
22. **Consequential damages**—This clause explains how the effects of consequential damages will be addressed if delays are caused during construction. Consequential damages are delays that happen to peripheral activities or succeeding activities when one or more activities are delayed during construction.
23. **Coordination with the work of others**—This clause outlines how multiple contractors on site simultaneously have to coordinate with each other in order to perform their work in such a manner as to not interfere with each other.
24. **Owner's right to supplement or complete work**—If this clause is included in a contract, it provides an owner with the legal right to perform work on the construction project even though a contractor is building the project. A contractor could not claim that an owner

is interfering with a project when the owner is performing work if this clause is included in a contract.

25. If a **termination for convenience** and **default** clause is in a contract, it gives an owner the right to cancel the contract at any time, even without a legitimate reason. Government contracts include this clause because funds to continue projects may not be available if they are not approved by taxpayers or the legislature.

    In addition to termination for convenience and default, there are many other standard termination clauses that could be included in contracts that allow parties to mutually consent to the termination of a contract. One example is from the Revised Codes of Montana (1997 ed., 28-2-1711):

    > 1) If the consent of the party rescinding, or of any party jointly contracting with him was given by mistake, or obtained through duress, menace, fraud, or undue influence, exercised by or with the connivance of the party as to whom he rescinds, or of any other party to the contract jointly interested with such party; 2) If, through the fault of the party as to whom he rescinds, the consideration for this obligation fails in whole or in part; 3) If such consideration becomes entirely void from any cause; 4) If such consideration, before it is rendered to him, fails in a material respect, from any cause; or 5) If all the other parties consent.

26. **Changes in the work** clauses explain how any deviations from the original scope of work will be dealt with during construction. This clause provides an owner with the legal right to make changes during construction using change orders if the contractor agrees to make them for appropriate compensation.
27. **Changed conditions**—This clause provides information on how a contractor should address conditions that are not as stated in the contract documents.
28. **Inspection of site and local conditions** and **work in place**—This clause indicates that the owner, or an owner's representative, has the right to inspect the work to determine whether it is being built in accordance to the contract documents.
29. A **price adjustment** clause provides information on how price changes will be handled during construction.
30. **Responsibility for safeguarding existing structures**—When a construction project is being built, neighboring structures could be compromised during the excavation stage since the adjacent structures had their foundations analyzed with the soil in place where the new project is being built and its attendant soil pressure. Either the owner or the contractor will be designated in this clause as being responsible for underpinning adjacent structures to ensure their structural integrity during construction of the new structure.
31. **Inspection**—This clause allows the owner's representative to order work to be removed to inspect work that has been covered up by the work with no recourse for the contractor, and at the expense of the contractor, for replacing the work.
32. **Effect of preliminary approvals**—This clause explains the consequences of any preliminary approvals that are issued during construction.
33. **Protests and disputes**—The procedures for addressing protests or disputes are discussed in this clause; they may include formal written or informal oral procedures.
34. **Claims**—This clause outlines how claims will be processed during construction and it could include a detailed formal, written process.
35. **Arbitration**—A clause could be included that states that either binding or nonbinding arbitration will be used to settle all claims that arise on a construction project.
36. A **wage** or **labor standard** clause explains the procedures for documenting wage and benefit rates that will be used during construction. For government contracts the Davis-Bacon Act will be invoked and prevailing wages and benefits will be paid during construction.
37. **Safety**—Instructions will be included on who is responsible for safety during construction and what it entails.
38. **First aid**—Normally this clause will indicate that the contractor is responsible for maintaining a first aid station during construction.
39. **Environmental protection**—The rights and responsibilities of a contractor are stated in this clause about what will be required to protect the job environment and the area surrounding the job site.
40. **Security and fire protection**—Contractors are normally responsible for providing security and fire protection during construction. This clause confirms this requirement or clarifies measures or options that will be utilized by the owner.
41. **Compliance with laws**—Requirements for complying with federal and state laws that affect construction such as labor laws, nondiscrimination laws, and equal employment opportunity laws.
42. **Utilities**—The contract will state whether the owner or the contractor is responsible for supplying temporary utilities during construction.
43. **Permits** and **licenses**—This clause indicates whether the owner or the contractor is responsible for securing permits or licenses.
44. **Subcontracting requirements**—Requirements are explained about how and what may be subcontracted during construction.
45. **Performance bonds**—Whether performance bonds are required and the specific format in which they must be submitted with the bid estimate.
46. **Payment bonds**—Whether payment bonds are required and the specific format in which they must be submitted with the bid estimate.
47. **Insurance** and **risk management**—Provides a list of the different types of insurance policies that are required to be maintained at all times during construction.

Contractors are normally responsible for obtaining insurance policies since the cost of insurance policies is based on the risk factors related to the contractor, not the owner.

48. **Indemnity**—If present, this clause will state that the contractor indemnifies the owner, which means that the contractor is taking on the liability of the owner in addition to his or her own liability.
49. **Acceptance** and **final payment**—The conditions under which the owner will accept the project as having met all of the contract requirements for completion of the project. This clause sets the dates for final inspection, certification of completion, acceptance of work, and issuance of the final payment, including retainage.
50. **Governing laws**—A statement about which state laws pertain to the construction project.
51. **Jurisdictional venue**—The court system that would have jurisdiction and that would try any case resulting from disputes during execution of the contract.
52. **Escalation**—In order to account for increasing prices in materials, contracts could include this clause, which could link the price of materials to some type of an index, such as the U.S. government inflation index. The contractor would be compensated for having to pay rising material prices during times of increasing inflation.

All of the above listed clauses are normally included in the general conditions of construction contracts. These clauses could be written in a variety of different ways yet they would have similar meanings. Legal terminology might be used to state the same information, which could make it difficult to interpret some of the general condition clauses. Some specific clauses were not mentioned in this section that require elaboration in order to provide a better understanding of their meaning; they are covered in the next section.

14.4 CONSTRUCTION CONTRACT CLAUSES REQUIRING INTERPRETATION

This section analyzes some of the construction contract clauses that are more difficult to interpret than boilerplate contract clauses. Sample clauses are included in this section that were extracted from actual construction contracts. Use of these clauses was granted with the provision that the actual documents not be cited or identified in any way.

Contractor's Warranties and Guarantees

Contractor warranty and **guarantee** clauses are confusing in that they may not have anything to do with the work or workmanship of the project and whether the contractor will repair defective work. This clause is used in a variety of ways to elicit agreement by contractors to obligations that could be potentially devastating if contractors are not aware of the consequences of not fully understanding what is being stated in this clause. It is a clause that should always be investigated and carefully interpreted to prevent contractors from becoming obligated to perform items of work when they do not fully understand the intent of the obligation.

Warranty clauses could be used to guarantee that contractors understand what is included in the work, or other statements of contractor knowledge. Typical examples of the wording used for this clause that are from construction contracts are the following:

1. The contractor is thoroughly familiar with all of the requirements of the contract.
2. The contractor has investigated the site and satisfied himself or herself regarding the character of the work and local conditions that might affect performance of the work.
3. The contractor accepts all risk directly or indirectly connected with the performance of the contract.
4. The contractor warrants that there has been no collusion.
5. The contractor warrants that he or she has not been influenced by any oral statements or promises by the owner or the engineer, but only by the contract documents.

Other statements that are included under this clause in actual contracts relate to:

- Contractor solvency.
- Contractor experience and competency.
- Statements in the proposal are true.
- The contractor is qualified and authorized to do work in the state.
- The contractor is familiar with laws, ordinances, and regulations.
- The contractor is familiar with tax, labor, and pay regulations.

Although the title of the clause is the same, as the above statements indicate, the contents of the clause could be vastly different in each construction contract; therefore, this is not a clause that should be skipped with the intention that it is merely a standard boilerplate clause.

The following legal case provides an example of how a contractor was obligated for performance by a certain date because of the requirements of a warranty clause (*Frank Briscoe Company, Inc. versus Clark County* (9th Cir. 1988) 857 F. 2d 606):

> Contractor versus County for cost overruns caused by delays. A $52.3 million contract for construction of an advanced wastewater treatment plant. The jury awarded $16.24 million in damages, which was reduced by $275,000 upon the contractor's acceptance of remitter. **AFFIRMED**. Project completed 15 months after adjusted contract deadline (original deadline 1,200 days after commencement of work). The major cause
>
> *(continued)*

of delay was from changes in the firms providing engineering services, making access to plans and specifications difficult. **HELD**: Jury instruction that the construction contract implied "a warranty by the Country that [the contractor], if it complies with the contract documents would be able to complete the job within the contemplated period of time," 857 F. 2d at 611, was proper. The instruction related to the duty imposed by law on the owner of a project who provides work instructions and specifications (the Spearin Doctrine). (The court instructed the contractor "at the close of its case-in-chief to prepare charts of the various breaches of contract underlying the major delay-causing events it claimed were caused by the County. The charts were to designate witnesses, trial transcript citations, and exhibit numbers supporting each alleged breach, and list the delay-causing event that resulted. [The charts were to] serve as master topical indexes for the transcripts and exhibits used by the jury during deliberations. The charts served as indexes to binders containing references to trial exhibits and witness testimony." *Id* 613. However, the contractor failed to take advantage of the opportunity to clarify its case to the jury.)

Defective Drawings

For a construction contract to not be slanted against the contractor it should include a statement that absolves the contractor of responsibility for checking the design drawings of the engineer. A typical statement that addresses **defective drawings** might be—if a contractor discovers an obvious error, he or she should:

1. Orally report it promptly to the engineer.
2. Confirm it in writing soon thereafter.
3. Not proceed with the affected work until directed to do so by the engineer.

One court case provided an interpretation of this clause in the following manner. A contractor is required to follow the plans and specifications. When he does, he cannot be held to guarantee that the work performed as required by the contract will be free from defects, or withstand the action of the elements, or that the completed job will accomplish the purpose intended. He is only responsible for improper workmanship or other faults, or defects, resulting from his failure to perform.

Approval of the Contractor's Drawings

When a contractor is responsible for generating shop drawings or working drawings, then the contract should state that the contractor should check them for accuracy, and then they will be checked for accuracy by the engineer, but the contractor is still responsible for performing work according to the contract. A typical clause that would cover **approval of the contractor's drawings** is the following: The approval by the engineer of any shop drawings or working drawings shall not in any way be deemed to release the contractor from full responsibility for complete and accurate performance of the work in accordance with the contract drawings and specifications.

Defective Work

One of the most difficult areas in construction contracting is the determination of who is responsible for work that is not completed in accordance to the contract documents and that is deemed to be **defective work**. If an engineer improperly labels a drawing, then they are responsible for the error. If a contractor follows the plans and specifications, he or she is protected as to his or her responsibility for proper performance. If a contractor does not follow the plans and specifications, then he or she is required to redo the work in the proper manner. A typical clause that protects the owner from defective work on the part of the contractor is the following: All material furnished and work done if not in accordance to the plans by the engineer shall be removed immediately and other material furnished or other work performed or a deduction from the contractor's compensation (it is impossible to remove the work in situations such as concrete work).

A clause that would protect a contractor from errors by the engineer might be: If correct on the drawings and a mistake is made in construction, the contractor is responsible. If the drawings are labeled improperly, the contractor is not responsible.

Conduct of the Work

The clause **conduct of the work** could have multiple meanings depending on what is inserted under the title by the engineers writing the contract. Examples of clauses that were found under this title are the following:

1. The owner, or the A/E, may issue instructions regarding the conduct of work if the owner feels something in the plans or specifications should be changed.
2. The contractor's operations should be carried on in a workmanlike manner and have a responsible party (superintendent) on duty at all times.
3. An engineer could order work to be speeded up, which may result in overtime and increase the cost of the job (this could be done if there is some type of threat such as a pending flood).
4. The owner, in the opinion of the engineer, assumes that the contractor would not be able to finish the work, could take over all, or a portion of, the work and have it finished by the contractor's forces or outsiders. This does not release the contractor or the surety from their obligation and liabilities.

As the above statements indicate, this clause could be used to obligate contractors to a variety of unknown requirements. The first clause does not indicate whether any compensation would be provided if the owner decides to change the work, only that the owner has the right to issue instructions when he or she wants to change the work.

The second clause could be interpreted to mean that the contractor is required to have a superintendent at the job site twenty-four hours a day. A more effective way to have stated the end of the clause would have been to say "during all times when work is being conducted at the job site" rather than "at all times."

The third clause would discourage contractors from undertaking a job where the contract contains this type of a clause. The problem with this clause is that it allows the engineer to speed up work without providing additional compensation if the contractor is required to pay workers overtime in order to complete the work on the accelerated schedule.

The fourth clause might cause some contractors not to bid on a project because the wording in that clause would allow the owner to take over a project for any reason or for no reason at all.

Order of Completion

Under an **order of completion** clause title the contract should state any requirements that there might be regarding the sequence of construction and the order in which items have to be completed during construction. The A/E should make sure that the requirements could be met by the contractor or the order of activities indicated could potentially add additional expenses to the contract, and result in higher bids by contractors, so that they would be able to perform the work in the order indicated on time.

Engineers normally prepare an outline that shows the expected sequence of work but the contractor is responsible for developing the project schedule. Owners do not want the liability of providing the schedule since it would allow subcontractors to sue the owner whenever there is a delay on projects.

If a contract does not indicate a specific time when a project is to be started or completed, the courts would interpret that it has to be completed in a reasonable time. A reasonable time would be determined by having expert witnesses provide testimony on similar types of projects and how long it took to complete them.

Inspections

Various types of inspections occur during construction projects. Contracts should include information on the right of the owner to inspect his or her project. The owner, or the owner's representative, would inspect the project to ensure that the contract is being followed during construction. If a project is large, the owner may have a separate inspection department or hire a separate firm to conduct inspections on the behalf of the owner. The inspectors would be checking materials, workmanship, shop fabrications, and verifying testing procedures and results. Owners should have a resident engineer or a staff of engineers in the field that monitors whether the contract and specifications are being properly executed during construction. A clause that would empower the owner to inspect the work is the following: All work, materials, processes or manufacture and methods of construction shall be subject to the inspection of the engineer.

Contracts could include a separate clause on the **duties of the inspector**. Inspectors are responsible for ensuring that the work is performed in accordance with the contract documents but it needs to be realized that if the contract is incorrect, it is not the responsibility of the inspector to change it, as they are not empowered to do this during construction. Inspectors are only allowed to assist the engineer as the engineer corrects the situation by issuing revisions, extras, or adjustments to correct the contract requirements.

Permits and Licenses

Most construction contracts indicate that it is the owner who is responsible for securing **building permits** for construction projects since it could take months or years to be issued a building permit. In addition to the building permit, other permits are required such as:

- A permit to close a street temporarily if dangerous work is being conducted in the area
- A permit to truck heavy steel, or other heavy materials, through streets or highways
- A permit to move existing structures to new locations
- A permit to temporarily shut off utilities

Labor Considerations

The **labor considerations** clause is used to include information pertaining to any federal, state, and local laws related to eligibility for employment or employment practices.

Underpinning

Underpinning is required when there are structures adjacent to construction projects because the structural integrity of the soil surrounding the existing structures is compromised when a new site is being excavated adjacent to them. Municipal ordinances or state laws require certain procedures to be followed to underpin adjacent structures if their safety is compromised during construction operations. Contracts should state that the owner is responsible for adjacent structures and their structural integrity during construction, but owners rely on the contractor to perform the underpinning or owners will hire a separate contractor that specializes in underpinning and have them perform the work.

Work Done by the Owner

In most construction contracts, owners reserve the right to perform some of the construction work using his or her own forces and directed by the engineer. Contractors are hesitant to work under this situation; therefore, a **work done by the owner** clause is usually included that indicates that the owner has to minimize interfering with the work of the general contractor.

Land and Facilities

A **land** and **facilities** clause indicates what the contractor is responsible for in relation to the job site and its facilities. Owners provide the land for construction jobs and they also provide access to the job site, space to work around the project as it is being constructed, and an area for storing materials. Contractors are normally responsible for providing the following:

1. Electric power
2. An access roadway
3. A railroad spur track if required (a track off of the main railroad line to the job site)
4. A warehouse for storing materials
5. Offices for the contractor and his or her employees and an office for the engineer
6. Utilities for all of the offices on site
7. Maintaining traffic around the job site

Archeological Remains

In states where there is a historical Native American Indian presence, construction contracts should include an **archeological remains** clause that states who is responsible if Native American Indian artifacts are discovered at the construction job site and the work is delayed due to a requirement for proper excavation of the artifacts. An archeologist may be required to canvas the job site to determine whether there are artifacts, and if any are discovered, they have to be excavated by archeologists. If the Native American Indian artifacts discovered contain any human remains, then representatives from the local Native American Indian tribe will exhume the remains and transport them to another location for a proper burial.

Hazardous Waste Mitigation

A **hazardous waste mitigation** clause should be included in all construction contracts. This clause indicates that the owner is responsible for the cost of hazardous waste mitigation and that the contractor will receive a time extension if previously unknown hazardous wastes are discovered during the excavation of job sites.

Performance

Once a contractor has completed all of the requirements contained in the contract, he or she has obtained **performance** of the contract. Performance is discussed in detail in Section 14.5.

Final Inspection and Acceptance

The contract, under the **final inspection** and **acceptance** clause, will state that the contractor will provide the owner with a written notice indicating that he or she has completed all of the contract requirements. After the engineer receives written notice, he or she will inspect the job to verify that every requirement in the contract has been properly completed by the contractor. If the engineer agrees that the contract is completed, or if there are deficiencies that surface during the verification process that are then corrected, the engineer indicates in writing that the work is satisfactory and that the work is accepted and the contract is fulfilled. A contractor will not receive their final payment until the engineer has approved all of the work performed by the contractor.

Miscellaneous Clauses

Construction contracts also contain a variety of miscellaneous clauses that might affect the cost of the final contract price. Examples of these types of clauses are:

1. **Employment**—A statement that the contractor is not allowed to hire the employees who are working for either the owner or the engineer during construction.
2. **Emergencies**—Specific procedures to follow in case of an emergency.
3. **Minimum wage rates**—A list is provided of the applicable rates that will be used during construction.
4. **Domestic versus foreign materials** (**Buy American**)—Public works projects could require that domestic products and materials be purchased in the United States if they are available; this is sometimes called a Buy American clause.
5. **Construction reports**—The owner will stipulate what types of reports are required, and how often they need to be produced, by the contractor for the owner.
6. **Payroll** and **bills for materials**—The contractor is required to supply the engineer with a copy of his or her payroll each month and with the bills for materials.
7. **Patents**—The contract could restrict the use of patented materials or products because it restricts competition and increases the cost of products and materials.
8. **Schedule of equipment**—The owner might require that the contractor guarantee that equipment installed by the contractor operates for a year or another designated time period.
9. **Dredging**—The contract could include specific instructions about how dredging will be performed, its sequence, when it will be performed, how traffic will be maintained, and the handling of obstructions during dredging.
10. **Borings**—The engineer will determine the types and number of borings required for the job.
11. **Medical facilities**—Information on whether the contractor is required to provide medical facilities during construction.

If contractors are able to carefully evaluate and analyze each and every contract clause, it reduces the opportunity for owners to take advantage of them and allows contractors the option of increasing the contract bid price to compensate for contracts that are slanted in favor of the owner.

14.5 PERFORMANCE AND BREACH OF CONTRACT

This section discusses the requirements for performance of a construction contract and the clauses that are typically used to accuse a party of **breach of contract** or as a defense when someone is accused of a breach of contract during construction.

Substantial Performance

A job is said to have reached **substantial performance** if it is completed except for small items that still need to be completed by the contractor. At this point the structure is fit for habitation and the contractor is paid for the work that has been completed to date but not for the remaining work. At this point a punch list is generated that includes all of the incomplete items; when these items are finished, the retainage is released by the engineer. A typical clause on substantial performance is: the accomplishment of all things essential to the fulfillment of the purpose of the contract, although there may be insignificant deviations from certain contract terms or specifications. Once a project has reached substantial completion, the engineer issues a **certificate of substantial completion**; a sample of this form is shown in Figure 14.1.

Partial Performance

A contractor is said to have only reached **partial performance** of the contract if the structure is not fit for habitation. This is considered to be a situation where the contractor is in default and in breach of contract. The owner could sue the contractor and receive damages but the contractor should still be able to recover some of their fee for services rendered to the owner if they invoke **unjust enrichment**, which means that one party should not benefit to the detriment of the other party if the other party has actually performed work. Partial performance occurs when there is no contract remaining. A sample clause on partial performance is: sufficient performance to confer appreciable benefit on the recipient, but falls short of the concept of substantial completion, or virtually complete performance.

Impossibility and Commercial Impracticability

There are two types of impossibility: (1) subjective and (2) objective. With **subjective impossibility** a situation exists where the skill level of the contractor is not high enough for them to be able to accomplish the work. For **objective impossibility** no one would be able to accomplish the work because of a situation that arises that either destroys the work or that prevents the contractor from being able to complete the work. Examples of where this would apply include when a project that is being built is destroyed by an act of God, if a contractor is a sole proprietorship and the owner passes away, fraud was committed in relation to the contract by the other party, or the other party prevents performance or breaches the contract. If a contracting firm is a corporation and the owner dies, then the firm would not be excused from performance due to impossibility. If a strike occurred during construction, an impossibility clause would not excuse the contractor from performance. A contractor would be excused from performance if he or she were able to prove that performing the work would be impossible by demonstrating without any reasonable doubt why it is impossible to the owner.

Another option is to use **commercial impracticability** as a defense against nonperformance. According to Smith, Currie, and Hancock in Common Sense Construction Law (2005, 208):

> The theory of commercial impracticability is related to the doctrine of impossibility. In the leading case of *Mineral Park Land Co. v. Howard*, the court described the concept as follows: *A thing is impossible in legal contemplation when it is not practicable; a thing is impracticable when it can only be done at an excessive and unreasonable cost....* We do not mean to intimate that the defendants could excuse themselves by showing the existence of conditions, which would make the performance of their obligations more expensive than they had anticipated, or which would entail a loss by them. But where the difference in cost is so great, as has the effect, as found, of making performance impracticable, the situation is not different from that of a total absence of earth and gravel (Emphasis added). In *Mineral Park Land*, the contractor was excused from performing a gravel excavation contract when the cost of performance proved to be twelve times that originally anticipated.

Destruction of Subject Matter

When a construction project is destroyed while it is being built, there needs to be a clause that protects contractors from having to perform. It seems obvious that if there is no longer a structure to work on that a contractor would be relieved from performance, but this is only the case if there is a clause in the contract that relieves a contractor from performance if the structure they are building is destroyed during construction. A clause that could be included in the construction contract is **destruction of subject matter**. Projects that would fall under this protection would have been destroyed by fires, unstable conditions, or violent natural acts. If this clause, or some type of similar clause, is not included in a construction contract, then a contractor is not relieved from further performance. If a contractor is not discharged from the project because of the destruction of the subject matter, then he or she is not able to recover for work that has already been performed on the project. It is also not equitable to require a contractor to start a project over at the original cost of the project, especially if it is several years into a project.

CERTIFICATE OF SUBSTANTIAL COMPLETION

Project ________________ Owner ________________ Owner's Contract No: ________________

Contract ____ Date of Contract ____________________

Contractor ____________________________________ Engineer's Project No: ________________

This (tentative) (definitive) Certificate of Substantial Completion Applies to:

☐ All work under the Contract Documents: ☐ The Following specified portions:

____

____

____

____

____

Date of Substantial Completion

The Work to which this Certificate applies has been inspected by the authorized representative of the owner, the Contractor, and the Engineer and was found to be substantially complete. The date of Substantial Completion of the project, or a portion of the project as designed above, is acknowledged and verified and also the date of the start of applicable warranties required by the Contract Documents, except as stated below.

A (tentative) (revised tentative) (definitive) list of items to be completed or corrected is attached to this document. This list may be all inclusive, and the failure to include any items on the list does not alter the responsibility of the Contractor to complete the work in accordance with the Contract Documents.

The responsibility between the Owner and the Contractor for security, operation, safety, maintenance, heat, utilities, insurance, and warranties shall be provided in the Contract Documents except as amended as follows:

☐ Amended Responsibilities ☐ Nonamended

Owner's Amended Responsibilities:

____

____

____

Contractor's Amended Responsibilities:

____

____

____

FIGURE 14.1 *(Continued)*

The following documents are attached and are included as part of this Certificate:

____

____

____

The Certificate does not constitute an acceptance of work not in accordance with the Contract Documents nor is it a release of the Contractor's obligation to complete the work in accordance with the Contract Documents.

____________________	____________
Executed by Engineer	Date
____________________	____________
Executed by Contractor	Date
____________________	____________
Executed by Owner	Date

FIGURE 14.1 Certificate of Substantial Completion Form.

Therefore, construction contracts should include a clause that allows for termination of the contract when the object of the contract has been destroyed during construction.

Each of the parties to a construction contract will insure their part of the project. Owners insure the work that is in progress and they would be compensated by their insurance company if the project were destroyed during construction. Contractors would have builder's risk insurance that would cover a project if it is completely destroyed. Builder's risk insurance is discussed in Chapter 19 under Risk Management.

Frustration of Object

For **frustration of object** to occur the value of the contract has to be almost completely destroyed by something occurring that was not reasonably foreseeable by the party who is claiming that they could no longer perform the contract.

Contract Fraud

No specific clause allows for protection from **fraud** in construction contracts. Either party to a construction contract could accuse the other party of committing fraud but fraud is difficult to prove unless hard evidence is available that proves that fraud occurred on the project. In most instances, fraud occurs prior to the start of construction when a contract is first being formed and one party provides false information to secure the contract. If it is discovered that fraud was used to secure a contract during construction, the parties to the contract could either void the contract and the party that was harmed by the fraud could recover damages based on the **legal theory of restitution** or the parties could agree to continue with the contract but with one party being paid damages to compensate them for however they were affected by the fraudulent act. In addition to actual damages, punitive damages could also be awarded to the party that suffered because of the fraudulent act.

Strikes and Labor Issues

Construction contracts should include a clause that helps protect contractors from situations when the workers go on **strike** and delay projects. Normally contracts allow for an extension of time when a strike occurs but not for additional costs related to the delay caused by the strike; contractors should be aware of this so that they do not anticipate that they will receive compensation when a strike occurs during construction. Most contracts do not have a provision for termination of the contract due to strikes, only time extensions.

Breach by the Other Party

If there is a **breach by the other party** clause when one party breaches a contract, this would allow the other party to terminate their work. This party would be able to collect damages if they are able to prove that they sustained a loss because of the other party breaching the contract.

Prevention by Other

When parties are performing a contract, they are obligated to not interfere with the work of the other party. If they do interfere, and their interference in any way keeps the other party from performing their part of the contract, and there is a **prevention by other** clause, then this relieves the party that was interfered with from performing their part of the contract. The person who interfered with the other party has breached the contract and could be liable for damages to the other party.

14.6 TYPES OF BREACH OF CONTRACT

There are three types of breach of contract:

1. **Total breach** of contract invalidates a contract and relieves the other party from performance of the contract.
2. **Partial breach** creates a situation where the parties would require the courts to determine the damages for breaching part of the contract.
3. **Anticipatory breach** occurs when one party realizes that the other party will breach the contract and they breach it in anticipation of this occurring during construction. Examples of situations that would lead to anticipatory breach are when one party to a contract is in jail, files for bankruptcy, or the law in some way makes it impossible for one of the parties to continue with the contract.

Nature of Breach

In situations where one party breaches the contract, it is not required that the other party demonstrate fault. The party that did not breach the contract would be able to recover their losses and any gains that they were not able to achieve if they are able to establish the **nature of the breach** and that the losses or gain prevention would have:

- occurred with reasonable certainty
- resulted from the breach of contract
- not been something that could have been anticipated
- been something that could have been avoided during construction

Remedies for Breach of Contract

When a breach of contract occurs, there are several different **remedies for breach of contract** that courts could award to compensate the party that was materially harmed by the breach of contract. The four types of damages are: (1) nominal, (2) compensatory, (3) pencitory, and (4) liquidated. The following sections explain these four types of damages (Bockrath 2000).

Nominal Damages

Nominal damages are awarded by courts in order to recognize that a breach of contract has occurred but no harm was caused by the breach of contract. The nominal damages would be a small sum that demonstrates that a party has breached the contract.

Compensatory Damages

Compensatory damages are used to provide compensation to the party that was materially harmed by the other party breaching the contract. They are not used as punishment but to compensate the harmed party for any losses they have incurred or gains prevented resulting from a breach of contract.

Pencitory Damages

Pencitory damages attempt to return the person that was materially harmed by the breach of contract back to the position they were in before the contract started. If this is not possible, they provide monetary compensation to the plaintiff.

Liquidated Damages

Rather than have a contractor be in breach of contract if he or she is not able to complete a project as scheduled, **liquidated damages** provide compensation to owners when this situation occurs at the end of a construction project. Liquidated damages could only be assessed if an owner is able to prove that he or she is losing a certain amount of money per day for each day that the completion of a project extends beyond the scheduled completion date. An example of a situation where liquidated damages could be assessed would be if an owner had preleased a building and the tenants could not move in on time; therefore, rent could not be collected until the building is complete and the owner could prove the monetary amount that he or she was losing per day in uncollected rent.

Most construction contracts contain a liquidated damage clause. If contractors are not aware of the requirement of owners having to prove an actual monetary loss per day, they might pay the liquidated damages assessed by owners even though they would not be required to if owners could not prove an actual loss.

Restitution

Restitution differs from the other forms of damages but it could also be awarded by the courts in breach of contract cases. Restitution is used to return plaintiffs to a position that they would have been able to achieve if the contract had been completed or they are awarded an amount of money that would be equivalent to the value of the project that was not completed at the time of the breach of contract.

14.7 WAYS TO TERMINATE CONTRACTS

The ideal method for terminating a contract is for both parties to perform their contractual obligations. Another method is for both parties to the contract to reach a *mutual agreement* to terminate the contract, which happens more frequently in contracts between owners and engineers than it does between owners and contractors. Owners work closely with engineers to generate contract documents and the designs for their projects, a process that potentially could require years to complete. If personality issues arise during the design process, it is sometimes easier for owners to terminate the contract and hire another engineer than to struggle through the design process with someone that they are having a difficult time working with on a regular basis.

Contracts are terminated when it is no longer possible to perform them, such as when there has been an act of God that destroys the project. The legal system could terminate a contract through the actions that occur when one of the parties to a contract files for bankruptcy. Some executives of firms file for bankruptcy in order to terminate their subcontractual obligations so that they would be able to negotiate more favorable subcontracts once the original subcontracts are no longer valid.

Contracts could also be terminated when there is a breach of contract. Owners try to prevent contractors from breaching the contract because the main objective of a project is to complete it. Owners might provide additional funds to contractors to prevent them from filing for bankruptcy and causing a breach of contract. Owners could assist contractors in other ways if it will lead to a satisfactory conclusion of the contract and prevent a breach of contract.

Chapter Summary

This chapter focused on some of the contract clauses that influence the successful completion of construction projects and on clauses that help provide defenses for breach of contract. The different types of damages that are awarded in court cases were also discussed including nominal, compensatory, pencitory, liquidated, and restitution to demonstrate the consequences of breaching a contract.

The duties and responsibilities of architects and engineers were listed to demonstrate how they are involved in construction contracts. A list of typical topics that are usually included or avoided in construction contracts was provided along with brief descriptions of each topic. The last part of the chapter explained methods for terminating contracts.

Key Terms

acceptance
access to the work site
Acts of God
adjustments for alternatives
anticipatory breach
approval of the contractor's drawings
arbitration
archeological remains
bill of materials
borings
breach by the other party
breach of contract
building permits
Buy American
certificate of substantial completion
changed conditions
changes in the work
claims
commercial impracticability
compensatory damages
compliance with laws
conduct of the work
consequential damages
construction reports
contractor is required to furnish
contractor warranty
coordination with the work of others
damages for delay
default
defective drawings
defective work
destruction of subject matter
domestic versus foreign materials
dredging
duties of the inspector
effect of preliminary approvals
emergencies
employment
environmental protection
escalation
facilities
final inspection
final payment
first aid
Force Majeure
fraud
frustration of object
general conditions
governing laws
guarantee
hazardous waste mitigation
incentive clause
indemnity
inspection
inspection of site and local conditions
insurance
interest
jurisdictional venue
labor consideration
labor standard
land
legal theory of restitution
licenses
liquidated damages
medical facilities
minimum wage rates
nature of the breach
nominal damages
objective impossibility
order of completion
owner will furnish
owner's right to suspend or complete work
partial breach
partial performance
patents
payment bond
payroll
pencitory damages
performance
performance bond
permits
prevention by other
price adjustment
progress payments
project schedule
protests and disputes
remedies for breach of contract
responsibility for safeguarding existing structures
restitution
retainage
risk management
safety
schedule of equipment
schedule of values
scope of work
security and fire protection
stored materials
strike
subcontracting requirements

subjective impossibility
substantial performance
suspend work
suspension of work
termination for convenience
time is of the essence
total breach
underpinning
unjust enrichment
utilities
wage
work done by the owner
work in place

Discussion Questions

14.1 Explain why construction contracts should include a destruction of subject matter clause.

14.2 Explain the difference between nominal, compensatory, and pencitory damages.

14.3 Explain why a contractor needs to understand the meaning of a contractor's warranty and guarantee clause.

14.4 Explain who is responsible and why when there are defective drawings in a construction contract.

14.5 Explain whether a constructor is released from liability if the owner's engineer approves the contractor's shop drawings.

14.6 Explain the difference between subjective and objective impossibility clauses.

14.7 How could commercial impractibility be used as a defense in a breach of contract case?

14.8 Explain whether a contractor would want an indemnity clause included in a construction contract. Why or why not?

14.9 If an owner orders work removed in order to conduct an inspection, who is responsible for the cost of replacing the work and why?

14.10 If an inspector discovers defects in the work, is he or she allowed to order that the work be changed? Why or why not?

14.11 What is a common procedure for safeguarding adjacent structures during construction?

14.12 How are owners allowed to make changes during construction? What clause provides them with this ability?

14.13 Explain what a termination for convenience clause allows owners to do during construction.

14.14 Is an incentive clause beneficial to contractors? Why or why not?

14.15 Why should contractors rather than owners provide project schedules?

14.16 Explain how an owner will benefit from a no damages for delay clause.

14.17 Under what circumstances would contractors be allowed a time extension if a force majeure clause is included in a contract?

14.18 Explain how and when liquidated damages could be assessed during construction projects.

14.19 What are consequential damages? How do they affect construction projects?

14.20 What are the ways in which a contract between an owner and a contractor could be terminated during construction?

References

Acret, J. 2002. *A Simplified Guide to Construction Law*. Washington, DC: BNI Building News Publishers.

Acret, J. 1990. *California Construction Law Manual*. New York: McGraw Hill, Inc.

Bartholomew, S. 2001. *Construction Contracting Business Law Principles*. Upper Saddle River, NJ: Prentice Hall.

Bockrath, J. T. 2000. *Contracts and the Legal Environment for Engineers and Architects*. New York: McGraw Hill Publishers.

Clough, R., G. Sears, and K. Sears. 2005. *A Practical Guide to Contract Management*. New York: John Wiley and Sons.

Frank Briscoe Company, Inc. v. Clark County (9th Cir. 1988) 857 F. 2d 606.

Smith, Currie, and Hancock. 2005. *Common Sense Construction Law: A Practical Guide for the Construction Professional*. New York: John Wiley and Sons.

Sweet, J., and M. Schneier. 2004. *Legal Aspects of Architecture, Engineering, and the Construction Process*. Toronto: West Publishing Company.

White, N. 2002. *Principles and Practices of Construction Law*. Upper Saddle River, NJ: Pearson Education, Inc.

Yohara, C., C. Turange, and F. Gatlin. 2001. *Fundamentals of Construction Law*. Washington, DC: American Bar Association.

SPECIFICATIONS

15.1 INTRODUCTION—DEFINITION OF SPECIFICATIONS

Specifications are an essential part of construction documents because they describe and define in detail the objectives of construction projects. Specifications are written to describe the characteristics of each item in a contract including materials and fabrication methods. Because it is difficult to portray every item legibly on plans, specification sections provide additional detailed information.

Engineering drawings show the orientation of structures, design details, and sections; specifications provide detailed descriptions about the materials and the production methods required to create structures. Specifications should also demonstrate the interrelationships that exist between each portion of a project and the rest of the project. They explain all of the materials and products and how they should be incorporated into a project.

Specifications also describe the desired quality of workmanship. They might also include precise procedures on how to measure quality. Official **quality control** and **quality assurance** programs could be included in specifications. These prescribe the exact processes for measuring and recording the results of quality testing processes.

Contract documents are not always written by the same person; therefore, items could potentially be omitted from the specifications that are in the plans or vice versa. Even if an item is only included in either the specifications or the plans, but not in both locations, contractors have to include that item in the project. Specifications are not always written in tandem with the plans even if they are being written by the same person who is preparing the plans. With interruptions, design changes, and drawing revisions, an engineer may not remember to change or update the specification sections that pertain to the drawings being altered or revised by him or her. Since specifications are lengthy and descriptive, many contracts will refer to standard specifications rather than having the actual text of standard specifications included in contracts.

This chapter describes the different types of specifications that are used in construction contracts, standard specifications, standards, and international standards. This chapter also includes information on the potential misinterpretation of specifications, vague terms that should not be used in specifications, and terms that might not be understood by foreign engineers and constructors. The last section of the chapter explains the Construction Specification Institute (CSI) Master Format, since it is utilized in a high proportion of construction contracts.

15.2 TYPES OF SPECIFICATIONS

Construction contracts contain a variety of different types of specifications. This section provides information on seven types of specifications:

1. Methods and Materials
2. Design
3. Performance
4. Functional
5. Purchase
6. Restrictive and Proprietary
7. Material and Product

Contracts do not contain one pure form of specifications and all of the seven types of specifications could be used within one construction contract document. An important aspect of understanding specifications is being able to identify which type of specification each specification section is in order to be able to properly perform what is described within each specification section. The following sections explain each of the seven types of specifications.

Method and Material Specifications

Method and material specifications describe the methods and/or materials that are required for a particular section of a construction project. Methods and material specifications are also referred to as design specifications; design specifications are covered in greater detail in the next section.

Design Specifications

Design specifications include detailed descriptions on how an item or a project is to be built by a contractor. They include specific standards that describe precise measurements, tolerances, descriptions of the materials to be used, inspection requirements, and the quality that has to be maintained during construction (Bockrath 2000). Some design specifications do not include the actual design because the product or structure is to be designed by the contractor following the details provided in the specifications.

Design specifications are more difficult to write than performance specifications since they have to include precise details that are created by the engineer writing the design specifications. Entire projects could be created using design specifications or only portions of projects might use design specifications. When a specification includes tolerances, inspection requirements, material requirements, exact measurements, and quality control requirements or a combination of these, it could be identified as a design specification. Examples of design specifications are the following (City of San Jose 1992, 73, 90, 101):

73.1.05 Curb Construction. Weakened plane joints shall be constructed at intervals of 10 feet or as shown on the plans. When a Portland cement concrete sidewalk or pavement is adjacent thereto, or to be constructed adjacent thereto, the joints shall coincide with score marks in the sidewalk or pavement.

The weakened plane joint shall be constructed by scoring the partially set concrete to a minimum depth of 2 inches by 1/4 inch with a tool that will leave the corners rounded.

Expansion joints shall be installed only where specifically called for on the plans or as directed by the Engineer.

All curb and gutter joints shall conform to City Standard Plan Details.

The batter of the curb face and lip of the gutter shall be constructed true to the dimensions as shown on the plans.

The use of existing asphalt pavement edge, as the lip of a gutter form will be allowed only upon express approval of the Engineer. The use of the excavated embankment for back forms will not be allowed, except for the bottom portions of A1 and B1 barrier curbs.

Defective curb shall be repaired by removing and replacing no less than 5 feet and leaving no less than 5 feet of a joint.

90-10.02 Materials. Minor concrete shall be Class A, 3000 psi concrete and shall contain not less than 564 pounds of cement per cubic yard unless otherwise specified on the plans or in the special provisions. The cement per cubic yard requirement may

(continued)

be waived by the Engineer in writing, provided the mix as designed consistently produces concrete whose compressive strength is in excess of 3000 psi and has a moving average of 3500 psi or more.

The compressive strength of the concrete will be determined according to the procedures and provisions of Section 90-9, "Compressive Strength" of Caltrans Standard Specifications, except that sampling and testing will be done according to ASTM C172 and C39.

Concrete samples for compressive strength requirements as a basis acceptance of minor concrete will be molded, cured, and tested as provided for in Section 90-9 of the Caltrans Standards Specifications, except that sampling and testing will be done according to ASTM C172, ASTN C31, and ASTM C39. The evaluation of compressive strength tests for minor concrete will be as provided for in Section 90-9 of the Catrans Standard Specifications or as specified herein.

The above examples of design specifications demonstrate the level of detail that is included in design specifications and their reliance on referring to standard specifications.

When design specifications are used rather than performance specifications, it could lead to claims against owners by contractors if the specifications are not accurate. One case that illustrated this issue was the case of *United States v. Spearin* (Supreme Court of the United States, 1919, 248 U.S. 132). The legal issue in this case, and similar cases (*MacKnight Flintic Stone Company v. The Mayor*, 160 New York; *Filbert v. Philadelphia*, 191, Pa.St. 53; *Bentley v. State, Wisconsin*, 416) involved defects in the specifications. The results of these and other cases indicate that "if the contractor is bound to build according to plans and specifications prepared by the owner, the contractor will not be responsible for the consequences of defects in the plans and specifications" (Sweet and Schneier 2004, 493). "The Spearin doctrine consists of two components: the owner's misrepresentation to the contractor through its issuance of the specifications, and the defective nature of those specifications. Both components must be satisfied for the contractor to recover"; therefore, the contractor has to demonstrate that he or she justifiably relied on the specifications and that the specifications were in some way defective (Sweet and Schneier 2004, 494).

Performance Specifications

Performance specifications are easier for contractors to implement than design specifications because they mainly focus on providing a description of the end result and not on how the end results have to be achieved during construction. Performance specifications provide contractors with latitude in the manner in which they achieve the end results. Performance specifications could also help lower the cost of a project, or a portion of a project since the actual design is created by the contractor and he or she might have methods for improving the construction process. Performance specifications state the criteria for achieving a desired level of performance of the final product. They do not include detailed measurements, or any other specific details, since the objective is to meet performance requirements not adhere to detailed specifications.

One example of a structure that could be built as a performance specification is a power plant. The specifications could stipulate that the power plant has to generate 3,000 megawatts of electricity per hour by using coal as the feedstock. Additional details would only be included if they are used to elaborate on the performance requirements for the project. The design and processes used to construct the power plant would be at the discretion of the contractor building the power plant. As long as the power plant meets, or exceeds, the performance requirement of generating 3,000 megawatts of electricity per hour, then the contractor will have fulfilled the contract requirements for the project.

Functional Specifications

Functional specifications are similar to performance specifications but they only set forth a particular function of the completed project rather than specific performance criteria. An example of a functional specification is a project that when completed would be able to store fifty recreational vehicles in a fully enclosed environment. The specification does not state how the objective has to be achieved; it only requires that the objective be accomplished such that the final structure performs the function it was built to perform.

Purchase Specifications

Purchase specifications are written in order to ensure that the correct materials and equipment are ordered and/or fabricated by vendors. Purchase specifications include the properties that are required for materials, equipment, and products. They could also include the required workmanship in order to fabricate, erect, and install materials and products. Specific manufacturers might be cited in purchase specifications, along with specific model numbers and part numbers, or a detailed product description; they might also include an "or equal" clause to prevent them from being restrictive specifications, which are described in the next section.

An example of the types of items listed in a purchase specification for a desktop computer that is to be ordered are the following:

- Hard drive capacity in gigabytes
- Processor speed
- Random-access memory (RAM) in megabytes
- Number of internal ports
- Compact disk drive speed
- Phone and/or Ethernet Internet connection
- Digital video disk drive speed
- Keyboard type
- Mouse type
- Operating system and version
- Color of the case
- Monitor size in inches
- Monitor resolution
- Maximum dimensions

Restrictive (Proprietary) Specifications

If a specification is written in a manner such that only one proprietary product would be accepted and the specification section does not include any provision for substitutions, then the specification section is a **restrictive specification**; this type of specification is not allowed in government contracts. One method for preventing restrictive specifications is to include the statement "or equal" at the end of the specification; this changes the specification into an **open specification**. Typical open specifications are written using a generic description, with no brand names or manufacturers listed, but they could include material quality, performance characteristics, or test values that have to be achieved for the product.

Proprietary specifications include specific brand names. When standard specifications are being referred to, it is important to ensure that they do not include restrictive or proprietary specifications if they are being referenced in government construction documents since it is illegal to use proprietary specifications in government contracts.

Material and Product Specifications

Federal specifications prescribe technical requirements for materials, products, and services that are provided by federal government agencies; these are referred to as **material and product specifications**. The **American Society for Testing and Materials** (ASTM) produces thousands of specifications that pertain to physical, chemical, electrical, thermal, performance, and acoustical properties of materials. The ASTM standards are referenced in construction specifications in order to control the quality of Portland cement, reinforcing steel, asphalt, insulation, and other products. The next section includes information on other organizations and societies that produce standard specifications that could be referenced in construction contract documents.

15.3 STANDARD SPECIFICATIONS

Writers of construction contract specifications take advantage of the hundreds of thousands of standard specifications that are developed and published by government agencies, professional societies, and professional organizations. Standard specifications are referenced in construction documents by organization name, specification number, title of the specification, and the date of issue.

Several categories of standard specifications include the following:

1. Basic Materials—The American Society for Testing and Materials (ASTM) standards
2. Product—The **American National Standards Institute** (ANSI) standards
3. Design—The **American Concrete Institute** (ACI), the **American Institute of Steel Construction** (AISC), and the **American Institute of Timber Construction** (AITC) standards

4. Workmanship—The ASTM standards
5. Test Methods—The ASTM standards
6. Codes—The ANSI standards, building codes, and government standard specifications

Table 15.1 contains a list of the acronyms and names of some of the other organizations that develop and publish commonly used standard specifications.

Table 15.1 Standard Specification Acronyms and Organization Names

Acronym	Name
AAES	American Association of Engineering Societies
ABMA	American Boiler Manufacturers Association
ABPA	Acoustical and Board Products Association
ACI	American Concrete Institute
AHMA	American Hardware Manufacturers Association
AIA	American Institute of Architects
AIAA	American Institute of Aeronautics and Astronomy
AIEE	American Institute of Electrical Engineers
AIQR	American Institute for Quality and Reliability
AIME	American Institute of Mining, Metals, and Petroleum Engineers
AISC	American Institute of Steel Construction
AIAI	American Iron and Steel Institute
AITC	American Institute of Timber Construction
AL	Associated Laboratories
ANSI	American National Standards Institute
API	American Petroleum Institute
ASME	American Society of Mechanical Engineers
ASQC	American Society for Quality Control
ASTI	American Society for Technology and Inspection
ASTM	American Society for Testing and Materials
BOCA	Building Officials and Code Administration
BSI	British Standards Institution
CFR	Code of Federal Regulations
COE	Corps of Engineers
CPSC	Consumer Product Safety Commission
CSA	Canadian Standards Association
CSI	Construction Specification Institute
DOC	Department of Commerce
DOD	Department of Defense
DOE	Department of Energy
DOT	Department of Transportation
ECSA	Exchange Carriers Standards Association
EPA	Environmental Protection Agency
FIEI	Farm and Industrial Equipment Institute
FS	Federal Specifications
ICBO	International Conference of Building Officials
IEC	International Electrotechnical Commission
MCAA	Mechanical Contractors Association of America
MIL	Military Standardization Documents
MSS	Manufacturers Standardization Society
NCMA	National Concrete Masonry Association
NDS	National Design Specifications
NIBS	National Institute of Building Sciences
NIOSH	National Institute for Occupational Safety and Health
NIST	National Institute of Standards and Technology
SBCCI	Southern Building Code Congress International
UL	Underwriters Laboratories

Members of many of the organizations listed in Table 15.1 also actively participate in developing standards that are referenced in the construction industry. In addition, international construction contracts might reference international standards rather than domestic standards; therefore, international standards are discussed in Section 15.4.

Definition of Standards

Standards as cited in codes, handbooks, or manuals are methods and procedures that are part of technical specifications; they set minimum requirements, exact measurements, processes, and other requirements that have to be satisfied in order to be in conformance with a contract. A standard is something that is accepted as a basis for comparison; it is a procedure, or product, used as a reference to determine the quality of similar procedures and products. It is established by determining the technical and nontechnical specifications of the procedure or product (McKechinie 2005).

The main basis for the technical requirements in building codes is the standards that are referenced and incorporated into contracts. Standards are used to evaluate qualities, characteristics, or properties of materials. They are also defined as "having the quality or qualities of a model, gauge, pattern, or type; generally recognized as excellent and authoritative; generally used and regarded as proper for use" (McKechinie 2005).

Standards could be required by law, as they are the common language that is used for communication between designers, constructors, vendors, fabricators, and regulatory agencies. Standard components, materials, and products are less costly; they provide savings in the cost of engineering materials and in time; and they foster improvements in quality.

Technical Standards

The basic intention of **technical standards** is to achieve a recognizable and acceptable level of quality in an industry. Standards help set common goals pertaining to quality that members of an industry strive to achieve, and in doing so facilitate communication between interacting parties. Enhancing communication to achieve quality in design and construction is one of the main purposes of standards. Factors of cost, materials of construction, and material specifications are the main elements of the requirements for standards, which in turn establish an acceptable level of quality.

According to the **U.S. Department of Energy (DOE) Technical Standards Program**, a technical standard is a "document that establishes uniform engineering and technical requirements for processes, procedures, practices, and methods. Standards may also establish requirements for selection, application, and design criteria of facility systems, components, and other items" (DOE 2003, 1). Standards document-proven methods and techniques to allow a better understanding between those who develop the standards and those who use them

(DOE 2003). Standards differ from specifications in that they are a long-term application, whereas specifications are used for unique or one-time applications.

Consensus Standards

A consensus approach is used when standards are developed by an organization. Consensus means "substantial agreement by concerned interests according to the judgment of a duly appointed authority, after a concerted attempt at resolving objections. Consensus implies much more than necessarily unanimity" (Gross 1989, 33).

In the United States **consensus standards**, as well as industry standards, are developed through a voluntary system that contains representatives from the industry, government, producers, consumers, institutions, and individuals. The system is referred to as voluntary since participants are not paid and the standards they produce are used voluntarily in building regulations and contracts (Gross 1989). The procedures used to develop standards in other countries are quite different from the methods used in the United States (Gross 1989). In countries other than the United States, there is greater governmental involvement and the processes are less rigorous than those used in the United States for the establishment of consensus. In many standards organizations, there is limited opportunity for participation, particularly by outsiders, no matter how affected an outsider might be by the development of standards.

Government Standards

In addition to standards that are developed by private organizations, governments also develop and adopt standards. Government standards account for 55 percent of U.S. standards. The largest developer of standards in the U.S. government is the **Department of Defense** with over 50,000 standards. ***Federal Specifications and Standards*** published by the **U.S. General Services Administration** (GSA) are used by federal and state agencies to procure common products. Federal test methods are used in many industries. It is estimated that there are over 100,000 standards developed by U.S. federal agencies such as the Occupational Safety and Health Administration (OSHA), the Environmental Protection Agency (EPA), the Postal Service, and others.

For many government standards it is difficult to distinguish whether they are truly standards or whether they are regulations (Breitenberg 1989). Although some federal standards are voluntary, many become mandatory when they are referenced in legislation or regulations, or they are invoked in contracts as a condition of sale to government agencies. Nongovernment standards could also be mandated in this manner.

Nongovernment Standards

According to the U.S. Department of Energy (DOE), **nongovernment standards** (NGS) are "a standardization document (also known as voluntary, consensus, and industry standards) developed by a private sector association, organization, industry association, or technical society which plans, develops, establishes, or coordinates standards and related documents" (DOE 2003, 2). Standards are put into practice by being referenced in contractual documents, specifications, drawings, and work authorization statements. Nongovernment standards are maintained by the technical society or trade organization that was involved in their development.

15.4 INTERNATIONAL STANDARDS AND THE INTERNATIONAL ORGANIZATION FOR STANDARDIZATION

Standards are an important aid in the process of homogenizing the world while incorporating its variations and international standards are a major component of global engineering and construction (E&C). This section discusses **international standards** and the **International Organization for Standardization** (ISO). The development process for standards is also explained along with a discussion on who participates in the development of international standards and how they are used in the E&C industry.

In the twentieth century, the United States developed many of the standards that were used throughout the world. In the United States, there are currently over 100,000 standards that are actively being maintained by different government agencies and private organizations. Countries in favor of international standards are now discarding some U.S. standards, which until a few years ago were being used internationally. Currently, the United States has limited representation in the international community for standardization. Other countries are now adopting the International Organization for Standardization (ISO) standards, which puts pressure on the United States to become more involved with the ISO standards. Clients throughout the world are stipulating international standards in their contractual arrangements with contractors; some of the countries in the European Union and the Pacific Rim are contributing millions of dollars to the development of international standards.

International standards are used to influence or regulate the types of goods and services that flow between nations. One of the main purposes of international standards is to work towards harmonizing the technical regulations of individual nations. Organizations, such as the American Society of Mechanical Engineers (ASME), the American Society of Testing and Materials (ASTM), the American National Standards Institute, the **British Standards Institution** (BSI), the **Canadian Standards Association** (CSA), and the **Deutsche Institute**, produce standards that are used in many parts of the world, but they are all national organizations. Many of the standards from these organizations are incorporated into international standards through the International Organization for Standardization (ISO). The technical work produced by the ISO is published as international standards and the ISO standards cover all areas except for those related to electrical and electronic engineering, which are covered by the **International Electrotechnical Commission** (ISO 2005).

The ISO has committees that develop guidelines, definitions, and documentation related to its standards; there are three main ISO publications (ISO 2005):

1. *ISO Guide 30—Terms and Definitions Used in Connection with Reference Materials*
2. *ISO Guide 31—The Contents or Certificates of Reference Materials*
3. *The ISO Directory of Certified Reference Materials*—This is an English and French guide to sources of certified reference materials throughout the world, with materials being classified according to domain.

International building standards (construction materials and building) account for 7 percent or about 1,000 of the more than 15,000 ISO standards, but many of the other ISO standards also have a direct, or indirect, influence on the E&C industry including testing, mechanical systems, fluid systems, electrical engineering, railway engineering, paint, materials handling equipment, chemical technology, petroleum, metallurgy, wood technology, environment, and civil engineering. Twelve percent of the standards being used internationally pertain to engineering and construction.

Despite the existence of the ISO and other such organizations, there is no international regulatory agency that oversees the implementation of and compliance to international standards. An engineering and construction client bidding internationally may or may not demand compliance with an international standard. Furthermore, registration for compliance to the ISO quality management and environmental management standards is performed by independent registrars and not internationally designated organizations or countries.

In the United States, in addition to the federal government, there are hundreds of organizations that develop and prepare standards. This is in contrast to countries that have one centralized agency responsible for preparing national standards. In the United States, there are over 100,000 standards that are maintained by the standardization community, of which 45 percent were developed by the private sector. Private organizations that develop standards could be categorized into three areas:

1. Scientific and professional
2. Trade associations
3. Standards developing organizations

International Organization for Standardization (ISO)

The International Organization for Standardization (ISO) was formed in 1947 to promote the development of standardization and to facilitate the international exchange of goods and services and to foster cooperation between intellectual, scientific, technological, and economic activities (ISO 2005).

The deficiencies caused by each country in the world using different standards are being recognized internationally and an immense effort to harmonize existing standards, and to develop new ones, is being done by the ISO. As long as the International Organization for Standardization (ISO) standards exist, the European Union is committed to using them. Therefore, some of the standards prepared by the **European Committee for Standardization** (CEN) are being put forth to the ISO for adoption as global standards. The U.S. government (or U.S. private firms) has no input into the European Committee for Standardization (ISO 1982).

As of 2009 there were 101 countries with ISO member status, most of which are funded by their governments, and 160 countries that are either member bodies, correspondent members, or subscriber members. Table 15.2 contains a list of the member countries that have voting rights. In the ISO, a one country/one vote rule prevails, but block voting is not uncommon.

There are many stages, or processes, that occur before a standard becomes an international standard. The processes for developing standards vary from country to country and from one standard-developing organization to another. Once an organization has developed a standard, they submit it to the ISO for review and possible inclusion as an international standard. A professional society, or association, is not able to directly submit their standards to the ISO for consideration. An official representative to the ISO has to submit standards from organizations within their country. The United States has only one official representative to the ISO, which is the American National Standards Institute (ANSI).

The ISO does have numerous committees and technical subcommittees that allow participation by individuals from different business sectors. Members of organizations either hold an observer (O) status on a task committee (TC) or a participating (P) member status. In order for members of organizations to register as a participating member on an ISO technical committee, the country's official representative of their country has to first designate an accredited standards developer as the **technical advisory group** (TAG) member. The United States has over 200 technical advisory groups.

When an organization is registered as a participating member of a technical advisory group, it is required to vote on **Draft International Standards** (DIS) and to participate in meetings. Societies and associations are allowed to comment on standards through their official representatives if they are registered as participating members on ISO technical committees. The designated member level is based on the following factors (ISO 1991, 2):

- The need for participation by a country
- Whether there is sufficient support indicated from those directly and materially affected to ensure effective participation

Table 15.2 ISO Member Countries

Algeria	Denmark	Lebanon	Serbia
Argentina	Ecuador	Libyan Arab Jamahiriya	Singapore
Armenia	Egypt	Lithuania	Slovakia
Australia	Ethiopia	Luxembourg	Slovenia
Austria	Fiji	Malaysia	South Africa
Azerbaijan	Finland	Malta	Spain
Bahrain	France	Mauritius	Sri Lanka
Bangladesh	Germany	Mexico	Sudan
Barbados	Ghana	Mongolia	Sweden
Belarus	Greece	Morocco	Switzerland
Belgium	Hungary	Netherlands	Syrian Arab Republic
Bosnia and Herzegovina	Iceland	New Zealand	Tanzania, United Republic of
Botswana	India	Nigeria	Thailand
Brazil	Indonesia	Norway	the former Yugoslav Republic
Bulgaria	Iran, Islamic Republic of Iraq	Oman	of Macedonia
Cameroon	Ireland	Pakistan	Trinidad & Tobago
Canada	Israel	Panama	Tunisia
Chile	Italy	Peru	Turkey
China	Jamaica	Philippines	Ukraine
Columbia	Japan	Poland	United Arab Emirate
Congo, Democratic Republic of	Jordan	Portugal	United Kingdom
Costa Rica	Kazakhstan	Qatar	Uruguay
Croatia	Kenya	Romania	United States
Cuba	Korea, Democratic People's	Russian Federation	Uzbekistan
Cyprus	Republic	Saint Lucia	Venezuela
Czech Republic	Korea, Republic of	Saudi Arabia	Viet Nam
Cote-d'Ivoire	Kuwait		

- Whether there is an acceptable, competent organization willing to serve as administrator for the technical advisory group

The ISO does not fund participation on their committees or subcommittees; therefore, committee participants must be self-funded or sponsored by firms, a consortium of firms, a professional society, or a trade organization. The extensive cost of funding a representative excludes many countries from participation in the ISO.

International Organization for Standardization Liaisons

Intergovernmental organizations help implement ISO standards when they use them in intergovernmental agreements. These organizations are closely associated with the work of the ISO by the following four mechanisms (ISO 1991, iii):

- International organizations are allowed to submit proposals for the preparation of ISO standards in a new field in the same way as ISO member bodies.
- International organizations may be granted **liaison status** with technical committees and subcommittees. Liaison status comprises two categories: "A" (effective contribution to the work) and "B" (provided with information only). Liaison A status confers the right to submit papers, attend meetings, and participate in discussions.
- International organizations are invited by the ISO to comment on relevant drafts of standards.
- Technical committees are instructed to seek full and, if possible, formal backing of the main international organization liaison for each ISO standard in which these organizations are interested.

There are over 500 international organizations that are liaisons to the ISO technical committees and subcommittees, which results in thousands of individual liaisons.

The Development Process for Standards

Figure 15.1 displays the organizational structure of the **ISO Central Committee** that is responsible for developing ISO standards.

There are ten main stages that standards developed by an organization must go through to be adopted as an ISO standard. Figure 15.2 shows the ten stages for the development of an ISO standard. Before a standard is submitted to the ISO for consideration, the standard has to first be created by an organization, society, or association. The following section outlines the steps required to have a standard reviewed and accepted by the ISO.

Standards are created through participation by individuals on standard committees within organizations and professional societies. International standards usually begin when there is a **New Work Item Proposal** (NP) for technical work in an area where there is no existing technical committee. If the proposed work is closely related to an existing technical committee, it is assigned to an existing committee by the **Committee of Action of the International Engineering**

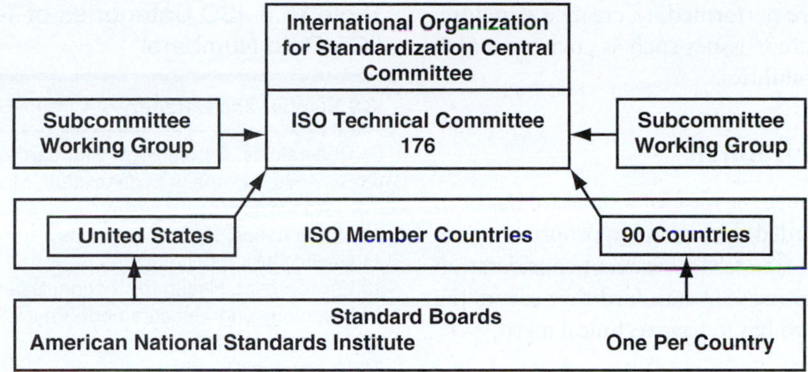

FIGURE 15.1 International Organization for Standardization Central Committee.

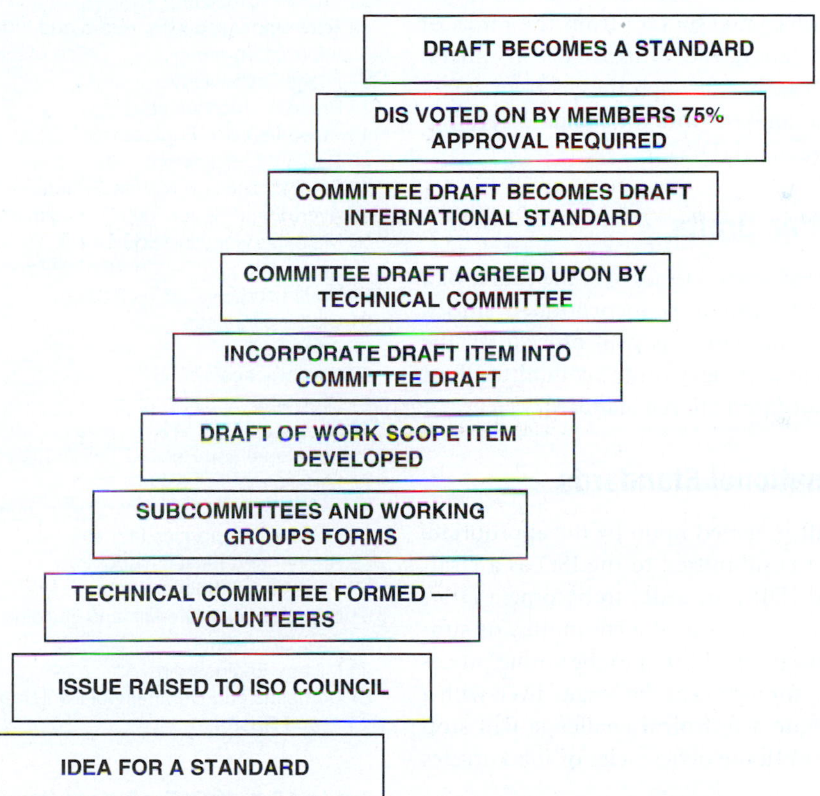

FIGURE 15.2 Steps for Developing an International Organization for Standardization Standard.

Council (IEC), or the Technical Board of the ISO. The IEC Central Office, or the ISO Central Secretariat, conducts a survey of the national member bodies to determine interest in the proposed standard.

If a two-thirds majority approves of forming a technical committee, and there are at least five member bodies willing to actively participate, the ISO creates a new committee. The committee is comprised of interested individuals, or individuals being sponsored by their firms. The committees meet only when it is necessary to discuss committee drafts (CD), or if there are other issues that cannot be handled through correspondence. The committees operate under the consensus procedures mentioned previously in the section on consensus standards.

The standards developed by committees should be evaluated for use based on their technical and quality performance (Termaat 1994, 5). To achieve technical performance, the standards should be written to the latest level of technology. The standards should not be a device to protect less efficient organizations or construction methods, or to assure business to a particular firm, standards developer, or professional assessor.

Quality performance standards set forth minimum levels of quality that must be achieved when products are produced, or processes are performed, to create a structure. Quality standards can include issues such as environmental, health, and safety responsibilities.

Issue Raised to ISO Council

As shown in Figure 15.2, once the idea for a standard has been fully developed it is forwarded through an appropriate representative to the ISO. For a standard to warrant consideration there must not be an existing ISO standard that covers the same item, and the standard has to have technical merit.

Formation of Technical Committees, Subcommittees, and Working Groups

Members of ISO technical committees, subcommittees, and working groups are nominated by ISO from the ranks of industry, government, labor, and individual consumers. These members work together through the different stages of the development of a standard until the standard becomes a Draft International Standard (DIS).

Technical Committee Drafts

Committee members meet several times over the course of a year to review an item until it is incorporated into a **committee draft** (CD). Through consensus procedures, the technical committee members agree on the committee draft before it becomes a Draft International Standard.

Approval of International Standards

Once a committee draft is agreed upon by the appropriate technical committee, it is submitted to the ISO as a Draft International Standard (DIS). In order to become a Draft International Standard, two-thirds of a committee or subcommittee must approve it. In addition to the voting procedures, ISO has a process for technical challenges. Even with a 75 percent favorable vote a technical challenge will stop approval of a standard until the deficiencies or inaccuracies have been cleared up.

Not all of the countries that are ISO members are allowed to vote on Draft International Standards. Only forty-one countries are voting members of ISO, with the remaining countries having only observer status. Countries with observer status may provide input into standards development through technical committees, subcommittees, and working groups.

15.5 INTERNATIONAL TECHNICAL STANDARDS

The ISO has forty categories of international technical standards; these categories are listed in Table 15.3. Each of the forty categories contain numerous subcategories. Table 15.4

Table 15.3 ISO Categories of Technical Standards (ICS Field Numbers)

ICS Number and Category
01 Generalities, Terminology, Standardization, Documentation
03 Services, Company Organization, Management, Quality, Administration
07 Mathematics, Natural Sciences
11 Health Care Technology
13 Environment, Health Protection, Safety
17 Metrology and Measurement, Physical Phenomena
19 Testing
21 Mechanical Systems
23 Fluid Systems and Components
25 Manufacturing Engineering
27 Energy and Heat Transfer Engineering
29 Electrical Engineering
31 Electronics
33 Telecommunications, Audio and Video Equipment
35 Information Technology, Office Machines
37 Image Technology
39 Precision Mechanics
43 Road Vehicles Engineering
45 Railway Engineering
47 Shipbuilding and Marine Structures
49 Aircraft and Space Vehicle Engineering
53 Materials Handling Equipment
55 Packaging and Distribution of Goods
59 Textile and Leather Technology
61 Clothing Industry
65 Agriculture
67 Food Technology
71 Chemical Technology
73 Mining and Minerals
75 Petroleum and Related Technologies
77 Metallurgy
79 Wood Technology
81 Glass and Ceramics Industry
83 Rubber and Plastics Industry
95 Paper Technology
91 Construction Materials and Building
93 Civil Engineering
95 Military Engineering
97 Domestic and Commercial Equipment, Entertainment, and Sports

Table 15.4 ISO Technical Standards for Civil Engineering—93 (ICS Numbers)

ICS Number and Category
93.010 Civil Engineering in General
93.020 Earthwork, Excavation, Foundation, Construction
93.025 External Water Conveyance Systems
93.030 External Sewage Systems
93.04 Bridge Construction
93.060 Tunnel Construction
93.080 Road Engineering
93.100 Construction of Railways
93.110 Construction of Ropeways
93.120 Construction of Airports
93.140 Construction of Waterways, Ports, and Dykes
93.160 Hydraulic Equipment

provides a sample of the subcategories under the major category of civil engineering standards to demonstrate the types of items that are included within categories.

Under each of the subheadings, there are specific standards, and they are available directly from the ISO at the ISO website at www.iso.org. The standards could be located by going to the appropriate categories and locating the desired standard or by going directly to the ISO catalog of standards at www.iso.org/iso/en/prods%2Bservices/catalogue/CatalogueListPage.CatalogueList. The telephone number for the ISO is +41 22 749 01 11 (Geneva, Switzerland) and their fax number is +41 22 733 34 30.

In the United States, the ISO standards could be purchased through national standards organizations such as the American National Standards Institute (ANSI) (phone: 212-642-4980, e-mail: ansioline@ansi.org), the American Society for Testing and Materials (ASTM) (phone: 610-832-9585, e-mail: service@local.astm.org), and Global Engineering Documents (phone: 800-624-3974, e-mail: globalcustomerservice@ihs.com).

15.6 OPPORTUNITIES FOR THE MISINTERPRETATION OF SPECIFICATIONS

The more precise a specification is, the higher the cost will be to comply with it. When writing specifications it is difficult to achieve a balance between specifications that are so precise that it would be difficult to achieve the exact results they prescribe and specifications that are so vague that a variety of results would meet their requirements. The engineers who write specifications attempt to translate into written words the intent of their design, which is a process that requires being able to verbally describe visual attributes or processes. Unfortunately, sometimes phrases are used in specifications that could be interpreted in a variety of different ways; this might lead to physical results that were not part of the intention of the person who designed the object and wrote the specifications. Therefore, it is crucial that after specifications are written they Should be reviewed to determine whether they contain vague words or phrases that could contribute to the specifications being misinterpreted by contractors.

Vague Terms to Avoid in Specifications

Table 15.5 contains a list of words and phrases that were included in thirty different contracts in the specification sections that should not have been included in them due to their vague nature and the potential for different interpretations.

If any of the terms listed in Table 15.5 are included in specifications, they could cause contractors to misinterpret their intent and lead to issues between owners and contractors that might result in contractors having to redo some

Table 15.5 Samples of Vague Terms to Avoid in Specifications

A few	Elements	Reddish
A lot	Encumbrance	Relevant
About	Enough	Responsible
Above	Excessive	Required
Above average	Expensive	Respectable
Acceptable	Expressly	Satisfying
Adequate	Extraordinary	Safe
Agreed upon	Extreme heat	Safe distance
Alleviate	Fair judgment	Short
Appropriate	Firm	Satisfactory
Almost	Familiar	Semi-significant
Anticipated	Feasible	Similar
Apparent	Few	Simple
Applicable	General	Small
Approximately	Fine	Secure
At least	Frequently	Slightly
Around	Glossy finish	Sometimes
Average attribute	Good	Something like
Bargain	Greater	Somewhat
Better	Hereunder	Just about
Big	Hard	Later
Bigger	Harmless	Local
Bright	High quality	Maybe
Clearly	Often	Old
Close	Perhaps	Overall
Comparable	Perfectly	Pleasant
Competent	Practically	Plenty
Correctly	Proficient	Possible
Damage	Proper	Practical
Decent	Qualities of	Precisely
Different	Proper	Really
Discrepancy	Proximity	Regularly
Earliest	Quiet	Semi-reflective
Easiness	Quick	Shortcoming
Efficient	Reasonable	Similar condition

items of work at either their own expense or at the expense of the owner. A protracted discussion would be required to determine the meaning of these types of terms unless the engineer who wrote the specification is available to explain the intention of a term.

Confusion in Specifications Related to Dimensioning

The numbering systems used on drawings and in specifications could be misinterpreted because most of the world uses a modern version of the **International System of Units**, the metric system, with the exception of the United States, although the metric system was legalized in the United States in 1866. References are available that contain conversion tables for units of weights and measures; scientific calculators and electronic translators have conversion functions. Examples of conversions are 2.5 centimeters per inch, and 1.094 meters per yard. Table 15.6 contains some of the more commonly used conversion factors.

Table 15.6 Examples of Conversion Factors

1 millimeter (mm)	1000 micrometers or microns
1 centimeter (cm)	10 mm
1 meter (m)	100 cm
1 meter (m)	3.2808 feet = 39.37 in
1 kilometer (km)	1000 m
1 inch	25.400 mm
1 foot	304.80 mm
1 sq. mm	0.00155 in squared
1 km squared	247.1 acres
1 hectare	2.471 acres
1 m squared	35.3 sq. ft
1 liter	0.264 U.S. gallons
1 gram (g)	0.035 oz.
1 kilogram (kg)	2.20 lb
1 ton	907 kg = 2.00 kips
1 m/sec	3.28 ft/sec
1 km/hr	0.911 ft/sec = .621 mph
1 kg/sq. m	0.0624 lb/sq.ft.
1 ton/sq. m	0.0328 lb/in

When dimensions are listed on drawings, engineers assume that they are in centimeters or meters if the metric system is their native numbering system, or they assume that they are in inches or yards if they are from the United States. If the plans do not state which system is being used, or if the inch symbol is not used, or if only the first page of the plans (the index) indicates that the metric system is being used and the plans are later separated, then errors could be introduced into projects.

Globally, dimensions are labeled in several different ways such as 7′6″, 7.5′, 7.5″, 7.5, 7 feet 6 inches, 7.5 feet, 7.5 inches, 0.75, 7 meters 6 centimeters, 7.5 meters, 7.5 centimeters, 0.75 meters, or 0.75 centimeters. Sometimes the foot and the inch symbols or decimal places and commas appear to be the same when blueprints are darker or lighter than the original plans, or if the plans have been copied on a copier, reduced in size, or printed on a printer that is low on toner. If the design engineer of record is not located in the country where a project is being built, it could be difficult to reach the engineer of record to clarify dimensions.

The dimension system used on drawings should be consistent throughout the plans and specifications and explained in several different places so that 7.5 is not interpreted as 7.5 inches, 7.5 feet, 7.5 centimeters, or 7.5 meters by different people working on the same project. To avoid confusion, dimensions could be listed as 7 feet 6 inches or 7 meters 6 centimeters, but using this system is a time-consuming process and too cumbersome on drawings. If the drawings have been reduced or reproduced on copiers, a note should be included that contains a scale that correlates to the reduced or copied versions. **Not to scale** (NTS) is a standard phrase that is used to indicate that the drawings have been altered in some manner from the original version.

One incident that occurred on one of the designs produced by the **National Aeronautics and Space Administration** (NASA) in the United States resulted in a $200 million error when an item was fabricated with the dimensions in inches when they were designed in centimeters.

The standard for lumber (2 × 4, 2 × 6, 2 × 8, 2 × 10, and so forth) also causes confusion since personnel might not understand that a 2 × 4 is actually 1 3/4 inches by 3 3/4 inches (nominal versus actual dimensions) as a result of dimension lumber being planed to obtain a smooth surface. Yet a piece of plywood is designated as a 4 by 8, and it is actually 4 feet by 8 feet.

Symbols Used in Drawings

Symbols that are familiar to engineers in one country could be completely foreign to engineers in other nations. One way to make sure that symbols are not misinterpreted is to provide handouts on standard symbols along with the plans and the terms used for symbols should be translated into local languages. Charts of standard symbols are available in drafting books and on drafting templates.

In some developing countries, people have never seen items such as: overhead showers or shower stalls, bathtubs, stoves, sinks, elevators, escalators, dishwashers, garbage disposers, central heating and air conditioning, self-flushing toilets, electronic garage door openers, revolving doors, or automatic door openers. Providing pictures of them along with plans and specifications helps workers have a better idea of what is supposed to be built.

Two examples of symbols that are not the same in Eastern and Western cultures are toilets and bathtubs. In Western cultures, toilets are used while sitting on a seat that is approximately eighteen inches high and that has a lid, or they are used in a standing position in front of the toilet with the lid on the toilet raised. Western toilets are flushed by depressing a handle on the front, top, or side of the toilet. Eastern toilets are flush to the floor, and they are used in a squatting position. They may be a porcelain bowl (or merely a hole in the ground) with ceramic footprints, or pieces of wood, that are used to position feet. If they do not have a Western flushing mechanism, water is poured into them from a separate container. In Western cultures, toilets are drawn as an oblong circle with a rectangular box attached to one end; in Eastern cultures toilets are drawn using a circle with a small oval on two sides of the circle. Figure 15.3 shows both a Western and an

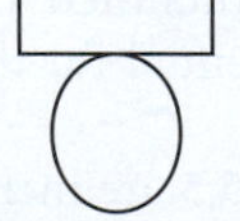

FIGURE 15.3 Sample of Western and Eastern Toilets on Design Drawings

Eastern version of toilets, as they would be displayed on design drawings.

In Western cultures, bathtubs could either be a fixture that a person lays in, or sits in. In Eastern cultures, bathtubs are a container that holds water that is ladled over the body using a pot while the user stands on the floor, which contains a drain.

Misinterpretation of Colors Used in Drawings and Specifications

Another example of an area that leads to numerous instances of confusion during construction projects that is related to specifications is the description of colors. If color samples are not provided in the specifications, then contractors have to use their judgment in order to determine what colors are supposed to be when painted on objects. This is especially difficult when the colors are labeled with terms such as canary yellow or battleship gray or other references to items that contractors or subcontractors may not be familiar with. Contractors must decide for themselves what shade the color is supposed to be when painted on a structure. This problem is even more prevalent when projects are being built in foreign countries where the objects being referred to in the specifications to identify a color are items that individuals in that country might not be familiar with if they have not seen the object or structure.

Another concern related to colors is that they are interpreted differently throughout the world. The color green represents parks and recreation in the United States, but it is associated with disease in countries with dense green jungles. Green is a favorite color in Saudi Arabia, but it is a forbidden color in parts of Indonesia. In Japan, green is used in the high-tech industry, but you do not see it used in the high-tech industry in the United States. The U.S. high-tech industry uses putty, black, and silver. For mourning, black is worn in the United States and Italy, but in Asia people wear white while they are in mourning. Other colors that signify mourning are: purple in Brazil, yellow in Mexico, and dark red in the Ivory Coast. Blue is a masculine color in the United States but red is masculine in the United Kingdom. Red is a sign of good fortune in China and the front doors of structures could be painted red. Pink is a feminine color in the United States but yellow is feminine in most of the rest of the world (Copeland and Griggs 2001). Since specifications stipulate specific colors, regional differences should be a consideration when designing projects and writing specifications.

Numerical Confusion in Drawings and Specifications

Numbers cause similar problems since the number four is considered bad luck in Japan. In China, four and fourteen mean death, three and thirteen mean life, and eight and eighteen mean good fortune, and if a building in China has eight (or eighteen) in the address, people are more likely to rent offices or apartments in the building on the eighth or eighteenth floors. Seven is unlucky in Ghana, Kenya, and Singapore, and the number thirteen is bad luck in the United States.

In the United States, in some buildings there is no elevator button for the thirteenth floor and offices and apartments on those floors are labeled as the fourteenth floor, but there is a thirteenth floor in the structure. In China, the fourth and fourteenth floors are not labeled in structures because four and fourteen are bad luck numbers as their pronunciation is similar to the word death in Mandarin. In the United States, some hotels and office buildings have the first floor labeled as the lobby, or a plaza level, and the second floor is the mezzanine, so the third floor is labeled as the first floor and the plans will be labeled in the same manner.

If **around-the-clock engineering** or **twenty-four hour engineering** is used to design a project, the electrical engineer might be in the United States, the mechanical engineer could be in India, the civil engineer in Lithuania, and the project is being built in South America. Engineers in each of these countries work on projects during their work hours and then they relay their work to engineers in a time zone that is eight hours earlier in the day, and they, in turn, relay their work to engineers in a time zone that is eight hours earlier. If twenty-four hour engineering is being used while a project is being designed, every engineer working on the project needs to know which standards are being used on the plans to designate floor levels and which numbering system for floors is being used on the project.

15.7 THE CONSTRUCTION SPECIFICAITON INSTITUTE (CSI) MASTER FORMAT

The Construction Specification Institute (CSI) Master Format is included in the ***CSI Manual of Practice*** (2008) and the ***CSI Master Format*** (2004); it is a process that is used to organize specifications. The CSI Master Format is used by over 85 percent of the large construction firms in the United States. If this format is not used, firms usually have their own unique format that is followed on all of their projects.

The CSI Master Format divides projects into sixteen divisions and each item in a project is included under one of the sixteen divisions. The sixteen CSI divisions are the following (Construction Specification Institute 2004, 3):

1. Division 1—General Requirements
2. Division 2—Site Work
3. Division 3—Concrete
4. Division 4—Masonry
5. Division 5—Metals
6. Division 6—Wood and Plastics
7. Division 7—Thermal and Moisture Protection

8. Division 8—Doors and Windows
9. Division 9—Finishes
10. Division 10—Specialties
11. Division 11—Equipment
12. Division 12—Furnishings
13. Division 13—Special Construction
14. Division 14—Conveying Systems
15. Division 15—Mechanical
16. Division 16—Electrical

The CSI Master Format uses these sixteen divisions to create a standard system for organizing specifications. Each of the sixteen divisions is divided into three subdivisions that are called: (1) **broadscope**, (2) **mediumscope**, and (3) **narrowscope**. According to the *CSI Manual of Practice*, these three subdivisions are defined as (Construction Specification Institute 2008, 3):

1. Broadscope titles are for broad categories of work and provide the widest latitude in describing units of work.
2. Mediumscope titles cover units of work of more limited scope.
3. Narrowscope titles are for use in covering extremely limited and very specific elements of work.

The *CSI Manual of Practice* describes the Master Format as (Construction Specification Institute 2008, 3):

> In Master Format the broadscope section titles are shown boldface capital letters with five digit numbers. The hyphenated three digit numbers with indented upper and lower case titles are recommended mediumscope sections. Unnumbered, indented titles, which in most cases follow a mediumscope title, are narrowscope section titles. These, when taken together with the mediumscope titles under a broadscope heading, describe the coverage of that broadscope section. For example:

02100	**SITE PREPARATION**	
02110	Site Clearing	
		Clearing and Grubbing
		Large Tract Tree Clearing
02115	Selective Clearing	
		Sod Stripping
		Tree and Scrub Removal
		Tree Pruning
02120	Structure Moving	

The *CSI Manual of Practice* and the *CSI Master Format* provides a list of all of the broadscope headings under each division and the medium and narrowscope headings are customized for each project. The actual specifications are available under each of the appropriate specification sections. The *CSI Master Format* makes it easier to locate information on specific items since they are consistently included in the same specification sections in any type of contract.

The *CSI Master Format* is also helpful when writing specifications because it provides a standard format for organizing specification sections. The *CSI Manual of Practice* also includes information on how to write contracts and specifications.

The last section of the *CSI Manual of Practice* includes an extensive list of key words for every item used in construction projects and a **CSI code** for each of the key words. This key word list could be used to locate the CSI code for an item and then the CSI code is used to locate the division and specification section where the item is described in the specifications.

Chapter Summary

This chapter discussed the different types of specifications, standards specifications, opportunities for misinterpretations of drawings and specifications, international standards, the ISO, international government standards, nongovernmental standards, the International Organization for Standardization (ISO) standards, international technical standards, consensus standards, along with the development process for standards. The *Construction Specification Institute Master Format* was also discussed along with how it is used in construction contracts and specifications.

Engineers and constructors need to understand the importance of the standards development process as it influences how and what they are able to design and build. The implementation of technical standards helps increase the quality of products, and costs are reduced due to designs being easier to interpret, better processes, and broader markets for products. Technical barriers to trade are reduced when technical standards are specified and universal technical communication and mutual understanding is also improved through the use of technical standards.

Key Terms

American Concrete Institute
American Institute of Steel Construction
American Institute of Timber Construction
American National Standards Institute
American Society for Testing and Materials
around-the-clock engineering
British Standards Institution
broadscope
Canadian Standards Association
committee draft
Committee of Action of the International Engineering Council
consensus standards
Construction Specification Institute Code

Construction Specification Institute Manual of Practice
Construction Specification Institute Master Format
Department of Defense
design specification
Deutsche Institute
Draft International Standards
European Commission for Standardization
federal specification
Federal Specifications and Standards
functional specification
International Electrotechnical Commission
International Organization for Standardization
international standards
International System of Units
ISO Central Committee
liaison status
material and product specification
mediumscope
method and material specification
narrowscope
National Aeronautics and Space Administration
New Work Item Proposal
nongovernmental standards
not to scale
open specification
performance specification
proprietary specification
purchase specification
quality assurance
quality control
restrictive specification
specifications
standard specification
standards
technical advisory group
technical standards
twenty-four hour engineering
U.S. Department of Energy Technical Standards Program
U.S. General Services Administration

Discussion Questions

15.1 Explain the difference between a technical standard and a quality standard.

15.2 Explain how international technical standards are implemented in the E&C industry.

15.3 Explain the difference between government and nongovernment standards.

15.4 Discuss who develops international standards and how they are developed by this organization.

15.5 Explain how international standards are enforced globally.

15.6 Explain how individuals could provide input into the development of international standards.

15.7 What are ISO Liaisons? How do they participate in the development of international standards?

15.8 Using Tables 15.3 and 15.4, discuss which categories of international standards would apply to the building of a twelve-story building.

15.9 Explain what technical standards are and how they are used in the E&C industry.

15.10 Discuss how engineers and constructors could overcome the potential problems that could be caused by using around-the-clock engineering.

15.11 What are the seven different types of specifications?

15.12 Explain why vague terms should not be used in specifications.

15.13 Discuss why the selection of appropriate colors is a concern in international contracts.

15.14 Discuss what types of errors could occur related to the dimensioning of drawings.

15.15 How are Construction Specification Institute codes used on construction projects?

15.16 Explain the difference between a performance specification and a design specification.

15.17 Explain how any type of a specification could be turned into an open specification.

15.18 What are the six types of standard specifications?

15.19 Explain restrictive or proprietary specifications.

15.20 Develop a list of ten vague terms that are not listed in Table 15.5 that should not be used in specifications.

References

Bockrath, J. T. 2000. *Contracts and the Legal Environment for Engineers and Architects*. New York: McGraw Hill Publishers.

Breitenberg, M. 1989. *Questions and Answers on Quality, the ISO 9000 Standard Series, Quality System Registration, and Related Issues*. Washington, DC: U.S. Department of Commerce.

City of San Jose. 1992. *Standard Specifications*. San Jose, CA: Department of Public Works, pp. 73, 90, and 101.

Construction Specification Institute. 2004. *Master Format 2004 Edition*. Alexandria, VA: Construction Specification Institute.

Construction Specification Institute. 2008. *The Project Resource Manual—CSI Manual of Practice*. Alexandria, VA: Construction Specification Institute.

Copeland, L., and L. Griggs. 2001. *Going International*. New York: Random House.

Department of Energy (DOE). 2003. *United States Technology Standards Program*. Washington, DC: Government Printing Office.

Gross, J. G. September/October 1989. International harmonization of standards: Done with or without us. *The Building Office and Code Administrator*. Washington, DC: 46–47.

The International Organization for Standardization (ISO). 1982. *ISO Benefits of Standardization*. Geneva, Switzerland: International Standards Organization Council Committee on Standardization Principles.

The International Organization for Standardization (ISO). 1991. Introduction. *ISO Liaisons*. Geneva, Switzerland.

The International Organization for Standardization (ISO). 2005. *The International Organization for Standardization Catalog.* Geneva, Switzerland: The International Organization for Standardization.

McKechinie, J. L. (ed.). 2005. *Webster's New Universal Unabridged Dictionary.* New York: Simon and Shuster.

Sweet, J., and M. Schneier. 2004. *Legal Aspects of Architecture, Engineering, and the Construction Process.* Toronto: Thomson Publishers.

Termaat, K. March 1994. Introducing SSM into both corporate and industry culture. *Proceedings of the American National Standards Institute.* Washington, DC: Annual Public Conference, 139–145.

U.S. Department of Defense. 2006. *Department of Defense Government Standards.* Washington, DC: www.dod.gov.

GOVERNMENT CONTRACTING

16.1 INTRODUCTION

Government contracting procedures are similar to the contracting procedures that are used in the private sector except that **government contracts** are administered by a **government contracting officer** that ensures performance of the contractual obligations. In addition, there are a variety of government regulations that have to be followed including the Federal Acquisition Regulations (FAR), the Contract Dispute Act, the Freedom of Information Act, and the Truth in Negotiations Act.

This chapter reviews government contracting requirements and government regulations that affect construction contracts, and explains what constitutes a claim against the government. Government requirements for bonds such as bid, performance, payment, and maintenance bonds are discussed along with how payment bonds are used to prevent mechanic's liens and how performance bonds are used to protect owners against contractor default. The role of sureties in government contracting procedures is discussed in relation to their involvement on construction projects and their requirements for issuing bonds to contractors. Stop notices are reviewed along with a discussion of what mechanic's liens are and how they are used on privately funded construction projects.

16.2 GOVERNMENT BONDING REQUIREMENTS

One of the differences between private and government construction contracts is in the requirements for **bonds**. Private owners have the option of either requiring or not requiring bonds to be secured by the contractors working on their projects. For government construction contracts bonds are required in order for contractors to work on government projects. Although it is owners, or government agencies, that require bonds, the bonds are issued by **surety companies** directly to contractors. In order to secure a bond, contractors have to be able to qualify for them based on a variety of criteria. Contractors pay for the surety bonds but they pass the cost of the bonds along to owners through their inclusion in their bid prices. When bonds are required on construction projects, it affords protection to owners for awarding of the contract, bid performance, and payment default. In the case of private projects it eliminates the ability of contractors and subcontractors to file mechanic's liens against the property and the structure that is being built.

A bond is a written contract that describes the legal obligations of the agreement that is formed between a surety and a contractor. Bonds guarantee all of the provisions of the construction contract, not anything that is in addition to the contract. Bonds are only invoked if a contractor breaches a contract, does not accept a bid that is awarded to them, or does not pay subcontractors or suppliers. Bonds are issued to qualified contractors at a cost that is a percentage of the total bond value such as 1 percent or 10 percent. The rate that contractors are charged for bonds is dependent on a number of factors, which are discussed in Section 16.3.

The requirement for bonds on government projects that exceed $100,000 was set forth in the **Miller Act** of 1935; many states have Little Miller Acts that set forth the bonding requirements at the state level (40 U.S.C. Section 3131 to 3134). The Miller Act requires contractors to post two bonds: (1) a performance bond and (2) a labor and material payment bond. Before work is awarded to a contractor for government construction projects, a contractor has to provide to the government the following (Miller Act 1935):

1. A performance bond in an amount that the contracting officer regards as adequate for the protection of the federal government.
2. A separate payment bond for the protection of suppliers and laborers. The amount of the payment bond shall be equal to the total amount payable by the terms of the contract unless the contracting officer awarding the

contract makes a written determination supported by specific findings that a payment bond in that amount is impractical, in which case the amount of the payment bond shall be set by the contracting officer. The amount of the payment bond shall not be less than the amount of the performance bond.

Subcontractors and suppliers that work directly with prime contractors are considered to be first-tier claimants. Second-tier claimants are the subcontractors and suppliers that have contracts with subcontractors rather than prime contractors.

16.3 BONDING PROCEDURES

This section discusses the procedures that occur during the bonding process on government contracts and private contracts that require bonds.

In contractual terms contractors are the principals in bonding procedures and contractors are the party that has an obligation to perform by constructing a project. The bonding company, which is called a **surety**, is the party that is guaranteeing the performance of the contract. Owners are the **obligees** and bonds are written for their benefit.

Surety bonds are the written instruments that are provided to owners by the surety and contractors. The surety is guaranteeing that owners will not suffer any losses if contractors do not perform as stipulated in the contract documents. Surety bonds constitute an indemnity agreement that guarantees that a surety will not incur any loss by providing bonds.

Surety Contractor Approval Process

In order for a surety company to provide bonds for government construction projects, the surety has to be approved and registered by the **U.S. Treasury Department**. The U.S. Treasury Department maintains a list of the sureties that are approved; therefore, contractors should verify that the surety firm they are using is government approved to issue bonds to contractors for public projects. The Treasury Department verifies that sureties have the ability to pay if the contractor they are bonding defaults on a project.

For private construction projects sureties do not have to be approved by the U.S. Treasury Department. Many major insurance companies also provide surety bonds. If a firm is large enough and it has adequate capital assets, it could provide its own bonds, but only a few of the largest construction firms are able to provide their own bonds.

Before a surety company will issue any type of a bond to a construction firm, they thoroughly review the construction firm and evaluate their character, their capital, and their capacity. The character of a firm is related to their standing in the industry and whether the firm has ever defaulted on a construction contract. Sureties investigate whether a construction firm has adequate working capital to undertake construction projects. The higher the level of working capital that a construction firm invests in a project, the less likely the company is to default on a project because the firm would risk losing the working capital invested in the project. The capacity of a firm includes whether the firm has the ability to perform the type of work they are proposing to build, including adequate qualified manpower, equipment, and other resources.

In addition to investigating the construction firm, sureties also evaluate the potential project and other accomplishments of the construction firm including:

1. The project risks, size, type, and nature.
2. The amount of work completed or bid by the contractor.
3. The amount of money **left on the table** every time the contractor has bid a project and been awarded the contract. This is the difference between the bid of the contractor and the next highest bid; it provides a measure of the ability of the construction firm to accurately estimate projects.
4. The largest project completed by the contractor to date. This indicates to the surety whether the contractor could potentially perform a project of an equal or larger size.
5. Area of expertise. Sureties are careful about bonding firms that are bidding on projects outside of their area of expertise. One way for construction firms to branch out into difference areas is to form a joint venture with a firm that has the appropriate expertise and if the joint venture successfully completes the project in that area, they would be eligible for bonding for the new type of work.
6. Length of the project. Sureties need to know the time period that will be required for a project in order to determine how long their funds will be committed to the project.
7. List of project subcontractors. Surety companies realize that it is not only the ability of the contractor that determines the success or failure of a project but that success is also influenced by the subcontractors that are performing work on a project.
8. Terms of the contract. Surety companies are supplied with copies of the contract documents. The contract is reviewed to determine whether the terms and the conditions of the contract are conducive to the contractor being able to fulfill the contract in a timely manner.
9. Payment measurement methods. Contractors are reimbursed for work completed by progress payments. Before progress payments are issued, the owner's representative verifies the amount of work performed using some type of progress measurement method such as:
    - Units Completed
    - Percent Complete
    - Level of Effort
    - Incremental Milestones
    - Start to Finish Percentages

If the progress measurement techniques being employed on a project do not result in progress payments that completely cover the amount expended to date by a contractor, there is a higher probability that a contractor might default on the project.

After reviewing all of the items listed above, surety companies determine a maximum amount that they will issue as a bond to a firm. In some instances, rather than having to review all of the documentation for each project that a contractor plans to bid on, the surety will provide a total bonding capacity figure and the firm could bid one project at the total value or multiple projects that when combined equal the total value.

Legal cases could arise if a surety company does not agree to provide a bond to a construction firm or if a firm loses its bonding capacity due to issues that have arisen on previous projects. The following legal case occurred when a firm alleged that the action of a bank impaired its ability to obtain a bond by a surety company and it affected their ability to obtain a construction contract [*S.C. Anderson Inc., v. Bank of America N.T. and S.A.* (1994) 24 Cap. App. 4th 529, 30 Cal. Rptr. 2d 286]:

> Plaintiff, a tenant improvement contractor, obtained judgment on a jury verdict of $248,004 compensatory and $300,000 punitive damages against the Bank of America, a foreclosing construction lender. Damages were based on a fraud cause of action. This appeal deals only with a nonsuit granted on the contractor's cause of action for profits lost from future jobs because of impaired bonding capacity. The California Supreme Court has recognized the cause of action for lost profits based on impaired bonding capacity. *Warner Construction Corporation v. Los Angeles* (1970) 2 C3d 285, 85 Cal. Rptr. 444, 466 P2d 996. The contractor showed that its bonding company refused to issue a bid bond because of unpaid receivables arising out of the job that was financed by the bank. The contractor showed it had figured the job and its bid would have been $2,900,000 including a 5 per cent profit of $140,588 and that the low bid was $3,027,036. Therefore, the contractor would have been the successful low bidder but for the lack of a bid bond. The nonsuit was **AFFIRMED**. Lost prospective net profits may be recovered if the evidence shows with reasonable certainty both their occurrence and extent. But the plaintiff has the burden to produce the best evidence available in the circumstances to establish a claim for lost profits. The contractor offered no evidence that would have enabled the jury to conclude it was reasonably probable that the company would have in fact earned a profit of $140,588. The contractor did not offer any evidence that its bid was accurate or reasonable or that it could have done the work for an amount equal to or less than its projects' costs. It is possible the bid contained errors or was based on overly optimistic takeoffs, and it is possible that the contractor would have performed inefficiently. **MOTION TO REOPEN.** The court required to grant contractor's motion to reopen if the testimony proffered would have cured the deficiencies in the case. The contractor offered the testimony of a bookkeeper as to the company's gross margins for 1983 through 1987. The offer of proof did not make it clear whether the testimony would have cured the deficiencies in the case since the offer did not specify whether the evidence was broken down on a job-by-job basis. Therefore, the trial court properly denied the motion to reopen.

This case study helps to illustrate not only the issue of damaging the bonding capacity of a firm but also the importance of providing complete and accurate testimony and documentation as evidence during trials. The contractor might have had a more viable case if he could have provided additional documentation for the claims he was asserting during this case.

16.4 TYPES OF BONDS

Government construction projects require four types of bonds:

1. Bid
2. Performance
3. Payment
4. Maintenance

These four types of bonds are discussed in the following sections.

Bid Bonds

Bid bonds are used to protect owners when the contractor that is the lowest, responsible, responsive bidder on a competitively bid project refuses to enter into the contract for the project. Bid bonds are either for a percentage of the total value of the project, such as 10 percent, or for a lump sum amount. When a bid bond is forfeited, the owner is entitled to the difference between the lowest bid amount and the next lowest bid up to the total value of the bid bond. The surety company pays the owner the difference between the lowest and the second lowest bid amount on behalf of the contractor. The contractor pays for the bid bond on a sliding scale that is based on their qualifications, as described in Section 16.3. Figure 16.1 shows a sample bid bond form that is used in construction contracts.

If the contractor with the lowest bid signs the contract, then the bid bond is not invoked but the surety does not refund the cost of the bid bond. Sometimes a surety will provide a bid bond for free, or at a minimum cost, if a contractor secures all of their bonds from the surety.

When a surety is required to pay an owner the bid bond when a contractor defaults on their bid, the surety might try to prevent having to pay the owner by invoking one of three possible defenses:

1. Proving that the low bidder made a mistake in their bid.
2. Claiming that the owner waited too long before awarding the bid, thus invalidating the time frame in which the bid bond was valid.
3. Claiming that the owner somehow changed or altered the terms and conditions of the contract before awarding the bid.

Bid bonds help prevent owners from having contractors bid on their projects that are not serious bidders, that do not intend to undertake the work, or that are awarded another job at the same time.

BID BOND

(Where applicable the references to Bidder, Surety, Owner or any other party are considered to be plural)

Bidder (Name and Address):

Surety (Name and Address of the Principal Place of Business):

Owner (Name and Address):

Bid
Bid Due Date:
Project (Brief description including location):

Bond
Bond Number:
Date (No later than bid due date):

Penal Sum: ______________________________ ____________
(Words) (Numbers)

Bidder:	**Surety**:
______________________________	______________________________
Bidder's Name and Corporate Seal	Surety's Name and Corporate Seal
By: ______________________________	By: ______________________________
Signature and Title	Signature and Title
Attest: ______________________________	Attest: ______________________________
Signature and Title	Signature and Title

FIGURE 16.1 Sample Bid Bond Form.

Performance Bonds

Performance bonds are guarantees by the surety that a contractor will perform the work as required by all of the conditions and terms of the contract. If a contractor breaches their contract, then the surety is obligated to complete the project or provide funds up to the original value of the contract plus all of the surety-approved change order amounts. Figure 16.2 provides a sample performance bond form that is used in construction contracts.

Performance bonds are limited to 100 percent of the original contract price. If any changes occur during construction that result in change orders for additional work, the amount of the contract that is increased by the change order is not covered by the bond unless the surety signifies its approval by signing the change order. Sureties will charge an additional fee over and above the original percentage of the contract price to cover the amounts listed in change orders. If performance bonds are used on construction projects, they eliminate **mechanic's liens** since the bonds guarantee payments for labor, materials, and equipment. The government does not allow mechanic's liens on public property; therefore, bonds are the only remedy for subcontractors and suppliers against nonpayment for their services on government contracts. The main purpose of bonds is to provide a financially solvent firm (the surety) to guarantee the completion of a project in case a contractor defaults during construction.

Before a surety will assume responsibility for a project, after a contractor has breached a contract, they might attempt to demonstrate that the owner in some manner delayed the contractor, failed to pay progress payments on time, or required excessive change orders. If none of these defenses are successful, then the surety has several options.

1. Finance the project by paying the subcontractor materials, labor, and equipment.
2. Wait until the owner terminates the contract because of the contractor breaching the contract and then the surety completes the contract with a different contractor and the balance of the contract is paid to the surety.
3. Require the owner to finish the project and pay the difference between the original bid amount and whatever amount is required to complete the project.
4. Complete the project with the original contractor by providing the contractor with additional capital.

PERFORMANCE BOND
(Any reference to the Contractor, the Owner, or any other party is plural where it is applicable)

Contractor (Name and Address): **Surety** (Name and Address of Principal Place of Business):

Owner (Name and Address):

Contract:

Date:
Amount:
Description (Name and Location):

Bond

Date (No earlier than the Contract Date):
Amount:
Modifications to this Bond Form:

The Surety and the Contractor are legally bound by the terms printed on the reverse side of this document to execute this Performance Bond on behalf of the authorized officer, agent, or representative.

Contractor as Principal: **Surety**:
Company:

Signature: ____________________ ____________________(Seal)
Name and Title: Surety's Name and Corporate Seal

By: ____________________
Signature and title
(Attach Power of Attorney)

(Additional space is provided below for other signatures if they are required)

Attest: ____________________
Signature and Title

Contractor as Principal: **Surety**:
Company:

Signature: ____________________ (Seal) ____________________(Seal)
Name and Title: Surety's Name and Corporate Seal

By: ____________________
Signature and title
(Attach Power of Attorney)

Attest: ____________________
Signature and Title

FIGURE 16.2 Sample Performance Bond Form.

5. Hire a contractor to complete the project on a cost plus basis.
6. Sue the owner.

Under any of these conditions, except for suing the owner, the surety will receive the total remaining contract amount for completing the work and the surety will only expend that amount while completing the project. The favored solution is for the surety to determine why the original contractor is in default and attempt to assist the contractor in completing the project since this would not delay the project.

If owners understand that they will lose control of their projects if the surety is required to assume control of the project, they will provide whatever assistance is necessary to help the contractor not default and to complete the contract as originally agreed upon in the contract.

Payment Bonds

Payment bonds, which are sometimes referred to as labor and material bonds, help protect against claims for payments by third parties such as subcontractors and vendors. Owners normally make progress payments to contractors and the contractors then pay their subcontractors and suppliers. On private construction projects if a contractor fails to pay their subcontractors and suppliers, then the subcontractors and suppliers would file a mechanic's lien against the property. Mechanic's liens are not allowed on public

PAYMENT BOND
(Any reference to the Contractor, the Owner, or any other party is plural where it is applicable)

Contractor (Name and Address): **Surety** (Name and Address of Principal Place of Business):

Owner (Name and Address):

Contract:

Date:
Amount:
Description (Name and Location):

Bond

Date (No earlier than the Contract Date):
Amount:
Modifications to this Bond Form:

The Surety and the Contractor are legally bound by the terms printed on the reverse side of this document to execute this Performance Bond on behalf of the authorized officer, agent, or representative.

Contractor as Principal: **Surety**:
Company:

Signature: ______________________ ______________________(Seal)
Name and Title: Surety's Name and Corporate Seal

By: ______________________
Signature and title
(Attach Power of Attorney)

(Additional space is provided below for other signatures if they are required)

Attest: ______________________
Signature and Title

Contractor as Principal: **Surety**:
Company:

Signature: ______________________ (Seal) ______________________(Seal)
Name and Title: Surety's Name and Corporate Seal

By: ______________________
Signature and title
(Attach Power of Attorney)

Attest: ______________________
Signature and Title

FIGURE 16.3 Sample Payment Bond Form.

property; therefore, payment bonds are required on government construction jobs and subcontractors and suppliers are paid by the surety company if they do not receive payments from the prime contractor. Subcontractors and suppliers are required to provide written notice of their claims to the contracting officer within one year of nonpayment. Figure 16.3 provides a sample of a payment bond form.

Maintenance Bonds

Maintenance bonds are used to protect owners from improper workmanship or defective materials after a structure is completed by a contractor. Maintenance bonds provide a guarantee to owners that contractors will correct defects that were caused by the contractor for a certain period of time after the completion of a project. Construction contracts provide guarantees for workmanship and materials for a stated period of time, but maintenance bonds shift liability to the surety company if defects are not corrected by contractors during the stated warranty period.

16.5 ADVANTAGES AND DISADVANTAGES OF BONDING

When a private owner is trying to determine whether to require bonds for their projects, they need to be aware of the advantages and disadvantages associated with requiring bonds on construction projects.

Advantages of Requiring Bonds

The advantages of requiring bonds include:

1. Bonds protect owners from contractors withdrawing bids after they have been evaluated and are about to be awarded to the lowest, responsible, responsive bidder. An owner would be awarded the bid bond if the lowest bidder does not agree to sign the contract. If there is a next lowest bidder, then the owner would only be awarded an amount equal to the difference between the lowest and the second lowest bidder rather than the total bid bond amount.
2. Payment bonds protect suppliers and subcontractors against not being paid by the prime contractor for their work or their products. Owners are protected because the surety will pay the subcontractors and suppliers out of the payment bond rather than the owner paying them since the owner has already paid the contractor for their services and products.
3. Performance bonds protect owners against the financial instability of contractors because sureties indemnify contractors and guarantee performance of contracts.
4. Bonds meet the requirements set for government projects.

Disadvantages of Requiring Bonds

Even with all of the advantages afforded by bonds it is important for owners to understand the negative aspects associated with requiring bonds including:

1. Requiring bonds may reduce the number of contractors that are able to bid on projects. With less competition the bid estimates could be higher than if there were more contractors bidding on a project.
2. If the bonding requirements are beyond the bonding capacity of individual firms, companies would be forced to form joint ventures in order to increase their bonding capacity. Joint ventures could be more difficult to administer, which could increase the potential for one of the joint venture partners to default on a project.
3. Owners might not completely understand the role of sureties and bonds. They could assume that the involvement of sureties guarantees that their project will be completed by the contractor awarded the project, which is not always the case after a contractor has defaulted during construction.
4. Owners have more control over completing their projects without a surety if a contractor defaults on a project.

16.6 GOVERNMENT CONSTRUCTION CONTRACTS

The contracts that are used for government construction projects are similar to the contracts used in private industry but they are administered by a government contracting officer. If there are disputes during construction, they are settled by the contracting officer. If a contractor is not satisfied with the decision of the contracting officer, then they have the option of "either appealing the decision to the Board of Contract Appeals with the opportunity for judicial review of the board decision in the Court of Claims or seeking direct judicial review of the decision of the contracting officer in the Court of Claims" (Cibinic, Nash, and Nagle 2006, 630). Additional information on government claims is provided in the section on what constitutes a construction claim against the government.

Government contracts for construction utilize **Standard Form 23-A** for the general provisions and **Standard Form 32** for supply contracts. The supplementary conditions and specifications are customized for each project.

The Government Services Administration (GSA) is the government agency that is responsible for administering government construction projects except for transportation-related projects, which are administered by the Federal Highway Administration and state Departments of Transportation.

Federal Acquisition Regulations

For guidance on contracting with federal agencies, working with government personnel on contracts, and information on the awarding and administration of federal construction contracts, refer to the **Federal Acquisition Regulations** (FAR) (United States Government, 2009). Another resource on government contracting is the book *Administration of Government Contracts* (2006) by John Cibinic, Jr., Ralph Nash, Jr., and James F. Nagle that is published by George Washington University.

The Federal Acquisition Regulations contains a procurement method whereby federal agencies have **indefinite-delivery/indefinite-quantity** (IDIQ) contracts. The following provides a description of IDIQ contracts (*Engineering News Record*, April 13, 2009, 12):

> IDIQ is a procurement method allowed under the Federal Acquisition Regulation for a range of architecture, engineering, and contracting services on federal projects. Unlike a stand-alone contract for a single project, an IDIQ approach allows the contract holder to perform several specific services for an agency within the term limits of the contract. Parameters within a contract include a specific contract period, a limit on the total value of work that can be awarded within that time frame, and the location where the tasks can be performed.
>
> Successful IDIQ contractors then compete against other contract holders for task orders. Agencies often use these contracts for easily defined, low-risk projects that need to mobilize quickly because contract holders are essentially pre-approved to execute the task.

In 2005, 52 percent of the total dollars spent on government orders processed through IDIQ, according to Federal Procurement Data System records (*Engineering News Record*, April 13, 2009). Construction projects utilizing IDIQ contracts are not as prevalent because projects are located in geographical diverse areas.

Government contracts could contain a Buy American clause that indicates that materials have to be purchased in the United States. For some materials and products it is no longer possible to buy products made in the United States since they are not manufactured in the United States. If this occurs on a construction project, the contractor is required to fill out a waiver application to receive authorization to buy the materials or products from a firm in a foreign country.

The Contract Disputes Act of 1978

The **Contract Disputes Act** (CDA) of 1978 was passed by the federal government to provide a method for reviewing contract claims that arise between contractors and the government. The Contracts Dispute Act applies to cases that involve contract disputes not tort claims for negligence acts. Section 605(a) of the Contract Disputes Act indicates that all contractual claims need to be submitted to the contracting officer who is administering the project and the contracting officer will render a decision on claims. If a contractor is not satisfied with the decision of the contracting officer, he or she could appeal the decision in the **U.S. Court of Claims** or the **U.S. Court of Federal Claims**. In addition, the U.S. Court of Appeals is also available for the federal circuit court. For breach of contract claims involving any federal agency the U.S. government has a **Board of Contract Appeals**, which is used mainly as an appellate court that hears appeals to cases that involve government contracting officers (United States Government, 1978).

The results of all of the previous decisions that have been rendered on claims against the government are available for review through the **U.S. Government Accountability Office Comptroller General**. Other decisions on cases involving claims against the government are available for review through:

1. Federal District Courts
2. Federal Circuit Court of Appeals
3. U.S. Court of Federal Claims
4. U.S. Court of Appeals

Contracting officers are responsible for addressing claims in a timely manner. In one legal case, the courts supported the requirement of contracting officers addressing claims quickly (*Pathman Construction Co. v. United States*, 817 F 2d 1573 Fed Cir. 1987):

> Under the Contracts Dispute Act, 41 U.S.C. 605 (c) (5), the failure of the contracting officer to issue a timely decision on a contract claim is deemed a decision by the contracting officer denying the claim, authorizing commencement of the appeal or suit.

The Freedom of Information Act (FOIA)

The **Freedom of Information Act** (FOIA) of 1966 provides contractors with a process for obtaining records kept by the government that pertain to projects that have prompted contractors to file claims. Information pertaining to project records is published in the federal register. This information could either be purchased or it is available for review in public reading rooms. Documents could be requested and the request will be honored if the individual requesting the record is able to adequately describe what is being requested for review.

To obtain FOIA records, a formal, written request needs to be sent to the person listed in the **Code of Federal Regulations** (CFR). The Code of Federal Regulations also provides time limits for requests; it describes the appeals process if an FOIA request is denied by the government.

Once the appropriate records have been secured, a contractor has to file a claim in order to be eligible for the dispute resolution procedures of the Contract Disputes Act. Claims that are over $100,000 have to be certified claims. Claims have to be filed within six years of the date of the occurrence of the activity that led to the claim. Contractors could also file a **Request for Equitable Adjustment** (REA) rather than a claim.

The Truth in Negotiations Act

The **Truth in Negotiations Act** of 1962 requires government contracting officers to issue a decision on claims filed by contractors within sixty days (10 U.S.C. 2306, 1962). After a decision is made, a contractor has ninety days in which to appeal a decision with the Board of Contract Appeals and one year to appeal to the Court of Federal Claims.

Alternative dispute resolution techniques could be used to settle claims or portions of claims including using:

1. A settlement judge
2. Minitrials
3. Summary trials with binding decisions
4. Other Alternative Resolution Techniques (ADR)

The four ADR techniques listed above are discussed in detail in Chapter 18.

In addition to a contractor being awarded a claim, the government will also pay attorney's fees associated with the preparation and presentation of claims.

16.7 WHAT CONSTITUTES A CONSTRUCTION CLAIM AGAINST THE GOVERNMENT

In order for a contractor to file a claim against the government, he or she has to be able to prove that there is a contractual basis that supports the right of the contractor to file a claim and that he or she is able to justify the monetary amount of the claim using schedules, cost accounting records, and supporting information from job estimates and other documentation. The two requirements of (1) proving a contractual basis for the claim and (2) providing supporting evidence for the claim amount are called **entitlement** and **quantum** (Cibinic, Nash, and Nagle 2006).

A claim against the government is a request that is filed in writing by a contractor detailing the justification for payment of a certain sum of money, or a request for clarification of contract terms, or some form of relief related to a contract.

16.8 STOP NOTICES

Stop notices could be used on both private and government construction jobs. A stop notice is a request issued by a contractor, subcontractor, or supplier to have funds withheld either by the lending institution or the government in the amount indicated in the written request to cover services rendered or materials. For private construction projects, stop notices are issued to the organization that is lending funds to the project or to the escrow company that is involved in the project. For government construction projects, stop notices are issued to the appropriate government agency at the city, country, state, or federal level.

On private construction projects prime contractors are allowed to file stop notices along with subcontractors, suppliers, vendors, and anyone else that provides services or supplies to the project who has not received payment. On public projects the same entities may file stop notices except for prime contractors. A stop notice bond is required when a stop notice is filed that covers 125 percent of the amount to protect the owner, contractor, and the lending institution from damages caused by the stop notice. A sample stop notice form is shown in Figure 16.4 and a stop notice bond is shown in Figure 16.5.

For private construction projects twenty days notice has to be provided before a stop notice is filed to allow time for the debts to be paid. Stop notices could be filed by subcontractors up to thirty days after a notice of completion is issued or sixty days if there has been no notice of completion. Prime contractors have sixty days after notice of completion in which to file a stop notice or ninety days if there is no notice of completion.

A lawsuit could be filed ten days after a stop notice has been filed and before ninety days after the end of the mechanic's lien period. Before filing a lawsuit, five days notice is provided to the defendant in the lawsuit. When a stop notice is released, the stop notice release bond is also released by the bondholder.

Public agencies could claim that a stop notice is not valid for one of four reasons (Cibinic et al. 2006):

1. The stop notice claim that was filed is not allowed under the stop notice law.
2. The person filing the stop notice is not listed as a potential claimant.
3. The amount that is being submitted in the stop notice is considered to be an excessive amount.
4. There is no basis in the law for the claim that is being filed by the plaintiff.

When a stop notice is being filed, it needs to include the following (Cibinic et al. 2006):

1. A description of the job site.
2. The amount of money due to the person filing the stop notice.
3. A statement requesting that the construction lender withhold enough of the construction funds to pay the plaintiff if his or her claim is valid.
4. The name of the person filing the stop notice.
5. A description of the work performed, or the materials supplied, for the project.
6. The name of the person who requested the work or materials for the project.
7. The total cost of the work or materials that were supposed to be supplied to the project.
8. The cost of the work or materials that have already been supplied to the project.
9. Verification of the amount being claimed in the stop notice.

16.9 MECHANIC'S LIENS

Mechanic's liens are used on private construction projects to protect against nonpayment for work or materials. Mechanic's liens are filed at the office of the county recorder in the county where the work was being performed and represent interest (an encumbrance) in the property in the amount the filer was not paid for services rendered during construction. Mechanic's liens may be filed by anyone who is working on a construction project who was hired to perform work, or to provide materials, by the owner or employed by someone who has been hired by the owner. Ten days notice needs to be provided to the owner before a mechanic's lien is filed to allow the owner time to pay the unpaid invoice.

Mechanic's liens are still valid even if the owner of the property goes bankrupt. Mechanic's liens are considered to be secured debts and have priority over unsecured debts of creditors. Liens could be filed up to ninety days after the job has been completed by the prime contractor. If a lawsuit, or a **notice of credit**, is not filed within ninety days after a lien was filed, then an owner may file for a **release decree** that removes the mechanic's lien.

Mechanic's liens include:

1. A description of the work performed or materials supplied to the project.

STOP NOTICE

LEGAL NOTICE TO WITHHOLD CONSTRUCTION FUNDS
(Public or Private Work)

To: ______________________ Project: ______________
(Name of Owner, Public body, or construction fund holder) (Name)

______________________ ______________
(Address. If directed to a bank or savings and loan, use the address of the branch holding the funds) (Address)

______________________ ______________
(City, state, and zipcode) (City, state, and zipcode)

TAKE NOTICE THAT ______________________
(Name of the person or firm claiming the stop notice)

whose address is ______________________
(Address of person of firm claiming stop notice)

has proclaimed labor, and furnished materials for a work of improvement as described as follows: ______________________
(Name and location of the project where the work or materials were furnished)

The labor and materials furnished by claimant are of the following general kind:
____
(General description of work and materials furnished)

The labor and materials were furnished to or for the following party : ______________
____
(Name of party who ordered the work or materials)

The value of the whole amount of labor and materials agreed to be furnished is: ________
____
(Total of everything claimant agreed or contracted to furnish)

The value of the labor and materials furnished to date is $ ______________
(Total value of everything actually furnished by claimant)

Claimant has been paid the sum of $ ______________ and there is due, owing, and
(Total amount which has been paid to claimant)

unpaid the sum of $ ______________ together with interest at the rate of
(Balance due to claimant of the project)

______________ % per annum from ______________, 20_.
(Interest as specified in contract, if none use legal rate) (Date when unpaid balance was due)

You are required to set aside sufficient funds to satisfy this claim with interest. Your are also notified that claimant claims as equitable lien against any construction funds for this project which are in your hands.

Firm Name: ______________________
(Name of stop notice claimant)

By: ______________________
(Owner or agent of stop notice claimant must sign here and verify below)

VERIFICATION

I, the undersigned say I am the ______________________ ,
(President of, Manager of, A Partner of, Owner of, etc.)

the claimant of the foregoing Stop Notice. I have read said Stop Notice and know the contents therefore, the same is true of my own knowledge.

I declare under penalty of perjury that the foregoing is true and correct.
Executed on ______________, 20_at ______________
(Date this document was signed) (Name of city and state where signed)

____
(Personal signature of the individual who is swearing that the contents of the stop notice are true)

FIGURE 16.4 Sample Stop Notice.

Project: ____________________
(Name)
____
(Address)
____
(City, state, and zip code)

STOP NOTICE BOND

Legal Bond for Stop Notice to Withhold Construction Funds
For Use with Stop Notices on Private Construction Projects
Not Required for Stop Notices on Public Works Projects

The undersigned principal and the undersigned sureties are held firmly bound unto
____ and
(Name and Owner of the Project)
____ and
(Name of Construction Fund Holder, if any)
____ and
(Name of Primer Contractor)
as obliges, in the sum of $ ______________________________
(One and one-half times the principal amount of the claim)
to be paid to the obliges, for which payment we bind ourselves.

THE CONDITION OF THE OBLIGATION IS SUCH THAT, WHEREAS, the principal has performed the work described on the stop notice accompanying this bond, and has filed a Stop Notice Claim on the project and

WHEREAS, in compliance with the provisions of the State Code of Civil Procedure, the principal is required to file a bond in the sum of one and one-quarter times the amount of the Stop Notice Claim, including interest.

NOW THEREFORE, if the defendant recovers judgment in an action brought on the Stop Notice, if the principal named below shall pay or cause to be paid all costs that may be awarded against the owner, contractor, construction fund holder, or other obligee named herein or any of them, and all damages that such owner, contractor, construction fund holder, or other obligee may sustain by reason of the equitable garnishment effected the Stop Notice or by reason of the lien (not exceeding the penal sum of this bond) then this obligation shall be void otherwise it will remain in full force and effect.

Date: ____________________

Surety: ____________________
(Signature of an individual who is over 21 years old)

Principal: ____________________
(Name of Firm which is the Stop Notice Claimant – exactly the same as it appears on the Stop Notice itself)

Surety: ____________________
(Signature of an individual who is over 21 years old)

By: ____________________
(Signature of an officer, partner, or the owner of the stop notice claimant – exactly the same as it appears on the Stop Notice itself)

FIGURE 16.5 Sample Stop Notice Bond.

2. A description of the property where the work was being performed for the job.
3. The name of the person who hired the person filing the mechanic's lien.
4. The name of the owner of the project.
5. The amount due to the person filing the mechanic's lien.

After a notice of completion is issued, the contractors have sixty days in which to file a mechanic's lien and subcontractors and suppliers have thirty days. The release of lien forms used for mechanic's liens are:

1. Conditional release for progress payment
2. Conditional release for final payment
3. Unconditional release for progress payment
4. Unconditional release for final payment

Forms that could be used for releasing mechanic's liens are included in standard contract documents.

Chapter Summary

This chapter explained some of the differences between private and government construction contracts including bonding requirements; bonding procedures; how sureties approve and provide bonds to contractors; the advantages and disadvantages of required bonds; and the different types of bonds such as bid, performance, payment, and maintenance.

The chapter also introduced Federal Acquisition Regulations, the Contract Dispute Act, the Freedom of Information Act, and the Truth in Negotiations Act and discussed how these affect government construction contracts. The later part of the chapter explained stop notices, which are used on both public and private construction projects, and mechanic's liens, which could only be used on private construction projects. Other resources are available on government contracting, two of which were mentioned in this chapter that describe government contracting procedures in detail.

Key Terms

bid bonds
Board of Contract Appeals
bonds
Code of Federal Regulations
Contract Disputes Act
entitlement
Federal Acquisition Regulations
Freedom of Information Act
government contracting officer
government contracts
indefinite-delivery/indefinite-quantity
left on the table
maintenance bonds
mechanic's liens
Miller Act
notice of credit
obligees
payment bonds
performance bonds
quantum
release decree
Request for Equitable Adjustment
Standard Form 23-A
Standard Form 32
stop notices
surety
surety bonds
surety companies
Truth in Negotiations Act
U.S. Court of Claims
U.S. Court of Federal Claim
U.S. Government Accountability Office Comptroller General
U.S. Treasury Department

Discussion Questions

16.1 Explain why an owner would require bid bonds on a construction project.

16.2 Explain how performance bonds protect owners.

16.3 What types of defenses could a surety claim in order to not pay an owner the bid bond when their client refuses to be awarded the project?

16.4 Why do government contracts require bonds?

16.5 What are bonds?

16.6 Where is additional information available that provides guidance for working on government contracts and with federal agencies?

16.7 How does the Freedom of Information Act help contractors when they are filing a claim against the government?

16.8 Under the Truth in Negotiations Act how long do contracting officers have to issue a decision on a claim?

16.9 What is a claim against the government?

16.10 How do mechanic's liens help contractors, subcontractors, and suppliers against nonpayment for services rendered to an owner?

16.11 Who is able to file stop notices on private or government projects? What do stop notices help to accomplish on construction projects?

16.12 Who has the legal right to approve a surety for use on a government construction project?

16.13 What does a surety check out before providing a bond to a contractor?

16.14 Why do contractors have to provide sureties with a set of construction documents before a surety will agree to issue the contractor bonds?

16.15 How are payment bonds used on government construction projects?

16.16 What are the disadvantages of requiring bonds on construction projects?

16.17 What are the advantages of requiring bonds on construction projects?

16.18 How is an owner compensated when there is a bid bond if the lowest responsible bidder refuses to be awarded a contract and the project is awarded to the second lowest bidder?

16.19 Why is it disadvantageous for an owner if a surety takes over a construction project after the contractor defaults on the project?

16.20 What is the main difference between government and private contracts?

References

Miller Act, Public Law 87-653. 1935. 40 U.S.C. Section 3131 to 3134, United States Congress. Washington, DC: Government Printing Office.

Cibinic, J., Jr., Nash, R. Jr., and J. F. Nagle. 2006. *Administration of Government Contracts.* Washington, DC: George Washington University.

Engineering News Record. April 13, 2009. Flexible approach is key to moving work out fast. Vol. 262, No. (12): 12–13. New York: McGraw Hill Publishing.

Pathman Construction Company v. United States (Fed Cir. 1987) 817 F 2d 1573.

S.C. Anderson Inc., v. Bank of America N.T. and S.A. (1994) 24 Cap. App. 4th 529, 30 Cal. Rptr. 2d 286.

Smith, Currie, and Hancock. 2005. *Common Sense Construction Law: A Practical Guide for the Construction Professional.* New York: John Wiley and Sons.

United States Government. 2009. *Federal Acquisition Register.* Washington, DC: Government Printing Office.

United States Government. 1978. Contract Disputes Act. Washington, DC: Government Printing Office.

United States Government. 1966. Freedom of Information Act. Washington, DC: Government Printing Office.

United States Government. 1962. Truth in Negotiations Act. Washington, DC: Government Printing Office.

CHANGE ORDERS AND CLAIMS

17.1 INTRODUCTION

Construction delay claims occur on every type of construction project. Construction professionals are mainly involved with planning, scheduling, monitoring, controlling, and administering construction projects, but they are also required to devote time to preparing for post construction resolution of claims. As-planned and as-built construction plans and schedules are compared in order to provide documentation for the **claims resolution process**.

During the claims resolution process project managers meet with company executives, company personnel, consultants, attorneys, and others. Project managers provide depositions, attend pre-trial conferences, and provide testimony at legal proceedings. All of these activities detract from the ability of project managers to successfully administer projects, commit time to new projects, and consume profits.

Since the frequency of delay claims is increasing in the United States construction industry architects, engineers, and specification writers are now required to prepare documents that are able to withstand the scrutiny of contract administrators, attorneys, arbitrators, and judges. This requires technical personnel to be versed in subjects that are not normally their fields of expertise. Designers and specification writers have to be as proficient in writing no damages for delay clauses as they are at designing structures.

Knowing about the types of issues that could become potential sources of litigation allows engineering and construction professionals to be able to implement measures to help prevent claims. Three suggestions that are useful to engineering and construction professionals in the area of claims management are:

- Be able to write clear and concise specifications and initiate preconstruction activities that address subjects that could potentially result in delay claims.
- Improve construction monitoring and administration techniques and work toward early resolution of potential disputes.
- Have streamlined, cost-effective, expeditious mechanisms for resolving delay claims including alternatives to costly litigation.

Because changes and claims will always be a part of the construction process, this chapter provides information on changes, change orders, addenda, construction claims, delay claims, construction, utilizing Critical Path Method schedules for claims analysis, and minimizing delay claims.

17.2 DEFINITION OF CHANGES

Construction documents are written by owners, or by the engineers who work for owners, in an attempt to communicate to contractors what the owner would like to have built. Unfortunately, it is difficult to completely capture all of the design requirements while construction contracts are being written; therefore, construction contracts allow for **changes** to be made to the design requirements that are included in contracts.

In addition to owner-requested changes, other circumstances might require that contracts be revised during construction such as:

- Inadequate design
- Material unavailability
- Budget changes
- Natural disasters

There are five types of potential alterations to construction contracts (Bockrath, 2000):

1. **Changes**—Owners order changes in the work from the original scope of work.

2. **Extras**—Owners request work that is in addition to the original construction contract and an extra amount is negotiated to compensate the contractor for the extra work.
3. **Deletions**—Work is removed from the original contract with an appropriate deduction in the price.
4. **Modifications**—The owner and the contractor both agree to change some of the work requirements of the contract with appropriate compensation paid to the contractor if it increases the scope of work.
5. **Waiver**—The owner in some way indicates that a certain portion of the original scope of work does not have to be completed but the contractor is compensated for the entire contract price.

The processes required to request changes or to initiate change orders are discussed in the next section.

17.3 CHANGE ORDERS

The purpose of **change orders** is to document in writing the terms agreed to by both the owner and the contractor for altering the scope or work. Change orders provide written instructions for contractors and they are signed by owners, the A/E, and the contractor. For projects where bonds are required, change orders have to also be signed by the surety. The role of the surety in change orders is discussed in Chapter 16.

Change orders authorize additions, deletions, or revisions to the scope of work. They indicate whether there will be a price or time adjustment or both for the proposed work. Most standard contracts include forms that could be used to issue change orders. The Engineers Joint Contract Documents Committee (EJCDC) *Document 1910-8* provides a **Work Directive Change Order Form**, as shown in Figure 17.1. Figure 17.2 shows a typical change order form.

Work Change Directive Number

Date of Issuance: ______________________ Effective Date: ______________

Project: Owner: Owner's Contract No.:

Contract: Date of Contract:

Contractor: Engineer's Project No.:

You are directed to proceed with the following change(s):

Item Number: Description:

Attachments (list documents supporting the change)

Purpose of Work Change Directive:

☐ Authorization for Work described herein is to proceed on the basis of the Cost of the Work due to:

☐ Nonagreement on pricing of proposed change

☐ Necessity to expedite Work described herein prior to agreeing to changes on Contract Price and Contract Time

Estimated Change in Contract Price and Contract Times:

Contract Price $ ________________ (increase/decrease)

Contract Time ________ (increase/decrease)
days

If the change involves an increase, the estimated amounts are not to be exceeded without further authorization.

Recommended Approval of the Engineering: Date:

Authorized for the Owner by: Date:

Accepted for the Contractor by: Date:

Approved by the Funding Agency (if applicable): Date:

FIGURE 17.1 Sample Work Change Directive Form.

CHANCE ORDER NUMBER: ______________

Date of Issuance: ______________ Effective Date: ______________

Project: Owner: Owner's Contract Number:

Contract: Date of Contract:

Contractor: Engineer's Project Number:

The Contract Documents are modified as follows upon execution of this Change Order:
Description:

Attachments: [List documents supporting the change(s)]

Change in Contract Price:	**Change in Contract Terms:** Working Days % Calendar Days
Original Price: $ ______________	Substantial completion (days or date): ______________ Ready for final payment (days or date): ______________
(Increase) (Decrease) from previously approved Change Orders No: ________ to No. ______________ $ ______________	(Increase) (Decrease) from previously approved Change Orders No.: ________ to No.: ______________ Substantial completion (days): ______________ Ready for final payment (days): ______________
Contract Price prior to Change Order: $ ______________	Contract Time prior to this Change Order: ______________ Substantial completion (days or date): ______________ Ready for final payment (days or date): ______________
(Increase) (Decrease) of this Change Order: $ ______________	(Increase) (Decrease) of this Change Order: Substantial completion (days or date): ______________ Ready for final payment (days or date): ______________
Contract Price incorporating this Change Order: $ ______________	Contract Time with all approved Change Orders: Substantial completion (days or date): ______________ Ready for final payment (days or date): ______________

RECOMMENDED:	**APPROVED:**	**ACCEPTED:**
By: ______________	By: ______________	By: ______________
Engineer (Authorized)	Owner (Authorized Signature)	Contractor (Authorized Signature)
Date: ______________	Date: ______________	Date: ______________

FIGURE 17.2 Sample Change Order Form.

The EJCDC *Document 1910-8* includes a **Field Order Change Form** that authorizes contractors to perform minor changes to the contract if changes will not alter the contract price or time. These do not have to be signed by the owner.

The American Institute of Architects (AIA) *Document A201*, **Construction Change Authorization Form**, is also used by some engineers to authorize changes and to instruct contractors to proceed with the changed work. It also indicates that adjustments will be made later to the contract price or time through an official change order.

Change orders should repeat whatever section in the original contract that pertains to the item being changed or cross-reference the original contract section. If work is being added by a change order, then a new section needs to be written and added to the specification. The original scope of work for the changed items should be reissued with it, marked to show the changed portion of the work. An asterisk is placed before and after the changed section in the margins. The change order number should be included and a means for highlighting changes should be used such as using different fonts, shading, different colors, or deleted words marked with a line or cross-hatching through them. Changes only need to be written once and then they are referenced in other locations in the contract.

Change orders are also used to correct construction documents if there are errors, omissions, or discrepancies. Some of the types of items that require change orders are the following:

- Additions and deletions of work
- Unforeseen site conditions
- Changes in building code supervision
- Changes in market conditions
- New products or product unavailability

If stipulated in the construction contract, owners, or their representative, are the ones who have the authority to execute change orders, but contractors are able to request changes using one of the following forms:

1. Change Order Proposal Request Form
2. Construction Change Authorization (AIA) Form
3. Work Directive Change (EJCDC) Form

The owner prepares an estimate for the proposed changed work and compares it to the estimate proposed by the contractor. Once both parties negotiate a price that is acceptable, a change order is issued before the work is performed by the contractor. If there is an increase or decrease in the cost of the item being changed, it could be determined using one of three methods:

1. Mutual agreement of a lump sum.
2. Using unit prices from the contract multiplied times the number of additional units or the number of units being deleted from the contract.
3. Actual cost plus a fixed percentage for overhead and profit.

Several areas could lead to delays in negotiating change orders, including:

- The overhead rate of the contractor, since some contractors might not want to disclose their overhead rate or the owner might assert that the overhead rate is excessive.
- Determining how to credit the owner for deductions in the work if it is only for a portion of a unit price item.
- The owner may not agree with the unit prices proposed by the contractor if they are not already included in the contract.
- There could be difficulties associated with verifying the costs of the contractor.
- Fluctuating equipment rental rates.
- Difficulties in determining the amount to be allocated for time extensions.

Change orders need to include a minimum amount of information in order to protect both the owner and the contractor. At a minimum change orders should include the following:

1. Change order number and date
2. Project name and job number
3. Name and address of the architect and/or engineer
4. Name and address of the owner
5. Name and address of the contractor
6. Original contract price
7. Statement that the change order modifies the original contract
8. Description of revisions with the cost of each item
9. Tabulation of the new contract amount
10. Time extension or reduction
11. Other contracts affected by the change order
12. Signature of the architect or engineer and the date of submittal
13. Signature of the owner and the date of acceptance
14. Signature of the contractor and the date of acceptance
15. List of the distribution of copies

Constructive Changes

In some situations, an engineer may determine that work being requested is included in the original scope of work but the contractor might not agree with this assertion and indicate that it is additional work. If the engineer refuses to issue a change order, the contractor should notify the engineer that he or she intends to file a claim to be compensated for what they interpret to be additional work. In this situation, the contractor is claiming that a **constructive change** occurred and that a change order should have been issued for the work. The contractor will still perform the work but a claim will be filed for additional compensation that will be reviewed during negotiations, arbitration, or litigation depending on which dispute resolution method is used for a project.

Cardinal Changes

In order to protect contractors from unreasonable demands for changes, contracts should include a limit on the value of the changes to the contract such as 10 or 20 percent of the total value of the contract. If no limit is included in the contract, court systems will evaluate the changes to determine whether they were within a reasonable limit or whether they exceeded the limits of normal change clauses.

A contractor has the legal right to walk off of a job, claim breach of contract, and recover a reasonable amount for the work already completed if an owner requests excessive changes, which are referred to as **cardinal changes**. Normally, before this happens the owner and the contractor would agree to void the original contract and renegotiate a new contract that incorporates the desired changes and that includes additional compensation for all of the requested changes. Another option is for the contractor to complete the job and then file a claim for the additional services, direct costs, overhead, and profit.

In some instances, a cardinal change occurs because of a series of small changes that add up to an amount that exceeds the maximum allowable value of changes. In order to determine whether this is occurring, it is important for contractors to maintain a change order log where change orders are entered by date and value and also includes a running total of the increases caused by change orders to the total contract price and indicates the percentage of contract price represented by the changes.

17.4 ADDENDA

Contract documents may have to be changed after they are issued but before the bid estimates are submitted by contractors. When a change is made to the contract documents prior to the bid opening by the engineer, the engineer is required to issue an **addendum**. Addenda are used to correct errors or omissions in the construction documents, or to make changes to the scope of work, such as:

- Change the date, time, or location of the bid opening
- Change the quantity of work
- Change the sequence of work
- Add to, make deletions to, or revise the contract
- Include additional products that could be used during construction

Addenda are numbered consecutively to allow contractors to be able to tell whether they have received all of the addenda. Whenever a new addendum is issued, an extension of time could be provided to allow contractors to incorporate the addenda into their bid proposals. If a change is minor, the extension might only be for a few days.

The basic components that are included in addenda are the following:

1. Addendum number and date
2. Project identification number
3. Name and address of the A/E
4. A list of who received a copy of the addenda
5. Opening remarks
6. Changes to prior addenda
7. Changes to bidding requirements
8. Changes to the agreement
9. Changes to the conditions of the contract
10. Changes to the specifications
11. Changes to the drawings

Once addenda have been issued, they become a legally binding component of the construction contract documents.

17.5 CONSTRUCTION CLAIMS

Whenever an owner changes contract requirements for the scope of work, contractors are able to file **claims** if they do not agree with performing the work for the amount proposed by the owner. Contractors could also file claims for other circumstances such as:

- An owner somehow causes a delay in the contractor starting the project.
- An owner mistakenly stops a contractor from performing their part of the construction project.
- An owner performs some type of an act that allows the contractor to justify stopping the work.
- The owner does not pay a contractor for work that the contractor has completed on a job.
- The owner in some manner causes the cost of construction to increase.

If owners and contractors are able to mutually agree on the amount of compensation for a change, then this process is formalized in a change order. If no agreement is reached on an adequate amount for a changed condition or for delays, contractors could file a claim against the owner. In some circumstances, contractors would file for damages in their claims. Under normal circumstances, claims are filed to cover additional costs incurred by the contractor that they were not compensated for by the owner or owners will file a claim for damages to compensate for work not completed by the contractor.

If a project were only partially completed due to the owner breaching the contract by not making progress payments, excessive changes, or not fulfilling other contractual obligations, the contractor would file a claim for restitution to be paid for the work performed to date including being compensated for materials, labor, overhead, and profit for the completed work. If a contractor breaches the contract, or its performance is defective, the owner could file a claim for restitution. If justified, the courts would award the owner an amount to compensate for the diminished value of his or her property.

In order to establish that a loss has occurred, the court system requires proof for any of the following:

- The costs associated with the work of the contractor.
- The cost for the owner or the contractor to complete the work.
- If the owner delayed the contractor, the additional costs incurred by the contractor.
- The amount of profit associated with the work performed and for the work not performed yet and any other consequential damages (see next paragraph).
- The profit lost by the owner if the project is delayed or the contractor provides defective work.

In addition to filing a claim for damages, an owner or a contractor could file a claim for **consequential damages**. These are damages that are beyond direct losses. Owners might claim diminished or lost profits because of delays. Contractors could claim they lost the opportunity for additional business or their bonding capacity was diminished or canceled by the surety.

17.6 CONSTRUCTION DOCUMENTATION TO SUPPORT CLAIMS

Information plays a key role in construction project management. For a construction project to be well managed, data from past projects, as well as data from current projects, has to be readily available. Construction projects use a variety of different types of documentation to determine the causes of claims and project delays. Being familiar with the different formats that could be used for scheduling and claims analysis provides insight into the most efficient methods for determining the causes of delays and for recording historical information.

Data and the accumulation and documentation of data are essential for project planning, control, and decision making. In each of these areas effective management of information is an integral part of any successful project management system where the primary objective is completing a project on time. In all three of these areas the documents that help to indicate the causes of delays are crucial.

Proving that a delay has occurred on construction projects is difficult; therefore, Table 17.1 includes a list of the different types of **construction documentation** that could be used to determine whether a delay has occurred during construction. Not all projects use the forty-three types of documents listed in Table 17.1 but the list could be used as a reference during the implementation of document retrieval systems or to help locate information that documents the causes of delays.

Table 17.1 Indicators of Project Delays

0) (O)	Other
1) (T)	Time cards
2) (IR)	Inventory reports
3) (S)	Schedule
4) (Q)	Quality control reports
5) (COR)	Correspondence
6) (COL)	Change order log
7) (PM)	Productivity reports
8) (DCR)	Daily construction reports
9) (DL)	Drawing log
10) (SDD)	Superintendent's log
11) (JL)	Job log
12) (CE)	Cost estimates
13) (PC)	Progress curves
14) (VA)	Variance analysis
15) (PR)	Procurement reports
16) (MR)	Manpower reports
17) (FLL)	Field labor reports
18) (SL)	Submittal log
19) (RFI)	Request for Info. Log
20) (E)	Experience
21) (PP)	Permit processes
22) (TS)	Time sheets
23) (DMR)	Design milestone reports
24) (ER)	Expediting reports
25) (CRMR)	Milestone review
26) (PML)	Project Manager's log
27) (FO)	Field observations
28) (DCN)	Design change notices
29) (BPR)	Bid package reviews
30) (DR)	Design reviews
31) (EXCR)	Exception reports
32) (DCM)	Daily contractor meetings
33) (VO)	Visual observations
34) (DS)	Delayed start
35) (DP)	Delayed permits
36) (MO)	Man. observations
37) (LCO)	Labor cost reports
38) (MCR)	Material cost reports
39) (SN)	Strike notice
40) (EUR)	Equipment usage reports
41) (TR)	Test reports
42) (NA)	Network analysis
43) (WL)	Weather log

Source: El-Nagar, H., and J. K., Yates. 1997. Construction documentation used as indicators of delays. *Journal of the American Association of Cost Engineers International*, 39(8): 33.

One example of how the information in Table 17.1 could be used is to relate the indicators of project delays to the technical causes of delays. One of the most prevalent technical causes of delays is design modifications. The primary indicators of delays that are used to determine that design modifications have delayed a project are (El-Nagar and Yates 1997):

- The Schedule (S)
- The Correspondence Log (COR)
- The Change Order Log (COL)
- The Request for Information Log (RFI)
- Daily Construction Reports (DCR)
- Progress Curves (PC)
- The Drawing Log (DL)

Another example is delays due to incomplete drawings, which uses the following indicators of project delays to determine whether incomplete drawings have caused a delay (El-Nagar and Yates 1997):

- The Request for Information Log (RFI)
- The Schedule (S)
- The Correspondence Log (COR)
- The Change Order Log (COL)
- Progress Curves (PC)
- Daily Construction Reports (DCR)
- The Drawing Log (DL)

For these two examples the top seven indicators of delay for both technical causes are the same, however, the order in which the indicators are used is different.

17.7 CONSTRUCTION DELAY CLAIMS

Owners require that their projects be completed by a specific date. A contract clause is normally included in construction contracts that indicates that **time is of the essence**. When this clause in not included in a contract, court systems will evaluate whether a project was completed within a reasonable amount of time. When a contractor delays a project, the owner might file a claim against the contractor for the delay in order to attempt to recover damages.

Four types of delays occur on construction projects: (1) noncompensable excusable, (2) compensable excusable, (3) nonexcusable, and (4) concurrent. These are discussed in the following sections (Bramble and Callahan 1987; Rubin, Guy, Maevis, and Fairweather 1983).

Noncompensable Excusable Delays

Noncompensable excusable delays are delays that are not caused by either the owner or the contractor. They are caused by Acts of God, or other unforeseeable causes, that are beyond the control of both parties. Delays that are caused by

Acts of God entitle contractors to receive time extensions, but not damages for the delay. Contracts usually contain a clause, sometimes referred to as a Force Majeure clause, that lists the causes of delays that are noncompensable excusable delays.

Litigation relative to the subject of establishing noncompensable excusable delays for unusually severe weather conditions is numerous, including cases that addressed (Yates and Epstein 2006):

- Formulating distinctions between the levels of frequency or severity that make conditions unusually severe as opposed to merely severe, such as, well above average rainfall deemed unusually severe.
- How to accurately and effectively document and evaluate actual weather conditions provided in contractor records, owner records, third-party records, National Oceanic and Atmospheric Administration records, National Weather Service records, and other records.
- The place where weather records are recorded, which should be as close as possible to the site. The impact of the weather on the actual work and establishing whether the weather delayed completion of the work.
- Lingering effects of weather, such as drying time for soils.
- Whether or not a contractor took reasonable precautions. If a contractor failed to provide protection for the site as specified in the contract, he or she may be denied a weather-based noncompensable excusable delays.

Compensable Excusable Delays

Compensable excusable delays are caused by owners and result in both a time extension and compensation being provided to contractors. Compensable excusable delays result from circumstances such as (Yates and Epstein 2006):

- Failure of the owner to have the work site available to the contractor in a timely manner.
- Owner initiated changes in the work.
- Owner delays in issuing a notice to proceed.
- Architect/engineer supplied designs that are defective.
- Owner not properly coordinating the work of other contractors.
- Owner not providing owner-furnished equipment in a timely manner.
- Owner providing misleading information.
- Owner interfering with the performance of the contractor.
- The owner, or the architect/engineer, delaying the approval of contractor submitted shop drawings.
- The owner, or the architect/engineer, using the shop drawing process as a means by which to change the contract requirements.
- The contractor encountering differing site conditions.

Nonexcusable Delays

Nonexcusable delays are attributable to the actions, or inactions, of contractors. When contractors cause delays to the completion of projects, such delays prevent contractors from obtaining time extensions for delays; therefore, the contractor might have to pay damages to owners. Some of the more common contractor-caused delays are (Yates and Epstein 2006):

- Failing to mobilize work crews and start the work in a timely manner.
- Failure to submit shop drawings and related materials to the owner for approval in a timely manner.
- Lack of adequate and sufficient construction equipment.
- Poor workmanship.
- Failure to perform the work properly.
- Improperly allocating labor, material, and other resources on the project.
- Lack of coordination of tradesmen and subcontractors.
- Failure to perform various portions of the work in a timely manner.

Concurrent Delays

Delays do not always fall into one of the three previous categories because there are often multiple factors that cause or contribute to delays. When more than one event delays a project, it is said to have been delayed by **concurrent delays** (James 1990). It is difficult to determine which delays are concurrent delays; if the owner and the contractor are not able to determine the effects of concurrent delays, legal proceedings might be required to resolve the issue.

Exculpatory Clause—No Damages for Delay

The **no damages for delay** clause has been utilized for a considerable amount of time. Yet the state of the law remains somewhat unsettled at this time and tends to vary from jurisdiction to jurisdiction. In an effort to relieve owners from liability for delay damages on construction projects, many contracts now contain a no damages for delay clause. This is an **exculpatory clause** (clears someone from an alleged fault) that purports to excuse an owner from contractual liability for damages due to delays caused by the owner. One example of such a clause is Article 13 of the City of New York's Public Works contract (Postner and Rubin 1992):

> The contractor agrees to make no claim for damages for delay in the performance of this contract occasioned by any act or omission to act of the City or any of its representatives, and agrees that any such claim shall be fully compensated for by an extension of time to complete performance of the work as provided therein.

A landmark case in New York that validates the application and enforceability of this clause is *Kalisch-Jarcho, Inc. v. City of New York*, 1983. In this case, the New York Court of Appeals held that (1) the provision of the contract under which the contractor agreed to make no claim for damages for delay was enforceable; (2) in order to recover, a contractor would have to show that the City's conduct amounted to gross negligence; and (3) the jury would have to find more than active interference on the part of the City.

While the court affirmed the enforceability of the clause, it did provide for broad exceptions. These exceptions include owner's bad faith; willful, malicious, or grossly negligent conduct by the owner; uncontemplated delays; delays so unreasonable they amount to abandonment by the owner; delays resulting from owner's breach of a fundamental obligation of the contract; and delays that were not reasonably foreseeable (*Corinno Civetta Construction Corp. v. City of New York*, 1986).

As a result of the exceptions created by the *Corinno* case, and similar exceptions in other jurisdictions, there has been an increase in litigation and controversy surrounding the enforcement of clauses of this nature. Subsequent cases have affirmed the principles set forth in *Corinno*, including *Port Chester Electrical Construction Corp. v. HBE Corporation and the Fireman's Fund Insurance Company*, 1992; *Earthbank Co., Inc. v. The City of New York, Department of Parks and Recreation*, 1991; *Novak and Co., Inc. v. Dormitory Authority of the State of New York*, 1991.

In order for an owner to strengthen a no damages for delay clause and reduce the likelihood that exceptions would be found, the clause should specifically list various types of delays for which damages will not be recoverable rather than stating that damages for delays will not be recoverable for any reason (Postner and Rubin 1992).

One option for countering a no damages for delay clause is to convince the court that a shocking or unconscionable result will happen if there is enforcement and to convince the court that the delays could not be contemplated and were not foreseeable and claim that active interference, hindrance, fraud, and unreasonable delays occurred on the project. The courts have to be convinced that hardship will result from the enforcement of the clause (Thomas and Wilshusen 1989):

Advocates of the clause indicate that (Postner and Rubin 1992):

- The clause allows fiscal stability by ensuring the owner knows the total price at the outset. In exchange for this stability, owners are willing to accept the possibility of higher bids.
- The clause could eliminate the need for lengthy delay claim litigation as damages are precluded by contract.
- The clause protects the bidding process as costly extras are averted. The low bid at the outset will remain the low bid, as there will be no delay claims.
- If contractors know there will be no delay claims on the job, they will not artificially delay progress to generate such claims.

Negative aspects of the clause include (Postner and Rubin 1992):

- This clause forces contractors to include high cost of delay contingencies in their bids. Contractors have to allow for delay damages that may not occur. This will artificially inflate bids.
- Some contractors will refrain from bidding since contractors prefer to bid on certainties. There is no way to estimate the contingency amount for delay damages.
- Strict enforcement of the clause would cause contractor losses and could cause contractor defaults.
- Litigation might increase as attempts are made by contractors to find exceptions to the harsh application of the clause.

17.8 UTILIZING CRITICAL PATH METHOD SCHEDULES FOR DELAY ANALYSIS

Several issues are often difficult to resolve when using **Critical Path Method Schedules** to analyze construction delay claims. They are discussed in the following sections.

Evaluation of Partial Suspensions or Partial Delays

Sometimes an owner will merely hinder or impede the performance of a contractor rather than completely delaying or suspending his or her efforts. However, when several separate hindrances delay overall project completion, Critical Path Method Schedule analysis could be utilized to demonstrate that these have occurred on a project. In order to establish that an entire project was delayed by **partial delays**, it needs to be shown that (Wickwire, Hurlbut, and Lerman 1989):

- Evidence that the original duration and person loading estimates were accurate.
- Evidence that the partial delay impacted the critical path.
- Evidence that the resources that were present on the project were capable of completing the work in accordance with the original schedule.

Originally the courts allowed contractors to have exclusive use of **float** in schedules. Thus, even where a contractor-caused delay occurred after an owner-caused delay used up all available float, some courts would still award delay damages to the contractor. In one legal case, the Contract Board of Appeals noted that float provides a contractor with latitude in scheduling noncritical activities to effect trade-offs of resources, to decrease costs, or to shorten the length of the project (Bennett, 1972). Float time was primarily considered to be a resource used in the scheduling of the work and belonged to the contractor.

The policy has changed on who actually owns or has use of float. It is no longer considered to be for the exclusive use of contractors (Zack 1992). In fact, many contracts have provisions stating that float is not for the exclusive use of any party to the contract. Contract language sometimes goes further and states that float is a project resource and could be used by whichever party first consumes it and that no time extensions will be granted until all of the float has been used.

As float is used up the critical path could change and new activities could become part of the critical path. Once activities that were originally on the critical path are delayed, a party is in a position to pursue a delay claim.

Time Extensions for Changes After the Contract Completion Date

Historically change orders that are issued after the contract completion date automatically entitle contractors to an extension of time to complete projects. These time extensions are arbitrarily granted without regard to evaluating the causes of the delays to the original work. This issue was eventually addressed by a Contract Appeal Board in the decision of Santa Fe that concerned a project that was in a state of negative float (VABCA, 1984). In this instance, the project was noted to be behind schedule, work was still in progress beyond the completion date, and no time extension was being issued for the delay by the owner. At that time, an additional change order was issued for which the contractor sought a time extension under the prior theory that any change order issued after the completion date should be the basis for a time extension. The Contract Appeal Board rejected the request of the contractor, noting that delays that do not affect the extended and predicted contract completion dates shown by the critical path in the network should not be the basis for a change to the contract completion date. Thus, if a change order issued after the original completion date does not delay the extended or predicted completion date, it does not serve as a basis for a time extension.

Baseline for Delay Claim Analysis

Another issue is at what point in time a delay claim should be analyzed since there are primarily three different times to consider (Wickwire, Hurlbut, and Lerman 1989):

- Evaluating delays when the events that are causing subsequent delays occur (estimate or project forward to assess delay).
- Evaluating delays concurrently with ongoing delays or just after they are completed.
- Evaluating delays after projects are completed.

There is a potential for conflict when evaluating delays by these three different methods. Developments in specification preparation, case authority, and government regulations appear to support the concurrent evaluation of delays. This is confirmed in the **Veterans Administration 1988 Master Specification** that states in paragraph 1.13:

> The contract completion time will be adjusted only for causes specified in this contract. Request for an extension of the contract completion date by the Contractor shall be supported with a justification, CPM data, and supporting evidence as the Contracting Officer may deem necessary for determination as to whether or not the Contractor is entitled to an extension of time under the provisions of the contract. Submission of proof based on revised activity logic durations and costs is obligatory to any approvals. The schedule must clearly display that the schedule has used, in full, all the float time available for the work involved in this request. The Contracting Officer's determination as to the total number of days of contract extension will be based upon the current computer-produced calendar-dated schedule for the time period in question and all other relevant information. Actual delays in activities which, according to the computer-produced calendar-dated schedule, do not affect the extended and predicted contract completion dates shown by the critical path in the network, will not be the basis for a change to the contract completion date. The Contracting Officer will within a reasonable time after receipt of such justification and supporting evidence, review the facts and advise the Contractor in writing of the Contracting Officer's decision.

In the legal case of Santa Fe, Inc., the Contract Appeal Board discussed the use of current CPM updates to measure delays and indicated that the use of current updated CPM schedules during the course of a project is the preferred method for evaluating project delays (VABCA, 1987). The owner used the information provided in the current CPM update to evaluate delays to the critical path for the project; the court accepted that analysis. This discussion and decision emphasizes the importance of using current updated CPM schedules to track the location of the critical path at a given point in time and the amount of float available for specific activities when analyzing construction delays.

Use of a *But For* Test for Extended Duration Claims

Courts have held that in order for contractors to recover additional costs for project delays, they must prove that such costs would not have occurred **but for** the action of the owner. In a decision on this subject the Contract Board of Appeals stated (Fishbach and Moore International Corp., ASBCA, 1977):

> It is axiomatic that a contractor asserting a claim against the Government must prove not only that it incurred the additional costs making up its claim but also that such costs would not have been incurred but for Government action.

The Use of Critical Path Method Schedules to Establish Early Completion Claims

Assume a contractor submits a reasonable CPM schedule showing completion in sixty days in a situation where the contract calls for one hundred days and the schedule is then approved by the owner. Legal cases indicate that if the contractor is delayed by the owner for thirty days, such recovery by the contractor is permissible. In Green Builders, Inc., the contractor's CPM schedule indicated completion would take seven months and the contractor's bid and related overhead costs were based on this estimate (ASBCA, 1988). The original contract completion date was set for a twelve-month period of construction. An owner-caused delay extended the project by three months and thus completion was achieved in ten months. The Contract Board of Appeals found that the contractor had a right to recover delay costs based on the scheduled early completion date because the contractor showed its performance plan was reasonable.

Fragnets

Fragnets are subnetworks used to break down one or more activities into additional levels of detail to develop the individual subactivities necessary for completion of the activity shown on the critical path (Wickwire, Hurlbut, and Lerman 1989). Fragnets also show additional (new) activities or logic revisions, due to delays. As such, fragnets are effective analytical tools for comparing portions of the as-planned CPM schedule with the actual events as impacted by changes. Fragnets are prepared for and used as part of a time impact analysis since they help visually display changes in critical portions of the sequence and duration of projects. Fragnets are then used to show the impact of changes on the overall project duration.

Contract Provisions for Notice Procedures

Typical notice provisions found in most construction contracts require that **notice of the occurrence of a delay** be provided within a fairly short period of time. Federal contracts require this notice within ten days and the standard American Institute of Architects contract forms require it to be within twenty days (Bramble and Callahan 1987). In respect to damages for delays, damages generally accrue when a project is substantially complete and accordingly notice of damages for delays are made after that point.

Generally, a contractor has to provide written notice of the delay setting forth the occurrence of a delay and the reasons for it. Some government contracts also require that after the initial notification of delay, a contractor must provide further information explaining the circumstances of the delay in greater detail, as well as documenting the cost impact. They may also require some form of critical path analysis (Bramble and Callahan 1987). Judges will usually look unfavorably towards notice provisions that they deem to be unreasonable, particularly clauses that require short time frames for notice and clauses that may also require contemporaneous documentation, or computation, of damages. In *E. C. Ernst, Inc. v. General Motors* (1973), a notice provision required that it contain an exact amount of damages. The court held the quantification aspect of the notice requirement to be unreasonable.

There are also some exceptions to compliance with the specified notice requirements. Failure to comply with written notice requirements may not be a bar to recovery if an owner has actual notice. When such actual notice is held by the owner, and it is coupled with some element of fault or causation on the part of the owner (such as defective specifications), the formal written notice requirements will frequently be deemed waived.

There are also a number of ways in which actual notice could be imputed to an owner. In some instances, monthly CPM updates provide sufficient notice of delays. Oral notice could be sufficient such as in Davis Decorating Service where the contractor advised the inspector working for the owner several times about a particular problem (ASBCA, 1973). In addition, another method of satisfying the notice requirement would be the existence of a written communication by the owner indicating an awareness of the operative facts.

Usually courts will seek to avoid the harsh forfeiture of rights brought on by lack of notice absent some prejudice to the owner. However, if an owner has been deprived of an opportunity to take corrective measures or to mitigate damages by a lack of notice, strict compliance with the notice requirements will be imposed by court systems.

Early Finish Critical Path Method Schedule Submissions

Early finish schedule submissions should be carefully evaluated since they might result in claims when a contractor, who does not meet its early finish completion date (as established by a contractually owner-approved CPM submission), as a result of the actions of the owner, seeks delay damages even when he or she is still able to complete work before the original contract completion date. Several court rulings have resulted in the following (Wigal, 1991):

> Where a contractor is delayed because the owner does not give the contractor the assistance necessary to meet the early scheduled completion date case law (*United States v. Blair*, 1944; *Pennor Installation Corp. v. United States*, 1950) holds that no compensation is due since the owner is under no obligation to make it possible for a contractor to complete early.
>
> Where a contractor is delayed because the owner unintentionally interferes with the contractor's scheduled early completion, case law (*Grow Construction Co. Inc. v. State*, 1977; *D'Angelo v. State of New York*, 1976) holds that such an owner caused delay will result in the contractor receiving compensation except where the delay was caused by a factor outside the control of the owner (*Sun Shipbuilding and Dry Dock Co., v. U.S. Lines, Inc.*, 1977).
>
> Where a contractor is delayed because the owner intentionally interferes with the contractor's early scheduled completion, such an owner caused delay will result in compensation to the contractor.

A variation on the theme of delayed early completion occurs when an owner explicitly, or implicitly, requires a contractor to expedite the scheduled completion date established by the formal contract. The contractor should recover additional costs based on the theory of directed, or constructive, acceleration. The elements that should be present to justify an action for constructive acceleration are (Gold 1977):

- Excusable delay.
- Notice by the contractor to the owner of such delay.

- An "order" by the owner to accelerate (could occur as a threat to terminate for default).
- Notice by the contractor that such an "order" is regarded as a constructive change.
- Incurring costs as a result of the acceleration.

17.9 QUANTIFICATION OF CONTRACTOR DELAY CLAIMS

Once entitlement to delay or disruption damages are established, the party seeking to recover has to substantiate its costs in order to receive payment. The types of delay damages that are awarded will vary from project to project. The categories of damages that could be recovered when a delay occurs are (Kutner 1989; Hohns 1979):

- Labor Escalation
- Material Escalation
- Increased Engineering and Supervision
- Loss of Productivity or Loss of Efficiency
- Interest
- Equipment Costs
- Impact Costs
- Field Office Overhead
- Main Office Overhead
- Insurance
- Bonding/Loss of Bonding

17.10 OWNER DAMAGES FOR DELAY

When delays occur that are caused by the contractor, the owner is entitled to damages for such delays. These damages may consist of actual damages for the losses incurred, or liquidated damages. The following are damages for delays that could be awarded to owners (Acret 1987):

- Loss of Use
- Additional Damages Related to Loss of Use
- Increased Interest
- Additional Professional Fees for architects and engineers

Cases that are illustrative of some of the above points are *Marshall v. Karl F. Schultz* (1983) where an owner recovers damages for delayed use of a facility, *United Telecommunications Inc. v. American Television and Communications Corp.* (1976) where an owner recovers damages for increased financing costs, and *Erecto Corp. v. State* (1968) where an owner recovers damages for additional fees.

Liquidated damages attempt to fix the limit of damages that would be incurred in the event that completion of a construction project is delayed by contractors. This predetermination of damages, which is based upon a precise dollar amount for each day of delay, eliminates the need to litigate actual damages.

With respect to assessing intent of the parties for the damages not to be penal, courts will generally allow parties the freedom to contract and to reach agreements, which will not be set aside as penal in nature. The legal theory for this is that when a contract is entered into that provides for liquidated damages, the parties have a right to set the amount by contract and, if reasonable, the courts will not allow such a provision by declaring them to be a penalty.

One problem with this view is that public construction contracts are frequently from documents that the contractor must take or leave. There is no freedom to negotiate the amount of liquidated damages. Nevertheless, this issue of unequal bargaining power among the parties is generally not considered by the courts in a typical evaluation of a liquidated damages clause.

In New York, the Court of Claims has sole jurisdiction to hear construction cases involving Public Construction Projects. With respect to disputes on federal projects, contractors have to comply with the Contract Disputes Act as required by the specifications. The procedures to be followed require that a claim has to be submitted to the contracting officer for a decision. The decision will be final unless the contractor appeals it to a Contract Appeals Board within ninety days of receiving it or unless a suit is brought in the United States Court of Claims within twelve months of receiving the decision.

The government has no opportunity to appeal the decision of a Contracting Officer. Where the decision of an officer is appealed to a Contract Board of Appeals by the contractor, and a board then renders a decision on that appeal, that decision may be appealed by either the owner or the contractor within 120 days of receipt. The decision of the board is final on questions of fact, absent fraud, bad faith, and so forth, but is not final on issues of law.

17.11 MINIMIZING DELAY CLAIMS—APPROPRIATE CLAUSES IN SPECIFICATIONS

Potential construction delay claims should be addressed at the earliest possible time. This includes the design and specification preparation stage for the owner and architect/engineer and the bid preparation period for the contractor. By focusing and concentrating on the potential for delay claims right from the inception of projects, prudent actions could be taken by the participants to significantly reduce the number and magnitude of delay claims. Efforts in this regard should continue from the pre-construction phase to the actual construction phase and through the post construction period should any unresolved claims remain outstanding.

Contract Completion Date

Realistic completion dates should be set in contracts. When a time frame is set that is unreasonably short, it will be difficult for a contractor to meet the completion date (Yturralde 1989).

A reasonable contract completion date will serve to eliminate many delay claims.

Excusable Delay—Weather

Clauses granting time extensions for weather delays should be written with more specific language with respect to excusable delays due to weather. The concept of awarding a time extension for **unusually severe weather** is vague; this has led to substantial amounts of litigation. Properly drafted contract documents should assign responsibilities and risks and reduce or eliminate uncertainties. To address this, a specification could be provided that does the following (Finke, 1990):

- Defines the daily severity of weather at which an impact on the work may be reasonable. One such threshold of severity could be given to apply to the entire project or separate levels could be assigned to different portions. As an example of the second option, daily severities could be tailored to the materials or type of work involved. The specifications should provide precise standards for such activities.
- Defines the number of days per month that each level of severity is foreseeable.
- Establish the source for actual weather records and site conditions (for lingering effects) and the required content of such records (National Weather Service, etc.).
- Defines and specifies an allowable number of additional work days that could be requested by the owner to make up for lost time due to weather.

The implementation of the above recommendations could assist in developing an objective procedure by which weather delay claims could be handled by the parties without the need for litigation.

Liquidated Damages

Liquidated damages should be set at reasonable and specific amounts. If they are too high, they will likely be declared by the courts as being penal in nature and therefore unenforceable. The setting of damages at unrealistic levels could cause contractors to generate delay claims. To reduce litigation regarding liquidated damage clauses, it would be advisable to provide in the contract a general description relative to how the amounts were calculated by the owner. During litigation, the courts will be guided by seeking fair compensation for monetary injuries sustained and will attempt to place the injured party in the position it would have been in if the delays had not occurred during construction. When the contracting parties keep this principle in mind in the setting of liquidated damages, the courts will generally allow the provisions to stand.

Chapter Summary

Construction delay claims are a major problem for the construction industry. Delay claims arise as a result of numerous factors including improperly drafted contract documents, erroneously prepared bids, owners failing in their responsibility to provide site access or to take other required action in a timely manner, and inadequate contract administration on the part of owners, contractors, architect/engineers, and other participants in the construction process. Claims could be made by contractors against owners and by owners against contractors and they could also involve A/E's, subcontractors, suppliers, and bonding companies.

This chapter provided a summary of the processes for filing change orders and claims including: changes, change orders, claims, construction delays, utilizing Critical Path Method schedules to justify claims, documentation used to justify claims, and methods for minimizing claims.

Key Terms

addendum
but for
cardinal change
change order
changes
claims
claims resolution process
compensable excusable delay
concurrent delay
consequential damages
Construction Change Authorization Form
construction documentation
constructive change
Critical Path Method Schedules
deletions
early finish schedule submissions
exculpatory clause
extras
Field Order Change Form
float
fragnets
liquidated damages
modifications
noncompensable excusable delay
nonexcusable delay
no damages for delay
notice of occurrence of a delay
partial delays
time is of the essence
unusually severe weather
Veteran's Administration Master Specifications
waiver
Work Directive Change Order Form

Discussion Questions

17.1 Explain how Critical Path Method schedules are used to justify claims by contractors.

17.2 Discuss how liquidated damage clauses could help prevent protracted claim negotiations at the end of a project that has been delayed by a contractor.

17.3 Explain what concurrent delays are and how a contractor could prove that they have occurred on a project.

17.4 How are noncompensable excusable delays addressed when they occur on projects?

17.5 Who benefits from nonexcusable delays and why?

17.6 If a project is delayed due to design modifications, what construction documentation could be used to prove it caused delays on a project?

17.7 Explain the difference between change orders and addenda.

17.8 What rights do contractors have if there is a cardinal change during a construction project?

17.9 What is required in order for a contractor to protect himself or herself when a constructive change occurs on a project?

17.10 Which standard construction documents provide forms for processing change orders?

17.11 Explain construction claims and how they are used during construction.

17.12 What are the disadvantages to having a no damages for delay clause in a construction contract?

17.13 What types of damages are owners entitled to for delay claims?

17.14 What types of damages are contractors entitled to for delay claims?

17.15 What are contractors entitled to when there are excusable weather delays?

17.16 How is a no damages for delay clause used during construction?

17.17 Explain notice procedures for delays.

17.18 Discuss how fragnets are used in construction.

17.19 What are the three times at which delays could be analyzed on construction projects?

17.20 Who is allowed to use float on construction projects and why?

References

Acret, J. 1987. *Construction Litigation Handbook*. Colorado Springs, CO: Shephard's/McGraw-Hill, Inc., 95–122.

American Institute of Architects. 2002. *Document 201—Construction Change Authorization Form*. Washington, DC: American Institute of Architects.

Baram, G. 1992. *Construction Claims—Documenting the Facts*. Transactions of the American Association of Cost Engineers Annual Conference, D.4.1–D.4.11.

Bennett, J. 1972. GSBCA No. 2362, 72-1, BCA (CCH) Decision No. 9764 at 43, 467.

Bockrath, J. T. 2000. *Contracts and the Legal Environment for Engineers and Architects*. New York: McGraw Hill Publishers.

Bramble, B., and M. Callahan. 1987. *Construction Delay Claims*. New York: John Wiley and Sons.

El-Nagar, H., and J. Yates. 1997. Construction documentation used as indicators of delay. *Journal of the American Association of Cost Engineers International* 39(8): 31–37.

Engineers Joint Construction Documents Committee. 2002. *Document 1910-8 Work Directive Change Order Form*. Washington, DC: National Society of Professional Engineers.

Finke, M. 1990. *Weather-Related Delays on Government Contracts*. 1990. Transactions of the American Association of Cost Engineers Annual Conference, F.3.1–F.3.8.

Gold, Harold. 1977. *The Problem of Extra Work and Extra Time*. The Masters Institute in Government Construction Contracting. Washington, DC: Federal Publications, Inc., B1–B127.

Hohns, M. 1979. *Preventing and Solving Construction Contract Disputes*. New York: Van Nostrand Reinhold Ltd.

James, D. 1990. Concurrency and apportioning liability and damages in public contract adjudications. *Public Contract Law Journal of 1990*, 490–531.

Kutner, S. 1989. *Project Management*. New York: Kimmons and Lawrence, 297–327.

Postner, W., and R. Rubin. 1992. *New York Construction Law Manual*. New York: McGraw-Hill, Inc.

Rubin, R., S. Guy, A. Maevis, and V. Fairweather. 1983. Construction Claims, Analysis, Presentation, Defense. New York: Van Nostrand Reinhold, Ltd.

Thomas, R., and F. Wilshusen. 1989. *How to Beat a "No Damage" Clause with Respect to Delay in Building or Construction Contract*. 74 ALR 3d 187, 187–264.

Wickwire, J., S. Hurlbut, and L. Lerman. 1989. Use of Critical Path Method techniques in contract claims: Issues and developments, 1978 to 1988. *Public Contract Law Journal*, 18(338), 338–391.

Wigal, G. January 1991. Interference with a contractor's early completion of a construction project. *The Construction Lawyer*, 10 (4): 17–24.

Yates, J., and A. Epstein. 2006. Avoiding and minimizing construction delay claims in relational contracting. *American Society of Civil Engineers Journal of Professional Issues in Engineering Education and Practice*, 132(2): 5–10.

Yturralde, D. October 7–11, 1989. *Mitigation of Construction Delay Claims*. Project Management Institute, Seminar/Symposium, Atlanta, GA, 289–291.

Zack, J. 1992. Schedule "games" people play and some suggested "remedies." *American Society of Civil Engineers Journal of Management in Engineering* 8(2): 138–152.

Legal Cases

Corinno Civetta Construction Corporation v. City of New York (Ct. of Appeals 1986) 502 NYS 2d 681.

D'Angelo v. State of New York (1976) 46 A.D. 2d 83, 984, 362 N.Y.S. 2d 283, Aff'd 39 N.Y. 2d 781, 385 N.Y.S. 2d 284, 350 N.E. 2d 615.

Earthbank Co., Inc., v. The City of New York (Department of Parks and Recreation) (A.D.1 Dept. 1991) 568 N.Y.S. 2d 101.

Erecto Corp. v. State (3rd Dept. 1968) 29 A.D. 2d 738, 286 N.Y.S. 2d 562.

E. C. Ernst, Inc. v. General Motors Corporation (5th Cir. 1973) 482 F.2d 1047.

Grow Construction Company, Inc. v. State (App. Div. 1977) 391 N.Y.S. 2d 726.

Kalisch-Jarcho, Inc. v. City of New York (Ct. of Appeals 1983) 461 N.Y. Supp. 2d 746, 448 N.E. 2d 413, 58 N.Y. 2d 377.

Marshall v. Karl F. Schultz, Inc. (Fla. Dist. Ct. App. 1983) 438 So.2d 533.

Novak and Company, Inc. v. Dormitory Authority of the State of New York (A.D. 2 Dept. 1991) 568 N.Y.S. 2d 453.

Port Chester Electrical Construction Corp. v. HBE Corporation and the Fireman's Insurance Company (2nd Cir. 1992) 894, F.2d 47.

Sun Shipbuilding and Dry Dock Co. v. U.S. Lines, Inc. (1977) 439 F. Sup. 671.

United States v. Blair (1944) 321 U.S. 730, 64 S. Ct. 820.

United Telecommunications, Inc. v. American Television Communications Corp. (10th Cir. 1976) 536 F.2d 1310.

Board of Contract Appeals Decisions

Capital Electric Co., GSBCA No. 3329, 1983, 83-2 BCA (CCH) Decision No. 16, 548.

Davis Decorating Service, ASBCA No. 17342, 1973, 73-2 BCA (CCH) Decision No. 10, 107.

Fishbach and Moore International Corp., ASBCA No. 18145, 1977, 77-1 BCA (CCH) Decision No. 12, 300 at p. 67, 874.

Green Builders Inc., ASBCA No. 35518, 1988, 88-2, BCA (CCH) Decision No. 20, 734.

Santa Fe, Inc., VABCA Nos. 1943–1946, 1984, 84-2 BCA (CCH) Decision No. 17, 341.

Santa Fe, Inc., VABCA No. 2168, 1987, 87-3 BCA (CCH) Decision No. 20, 104.

CONTRACT DISPUTE RESOLUTION TECHNIQUES

18.1 INTRODUCTION

In spite of all the efforts of owners, contractors, and subcontractors, contract disputes still occur on construction projects. It is difficult, if not impossible, to draft construction contracts that address every potential problem that could arise during construction; therefore, one or both parties to construction contracts could file claims against each other. In order to be able to settle contractual disputes, engineers and contractors should be familiar with contract **dispute resolution techniques**. Several types of dispute resolution techniques are used in the construction industry and this chapter introduces traditional and **alternative dispute resolution** (ADR) methods. Traditional dispute resolution methods include litigation and negotiation; alternative dispute resolution (ADR) techniques include:

1. Arbitration
2. Mediation
3. Mediation/Arbitration (Med/Arb)
4. Early neutral evaluation
5. Rent-a-Judge
6. Court-annexed arbitration
7. Summary jury trials
8. Dispute review boards

Litigation and arbitration are both methods that use **adversarial processes**, which means that one party will win and one party will lose. Using adversarial processes could lead to deteriorating relationships between the parties involved in construction contracts; that is why other dispute resolution techniques have been evolving during the past few decades. Alternative dispute resolution techniques rely on voluntary, **nonadversarial processes** to create a win/win resolution to disputes that helps to preserve working relationships.

The traditional dispute resolution process is shown in Figure 18.1 in which a box indicates where ADR techniques could be attempted before resorting to litigation or arbitration.

18.2 CONTRACT CLAIMS

Contract claims could occur at any stage in the execution of a construction contract. Typical causes of claims include:

- Differing site conditions
- Delays
- Design errors or changes
- Interpretation differences
- Acceleration or suspension of work
- Construction failures
- Additional work
- Deleted work

Contract claims arise when there are disputes between the contracting parties or when the owner and the contractor

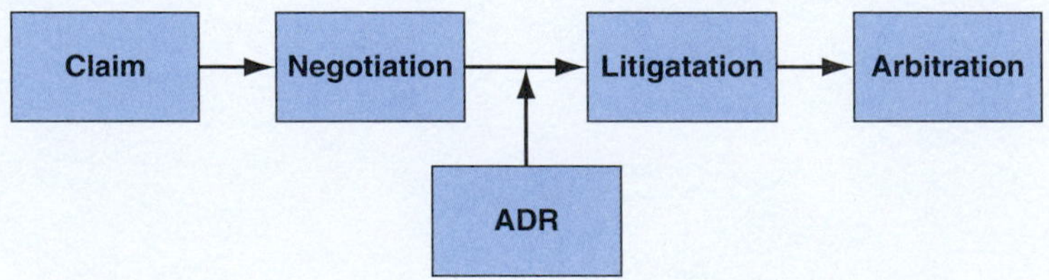

FIGURE 18.1 Traditional Dispute Resolution Process with ADR Inserted.

are not able to agree that there has been a change in the original scope of work or on the degree of the change.

When a dispute arises and the contracting parties are not able to quickly resolve it, one of the parties will file a change order or file a claim. Change orders and claims are discussed in detail in Chapter 17 in Sections 17.3 and 17.5.

18.3 INFORMATION TECHNOLOGY AND THE ELECTRONIC MARKETPLACE

In addition to the typical issues that plague construction, when one or both of the parties to a contract file claims, the introduction of **information and communication technologies** (ICT) is complicating construction claims and dispute resolution processes. Many of the new information and communication technologies have not been addressed yet by legal and contractual practices; therefore, potential precedent legal cases are only now just starting to work their way through the legal system. In the meantime, the use of ICT during construction projects could complicate legal proceedings when there are contractual disputes. The lack of legal validity for ICT might pose a barrier for further implementation of ICT. The following sections address three of the legal ICT issues: (1) the admissibility of e-mails in court proceedings, (2) Electronic Signatures in Global and National Commerce Act, and (3) the eLEGAL European Commission Research.

Admitting E-mails as Evidence in Dispute Resolution Proceedings

The use of **e-mails as evidence** in court proceedings is a difficult area for the court system. Because the dates on computer systems could be altered by setting them to earlier or later dates, the dates shown on e-mails may not be valid as evidence. E-mails could also be sent from computers that do not belong to the person sending them. Therefore, some court systems are reluctant to allow e-mails to be admitted as evidence if it is time sensitive or sender sensitive evidence. If an e-mail was printed and then signed and dated by the sender, then it may be allowed as evidence in the same manner that signed and dated memos are permitted as evidence.

Electronic Signatures in Global and National Commerce Act of 2000

The **Electronic Signatures in Global and National Commerce Act of 2000** was passed by the 106th Congress of the United States to facilitate interstate and global electronic commerce. Prior to the passing of this act, each individual state was regulating the use of electronic and digital commerce within each state. The act only covers transactions that are related to business, commercial (including consumer), and governmental matters. This act states that (U.S. Congress S.761-1, 2000, p. 1):

> with respect to any transaction in or affecting interstate commerce or foreign commerce—
>
> (1) a signature, contract, or other record relating to such transaction may not be denied legal effect, validity, or enforcement solely because it is in electronic form; and
>
> (2) a contract relating to such transactions may not be denied legal effect, validity, or enforceability solely because an electronic signature or electronic record was used in its formation.

The act does require informed consent on the part of the consumer to allow the use of electronic transfer of information and their digital signature. Consumers have to also be informed about the procedures for withdrawing their informed consent.

The provisions of the act will not apply to contracts or records that are included in the exception list to the act; they are the following (U.S. Congress S.761-5, 2000, p. 5):

> (1) a statute, regulation, or other rule of law governing the creation and execution of wills, codicils (postscript to a will), or testamentary trusts (trusts created by wills);
>
> (2) a State statute, regulation, or other rule of law governing adoption, divorce, or other matters of family law; or
>
> (3) the Uniform Commercial Code, as in effect in any State, other than sections 1-107 and 1-206 and Articles 2 and 2A.

Additional exceptions to the act include (U.S. Congress S.761-5, 2000, p. 5):

> (1) court orders or notices, or official court documents (including briefs, pleadings, and other writings) required to be executed in connection with court proceedings;
>
> (2) any notice of—
>
> a. the cancellation or termination of utility services (including water, heat, and power);
>
> b. default, acceleration, repossession, foreclosure, or eviction, or the right to cure, under a credit agreement secured by, or a rental agreement for, a primary resident of an individual;
>
> c. the cancellation or termination of health insurance benefits or life insurance benefits (excluding annuities);
>
> d. recall of a product, or material failure of a product, the risks endangering health or safety; or
>
> (3) any document required to accompany any transportation or handling of hazardous materials, pesticides, or other toxic or dangerous materials.

The electronic signature component of the act refers to any electronic means such as "an electronic sound, symbol, or process attached to or logically associated with a contract and other record and executed or adopted by a person with the intent to sign the record" (U.S. Congress S571-9, 2009, p. 9). Signatures have to be attached or logically associated with the document to which they pertain.

eLEGAL European Commission Research

The **European Commission** has funded an international research project called **eLEGAL** that is developing "a framework for legal conditions and contracts related to the use of ICT in the construction sector." According to the eLEGAL assessment and workshops with international experts, a number of key issues have to be addressed in order to facilitate the application of ICTs to the construction industry, including (Nielsen, Hassan, and Ciftci 2007, 255):

- Making e-mail and exclusive electronic communication legally admissible and contractually valid.
- Avoiding a "media-breach" between project principles.
- Establishing a contractual framework for the use of application service providers (ASPs) governing the relationships between the client, the SP, and end users.
- Liability for errors resulting from conversions between different computer-aided design (CAD) software programs—or preferably avoidance of errors.
- Managing hard/soft copy discrepancies.
- Enabling project adoption of digital signatures.

In some countries, the use of e-signatures is governed by electronic signature laws, such as the **Turkish Electronic Signature Law** passed in 2004 that indicates (Nielsen, Hassan, and Ciftci 2007, 261):

> The law of e-signatures stipulates an electronic signature shall be considered equivalent to a handwritten signature in all of the Member States (in Europe), provided that it meets the functional requirements of the Advanced Electronic Signature; based on a Qualified Certificate (QC); and created by a Secure Signature Creation Device. Further to the recognition of the legal equivalence the law reaffirms that the Qualified Electronic Signature is admissible as evidence in court proceedings.

Signatures that require verification by a notary republic could also be verified by a **digital notary service** where the notary republic attaches their digital signature to the document including the date the signature was verified and the time when the verification was certified by the notary republic (Nielsen, Hassan, and Ciftci 2007).

Intellectual property rights (IPR) need to be respected when customers and suppliers are exchanging data and this requires some type of assurance that technical data will not be compromised by other parties when it is being used electronically. "A legal statement or consortium agreement on the ownership and licensed use of IPR by the parties for the purpose of fulfilling their obligations should be defined in the contract" (Ren and Hassan 2007, 248).

18.4 BUILDING INFORMATION MODELING (BIM)

Some owners and architects are beginning to require the use of **Building Information Modeling** (BIM) software on their projects and engineering and construction firms are being contractually required to utilize these three-dimensional modeling technologies. The leaders in BIM technology in 2010 were *Autodesk* with its *Revit* suite of programs and *Bentley Systems.* Both of these firms provide software that allows the importation of different platforms and formats for design drawings and three-dimensional models are generated that incorporate the contributions of all of the designers. Other software systems that are **aggregate model viewers** and conflict resolution tools are *Autodesk's NavisWorks, Bentley's ProjectWise Navigator, VICO Contractor, ArchiCAD 12's Virtual Building Explorer,* and *Tekla Structures.*

Building Information Modeling software also provides the capability of creating four-dimensional schedules that generate the three-dimensional models in predefined scheduling sequences and **Quantity Take-Offs** (QTO) could be automatically generated from three-dimensional models. One of the features of BIM software that benefits contractors is **clash detection** or the **highlighting of interferences** between building elements and the generation of clash detection reports.

Projects that have utilized the clash detection feature in BIM software experience a significant decline in change orders during construction because construction interferences are discovered during the design phase and construction methods and processes could be modeled using the BIM software to determine their viability. With a reduction in the number of change orders issued during construction, the number of claims are also reduced, thus saving owners and contractors money. Even though the initial cost of BIM software seems prohibitive to small- and medium-size firms, the cost savings that are being realized through its use is leading to its adoption by members of firms in the engineering and construction industry.

Integrated Project Delivery (IPD) has led to the development of **Collaborative Agreements** that are now available through the American Institute of Architects, the Army Corps of Engineers, and other entities, including agreements from Collaborative Agreements:ConsensusDOCS, LLC, such as *ConsensusDOCS 300 Standard Form of Tri-Party Agreement for Collaborative Project Delivery* (2007). Agreements that are specific to Building Information Modeling applications include:

1. The ConsensusDOCS 301-2008 BIM Addendum
2. The American Institute of Architect's E202-2008 Building Information Model Exhibit and Protocol
3. The Army Corps of Engineers Building Information Modeling Road Map (October 2006)
4. The Associated General Contractors Guide to BIM

Since BIM technology has only been available for a few years, the legal ramifications of its use are not known and the lawsuits that will eventually set legal precedents have not yet materialized in the legal system. Potential areas for disputes include (Salmon 2009):

1. Liability issues related to generating three-dimensional models using drawings from several different firms and who would be responsible for errors.
2. Disputes arising from software incompatibility and inaccurate data being uploaded into the BIM software, as a result of incompatibility.

3. Disputes arising from changes being made after drawings are uploaded into the three-dimensional models on the two-dimensional drawings.
4. Disputes from changes resulting from clashes between construction elements designed by members of different firms.
5. Disputes resulting from how construction elements are displayed in a three-dimensional model versus how they are shown on individual engineering or shop drawings.
6. Disputes prompted by the engineering drawings not containing all of the changes that were made on the three-dimensional model, although the three-dimensional modeling software is supposed to translate the three-dimensional model back into accurate two-dimensional drawings.

Any and all of the above potential disputes could arise on projects that utilize BIM software programs. Because there are currently no precedent law cases, each new dispute will have to be decided on its own merits until a body of precedent legal cases is established through the settlement of lawsuits.

18.5 CONTRACT NEGOTIATIONS

Claims could either be resolved during construction, or at the end of a construction project, by **negotiations**. Negotiations require both parties to compromise in order to settle claims. Contract claim negotiations are conducted by representatives of the conflicting parties without assistance from neutral third parties.

In construction, 90 to 95 percent of construction claims are settled through negotiations rather than through court settlements, but this statistic also reflects the situations where a negotiation is concluded right before a case is about to be heard in a court of law. Even if a case is settled right before it was about to be tried in court, both parties would have already expended funds on lawyers, discovery, depositions, expert witnesses, and other associated legal costs. The earlier a dispute is settled, the less it costs all of the parties involved in the dispute.

Negotiations could either be conducted one claim at a time or by combining all of the claims and negotiating them all at the same time. Contractors tend to file claims for excessive amounts knowing that they will be able to negotiate down to the amount that they are hoping to settle for at the end of the negotiation. Having appropriate documentation that helps support claims leads to realistic settlements, but if consequential damages are being claimed, it is more difficult to prove losses or delays. The documentation from some of the sources listed in Chapter 17 in Table 17.1 strengthen negotiations and help by providing background information for proposed time and cost adjustments to projects.

Negotiators will try to persuade the opposing party to relinquish one or several claims if they are awarded another large claim; this is a typical process during negotiations. Negotiators have to be prepared to bargain with the opposing party, as both sides need to think they have gained something during negotiations. If one of the parties does not achieve their goals, they could forego further negotiations and switch to trying to settle the disputes through arbitration or litigation. If negotiations are held at the end of construction projects, it consumes valuable time of the key project team members and prohibits them from starting to work on new construction projects. This is an impetus for the negotiation team members to reach a settlement quickly since it would save both time and money.

Sometimes it is difficult to initiate negotiations during or after construction projects because adversarial (enemy or rival) relationships have developed due to the opposing team members having had negative experiences while working with each other. Often the negotiations turn into a continuation of disagreements that develop during construction rather than being about the stated disputes. In these types of situations, it might be beneficial to have other people involved in the negotiations rather than the principal team members. The disadvantage of this is losing the **institutional memory** (familiarity with) the project that the principal team members possess and that they bring to the bargaining table.

Negotiators

One of the most difficult and crucial steps in negotiations is selecting the proper person to participate in negotiations for a firm. Project managers are usually involved in negotiations not only because of their expertise but also because they have the authority to make decisions or agree to settlements. If someone without proper authority is negotiating for a firm, it could slow down negotiations if they have to interrupt the negotiations to transmit settlement offers to whoever in the firm has the authority to authorize settlements.

Just because someone is technically competent or is the most knowledgeable about a dispute, does not necessarily mean that they are the best person to conduct negotiations. Some of the traits required to be able to effectively conduct negotiations include (ACEC Guidelines to Practice 1988, 34):

- Preparation and planning skill
- Knowledge of subject matter
- Ability to understand the true interests of the firm
- Ability to think clearly and rapidly under pressure
- Ability to express thoughts verbally
- Good listening skills
- Patience
- Ability to persuade others
- Ability to understand others
- Ability to control emotions
- Ability to maintain flexibility

When negotiations break down due to either personality conflicts, impatience, lost tempers, a stalemate, or slow progress, then the negotiator should try to convince the opposing parties that the process would benefit from switching to another form of dispute resolution.

18.6 CONTRACT MEDIATION

Mediation is a variation on negotiation that inserts a neutral third party into the negotiation process. The mediator helps the opposing parties to reach a settlement. Mediation sessions might take the form of conferences or they could be held in private sessions. Sometimes if the negotiations have deteriorated into an explosive situation, a mediator will have the opposing parties in separate locations and bring settlement proposals from one location to the other location. Mediation sessions could be conducted over several days and nights without adjourning until a settlement is reached. This method is used during union contract mediation sessions. Mediation sessions for construction contract disputes are not usually as intense as union sessions; therefore, they may be conducted during regular business hours only. Mediation is used for several reasons including:

- Controlling losses
- Containing damages
- Preserving relationships
- Clarifying issues
- Securing agreements

Mediation does not replace negotiation, rather it is used to augment negotiation and is only attempted if both parties agree to use it. Mediators do not force either party to agree on a settlement. They only provide a neutral observation of the situation. Mediators help translate information to ensure that both parties understand it, they may provide advice concerning the other party and their objectives, they help direct the negotiations, and they offer suggestions. In order for mediation to be successful, the parties involved in it need to genuinely want a resolution of their differences and be committed to not having the discussions end in failure.

Arbitration clauses in construction contracts could also include stipulations that require mediation before arbitration when there are contract disputes as a way of ensuring that mediation will be explored before using arbitration.

18.7 ARBITRATION

Many construction contracts contain **arbitration** clauses that require either **binding** or **nonbinding arbitration**. If binding arbitration is required, then the decisions of the arbitrators are final and may not be appealed in a court of law. If nonbinding arbitration is required, then if either of the parties involved in an arbitration are not satisfied with the outcome of the arbitration proceedings, they could sue the other party to try to obtain a different decision through the court system.

Many different types of contracts contain arbitration clauses including utility, telecommunication, purchase agreements, real estate, and insurance. For these types of contracts it is important to note whether the arbitration requirements are binding or nonbinding, because if they are binding, then signing these contracts waives the right to judicial legal proceedings. The **Federal Arbitration Act** applies in situations where there are contracts for interstate commerce including materials or the services of architects or engineers. The **American Arbitration Association** provides *Guidelines on Construction Industry Arbitration Rules and Mediation Procedures*. This document is provided in Appendix J.

Arbitration is a process whereby a neutral third party or several people are hired to evaluate evidence, listen to the arguments of both parties, and provide a binding or a nonbinding decision and settlement award. For construction disputes either one arbitrator is used or the arbitrators are selected by a process where the owner chooses one arbitrator, the contractor chooses another, and the two arbitrators selected decide on the third arbitrator.

Many different organizations maintain lists of arbitrators such as the American Arbitration Association and state and local arbitration organizations. Arbitrators could be architects, owners, engineers, or constructors and having one of each of these occupations represented helps provide more balanced decisions. Arbitrators used in construction that are architects, engineers, or constructors have a better understanding of the technical merits of arbitration proceedings than judges who would only have a legal background.

Arbitrators may not have, or have had, any type of affiliation with either the owner or the contractor involved in the case they are evaluating, which could be difficult in areas with small populations. Arbitrators might have to be brought in from other parts of the state, or from other states, in order to guarantee the impartiality of the arbitrators. When arbitrators are selected, they have ten days in order to disclose information on cases they have arbitrated in the past. If they have not provided this information after fifteen days, they are disqualified as an arbitrator.

The parties involved in arbitration proceedings are responsible for supplying the arbitrators with the construction contract and drawings and other documentation prior to the arbitration proceedings to allow adequate time for the arbitrators to review the documents.

In construction, arbitration proceedings are preferred over litigation for a variety of reasons. The amount of time required for arbitration is minimal compared to litigation. Depending on how crowded the court dockets are, a lawsuit might not be tried in court for years. In the construction industry, risking having to wait that long to settle a dispute is not acceptable; therefore, arbitration is an alternative that saves time since arbitration proceedings take months, or less than a month, rather than years.

Not having to wait for a case to be heard in the court system speeds up the dispute resolution process. Arbitration proceedings could commence as soon as both parties are ready to present their evidence. The arbitrators have reviewed the construction contract and documents.

Arbitration is also less expensive than litigation since lawyers are not required and members of firms may present their own evidence. Instead of paying for attorneys, the disputants pay the arbitrators by the hour to review the construction documents and to conduct the arbitration proceedings.

One primary reason that contractors prefer arbitration rather than litigation is because they are private proceedings, unlike court cases that are part of the public record. Having private proceedings helps preserve the reputation of a firm, since others will not know about the disputes and how they are settled by the arbitration panels.

Another benefit of arbitration proceedings is that their location, time, and the arbitrators are all selected by the disputing parties not the court system. This allows arbitration proceedings to be conducted at construction job sites where the arbitrators could physically review the items under dispute.

In arbitration, the rules of evidence that are used in the court system, which require someone to refer to something before it is allowed to be entered as evidence, do not apply and anyone or anything could be presented as evidence. This allows all types of construction documentation to be presented in any format. Discovery could also be included in arbitration proceedings if it is stipulated in the arbitration clause in the construction contract. Discovery allows each side to review the evidence that will be presented by the other side in hopes of fostering a settlement at that stage of the proceedings.

Arbitration proceedings do not follow precedent law, which is required in court proceedings; therefore, each case is evaluated on its own merits not on how previous cases were settled by the courts. This allows for more equitable settlements since disputes are not evaluated based on previous case law disputes that may or may not be similar to the current dispute. Not having to follow precedent law allows cases to focus on technical issues rather than legal issues.

After arbitration proceedings are complete, the arbitrators have thirty days to provide their decisions and the amount of awards. Settlement awards are based on actual losses. There are no awards for damages unless punitive damages are awarded, because a claim was unrealistic or they would have been awarded if the dispute had been settled in a court of law. If attorneys are used during arbitration proceedings, then the attorney's fees for the party that wins the settlement may also be included in the settlement award.

If one party does not attend the arbitration proceedings, the arbitrators could still award a settlement if there is a contract clause requiring arbitration to settle disputes. Even if one party breaches a contract, either party could still demand that all of the disputes that occurred before the breach be settled through arbitration if there is an arbitration clause in the contract.

If an award is made against a party and they do not pay it, the other party would have to file a petition with the courts. If there is no response from the other party within ten days, then the award becomes a judgment and it is processed in the same manner as court judgments. Without there being a court-awarded judgment, a sheriff is not able to attach property nor could the court order a receiver to take possession of property.

There are five grounds for invalidating arbitration awards:

1. If awards are procured through fraud.
2. Misconduct on the behalf of the arbitrators.
3. The arbitrators make awards for disputes that were not part of the disputes submitted to the arbitration panel.
4. An arbitrator was not listening to important evidence.
5. An arbitrator refused to postpone the arbitration proceedings to allow one party to collect evidence needed for the proceedings.

Appeals are not allowed for binding arbitration decisions unless it could be proven by the courts that fraud occurred during the proceedings. Arbitrators have immunity from being sued by the disputants, since they are performing in a capacity that is similar to jurors in court cases.

One additional complication to the use of arbitration is that in some states sureties are not bound by the arbitration awards of their clients. If a surety is involved in arbitration proceedings, a separate case may have to be filed against the surety. In some states, sureties are either bound by the arbitration awards or they are at subject to a motion for summary judgment.

18.8 LITIGATION

Litigation involves one party suing another party in the court system. The fundamentals of how lawsuits are initiated and conducted are discussed in Chapter 3. Contractors or owners would only resort to litigation, rather than one of the other available dispute resolution techniques or alternative dispute resolution techniques, when large sums of money are involved in the claims or if there has been an egregious (notably bad) breach of contract.

The reasons that owners and contractors would not use litigation to settle claims include:

1. The results of lawsuits become part of the public record, which could be viewed by anyone. This could be damaging to the reputation of a firm or produce unwanted publicity in newspapers, on television, or on the Internet.
2. The cost of litigation could be prohibitive because there are filing fees, court costs, and legal fees.
3. Lawsuits could take years before they reach the **court docket**.
4. Lawsuits are tried in courtrooms where it is difficult to demonstrate engineering or construction concepts without being able to see the actual items being described in court cases.
5. Court proceedings require that cases follow rules of evidence where items may only be introduced as evidence if they are first introduced during the testimony of witnesses.
6. Cases are argued based on precedent law. This requires the investigation of previous legal cases in order to locate similar cases that could be used to argue the merits of the current case. This requires that cases focus on legal not technical issues.

7. The decisions rendered by the court could be appealed if one of the parties is able to prove that the trial was not conducted properly.
8. Disputants risk settlements that might include damages in addition to actual losses.

Litigation and legal settlements are also unpredictable because they are influenced by the judge trying the case and the ability of lawyers to present a case involving technical issues that are outside of their area of expertise. Owners and contractors prefer to have their cases decided by arbitrators who are knowledgeable about engineering and construction.

18.9 DISPUTE REVIEW BOARDS

Dispute review boards (DRB) have been used in Europe for the past thirty-five years and they are gaining acceptance in the United States in the twenty-first century. In order to try to prevent disputes from escalating to a point where disputes could only be settled by arbitration or litigation, dispute review boards could be used on construction projects. Dispute review boards are a panel of independent arbitrators who are chosen by contractors and owners. The dispute review board members are hired at the beginning of a project to review the contract and make recommendations for settling disputes and claims on a regular basis. Decisions of the DRB are not binding. Contractors and owners may still use other dispute resolution methods to settle disputes and claims (Vorster 1993). One definition for dispute review boards is the following (Vorster 1993, 3):

> A dispute review board (DRB) is a small group of independent, knowledgeable and respected individuals selected by the owner and contractor and appointed under the contract to review and make recommendations on disputes that arise on the project. The board is appointed at the outset of the project. The board visits the project regularly and has a long-term perspective on any issue it is asked to address. The recommendation it makes is the first step in the dispute resolution process. The recommendations are not binding and they do not preclude either party to the contract from exercising any of the other mechanisms for resolving claims.

In addition to analyzing contract documents before they render a decision, dispute review board members also conduct site visits. On some construction projects, DRB members have offices at the job site and conduct regularly scheduled weekly meetings to settle disputes before work progresses any further. Presentations are given by contractors and representatives of owners (engineers) to DRB members before the board makes their decisions.

The American Arbitration Association (AAA) publishes a ***Dispute Review Board Guide Specifications*** (*DRB Guide Specifications*). The AAA also provides a **Three-Party Agreement** and a list of potential DRB members (Jenkins and Stebbings 2006). The **International Chamber of Commerce** publishes **Dispute Board Clauses** (ICC DB Clauses), **Dispute Board Rules** (ICC DB Rules), and a **Model Dispute Board Member Agreement** (DBMA) (Jenkins and Stebbings 2006).

In a case study that was conducted while a project was using DRB to settle disputes, it was noted that whoever provided the most polished, well-documented presentation usually had DRB decisions made in their favor. Recommendations that were derived from the case study project for improving DRB techniques include (Yates and Duran, 2000, 36):

- If a project is using a construction manager, then special meetings between the contracting parties and the construction manager are beneficial whenever there is sufficient evidence about potential disputes.
- If a project management oversight constructor is used, they should audit the effectiveness of the dispute review board and tabulate records of the final costs of settled disputes compared to original requested costs. Final costs should be compared to costs arrived at through negotiations by both sides rather than the cost of potential litigation or of unsettled disputes.
- The DRB members should be cautious about allowing subcontractors to participate in hearings. The three-party agreement is with the owner, the constructor, and the board members, not with subcontractors. The master construction agreement is executed between owners and contractors, not with subcontractors.

The *Technical Committee on Contracting Practices of the Underground Technology Research Council* (UTRC) (1991) discusses the DRB process and outlines the different perspectives of owners, attorneys, contractors, and members of the DRB. Two of the most important elements of the DRB process are described—escrow bid documents and geotechnical design summary reports. In addition, several case histories are presented along with specification clauses for implementing the DRB, examples of dispute review board recommendations, user comments, and a report outline for the geotechnical design summary.

The ***Construction Dispute Review Board Manual*** (1996) is a do-it-yourself guide to understanding the DRB and to setting up the mechanisms for implementing a successful DRB process. A comprehensive DRB methodology checklist with twenty-two statements is included and intended to be used to assess compliance of a particular DRB with the classic methodology developed by American Society of Civil Engineers and the Underground Technology Research Council. A positive response to all twenty-two statements should reassure contracting parties that the DRB process should be used on a particular project. The manual also includes useful information on sample contract specifications, practices of the DRB, several case histories, and a comprehensive table that describes projects that have used the DRB process.

The report called *Avoiding and Resolving Disputes during Construction* explains the history of the development of the DRB as an ADR technique (Technical Council on Contracting Practices of the Underground Technology Research Council 1991). In this publication, the author states that the

DRB does not supplant existing dispute settlement methods being used by an owner, but rather it is an earlier, nonbinding, intermediate step directed at avoiding the need to resort to other more expensive, more time-consuming, and less satisfactory procedures. In this reference, following the description of the DRB process, there is a comprehensive discussion on the use of the DRB followed by perspectives of the owner, contractor, attorney, and board member. In the section on the perspective of the owner, several reasons are provided for using the DRB on risky projects. Risky projects may involve some of the following: subsurface excavation, high-technology facilities, hard-to-implement construction methods, new construction methods, or large complex projects.

Problems and disagreements are an inherent part of all construction projects. When contractors, owners, or construction managers do not deal directly with these problems, often they escalate into major conflicts. The DRB is an active, real-time method that allows contracting parties to plan a combined approach for discussing potential problems, and proposing solutions to these problems, before the problems escalate to the point where they are being improperly processed by one or both parties to the contract.

Using the dispute review board as a dispute resolution method on public works, construction projects could help owners and contractors resolve disputes rapidly. In addition to the use of the DRB during a project, there should also be "preventative" dispute resolution methods. These preventative methods could be implemented prior to the commencement of a project, as well as being practiced during construction.

Several methods for mitigating the weaknesses of dispute review boards are (Yates and Duran, January 2007):

- Potential DRB board members selected by the owner should proactively guide the owner and highlight any potential imbalance of experience in favor of the contractor or the owner.
- If the contractor is unresponsive to either the owner or the construction manager, the DRB members should inform the contracting parties and hold special meetings to correct the situation, because continuance of this behavior by the contractor could result in a difficult dispute resolution process.
- Contract documents should always be reviewed for areas where they might lead to potential disputes. The DRB members should actively perform this task whether or not they are formally asked to do so by the contracting parties.
- The contracting parties should measure the effectiveness of using the DRB process at scheduled time intervals during the project. Measuring the effectiveness of the DRB process provides the contracting parties with insight into whether the credibility of the DRB is being eroded. If there is erosion of credibility, immediate action is needed by the contracting parties to restore the credibility of the DRB.
- Special meetings between the contracting parties and the construction manager are necessary whenever there is sufficient evidence about potential disputes.
- Once disputes begin, a risk assessment and an allocation plan should be prepared for anticipated potential disputes.

18.10 ALTERNATIVE DISPUTE RESOLUTION TECHNIQUES

There are several **alternative dispute resolution** (ADR) techniques that are also used to help settle construction disputes. The following sections describe some of the more common ADR techniques.

Mediation/Arbitration (Med/Arb)

Mediation/Arbitration (Med/Arb) is a composite of both standard mediation techniques and arbitration. Med/Arb agreements stipulate that a mediator will be appointed prior to the disputes and the mediator in turn switches roles and becomes an arbitrator if the disputing parties are not able to reach an agreement through mediation.

The advantage of using this process is that the mediator will already be familiar with the issues surrounding the disputes by the time they change roles and arbitrate the disputes. If a different person were brought in to arbitrate the disputes instead of the mediator, then all of the evidence presented during the mediation process would have to be repeated for the new arbitrator.

The disadvantage to Med/Arb is that having someone who has the authority to make binding decisions involved in the mediation process might cause the disputants to withhold information that might help achieve a settlement during mediation in order to wait and disclose it during arbitration.

Early Neutral Evaluation

In standard mediation proceedings, the mediators do not provide opinions to either party on the merits of their cases because they are only acting as a facilitator during mediation. In **early neutral evaluation** the mediator is empowered to provide an evaluation of the merits of the case of each party and provide their evaluation to both parties. This process provides each party with an evaluation of the facts of their case and how their case would be viewed by neutral parties. Another option is for a neutral evaluator to be hired to only evaluate the cases being presented by the disputants rather than also being involved during the mediation process.

Minitrials

Minitrials are used along with negotiation and are not actually trials but a way of eliciting additional facts that were not disclosed during regular negotiations. If one of the parties to a dispute thinks they would prevail during binding arbitration, they may withhold key facts during negotiations. Minitrials are used to try to uncover all of the key facts to encourage settlements during negotiations rather than waiting for arbitration.

Minitrials are dry runs of the actual proceedings that would be used during arbitration that are presented to company executives with the authority to authorize settlements. The premise of minitrials is that if the executives with authority to settle disputes have witnessed both sides of the case, they will be in a better position to provide a realistic settlement offer.

Rent-a-Judge

When arbitration is being used and the disputing parties are not able to agree on the selection of an arbitrator or multiple arbitrators, their first option is to have one of the arbitration associations assign someone. Unfortunately, this could lead to a case being arbitrated by an individual that the disputants are not satisfied with or that does not have the appropriate background to review the dispute. Another alternative for the disputants is for them to hire a former judge, a lawyer, or a private expert in their field to hear the case and render either a binding or a nonbinding decision, a process called **Rent-a-Judge**. Dispute resolution organizations could provide a list of names of potential experts and the disputants are responsible for directly hiring the individual to arbitrate their case.

Court-Annexed Arbitration

Court-annexed arbitration is a dispute resolution procedure that is available within the court system. It is similar to a minitrial in that both sides present their case but the arbitrators who hear the case are lawyers who are volunteering their time and have been appointed by the court to hear the case. The decisions reached in these proceedings are not binding unless both parties have waived the appeal process before the proceedings commence. If an appeal is not waived and one of the disputants would like to appeal the decision, then the case would be tried in a court of law.

Summary Jury Trial

Summary jury trials are used while disputants are waiting for their case to be tried in the court system; they are considered to be **mock trials**. Attorneys are used to summarize the cases and the evidence and present them to a jury that renders an opinion. The opinion of the jury is then passed along to a judge. The judge presides over the negotiations until a settlement is agreed upon by both parties. After the mock trial, both of the disputants will know how the case might be settled by a jury and this knowledge is used by the judge to influence the disputants to settle the case during negotiations.

Chapter Summary

This chapter discussed some of the issues that are emerging as a result of the implementation of new information technology and communication methods into the construction arena such as: the admissibility of e-mails as evidence in dispute resolution proceedings, liability for computer-aided design (CAD) software format conversions, electronic signature laws, discrepancies between electronic and paper documents when using Building Information Modeling (BIM), and electronic signatures.

This chapter also presented some of the options that are available in the construction industry for setting construction contract disputes. The traditional dispute resolution techniques that were presented were: mediation, negotiation, and litigation. In addition to the traditional dispute resolution techniques, this chapter also presented information on alternative dispute resolution methods including: arbitration, dispute review boards, Med/Arb, early neutral evaluation, minitrials, Rent-a-Judge, court-annexed arbitration, and summary jury trials. Alternative dispute resolution methods are not limited to the ones presented in this chapter since any type of dispute resolution process could be used as long as both of the disputants agree to use it.

Key Terms

adversarial processes
aggregate model viewers
alternate dispute resolution
alternative dispute resolution
American Arbitration Association
arbitration
binding arbitration
Building Information Modeling
clash detection
Collaborative Agreements
Construction Dispute Review Board Manual
contract claims
court-annexed arbitration
court docket
digital notary service
Dispute Board Clauses
Dispute Board Rules
dispute resolution techniques
Dispute Review Board Guide Specifications
dispute review board
early neutral evaluation
eLEGAL
Electronic Signatures in Global and National Commerce Act of 2000
e-mails as evidence
European Commission
Federal Arbitration Act
highlighting of interferences
information and communication technologies
institutional memory
Integrated Project Delivery
intellectual property rights
International Chamber of Commerce
litigation
Mediation/Arbitration
mediation
minitrials
mock trials
Model Dispute Review Board Member Agreement
negotiations
negotiators
nonadversarial processes
nonbinding arbitration
Quantity Take-Offs
Rent-a-Judge
summary jury trials
Three-Party Agreement
Turkish Electronic Signture Law

Discussion Questions

18.1 Explain what leads to the necessity of having to have dispute resolution techniques available during construction.

18.2 Discuss whether e-mails could be used as evidence in dispute resolution proceedings.

18.3 Explain *qualified electronic signatures* and how they are used in construction.

18.4 Under what circumstances would disputants use mediation to settle construction claims?

18.5 What characteristics are important for negotiators to possess?

18.6 How would disputants locate arbitrators to hear their arbitration cases?

18.7 What grounds could be used to invalidate arbitration awards?

18.8 Under what circumstances would arbitrators award damages in addition to actual losses as part of an arbitration settlement?

18.9 What types of claims could arise related to the use of Building Information Modeling software on construction projects?

18.10 What aspects of Building Information Modeling are the most useful to contractors?

18.11 What are the legal issues that are currently being addressed in the eLEGAL research?

18.12 How are notary republics being used online to verify signatures?

18.13 What are the advantages of using arbitration over litigation?

18.14 If a project manager is the person who is negotiating claims at the end of a project, what negative impacts does this have on the firm that employs the project manager?

18.15 What are the primary reasons that negotiations are not successful and the disputants have to use arbitration or litigation to settle disputes?

18.16 Why is it beneficial to hold arbitration proceedings at construction job sites where the disputes have occurred that are being arbitrated in a claims case?

18.17 What are the advantages of using dispute review boards?

18.18 What are the disadvantages of using dispute review boards?

18.19 What would be the benefit of using early neutral evaluations to help settle construction disputes?

18.20 Why would summary jury trials influence the disputants to settle their claims before their case is heard in a court of law?

References

American Council of Engineers Contractors Guidelines to Practice. 1988. *Alternative Dispute Resolution for Design Professionals*. American Council of Engineering Companies, 1(7): 38–44.

Dispute Review Board Manual. 1996. New York: McGraw-Hill Publishing Company.

Jenkins, J., and S. Stebbings. 2006. *International Construction Arbitration Law*. The Netherlands: Kluwer Law International.

Nielsen, Y., T. Hassan, and C. Ciftci. July 2007. Legal aspects of information and communication technologies implementation in the Turkish construction industry. *American Society of Civil Engineers, Journal of Professional Issues in Engineering Education and Practice* 133(3): 210–215.

Ren, Z., and T. Hassan. July 2007. Legal requirements and challenges for e-Business within the single electronic european market. *American Society of Civil Engineers, Journal of Professional Issues in Engineering Education and Practice* 133(3): 216–220.

Salmon, J. Spring 2009. The legal revolution in construction. *Journal of Building Information Modeling*, 18–19.

The Technical Committee on Contracting Practices of the Underground Technology Research Council. 1991. *Avoiding and Resolving Disputes During Construction—Successful Practices and Guidelines*. New York: American Society of Civil Engineers.

Vorster, M. C. 1993. *Dispute Prevention and Resolution: Alternative Dispute Resolution in Construction with Emphasis on Dispute Review Boards*. Austin, TX: Construction Industry Institute.

Yates, J., and J. Duran. January 2007. Dispute review boards strength and weaknesses. *American Association of Cost Engineers International Journal, Cost Engineering Journal* 42(1): 31–37.

Yates, J., and J. Duran. October 2006. Utilizing dispute review boards in relational contracting: A case study. *American Society of Civil Engineers, Journal of Professional Issues in Engineering Education and Practice* 132(4): 334–341.

CHAPTER NINETEEN

RISK MANAGEMENT AND CONSTRUCTION INSURANCE

19.1 INTRODUCTION

Construction projects are inherently risky endeavors that create many opportunities for things to go wrong, or for people to be injured or killed during daily operations. Since the United States is a highly litigious society, with high numbers of law suits per capita, owners and contractors attempt to locate methods for reducing their exposure to losses; this is called **risk management**. Prior to the start of construction, owners and contractors try to identify potential risks and their seriousness and to categorize them into (1) risks that they could try to eliminate through implementing safety programs or modified construction operations or as (2) risks where they could try to reduce their financial exposure through insurance, sharing risks, or self-insuring against risks. When contractors agree to perform a construction contract, they are obligated by the terms of the contract to obtain the types of insurance required in the contract. The cost of the required insurance policies is passed onto owners in the bid price. Insurance policies are legally binding contracts that obligate insurance companies to provide financial compensation for incidences and specified losses, as described in insurance policies.

This chapter provides information on risk management and discusses risk mitigation strategies such as professional liability insurance, the types of commercial insurance available, builder's risk insurance, and liability insurance.

19.2 INDEMNITY

In an attempt to protect themselves from **open-ended liability**, owners normally include a clause in construction contracts that indemnifies and holds harmless the owner against any acts or actions of the contractor that result in losses, injuries, or deaths. Even though **indemnity clauses** are a standard practice in the construction industry, sometimes the judicial system has invalidated these types of clauses by indicating that owners are not able to use a contractual clause to waive their liability. In addition to indemnity clauses being included in contracts, the legal system also provides some measure of indemnity in certain cases.

19.3 PROFESSIONAL LIABILITY INSURANCE

In addition to there being contractual requirements for various insurance policies during construction that are required for contractors, engineers also secure insurance policies to try to protect themselves from liability related to either errors or omissions in the construction documents, including the design, plans, and specifications. **Professional liability insurance** is not as common as the other types of insurance required for construction projects because professional liability insurance policies are expensive.

One example of the prohibitive cost of professional liability policies is a three-person structural design firm that is operating in the state of California, and all three of its members have professional engineering licenses as structural engineers and as professional engineers. The yearly cost of the professional liability insurance premium for the firm is $75,000. In order for the firm to cover the salaries of the principals of the firm, along with the cost of the professional liability insurance premiums, the firm has to either perform a high volume of work per year or charge noncompetitively high fees. As a consequence of the high cost of professional liability insurance, many small to medium-size engineering firms may not have professional liability insurance.

One option for small design firms that do not have professional liability insurance is to form a corporation and ensure that none of their personal assets are commingled with the assets of the firm. Only the assets of the firm would be in jeopardy if the firm were sued for negligence. If a firm does not own many assets, then it is unlikely that someone will sue the firm since they would not be able to receive much in the way of compensation if they win the lawsuit. If a firm is not incorporated, the same type of strategy could be used if the assets of the owner were transferred to someone else. If the owner of the firm is sued and the plaintiff wins the case, there would be no assets to pay the judgment.

19.4 COMMERCIAL INSURANCE

This section describes the basic types of **commercial insurance** that are required for construction projects including:

1. Worker's Compensation
2. Employee Liability
3. Comprehensive General Liability
4. Property
5. Liability Insurance to Protect the Owner
6. Motor Vehicle
7. Unemployment
8. Social Security and Medicare

When an insurer provides any type of commercial insurance policy, the insurance firm is assuming financial responsibility for specific types of loss, as stated in insurance policies. Although there are standard types of insurance policies, insurance companies also provide nontraditional insurance policies for items such as body parts, if body parts are instrumental in the profession of the policy holder such as arms for professional football quarterbacks or noses for singers. These types of policies are usually underwritten by Lloyds of London and the premiums for these policies are expensive compared to traditional insurance policies. As long as a policyholder is willing to pay the cost of the premiums for insurance policies, companies will underwrite policies for unusual insurance.

For construction projects, insurance companies evaluate the insurability of individual contractors and the policy premiums charged for insurance policies are based on the historical safety and liability records of the contractors. If a construction firm has a history of never having a claim or a minimal amount of claims against their insurance policies, then the insurance premiums they will be charged will be lower than the premiums charged for firms that have filed claims, thus providing this contractor with a competitive advantage because the contractor is able to pass on the cost savings for premiums to the owner in their bid price.

Unfortunately, the prevention of having claims on policies could also lead to questionable practices at construction job sites. A worker may be asked to wait until the end of the day to go to see a doctor after they have been injured on the job, because if they complete the day, the injury is not considered a **loss time accident**. It is not counted against the firm by the insurance company. For some minor injuries, this practice may not lead to any type of permanent injury or harm to the worker, but for more severe injuries it could have lasting consequences. At one construction job site a worker was asked to hold onto the tip of his finger that had been severed by a saw and to wait thirty minutes until the end of the day to leave to see a doctor to have the tip of his finger surgically reattached to his finger.

Contractors should be aware that no matter which type of insurance policies they secure, the policies do not cover deliberate actions that cause a claim, since insurance policies only cover accidental incidents. Negligence and oversight could be covered by insurance policies, as long as the negligence is not willful negligence. Each case is reviewed by the insurance company on its own individual merits before there is a determination on whether the insurance company will cover the damages that result from a claim against the insurance policy.

The following sections describe some traditional insurance policies that are required for construction projects.

Worker's Compensation Insurance

Worker's compensation insurance is required in all of the fifty states, as set forth by Worker's Compensation Acts. Worker's compensation is paid by employers based on a different multiplier for each craft times the wages earned by the workers in each craft into a government fund. It covers employees when they sustain a personal injury or contract a disease and there is a direct causal relationship between the work environment and the disease contracted by the employee. Worker's compensation benefits are not based on proving that an employer caused an injury or disease, as claims are only based on proving a connection between the work performed and the injury sustained or the disease contracted by the employee.

If an employee accepts worker's compensation benefits, they should be aware that by accepting them the employee is waiving their right to sue their employer. Since worker's compensation benefits start shortly after an injury occurs, an employee may not yet know the long-term consequences of their injuries and therefore will not have the option to sue their employer at a later date when they are more aware of the severity of their injuries. Worker's compensation benefits include:

- Lost wages
- Economic loss
- Past and future medical expenses
- Benefits to dependents if someone is killed

With workmen's compensation claims there is no punitive damages or compensation for pain and suffering. That is why some claimants file lawsuits rather than merely accepting what they are entitled to through their workmen's compensation insurance policy.

If an employee causes their own injury, or the injury results from willful misconduct on the part of the employee, then the employee would not be eligible for worker's compensation benefits.

Unemployment Insurance

Employers are responsible for paying unemployment premiums to both the federal and state governments for each employee. The amounts paid are based on the salaries of their employees. **Unemployment insurance** rates, which were first implemented in 1935 at the federal level, were 6.2 percent of wages in 2009 and state payments receive a credit of up to 5.4 percent; therefore, the federal portion would only be .8 percent if 5.4 percent is paid at the state level. Employees are not eligible for unemployment premiums if they voluntarily separate from their employer. Unemployment benefits could only be collected if an employee is terminated by their employer either by being laid off or fired from their job.

Unemployment premiums vary based on the salary that an employee was being paid when they were terminated from work and the unemployment premium rates set by each individual state. Unemployment premiums are only paid to workers for a limited amount of time, such as six months, after they are terminated from their jobs. Workers are required to demonstrate that they are actively seeking employment while they are receiving unemployment benefits to remain eligible for collecting benefits.

The unemployment rates published by the government are based on the number of workers collecting unemployment not on the number of workers who are no longer eligible for benefits because it is past the allotted time in which they are able to collect benefits. Therefore, the unemployment rates published by the government may be artificially lower than the actual number of workers who are still seeking employment. One example of this occurred in June 2009 when the published unemployment rate was 9.0 percent and the rate for people actually seeking employment who were no longer eligible for unemployment insurance benefits or who were underemployed was 15.8 percent (*U.S. News and World Report*, July 2009).

Social Security

Employers and employees both pay into the **social security** system, which was established in 1935, on an equal basis at a rate of 7.65 percent for a total of 15.30 percent (2009 data), of the salary of an employee up to the maximum yearly earning limit of $106,800. Of the 7.65 percent paid to the government, 6.255 percent is applied toward social security benefits and the remaining 1.45 percent is for Medicare (U. S. Social Security Administration 2009). After an employee reaches the maximum limit set by the government, which was $108,600 in 2009, for payment for social security each year additional payments are not made for the remainder of the calendar year.

Monthly social security benefits are paid to employees or their survivors when (U. S. Social Security Administration 2009):

1. An employee reaches retirement age. Employees are eligible for reduced monthly social security benefits at sixty-two years old or their entire monthly premium at a later age that depends on the year when they were born. For some it is sixty-five and for others it is sixty-seven and a half and for others it is later. The age at which someone is allowed to file for social security benefits for retirement could be changed by an act of Congress.
2. An employee becomes permanently disabled.
3. An employee passes away.

Liability Insurance

Liability insurance is used to compensate victims for injuries, death, or for loss of property, but there are also other forms of liability insurance. During construction, contractors have **contingent liability** because they are responsible for the actions of their subcontractors; therefore, they could have contractor's **public liability and property damage insurance** to protect them from negligent acts by their subcontractors. Contractors could also have **contractual liability insurance** to protect them if they are sued by owners. Contractors normally have **comprehensive general liability insurance**, which covers other incidents that could arise during construction.

Public liability and property damage insurance is used to protect contractors when third parties are injured at construction job sites or personal property is damaged during construction activities. When children enter jobs sites and they are injured or killed, the legal system considers job sites to be **attractive nuisances** and contractors are responsible for preventing access to their job sites by fencing them and/or providing security.

There is increasing concern about the liability of firms for material posted on websites by their employees. If employees post material anonymously, then Section 2003 of the **Communications Decency Act of 1996** protects websites from liability for the postings (United States Government 1996). If the postings are not anonymous, then firms could be liable for the materials posted online by their employees; this has led to some firms in the United States (approximately 40 percent) monitoring the e-mails and web searches of their employees to try to prevent situations from arising that would involve company liability.

Builder's Risk Insurance

Contractors invest in **builder's risk insurance** because it covers the project risks named in the policy and also temporary facilities including materials and supplies while they are in temporary storage, during transit, and after they are delivered to the job site. It also covers the tools and equipment that are located at job sites. Builder's risk policies provide insurance against direct physical damage or loss from external causes.

In addition to builder's risk insurance, contractors could also obtain riders to cover any or all of the items that are excluded from the policy. Riders are additional policies that cover specific items; an extra premium is paid to obtain them. The typical exclusions to builder's risk policies are **XCU exclusions**—explosions, collapse, and underground damage. Other exclusions might include:

1. Freezing
2. Glass breakage
3. Earthquakes
4. Subsidence
5. Floods
6. Nuclear radiation
7. Cost of correcting defective workmanship
8. Losses during testing
9. Loss or damage due to errors or omissions

How items are covered by builder's risk insurance policies varies, as policyholders could be reimbursed for actual value or replacement costs. Compensation could be computed on work in place, completed value, or reported value or it could be calculated differently for each project. Builder's risk policies might contain a **subrogation** clause that allows the insurance company to sue for recovery, as if they were the actual insurance policyholder.

Umbrella Excess Liability Insurance

Contractors may also obtain **umbrella excess liability insurance** that increases their insurance policy coverage limits to a higher value and also covers all of the risks that are not accounted for in standard insurance policies.

Owner's Protective Liability Insurance

Owners may elect to obtain **owner's protective liability insurance** to insure against accidents that occur because of something the owner did or did not do and for which the liability cannot legally be passed on to the contractor.

Miscellaneous Insurance

Engineers and constructors may also provide insurance to themselves and their employees; the most common types of insurance are medical and life insurance. Other types of insurance that they might choose to offer their employees include dental, vision, and disability insurance. Disability policies usually pay up to 60 percent of the salary of an employee for **long-term disability**. **Short-term disability policies** pay wages, or partial wages, until an employee reaches the allotted amount of time to be eligible to collect long-term disability.

19.5 GREEN BUILDING TECHNOLOGY LIABILITY

In the emerging area of green building, there are currently no new types of liability insurance that would insure against green liability such as (ENR, July 14, 2008):

- Failure to obtain the promised green certification for a project
- Failure of a green design to deliver expected results
- Problems with new products and designs
- New products not providing the green savings as advertised
- Delays from lack of green-product availability

"There are nearly 70 jurisdictions in 28 states that call for some form of green building. Many legal experts believe that if a building falls short of these requirements, the designer and contractor could be subjected to negligence claims" (ENR, July 14, 2008, 11).

With the advent of the implementation of **Green Building Technologies**, the incorporation of sustainable design and construction into projects could increase the potential for litigation during or after construction. Because green building projects require a highly coordinated team effort, there is not one entity that controls the entire process. If any of the team members fails in their assigned task and the structure is not awarded Leadership in Energy and Environmental Design (LEED) certification, or it does not obtain the levels of compliance mandated by law for sustainable structures, then issues of responsibility for this failure could lead to legal battles over who is liable (ENR, July 14, 2008).

In order to address the risks and responsibilities associated with sustainable design and construction, new types of contracts are being investigated that would help define the roles of each party and provide an equitable allocation of the contractual risk. In the future, new types of liability insurance may arise that protect the various team members from liability for failure to achieve the predetermined sustainable goals.

Chapter Summary

This chapter provided insight into the types of risks that owners and contractors attempt to manage during construction projects. Although it is impossible to completely reduce risks, safety procedures and carefully monitoring construction operations could help to reduce the incidence of accidents. For the risks that cannot be completely managed, insurance policies are obtained by owners and contractors to try and protect the firm when workers or third parties are injured or killed while at construction job sites. Insurance policies also help protect contractors from damage sustained during construction, loss of materials and equipment, motor vehicle accidents, and the other perils of construction.

This chapter covered some of the basic insurance policies that could be required on construction projects such as builder's risk, comprehensive general liability, professional liability, contractual liability, owner's protective liability, and umbrella excess liability. In addition other forms of risk management were presented including worker's compensation, social security, and unemployment insurance.

Key Terms

attractive nuisance
builder's risk insurance
commercial insurance
Communications Decency Act of 1996
comprehensive general liability insurance
contingent liability
contractual liability insurance
Green Building Technology
indemnity clauses
liability insurance
long-term disability
loss time accident
open-ended liability
owner's protective liability insurance
professional liability insurance
public liability and property damage insurance
risk management
short-term disability policy
social security
subrogation
umbrella excess liability insurance
unemployment insurance
worker's compensation
XCU exclusions

Discussion Questions

19.1 Why is it important for construction firms to provide worker's compensation insurance to its employees?

19.2 Which types of insurance protect third parties when they are injured or killed while at construction job sites?

19.3 Who benefits from an indemnity clause in a construction contract and how do they benefit?

19.4 Is it possible to obtain insurance for nontraditional items such as body parts? If so, why might someone pay for this type of policy?

19.5 How does unemployment insurance help workers who are terminated? How does someone become eligible for it?

19.6 How does professional liability insurance protect engineers?

19.7 When is an employee eligible to collect social security?

19.8 What are the different types of liability insurance?

19.9 Explain attractive nuisances at construction job sites.

19.10 What types of items are excluded from builder's risk insurance policies?

References

ENR. July 14, 2008. Insurers worry about green building risk. *Engineering News Record*, 26(1): 10–11.

U.S. Government. 1996. Communications Decency Act. Washington, DC: Government Printing Office.

U.S. News and World Report. July 2009. A lot can still go wrong, 79–80.

U.S. Social Security Administration. 2009. *Trust Fund Data.* www.ssa.gov/OACT/ProgData/taxRates.html.

CHAPTER TWENTY

INTERNATIONAL LAW, CONTRACTS, AND ARBITRATION

20.1 INTRODUCTION

When working on global engineering and construction projects, it might be difficult to determine which legal system has jurisdiction over contractual disputes. Therefore, international contracts will indicate that international arbitration or a specific legal jurisdiction will be used to settle claims and disputes.

Examples of issues that increase the legal complexity of global engineering and construction (E&C) projects include: regime changes that create instability and uncertainty on engineering and construction projects; injured workers being from several foreign countries; foreign government restrictions and laws; construction material delivery problems when clearing customs; terrorist attacks; and kidnappings for ransom or for political reasons.

This chapter discusses global contractual legal issues including: regional legal issues, international contract clauses, claims and change orders, international arbitration, terrorism and kidnapping insurance, regime changes, and liability issues. International contract clauses are discussed in relation to how specific clauses might impact global engineering and construction projects and the ramifications of contract clauses when they are interpreted by judges in foreign legal systems.

20.2 INTERNATIONAL CONVENTIONS

The *Legal Guide for Drawing Up International Contracts for the Construction of Industrial Work*, provided by the **United Nations Commission on International Trade and Law** (UNICITRAL), is frequently utilized as a reference during the development of international contracts (UNCITRAL 1988). The UNCITRAL book provides an analysis of international contract clauses and discusses global legal issues related to contracts.

The **Martindale-Hubbell International Law Digest**, which is updated yearly, contains information on international legal conventions and laws used throughout the world (Martinale-Hubbell 2000). Examples of legal conventions include the following:

1. The **Convention on the Recognition and Enforcement of Foreign Arbitral Awards**, which is also known as the **New York Convention**, requires foreign courts to honor international arbitration awards (UNCITRAL 1988).
2. The **United Nations Convention on Contracts for the International Sale of Goods** provides information on how to draft international contracts (O'Hare 1980).
3. The **International Commercial Terms** (INCOTERMS), which is published by the **International Chamber of Commerce**, defines terms that are used in international contracts (Murphy 2005).

20.3 REGIONAL LEGAL JURISDICTIONS

Common law, civil law, Shari'a (Islamic) law, and Asian legal systems are discussed in Chapter 2. Common law systems are used in the United States, the United Kingdom, and former British colonies. Civil law systems are used in Europe, South America, Scotland, Quebec, and Louisiana.

20.4 INTERNATIONAL ENGINEERING AND CONSTRUCTION CONTRACTS

International engineering and construction (E&C) contracts are similar to the types of E&C contracts that are used in the United States but they also contain additional clauses that address situations that arise that are related to the global nature of international projects. In the international arena, the construction contracts used most frequently are the standard forms published by:

1. The **Federation Internationale des Ingenieurs-Conseils** (FIDIC) (**International Federation of Consulting Engineers**).
2. The **Engineering Advancement Association of Japan** (ENAA).
3. The **Institution of Civil Engineers** (ICE).
4. The **American Institute of Architects** (AIA).

The Federation Internationale des Ingenieurs-Conseils (FIDIC)

The FIDIC standard forms that are available are the following (Jenkins and Stebbings 2006, 14):

- *Conditions of Contract for Building and Engineering Works Designed by the Employer* (Red Book)
- *Conditions of Contract for Plant and Design-Build for Electrical and Mechanical Plant, and for Building and Engineering Works, Designed by the Contractor* (Yellow Book).
- *Conditions of Contract for EPC/Turnkey Projects* (Silver Book)
- *Short Form of Contract* (Green Book)

The Red Book is "the first choice of all major international development banks and agencies for use as general conditions of contract on a construction project financed by them. These institutions include" (Jenkins and Stebbings 2006, 14):

- Asian Development Bank (ADB)
- African Development Bank (AFDB)
- Caribbean Development Bank (ADB)
- Commission of the European Communities (CEC)
- European Bank for Reconstruction and Development (EBRD)
- European Investment Bank (EIB)
- Inter-American Development Bank (IBD)
- International Bank for Reconstruction and Development (IBRD)
- United Nations Development Programme (UNDP)

The Engineering Advancement Association of Japan

The Engineering Advancement Association (ENAA) of Japan provides a contract that is used for process plant construction called the *Model Form International Contract for Process Plant Construction* (ENAA Process Plant Model Form 1992). This contract contains five volumes (Jenkins and Stebbings 2006, 15):

1. Volume 1—Agreement and General Conditions
2. Volume 2—Samples of Appendices
3. Volume 3—Guide Notes
4. Volume 4—Work Procedures
5. Volume 5—Alternative Form without Process License

The ENAA also publishes the *Model Form International Contract for Power Plant Construction* (ENAA Power Plant Model Form 1996) that consists of three volumes (Jenkins and Stebbings 2006, 15):

1. Volume 1—Agreement and General Conditions
2. Volume 2—Samples of Appendices
3. Volume 3—Guide Notes

The Institution of Civil Engineers

The Institution of Civil Engineers (ICE) publishes two standard forms: (1) the *ICE Conditions of Contract* and (2) the *New Engineering Contract* (NEC). The ICE has over 70,000 civil engineers as members that are from the United Kingdom, China, Russia, India, and 140 other countries (Jenkins and Stebbings 2006). The *ICE Conditions of Contract* were developed by the Conditions of Contract Standing Joint Committee (CCSJC) that includes civil engineers from the **Civil Engineering Contractors Association** (CECA) and the **Association of Consulting Engineers** (ACE). The *ICE Conditions of Contract* include (Jenkins and Stebbings 2006, 15):

- Measurement Version
- Design and Contract
- Term Version
- Minor Works
- Partnering Addendum
- Tendering for Civil Engineering Contracts
- Agreement for Consultancy Work in Respect to Domestic or Small Works

The ICE also publishes the *Engineering and Construction Contract* (ECC) that includes "the ECC and a subcontract, a professional services contract, an Adjudicator's (arbitrator) contract, a short contract, a short subcontract, and the NEC partnering option" (Jenkins and Stebbings 2006, 16).

The American Institute of Architects

The American Institute of Architects construction contract documents are discussed in Chapter 12 under Engineering Contracts.

20.5 INTERNATIONAL CONSTRUCTION CONTRACT CLAUSES

This section includes information on the types of clauses that are included in international construction contracts. It also provides information on how and why some of these clauses affect international projects.

In order to keep funds in the economy of a host country, international construction contracts might be divided into two contracts: (1) for work performed within a host country and (2) for work performed outside of a host country (Lantis 2005). Table 20.1 provides a summary of some of the clauses that are typically included in international contracts

Table 20.1 Synopsis of Common International Contract Clauses

Clause Title	Highlights of Clauses	
Technical Standards and Inspections	Safety	Local regulations.
	Environment	Local regulations.
	Technical	International or local or other country.
	Testing	Can use national or international testing institutions.
	Inspections	Municipal governments, for safety, health, and environmental compliance.
Confidentiality	Local governments might require all documents for a project, which compromises confidentiality. Hard to enforce as parties to the contract are subject to the laws of different nations.	
Patents and Trademarks	Only protected in the country issued.	
Currency Clauses	Contractor paid in the local currency with a multiplier for the currency exchange rate. Can specify payments in a different currency. Subcontract may have to be paid in local currency.	
Language	May be different versions of the contract in different languages. Original version controls.	
Local Subcontractors and Suppliers	Contract may require the use of local subcontractors and suppliers. Use local subcontractors and suppliers to maintain goodwill.	
Transporting Equipment	May be size and weight restrictions; permits may be required.	
Laws of Host Nation	Owner provides local building permits; owner helps contractor get visas and local work permits for personnel. Customs duties on imported equipment. May be duties on some exported items. Duties on construction equipment required, even if it will be exported at the end of the job. Contract should state who pays fees. May be transit taxes on items shipped. Some things are illegal to import or export—check with respective governments. If duties increase after a contract is signed, a contractor may file a change order.	
Liens Subcontracting	Owner may require proof of payment to subcontractors and suppliers. Find out if the government allows liens to be placed on property if subcontractors or suppliers are not paid. May be local limits on how much work is done using subcontracts. Government may require a certain portion of the contract to be performed by local nationals. If used, foreign subcontractors may have to pay taxes locally and in the other country.	
Bankruptcy of the Contractor	Owner may be required to make back payments to subcontractors if they take over the subcontract.	
Liquidated Damages	Clause may be voided in some countries if it is being used to punish a contractor. The amount could be reduced by the courts, if could prove that there was partial performance.	
Hardship	Allows a renegotiation of the contract based on hardship.	
Index and Currency	If prices for products rise faster than the inflation rate, this clause allows for renegotiation of the contract.	
Exemption	One party is not able to seek damages if the other party does not perform due to something in the exemption clause occurring. Examples include: severe weather, natural disasters, civil strife, riots, war, destruction of the subject matter, fire, flooding, and the failure of a government to approve a project.	
Termination Clause	Owners may terminate a contract for any reason if the clause is included in a contract. Some governments do not allow termination unless it is stated in the contract. Should state the specific conditions for termination. A contract may allow a contractor to terminate a contract if the owner does not pay, orders a suspension of work, or files for bankruptcy. Some countries do not allow termination due to bankruptcy. Proceedings similar to bankruptcy may be called: receivership, liquidation, insolvency, assignment of assets, reorganization. If a contract is terminated, it may void contractual obligations, such as dispute settlement and confidentiality.	
Spare Parts and Maintenance after Construction	Contractor provides spare parts and repairs facilities for a set period of time. Check whether owner or contractor pays if equipment is shipped out of a country for repairs.	
Choice of Law and Choice of Forum	Allows parties to select which legal system will be used to litigate contractual disputes. If there is no choice of law clause but an arbitration clause, arbitrators decide which laws apply. May be limited to a country with a connection to the contract. Could be different legal systems for different legal issues. May stipulate an exclusive jurisdiction and a specific court.	
International Conventions	See subheading on international conventions.	

References: UNCITRAL 1988; Bonell 2000; Stokes 1978.

and also includes a brief explanation of why each of the clauses listed is important to international contracting.

The following sections elaborate on some of the essential international contract clauses.

Technical Standards and Inspections

Host country (country in which an engineer or contractor is working) legal jurisdictions often regulate the technical aspects of E&C projects, as well as safety and pollution, by enforcing local standards. Contracts for construction services in a foreign country benefit from using technical standards that are internationally accepted if they are familiar to local contractors and suppliers. If local inspection and testing institutions are used on projects that are in unfamiliar locations, local contractors will understand the testing procedures and processes. If governments restrict access to facilities where equipment and machinery are being manufactured, then owners should specify testing agencies that have access to the restricted facilities.

Government authorities usually inspect construction and test facilities to ensure that they comply with safety, health, and environmental regulations; therefore, it is important to be familiar with host country standards that could be enforced on local construction projects. If governments change inspection and testing requirements after a contract has been signed, then contractors are entitled to change orders that reflect the new cost of complying with government regulations.

Confidentiality

If the governments of host nations require that all project documentation be submitted to them, this limits the level of confidentiality that may be maintained on projects. **Confidentiality agreements** should be entered into before contract negotiations commence, but these agreements are difficult to enforce if parties to the agreement are from different nations. In the global arena, attorney-client confidentiality is not guaranteed, which creates problems for members of foreign and domestic firms that are competing for projects in foreign countries.

Patents and Trademarks

Patents and **trademarks** are only protected in the country where the patent or trademark is registered since there are no global patents and patent and trademark laws of one nation do not extend into other countries. Engineering and construction personnel may be reluctant to use proprietary processes, or products, in foreign nations unless they have had a previous (positive) relationship with a particular client.

The Use of Local Suppliers

Contractors could be required to use local subcontractors or suppliers to generate employment in a particular nation or owners might specify the use of local subcontractors and suppliers to earn and maintain goodwill in a local community.

Packaging and Transportation of Equipment and Materials

Shipping agencies have regulations restricting the size and weight of items to be shipped, which could preclude the use of large construction equipment on some projects. The shipping of hazardous materials could also be restricted by government regulations or **intragovernmental treaties**. **Transportation permits** are required for shipping large construction equipment, materials, and machinery. Contracts should specify who is responsible for obtaining and paying for these permits.

Laws Governing Engineering and Construction Personnel

Engineering and construction personnel are subject to the laws of host nations, as well as the federal laws of their native country, such as laws governing the bribery of foreign government officials. International treaties such as the Kyoto Protocol Treaty, which stipulates the amount of greenhouse gas emissions a country is allowed to produce per year, govern transactions between members of the E&C industry if the treaty has been ratified by host country governments prior to or while a project is being designed and built by an owner.

Local Construction Permits, Visas, Customs, and Duties

International contracts should indicate who is responsible for securing permits and licenses and who will pay for **customs duties** and **transit taxes** if they are required for materials used on construction projects. Customs duties are normally levied on materials and equipment being imported into a country and duties are imposed on items being exported to other countries. Whoever receives the items being shipped usually pays for customs and transit taxes.

Subcontracting

In legal systems where contractual relationships are only between contractors and subcontractors, not subcontractors and the owner, owners will not be able to recover losses from subcontractors, only from contractors. Owners should protect themselves from liens by requiring contractors to submit written proof when they pay subcontractors. If contractors order materials from foreign suppliers, they might have to pay taxes in the host country and also pay taxes in the country of origin of the materials.

Host country governments might require a certain percentage of participation by national contractors on domestic projects even if construction projects have foreigner owners. Owners could set requirements on the maximum amount of work that may be performed by subcontractors. Owners could also specify that a specific subcontractor be used on a project and if this situation arises, prime contractors should request that their liability be limited if they use subcontractors specified by owners (UNICITRAL 1988).

Liquidated Damages

The assessment and enforceability of **liquidated damages** varies throughout the world. Liquidated damage awards could be reduced, or eliminated, if they are merely being used to punish a contractor for not completing a project on time. In order for owners to be awarded liquidated damage, they have to prove that they sustained an actual monetary loss due to their project being delayed by the contractor. Reviewing previous court decisions from a local court provides insight into how the local legal system has ruled on prior cases involving liquidated damages.

Hardship Clauses

The United Nations Commission on International Trade and Law (UNCITRAL) defines **hardship** as "a change in economic, financial, legal or technological factors that cause a serious adverse economic consequence to a contracting party, thereby rendering more difficult the performance of his contractual obligations," which means that some type of hardship makes performing a contract more difficult, but not impossible to perform (UNCITRAL 1988, 242). A hardship clause allows owners and contractors to renegotiate a contract based on the conditions that create hardship. Hardship clauses are not allowed in all legal systems, and some legal systems may not uphold the clause when it is challenged in a court of law. Contractual parties have the right to agree to renegotiate a contract even if a hardship clause is not in a contract, but the presence of a hardship clause makes it easier for contractors to renegotiate a contract.

Index and Currency Clauses

When prices for certain products increase faster than the average inflation rate, having an **index clause** allows for automatic adjustments to payments when there are fluctuations in exchange rates. **Currency clauses** are also used to account for fluctuations in exchange rates.

Exemption Clauses

In international contracts, **exemption clauses** might be used in a manner similar to impossibility clauses. Items that could be included in an exemption clause include: abnormally severe weather or natural disasters; civil strife, riots, or war; destruction of subject matter; fire; and flooding. Exemption clauses could also include interference by host governments, acts of terrorism, coups, or host government agents not approving projects.

Termination Clauses

In some regions of the world, owners are allowed to terminate contracts merely for their convenience even if there is no specific clause in a contract. Other governments do not allow such **unilateral termination** unless there is a **termination clause** in the contract. Some legal systems require judicial consent for owners to terminate contracts without termination clauses. If termination clauses are used, contracts should state the conditions under which an owner is allowed to terminate a contract. Some contracts state that contractors are allowed to terminate contracts if owners do not provide prompt payment, if an owner orders a long suspension of work, or if an owner files for bankruptcy. Some judicial systems do not allow parties to terminate contracts solely on the basis of the bankruptcy of the other party to a contract. Bankruptcy proceedings exist throughout the world, except in India, and they may be called: receivership, liquidation, insolvency, the assignment of assets, or reorganization (UNCITRAL 1988).

In some legal systems, if a contractor is terminated, they could lose their legal right to use dispute settlement techniques, and also waive their rights as stated in their confidentiality agreement, unless a contract states that these rights are valid even when either party terminates a contract.

Choice of Law, Choice of Forum, and Exclusive Jurisdiction Clauses

Choice of law clauses, **choice of forum clauses**, and **exclusive jurisdiction clauses** indicate which legal system has jurisdiction over contracts. If contracts do not have any of these clauses, but they contain an arbitration clause, then arbitrators have the right to choose which laws will apply to a contract, and these laws might be a combination of laws from different legal systems (Stokes 1978). The legal system selected by arbitrators does not have to be from a host nation or the native country of any of the parties to the contract. The parties writing the contract should make sure that the legal system used in the country where the project will be built will uphold an exclusive jurisdiction clause.

There could be local legal restrictions imposed on contracts due to the choice of law that governs items such as: job site safety standards, environmental protection, the importing and exporting of materials, taxes and duties, and the transfer of ownership of property. A local legal system might restrict the choice of law to a place that has a connection to a contract, such as a host nation or the legal system of the native country of an owner or a contractor (UNCITRAL 1988). Choice of law clauses cause confusion when contracts specify different choices of law for each legal issue that arises during the performance of a contract (Rubino-Sammartano 2001).

A **Calvo clause**, which is named for Carlos Calvo who was a nineteenth-century Argentinean diplomat, is an exclusive jurisdiction clause that specifies that legal disputes will be litigated by the court of a host nation (Bondzi-Simpson 1990). Calvo clauses were developed in 1927 by the **United States-Mexican General Claims Commission**. They stipulate that, when a plaintiff has a legal claim, they must first seek legal recourse in the local jurisdiction before they seek restitution for their claim in a foreign court (Stokes 1978).

Legal Issues in International Joint Ventures

International joint venture (IJV) agreements should include a warranty clause that waives the liability of partners if there is any pending litigation, or omission of facts that might cause litigation. International joint venture bylaws are frequently used in common law countries, but not in civil law countries. How the IJV will be managed should be stated in writing along with who has the authority to make decisions and who has veto power (Wolf 2000).

When forming an IJV, potential partners should verify that contracts with suppliers and customers that existed before an IJV is formed would still be valid and remain in force if a firm that has a contract with the suppliers and customers enters into an IJV. It is possible that licenses could be terminated when a company is sold or if it enters into an IJV to prevent the transfer of proprietary data, or information, to competitors (Wolf 2000).

20.6 CLAIMS AND CHANGE ORDERS

Legal claims become complicated when the governments of the home nation of employees and owners become involved in contractual disputes in addition to the nation where a project is being built and the government where a firm is incorporated.

Adjustment and Revision Index Clauses

Adjustment of price is when a contractor is paid a different amount from what is stated in the contract when the scope of work on a project changes. A **revision of price** clause is used to compensate contractors for changes due to fluctuating exchange rates and inflation. A **revision index clause** allows changes to be made to the payments to contractors based on a specific index when there is inflation. If an index clause is included in a contract, an algebraic formula should also be included for revising payments based on a price index (UNCITRAL 1988).

The Defense of Contractors by Home Nations in Contractual Disputes

The legal theories of **denial of justice** and **expropriation** are involved when the government of one party to a contract acts on the behalf of a party from their nation based on the concept that a nation is damaged if a company from that country is damaged in a lawsuit.

To avoid denial of justice, governments try to ensure that their citizens receive due process of law, including access to the legal system and a fair trial, when they are involved in a lawsuit in another country (Stokes 1978). For a denial of justice claim to be valid, someone has to demonstrate that the judicial process is being delayed or that there is prejudice or hostility involved in the legal proceedings making it impossible for a foreign national to receive a fair hearing. Foreign nationals may only seek help from their native country prior to using a local legal system (Stokes 1978). Usually foreign ministers or state department officials negotiate a settlement instead of taking cases to an international court. In the United States, the **Hickenlooper Amendment** allows the U.S. government to suspend aid to a foreign country that seizes property that is owned by U.S. citizens, but this law is primarily used as a negotiating threat (Stokes 1978).

Expropriation occurs when a foreign nation seizes the property of a foreigner, such as construction equipment, without proper compensation. Foreign governments do not even have to seize or transfer the title of property for a case of expropriation to arise, since they only have to prevent an owner from using their property (Stokes 1978).

Typical Causes of Claims and Change Orders

Some examples of issues that could arise on international contracts that might lead to claims being filed for extra payments include the following:

- Local regulations for testing structures could be changed after a contract has been executed. If this occurs, a contractor may submit a claim for any additional costs associated with the new testing requirements for the project.
- Contractors are not able to finish projects in developing countries on schedule because they do not have enough laborers, equipment, technically competent employees or supervisors, or delays by subcontractors (Scott 1993).
- Owners could file a claim if projects are delayed due to contractors not importing enough employees or arranging for subcontracts in a timely manner.
- Guarantees could be voided if contracts do not specify how change orders affect guarantees (UNCITRAL 1988).
- Owners do not provide prompt payment; therefore, contractors are entitled to submit a claim for interest on late payments.
- Interest rates for late payments are specified in contracts, and applicable local laws might also stipulate interest rates, and the two interest rates could be in conflict. Some legal systems do not allow interest to be paid on late payments, such as is the case in some Islamic countries.
- Local, regional, or national government officials interfere with the progress of a project.
- Materials or supplies are not released from customs in a reasonable amount of time.
- An owner substantially changes the scope of work.
- An owner requires a contractor to use methods that are different from the original methods, as stated in the contract.
- Materials that are required in the specifications are no longer available.
- When concurrent delays are caused by an owner when they are self-performing part of the construction work.
- When an owner does not provide materials or supplies that they are required to provide for a project.

Another issue that could lead to claims being filed for extra payments in the global arena includes miscommunication between owners and contractors. Therefore, it is important for owners to provide instructions in writing and to make sure that their instructions are properly translated into the language of the contractor.

Cardinal Changes to Contract

If contracts are changed more than a certain percentage, usually 10 percent in the United States, this is considered to be a **cardinal change** and it could void a contract. Before doing work in a country, it is important to know the extent to which an owner is allowed to change the scope of work without voiding a contract because contractors could end up performing an unreasonable amount of extra work if there is no limit to the number of changes allowed during the execution of a contract.

20.7 GLOBAL CONTRACT DISPUTE RESOLUTION TECHNIQUES

In the global arena, various methods are used to resolve disputes. Some of the primary techniques are negotiation, mediation, conciliation, arbitration, litigation, and dispute review boards. Chapter 19 covers both traditional and alternative dispute resolution techniques. The most frequently used contract dispute resolution techniques in the global arena is international arbitration, which is discussed in the next section.

20.8 INTERNATIONAL ARBITRATION

International arbitration provides a nonjudicial international forum in which to settle disputes. In order to settle claims, three arbitrators are selected by the disputing parties; the arbitrators are invested with the power to settle disputes. Many different arbitration associations provide lists of arbitrators and in the global arena the **International Arbitration Association** is one of the most frequently used arbitration organizations.

During international arbitration some legal systems allow parties to obtain interim relief while they are waiting for the settlement of a dispute. Arbitration awards have to be enforceable, but the enforceability of awards is affected by government restrictions.

There are several reasons why arbitration is a more efficient dispute resolution technique than litigation, including that it are less formal than court hearings; parties are allowed to select arbitrators who are experts in the fields of architecture, engineering, or construction; and arbitrators usually respect the contract choice of law, whereas court systems may not always honor the contract choice of law. Other advantages to arbitration in the international construction environment are that a contract may specify the location of arbitration proceedings and the language of the proceedings. Arbitration proceedings are less disruptive than litigation, and the results of the proceedings are confidential. If legal proceedings are used to settle a dispute, there is no confidentiality. Arbitration is usually faster and less costly than judicial proceedings due to international conventions that apply to international arbitration.

Legal systems might allow judicial challenges to arbitration clauses and to arbitration awards. Arbitration awards could be overturned if the arbitrators were incompetent, if a party was unable to thoroughly explain their case, if the **arbitral tribunal** was not conducted in accordance with the contractual requirements, or if the arbitration award does not follow public policy (UNCITRAL 1988).

In the People's Republic of China (PRC), arbitration proceedings are conducted according to the *Beijing Arbitration Commission's Arbitration Rules* (BAC Rules). Arbitration awards are enforceable according to the *1958 Convention on the Recognition and Enforcement of Foreign Arbitral Awards,* the New York Convention, because the PRC adopted the New York Convention in 1987. Arbitral awards from Hong Kong could be enforced in the PRC because of the *Supreme People's Court's Arrangement for Mutual Enforcement of Arbitral Awards between the Mainland and Hong Kong Special Administrative Region*. Arbitration awards in Taiwan may be enforced under the terms of the *Supreme People's Court's 1998 Directive for Recognition of Civil Judgments of the Courts of Taiwan Region*. Additional information is available related to arbitration in the PRC in the book PRC Construction Law—A Guide for Foreign Companies, published by the Commerce Clearing House, Asia Pte Limited (2006).

Agreement to Arbitrate and Arbitration Clauses

In order for arbitration to take place, all parties to an agreement have to agree to arbitrate either by including an agreement to arbitrate in the original contract or by an agreement that is created when a claim has not been settled through negotiations. Some legal systems make it a requirement that all parties have to agree to arbitration when a dispute is not settled through negotiation (UNCITRAL 1988). Parties to a contract should provide a written agreement to arbitrate at the beginning of the arbitration hearing even if they have already agreed to arbitrate because there is an arbitration clause in the contract.

Contracts usually specify an odd number of arbitrators, such as three, because when an odd number of arbitrators are used, there will not be split decisions and having three, or more, arbitrators provides arbitrators with diverse backgrounds.

The contract or an agreement to arbitrate should specify the location of the arbitration proceedings, since local jurisdictions regulate arbitration tribunals within their jurisdictions. Arbitration proceedings should be located in a jurisdiction where awards will be enforced, such as in a jurisdiction where one of the parties or their assets are located; or in a jurisdiction within a nation that has an international convention with one or more of the nations involved; or in a jurisdiction with arbitration regulations that are well suited for international cases. Other issues that should be considered in selecting the location for arbitration proceedings are whether a location is convenient for the parties involved in the dispute, the availability of facilities, and the availability of support from members of international arbitration institutions.

If arbitration proceedings are in the home jurisdiction of a defendant, then the award may be more readily enforced than if the proceedings are in the home jurisdiction of a plaintiff. Specifying that more than one language be used in the arbitration hearings is useful but paying for translators increases the cost of the proceedings. If arbitration awards are not paid, it could result in a breach of contract claim (UNCITRAL 1988).

Authority of Arbitral Tribunal to Order Interim Measures

Some jurisdictions allow arbitrators to order interim measures before issuing an award. In jurisdictions where interim measures are not set by arbitrators, court systems might order interim measures even if the case is going be settled by arbitration (UNCITRAL 1988). However, if a party sues in a court of law to secure an interim measure, it could void their right to use arbitration proceedings (Knutson 2005).

The New York Convention

The *Convention on the Recognition and Enforcement of Foreign Arbitral Awards* is commonly known as the New York Convention. The New York Convention was written in 1958, but it was not widely used until the 1970s. The New York Convention requires defendants to prove that an award is invalid as opposed to the 1927 **Geneva Convention**, which requires plaintiffs to prove that an award is valid. The New York Convention only applies to the enforcement of an award and not to the hearing stage of a dispute. Arbitrators have to follow the procedures outlined in arbitration agreements; if they are not followed, then awards may be invalidated based on the New York Convention (Chukwumerije 1994).

Ad Hoc versus Institutional Arbitration

Under **ad hoc arbitration** contractual parties write their own rules for the arbitration process. **Institutional arbitration** occurs when parties agree to follow the rules of a particular arbitration institution and the rules usually are not altered by either party. An advantage of institutional arbitration is that the institution being used will provide staff members to conduct arbitration hearings (Redfern and Hunter 1986). Some examples of arbitration institutions include: the American Arbitration Association (AAA), the Inter-American Commission of Commercial Arbitration (IACAC), the International Centre for the Settlement of Investment Disputes (ICSID), the International Chamber of Commerce (ICC), the International Arbitration Association, the London Court of International Arbitration (LCIA), and the Stockholm Chamber of Commerce (SCC).

20.9 ANTICORRUPTION LEGISLATION

In the United States, the federal **Foreign Corrupt Practices Act** of 1977 prohibits questionable and illegal payments to foreign officials by U.S. citizens (National Law Journal, July 12, 2004). As of May 2005, less than one hundred cases had been prosecuted under this law in the United States since the act became law in 1977 (*New York Times*, May 17, 2005).

The U.S. **Sarbanes-Oxley Act** of 2002 compliments the *U.S. Foreign Corrupt Practices Act* in that the Sarbanes-Oxley Act also pertains to bribes in addition to accounting fraud (*New York Times*, May 17, 2005). In 1996, the World Bank Group enacted a process to disqualify corrupt companies from participation in bank-financed projects (*Broward Daily Business Review*, January 6, 2005).

The *Convention on Combating Bribery of Foreign Public Officials* of the *Organization of Economic Cooperation and Development's* (OECD) is a thirty-five nation treaty designed to reduce international bribery. Under this convention, Australia, Japan, Mexico, New Zealand, South Korea, and many European countries have stopped allowing companies to write off their bribes in foreign lands (*New York Times*, May 17, 2005). The *United Nations Convention Against Corruption* is another attempt at stopping international bribery; it was signed by members from ninety-five nations in a meeting in Marida, Mexico, on December 19–21, 2003 (*Broward Daily Business Review*, January 6, 2005).

The *American Society of Civil Engineers International Activities Committee*, Task Committee on Global Principles for Professional Conduct, which has 137,000 members, 15,000 of which are outside of the United States, in conjunction with the *International Federation of Consulting Engineers*, the *Pan American Federation of Consultants Institution of Engineers*, the *Japanese Society of Civil Engineers*, the *National Institute for Engineering Ethics* (U.S.), the *United Kingdom Institution of Civil Engineers*, and the *World Federation of Engineering Organizations*, along with sixty-six other engineering societies have been addressing issues of bribery, fraud, and corruption with worldwide agreements of cooperation related to "mutually acceptable principles and guidelines for the procurement of services and execution of work worldwide—zero tolerance for bribery, fraud, and corruption" (American Society of Civil Engineers, December 2004, 1).

20.10 KIDNAPPING AND RANSOM INSURANCE

When foreign expatriates are kidnapped, their employers might have **kidnapping and ransom** (K&R) **insurance**. Kidnapping and ransom insurance is an important consideration for members of E&C firms because there are 15,000 reported kidnappings per year, 70 percent of which are resolved through the payment of ransom. However, an estimate for the number of kidnapping victims who have the protection of K&R insurance is less than 1 percent (*Insurance Day*, October 22, 2003).

One estimate of the total value of K&R insurance premiums paid in 2005 was $150 million (U.S. dollars) and another estimate was between $120 and $130 million (U.S. dollars) per year (*Insurance Day*, October 22, 2003; *Financial Times*, July 23, 2003). The amount of money that kidnappers demand for ransom payments is between $19.5 million and $77.5 million (U.S. dollars) (*Lloyd's List*, December 19, 2003).

Executives in engineering and construction firms do not admit to having K&R insurance for fear that their employees would be targeted for kidnapping. Public disclosure of a K&R insurance policy typically voids the insurance policy (*Insurance Day*, September 2, 2004). Kidnappers could also be prosecuted on charges of money laundering in addition to kidnapping in some nations such as the United States and Japan (*Japan Economic Newswire*, September 30, 2004). Governments of host nations are reluctant to pay ransom because it encourages more kidnappings.

20.11 CHANGING GOVERNMENTS

Changing government regimes in a violent or dramatic manner presents particular risks when millions of dollars in heavy construction equipment are being used on projects. For example, the violent Iranian revolution that resulted in regime change in 1979 resulted in 6,000 claims, worth an aggregate total of more than $65 million (U.S. dollars) (Amin 1983). The international claims tribunal had difficulty processing all of the arbitration claims because the Iranian contracts included exclusive jurisdiction clauses specifying Iranian courts (Amin 1983). The **United Nations Compensation Commission** was set up after the Iraq invasion of Kuwait; it was a "political organ that performs an essentially fact finding function of examining claims, verifying their validity, evaluating losses, assessing payment and resolving claims" (Rubino-Sammartano 2001, 160–161).

Force Majeured and Acts of God or beyond anyone's control clauses may be used by contractors during periods of political instability. Another possible defense for contractors is the legal theory of impossibility, which are clauses that state that no one could perform the work due to its destruction or no one has the ability to perform the work in its present state (Murphy 2005).

20.12 LIABILITY ISSUES

This section discusses liability issues in the international contracting arena including sources of risk, risk of loss, taxation within a host country, insurance, professional liability and structural defect insurance, political risk insurance, multiple pledge of shares problems, worker's compensation insurance, contractor's guarantee and performance bonds, risk associated with currency variations, local labor laws, and risks associated with using the local workforce.

Sources of Risks

Risks may arise from uncertainties, technical issues, financial issues, political issues, and uncertainties in contracts (Williams 1992). Decision risk is highest during contract negotiations and as construction projects progress the risks associated with decisions declines (Wearne 1992).

The laws applicable to the **transfer of ownership of property**, including real estate, are usually specified by the local jurisdiction where the property is located at the time of the transfer. In most legal systems, building materials become a part of a project at the time the materials are installed, since it would be difficult to move them and reinstall them in the same manner and with the same quality. However, legal systems vary as to the specific time that equipment becomes part of a structure. Some legal systems require that equipment be permanently installed, yet other legal systems only require temporary installation for the transfer of ownership to occur. If an applicable law is not clear about when transfer of ownership takes place, then the contract should state when it occurs. For example, if International Commercial terms (INCOTERMS) are specified in the contract, then they will govern the transfer of risks associated with ownership (UNCITRAL 1988).

Risk of Loss

Legal systems will either assign risk of loss to the party in possession of the property or to the party that owns the property. In legal systems that assign risk, contractual parties are allowed to specify their own terms regarding transfer of risk with ownership. The risks associated with loss of or damage to the tools and equipment of a contractor are normally borne by contractors; this risk does not usually extend to owners (UNCITRAL 1988).

Taxation within a Host Country

To avoid having contractors inflate their overhead to avoid **taxes in a host country**, some countries require a **gross receipts tax** on international contracts because auditing international construction overhead is difficult. Requiring joint ventures with local companies helps track income taxes of firms from foreign countries. Host country taxes could be reduced if money is not transferred out of the host country. Contractors could also become incorporated in a nation that has a tax-reducing treaty with a host nation where they are conducting work (Stokes 1978).

Insurance

Lending institutions might require contractors to seek advice from insurance companies before undertaking overseas projects. If multiple insurance policies are being used on a project, the policies should be coordinated so that there are no gaps in coverage for some risks and double coverage for other risks. Host nations could mandate that the insurance companies insuring international construction projects be located within the host country and that they pay insurance claims in the local currency (UNCITRAL 1988).

Designers might have to provide insurance against errors and omissions in their designs when they perform work in certain countries. Some nations might hold contractors responsible for defects ten years after the project is completed; **professional liability and structural defect insurance** is available to cover this risk (UNCITRAL 1988).

Political risk insurance is available through private companies such as Lloyds of London and some governments such as the U.S. government (Kangari and Lucas 1997). The U.S. **Overseas Private Investment Corporation** (OPIC) assists with feasibility studies and loans, as well as political risk insurance to protect against the "inconvertibility of local currency; loss due to expropriation or confiscation, and loss due to war, revolution, or insurrection" (Stokes 1978, 55).

Insurance companies also offer political risk insurance to protect multinational investors against expropriation, the seizure of someone's assets by another person or government, but most insurance companies offering this protection do so under the premise that they will attempt to recover the cost of a claim from the government responsible for the expropriation. To strengthen an insurance company's claim against a nation, insurers want a clear title without liens to the investor's equity shares at the time they honor an investor's claim. However, investors will typically have a lien from a financing institution. This creates a **multiple pledge of shares problem** where insurance companies need clear title to pursue their claims, while lending institutions also have a lien (Moran 2001, 9). To solve this type of a problem, several solutions exist:

1. The lending institution may secure their own political risk insurance.
2. Lending institutions may agree to release their liens under an expropriation situation in exchange for charging a higher fee to the investor borrowing the money.
3. Lending institutions could temporarily suspend their lien or completely release their lien in exchange for a portion of the insurance settlement.
4. The lending institution could release their lien and proceed with a simultaneous claim against the government at the same time as the insurance company proceeds with their claim against the government (Moran 2001).

Companies could be required by governments to provide insurance to compensate employees who are injured on the job. Governments that provide **worker's compensation insurance** pay either a lump sum or a monthly stipend, which is paid to injured workers or the families of deceased workers. Owners could require worker's compensation insurance to protect them from being sued if an employee is injured on a job site. Expatriate engineering and construction personnel should check to see if they are covered by worker's compensation insurance of their host nation if they are working overseas.

Contractor's Guarantee and Performance Bonds

Governments could mandate that a contractor provide a **standby letter of credit** instead of a performance bond. If an owner provides documents proving that a contractor failed to perform, such as an arbitration award, to the bank holding the standby letter of credit, the bank will then issue funds to perform corrective actions up to a predetermined limit (UNCITRAL 1988).

Risk Associated with Currency Variations

Fluctuations when governments devalue their currency create problems for owners and contractors; therefore, it is important to monitor currency fluctuations and the events that cause them. Daily television news broadcasts, such as the **British Broadcasting Company** (BBC) News, the **Deutsch Weller** News, a German news broadcast in English, and weekly financial newspapers are sources of information on world events and world economic news.

If a contractor is paid in their native currency, then the owner is assuming the risk of exchange rate fluctuations. If a contractor is paid in the currency of a host nation, then the contractor is assuming currency exchange risk. If a contractor pays subcontractors using their native currency, the contractor is assuming exchange rate risks. Both owners and contractors assume exchange rate fluctuation risk if the contractor is paid in the currency of a different nation. One option for avoiding these types of risks is for owners to pay contractors with the currency of the financing institution supporting a project.

In lump sum and unit-price payment contracts risks for contractors are reduced if they are paid in the same currency that they use to pay their subcontractors and suppliers; therefore, contractors might request that they be paid in multiple currencies. When a contractor is paid in a cost-reimbursable system, they might be paid in the same currency that they use to pay their costs. Contracts should not specify two currencies and then allow one party to choose which one will be used without a currency adjustment because the party with the choice could be **unjustly enriched**, which means that they receive more money than they are entitled to from the other party to the contract.

It might be difficult for contractors to export their profits back to their home nation because of host country laws limiting the international transfer of funds. One alternative is to export goods out of the host nation instead of money, but this could be risky if there is no market for the goods being exported or prices decline for the products exported back to a home country (Stokes 1978).

Local Labor Laws and Risks Associated with Using the Local Workforce

The hiring or firing of members of a local labor force is controlled by national and local labor laws; these laws could restrict the firing of employees. In Germany, employers are required to pay two years worth of wages for each worker that is laid off. One method for reducing this risk is to use local companies that provide temporary workers who remain employees of the original firm and not the contractor.

In some countries, there might be a mandate to use a certain amount of local labor and the local labor may be unskilled labor (Stokes 1978). Governments, or owners, might require foreign companies to train local labor, but if the owner is the government, contractors should try to negotiate with members of the government to be compensated for some of their training costs because the host nation benefits from having a more skilled workforce. Using unskilled managers is risky for a contractor because managers affect the workers under their span of control. If foreign managers are used, they could insist that they be allowed to hire their own personnel (Stokes 1978).

Chapter Summary

As with all construction projects, legal issues and claims in the global arena could sometimes be the determining factors on whether a project makes a profit or not. Having a basic understanding of different legal systems is essential when operating in a global environment. Engineering and construction personnel should also be familiar with whether the legal system they will be operating under will be similar or different from their native country in terms of whether it is a common law legal system or a civil law legal system or some other type of legal system. International contracts are similar to domestic contracts but they include additional clauses that address international issues.

This chapter presented information on the types of legal and contractual issues that arise on international contracts. If a person knows which legal system is being used, it helps when they are involved in settling contractual disputes. This chapter also provided information on resources that address international contracts and international conventions.

International contract clauses were discussed in this chapter in terms of the clauses that would differ in an international contract versus domestic contracts. In addition, a table that includes standard international contract clauses, along with a brief explanation of what they mean to someone involved in an international contract, was also provided in this chapter. A section on claims and changes to international contracts was included, along with some examples of situations that could cause claims to arise on international contracts such as labor situations and payment delays. Information was also provided on the involvement of home nations in contractual disputes in cases of expropriation or denial of justice.

International arbitration, agreements to arbitrate, and arbitration clauses were reviewed, along with the authority of arbitration tribunals to order interim measures in arbitration cases. Anticorruption laws were briefly discussed and they vary from country to country. A discussion on kidnapping and ransom insurance was also included, along with information on global liability issues, sources of risk, insurance, and changing governments and the effect they have on construction projects.

Key Terms

ad hoc arbitration
adjustment of price
American Institute of Architects
arbitral tribunal
Association of Consulting Engineers
British Broadcasting Company
Calvo clause
cardinal change
choice of forum clauses
choice of law clauses
Civil Engineering Contractors Association
confidentiality statements
Convention on the Recognition and Enforcement of Foreign Arbitral Awards
currency clause
customs duties
denial of justice
Deutsch Weller
Engineering Advancement Association of Japan
exclusive jurisdiction clauses
exemption clauses
expropriation
Federation Internationale des Ingenieurs-Conseils
Force Majeure
Foreign Corrupt Practices Act
Geneva Convention
gross receipts tax
hardship
Hickenlooper Amendment
host country
index clause
Institution of Civil Engineers
institutional arbitration
International Arbitration Association
international arbitration
International Chamber of Commerce
International Commercial Terms
International Federation of Consulting Engineers
international joint venture
intragovernmental treaties
kidnapping and ransom insurance
liquidated damages
Martindale-Hubbell International Law Digest
multiple pledges of shares problems
New York Convention
Overseas Private Investment Corporation
patents
political risk insurance
professional liability and structural defects insurance
revision index clause
revision of price
Sarbanes-Oxley Act

standby letter of credit
taxes in a host country
termination clauses
trademarks
transfer of ownership of property
transit taxes
transportation permits
United States-Mexican General Claims Commission
unilateral termination
United Nations Commission on International Trade and Law
United Nations Compensation Commission
United Nations Convention on Contracts for the International Sale of Goods
unjustly enriched
worker's compensation insurance

Discussion Questions

20.1 Why would a firm require that its employees not divulge whether it has kidnapping and ransom insurance policies?

20.2 What are the most commonly used construction contract documents in the global arena?

20.3 Explain whether patents or trademarks are protected in the global marketplace. Why or why not?

20.4 Explain how termination clauses are used on construction projects.

20.5 Discuss Calvo clauses and how they could affect construction projects.

20.6 Which five clauses from Table 20.1 should always be included in international contracts? Why are the five clauses selected more important to include than the other clauses in Table 20.1?

20.7 How could confidentiality agreements be enforced on international contracts?

20.8 When engineering and construction personnel are working in foreign countries, which legal system governs their actions, and why?

20.9 Explain how a hardship clause benefits owners and contractors if it is included in international contracts.

20.10 Explain how exemption clauses are similar to impossibility clauses.

20.11 Explain the conditions under which owners and contractors are allowed to terminate contracts if a termination clause is included in contracts.

20.12 Explain choice of law clauses.

20.13 Explain the difference between an adjustment of price and a revision of price.

20.14 Explain the difference between denial of justice and expropriation.

20.15 What would be a reasonable percentage that a contract could be changed by an owner without it being a cardinal change? Why?

20.16 Why is international arbitration a better way to settle international contract disputes than litigation?

20.17 How might the organizations listed in the section on anticorruption legislation be able to reduce corruption on engineering and construction projects?

20.18 Why was it so difficult for contractors to settle their contractual disputes in Iran after the Iranian revolution in 1979?

20.19 Explain political risk and how it affects engineering and construction contracts.

20.20 Discuss which currency should be used to pay a foreign contractor if they are performing work in a host country using subcontractors that are from a country that is not the host nation and explain why.

References

American Society of Civil Engineers. December 2004. ASCE members work to reduce corruption worldwide. *American Society of Civil Engineers Magazine*. Reston, VA.

Amin, S. 1983. Iran-United States claims settlement. *The International and Comparative Law Quarterly* 32(3): 750–756.

Bondzi-Simpson, P. 1990. *Legal Relationships Between Transnational Corporations and Host States*. Westport, CT: Greenwood Publishing Group, Inc, Quorum Books.

Bonell, M. 2000. *UNILEX: International Case Law and Bibliography on the UNIDROIT Principles of International Commercial Contracts*. Ardsley. NY: Transnational Publishers.

Broward Daily Business Review. January 6, 2005. World joins U.S. in cracking down on businesses offering bribes overseas. Legal Events Section 46(20): 10.

Chukwumerije, O. 1994. *Choice of Law in International Commercial Arbitration*. Westport, CT: Quorum Books.

Commerce Clearing House. 2006. *People's Republic of China Construction Law—A Guide for Foreign Companies*. Singapore: Wolters Kluwer Business, Asia Pte Limited.

Financial Times (London Edition). 2003. Economic growth boosts the demand for kidnap insurance: Services are for more than Colombians and oil companies. Section: Companies in the Americas: 28.

Insurance Day. October 22, 2003. Kidnap risk on agenda again for multinationals. Informa Publishing Group Ltd., Section: News Part I.

Insurance Day. September 2, 2004. Kidnap risk on agenda again for multinationals. Informa Publishing Group Ltd., Section: News Part II.

Japan Economic Newswire. September 30, 2004. Suspected kidnapper of japanese charged with moneylaundering. Manila, Philippines: Kyoto News Service, International News Section.

Jenkins, J., and S. Stebbings. 2006. *International Construction Arbitration Law*. The Hague, The Netherlands: Kluwer Law International.

Kangari, R., and C., Lucas. 1997. *Managing International Operations*. New York: American Society of Civil Engineering.

Knuston, R. 2005. *FIDIC: An Analysis of International Construction Contracts*. The Hague, The Netherlands: Kluwer Law International.

Lantis, M. May 9, 2005. Telephone Interview. Clark Construction Company, San Jose, CA.

Lloyd's List. December 19, 2003. Baghdad kidnappings push up premiums. *Lloyd's List International*, 14.

Martindale-Hubbell. 2000. *Martindale Hubbell International Law Digest*. Chicago: R.R. Donnelley and Sons Company.

Moran, J. 2001. *International Political Risk Management: Exploring New Frontiers*. Washington, DC: The World Bank Group Multilateral Investment Guarantee Agency (MIGA).

Murphy, O. 2005. *International Project Management*. Mason, OH: Thomson Higher Education.

National Law Journal. July 12, 2004. Oil company bribery suit settles: Government lays out blueprint for successor liability in resolving case. News Section: 5.

New York Times. May 17, 2005. Antibribery efforts fail to stop executives from greasing palms. The International Herald Tribune, Finance Section: 13.

O'Hare, C. April–July 1980. Cargo dispute resolution and the Hamburg Rules. *International and Comparative Law Quarterly* 29 (Parts 2 and 3): 219–237.

Redfern, A., and M. Hunter. 1986. *Law and Practice of International Commercial Arbitration*. London: Sweet and Maxwell.

Rubino-Sammartano, M. 2001. *International Arbitration: Law and Practice*. The Hague, The Netherlands: Kluwer Law International.

Scott, S. August 1993. Dealing with delay claims: A survey. *International Journal of Project Management* 11(3): 143–153.

Stokes, M. 1978. *International Construction Contracts*. New York: McGraw-Hill Publications.

UNCITRAL. 1988. *Legal Guide on Drawing Up International Contracts for the Construction of Industrial Works*. Vienna, Austria: United Nations Commission on International Trade Law.

Wearne, S. February 1992. Contract administration and project risk. *International Journal of Project Management* 10(1): 39–41.

Williams, T. November 1992. Risk management infrastructure. *International Journal of Project Management* 10(4): 5–10.

Wolf, R. 2000. *Effective International Joint Venture Management*. Armonk, NY: M.E. Sharpe.

INTERNATIONAL SUMMARY

During the twenty-first century, engineering and construction (E&C) professionals are dealing with challenges that they did not have to address in the twentieth century, which are being created by the rapidly increasing globalization of the world. Members of countries throughout the world are now irrevocably linked by telecommunications systems and the media. In addition to innovative technical processes political, social, and economic events are changing how engineers and constructors have to interface with the rest of the world. Engineering and construction professionals receive a solid scientific and engineering background during college, but during the twenty-first century, they also should be familiar with the nuances of other cultures, especially legal issues, and how to effectively work in the global arena. Another major influence that is complicating the legal issues that E&C personnel are required to address is increasing foreign investment in domestic firms. It is now becoming harder to distinguish between domestic and foreign firms because many previously domestic firms now have foreign stockholders (ownership).

Some of the major components of construction materials such as steel are being produced in only a few countries in the world; therefore, engineers and constructors have to rely on purchasing materials in the global marketplace, which requires an understanding of how to negotiate and implement international contracts. Engineering and construction firms are no longer competing only against other domestic firms for construction materials, since they now must compete against foreign firms for scarce materials. The price of structural steel members doubled during 2005, along with similar price increases for timber products. Steel is now being produced mainly in Pacific Rim nations. The virgin forests in North America, as well as the forests in many other countries, have been denuded and building materials are now being made of compressed wood chips (laminated wood), which are replacing virgin wood as a structural building component. The quality of construction materials is steadily declining due to relaxed safety requirements in the nations where construction materials are being produced or fabricated, so designs have to be adjusted to accommodate these lower quality materials.

Since firms are forced to compete for scare resources that are produced by firms in foreign countries, they are influenced by global legal issues even if they are only domestic E&C firms. When domestic firms no longer produce or process the raw materials and components used in construction projects, domestic E&C firms have to compete for scarce materials in the global marketplace, which requires an understanding of the legal issues and contract implementation processes that surface when working with global suppliers, fabricators, and foreign E&C firms.

When E&C personnel work on construction projects in foreign countries, they work with personnel from the host nation, as well as with personnel from other countries, because clients usually hire foreign nationals from several nations to obtain the technical expertise required to design or build projects. Global project managers and project management team members might be managing multicultural personnel and laborers, and in some situations they might be managing workers from several different regions of the world. In these types of situations, project personnel may be required to oversee E&C contracts that are with firms from various countries and that are being administered by courts from different legal jurisdictions.

Engineering and construction professionals are required to work within societal and cultural constraints when they are performing work for global clients in foreign countries or merely working with foreign nationals within their own nation. Global variations also account for the vast differences seen in the construction techniques and processes used to construct projects throughout the world. The construction techniques and processes that are used within a particular country also influence how contracts are written and implemented on construction projects because indigenous legal systems are the major influencing factor on how contracts are written and applied to construction projects.

The twenty-first century is being shaped by global political, social, and economic events that are no longer concentrated in Western nations as Eastern nations are moving to the forefront of global visibility. Forty percent of the working age population is concentrated in India and China. Projections are that by 2032 three of the four largest economies in the world will be Japan, China, and India (Zakaria 2005). The legal systems used in Eastern nations differ from Western legal systems in that they stress collaborative approaches to settling contract disputes rather than adversarial relationships.

The definition of global competitiveness includes *firms competing for work in foreign countries,* but it also requires an analysis of the affects of global competition on domestic E&C markets. Diverting personnel and resources onto foreign projects creates more opportunities in domestic markets for native firms. Monitoring the trends and events that transpire in the global E&C arena is essential, since members of domestic E&C firms are affected by price fluctuations in the cost of construction materials, the increasing cost of transporting materials, emerging innovative designs, the availability of technical personnel, foreign ownership of domestic E&C firms, and changing global economic and political climates.

In order for E&C firms to remain competitive in the global marketplace, their personnel have to be able to quickly adapt to working with people from other cultures. They have to develop a global perspective that is incorporated into their designs, into the techniques and processes used to construct a facility, into the way they manage construction projects, and most importantly into international contracts. Engineers and constructors that are only familiar with common law legal systems will have to develop an understanding of the differences between common law, civil law, Shari'a law, and Asian law legal systems in order to operate in countries where common law legal systems are not the prevailing legal system.

The competitiveness of domestic E&C firms is also affected when firms are no longer able to locate qualified personnel in their native country to hire to fill technical positions. Increasing competition for highly qualified E&C professionals has led to firms hiring foreign nationals to fill positions, because the country of origin is not as important as a firm being able to perform the work required for the fulfillment of an E&C contract. This process creates a situation where E&C personnel may be familiar with only one type of legal system—that of their native country. In order for firms to operate effectively in countries with legal systems that are different from the legal system of their native country, E&C personnel need to have a basic understanding of the differences between their native legal system and the legal system of the host country where they are working on projects.

Unless E&C higher educational institutions introduce courses that cover common law, civil law, and Eastern legal systems, E&C professionals will have to seek out additional references related to foreign legal systems and educate themselves on the differences between legal systems throughout the world. Having a basic understanding of worldwide legal systems helps to prepare engineers and constructors to work on global projects anywhere in the world.

References

Zakaria, F. 2005. The rise of China, India, and Brazil. *Newsweek* pp. 27–31.

UNITED STATES CONSTITUTION

From the U.S. Government Archives

We the People of the United States, in Order to form a more perfect Union, establish Justice, insure domestic Tranquility, provide for the common defense, promote the general Welfare, and secure the Blessings of Liberty to ourselves and our Posterity, do ordain and establish this Constitution for the United States of America.

ARTICLE 1

Section 1

All legislative Powers herein granted shall be vested in a Congress of the United States, which shall consist of a Senate and House of Representatives.

Section 2

The House of Representatives shall be composed of Members chosen every second Year by the People of the several States, and the Electors in each State shall have the Qualifications requisite for Electors of the most numerous Branch of the State Legislature.

No Person shall be a Representative who shall not have attained to the Age of twenty-five Years, and been seven Years a Citizen of the United States, and who shall not, when elected, be an Inhabitant of that State in which he shall be chosen.

Representatives and direct Taxes shall be apportioned among the several States which may be included within this Union, according to their respective Numbers, which shall be determined by adding to the whole Number of free Persons, including those bound to Service for a Term of Years, and excluding Indians not taxed, three-fifths of all other Persons.

The actual Enumeration shall be made within three years after the first meeting of the Congress of the United States, and within every subsequent term of ten Years, in such a manner as they shall by law direct. The Number of Representatives shall not exceed one for every thirty Thousand, but each State shall have at Least one Representative; and until such enumeration shall be made, the State of New Hampshire shall be entitled to choose three, Massachusetts eight, Rhode Island and Providence Plantations one, Connecticut five, New York six, New Jersey four, Pennsylvania eight, Delaware one, Maryland six, Virginia ten, North Carolina five, South Carolina five and Georgia three.

When vacancies happen in the Representation from any State, the Executive Authority thereof shall issue Writs of Election to fill such Vacancies.

The House of Representatives shall choose their Speaker and other Officers; and shall have the sole Power of Impeachment.

Section 3

The Senate of the United States shall be composed of two Senators from each State, chosen by the Legislature thereof, for six Years; and each Senator shall have one Vote.

Immediately after they shall be assembled in Consequence of the first Election, they shall be divided as equally as may be into three Classes. The Seats of the Senators of the first Class shall be vacated at the Expiration of the second Year, of the second Class at the Expiration of the fourth Year, and of the third Class at the Expiration of the sixth Year, so that one-third may be chosen every second Year; and if Vacancies happen by Resignation, or otherwise, during the Recess of the Legislature of any State, the Executive thereof may make temporary Appointments until the next Meeting of the Legislature, which shall then fill such Vacancies.

No person shall be a Senator who shall not have attained to the Age of thirty Years, and been nine Years a Citizen of the United States, and who shall not, when elected, be an Inhabitant of that State for which he shall be chosen.

The Vice President of the United States shall be President of the Senate, but shall have no Vote, unless they be equally divided.

The Senate shall choose their other Officers, and also a President pro tempore, in the absence of the Vice President, or when he shall exercise the Office of President of the United States.

The Senate shall have the sole Power to try all Impeachments. When sitting for that Purpose, they shall be on Oath or Affirmation. When the President of the United States is tried, the Chief Justice shall preside: And no Person shall be convicted without the Concurrence of two-thirds of the Members present.

Judgment in Cases of Impeachment shall not extend further than to removal from Office, and disqualification to hold and enjoy any Office of honor, Trust or Profit under the United States: but the Party convicted shall nevertheless be liable and subject to Indictment, Trial, Judgment and Punishment, according to Law.

Section 4

The Times, Places and Manner of holding Elections for Senators and Representatives, shall be prescribed in each State by the Legislature thereof; but the Congress may at any time by Law make or alter such Regulations, except as to the Place of Choosing Senators.

The Congress shall assemble at least once in every Year, and such Meeting shall be on the first Monday in December, unless they shall by Law appoint a different Day.

Section 5

Each House shall be the Judge of the Elections, Returns and Qualifications of its own Members, and a Majority of each shall constitute a Quorum to do Business; but a smaller number may adjourn from day to day, and may be authorized to compel the Attendance of absent Members, in such Manner, and under such Penalties as each House may provide.

Each House may determine the Rules of its Proceedings, punish its Members for disorderly Behavior, and, with the Concurrence of two-thirds, expel a Member.

Each House shall keep a Journal of its Proceedings, and from time to time publish the same, excepting such Parts as may in their Judgment require Secrecy; and the Yeas and Nays of the Members of either House on any question shall, at the Desire of one-fifth of those Present, be entered on the Journal.

Neither House, during the Session of Congress, shall, without the Consent of the other, adjourn for more than three days, nor to any other Place than that in which the two Houses shall be sitting.

Section 6

The Senators and Representatives shall receive a Compensation for their Services, to be ascertained by Law, and paid out of the Treasury of the United States. They shall in all Cases, except Treason, Felony and Breach of the Peace, be privileged from Arrest during their Attendance at the Session of their respective Houses, and in going to and returning from the same; and for any Speech or Debate in either House, they shall not be questioned in any other Place.

No Senator or Representative shall, during the Time for which he was elected, be appointed to any civil Office under the Authority of the United States which shall have been created, or the Emoluments whereof shall have been increased during such time; and no Person holding any Office under the United States, shall be a Member of either House during his Continuance in Office.

Section 7

All bills for raising Revenue shall originate in the House of Representatives; but the Senate may propose or concur with Amendments as on other Bills.

Every Bill which shall have passed the House of Representatives and the Senate, shall, before it become a Law, be presented to the President of the United States; If he approve he shall sign it, but if not he shall return it, with his Objections to that House in which it shall have originated, who shall enter the Objections at large on their Journal, and proceed to reconsider it. If after such Reconsideration two-thirds of that House shall agree to pass the Bill, it shall be sent, together with the Objections, to the other House, by which it shall likewise be reconsidered, and if approved by two thirds of that House, it shall become a Law. But in all such Cases the votes of both Houses shall be determined by Yeas and Nays, and the names of the persons voting for and against the bill shall be entered on the journal of each House respectively. If any Bill shall not be returned by the President within ten Days (Sundays excepted) after it shall have been presented to him, the Same shall be a Law, in like Manner as if he had signed it, unless the Congress by their Adjournment prevent its Return, in which Case it shall not be a Law.

Every Order, Resolution, or Vote to which the Concurrence of the Senate and House of Representatives may be necessary (except on a question of Adjournment) shall be presented to the President of the United States; and before the Same shall take Effect, shall be approved by him, or being disapproved by him, shall be repassed by two-thirds of the Senate and House of Representatives, according to the Rules and Limitations prescribed in the Case of a Bill.

Section 8

The Congress shall have Power to lay and collect Taxes, Duties, Imposts and Excises, to pay the Debts and provide for the common Defense and general Welfare of the United States; but all Duties, Imposts and Excises shall be uniform throughout the United States;

To borrow money on the credit of the United States;

To regulate Commerce with foreign Nations, and among the several States, and with the Indian Tribes;

To establish an uniform Rule of Naturalization, and uniform Laws on the subject of Bankruptcies throughout the United States;

To coin Money, regulate the Value thereof, and of foreign Coin, and fix the Standard of Weights and Measures;

To provide for the Punishment of counterfeiting the Securities and current Coin of the United States;

To establish Post Offices and Post Roads;

To promote the Progress of Science and useful Arts, by securing for limited Times to Authors and Inventors the exclusive Right to their respective Writings and Discoveries;

To constitute Tribunals inferior to the Supreme Court;

To define and punish Piracies and Felonies committed on the high Seas, and Offenses against the Law of Nations;

To declare War, grant Letters of Marque and Reprisal, and make Rules concerning Captures on Land and Water;

To raise and support Armies, but no Appropriation of Money to that Use shall be for a longer Term than two Years;

To provide and maintain a Navy;

To make Rules for the Government and Regulation of the land and naval Forces;

To provide for calling forth the Militia to execute the Laws of the Union, suppress Insurrections and repel Invasions;

To provide for organizing, arming, and disciplining the Militia, and for governing such Part of them as may be employed in the Service of the United States, reserving to the States respectively, the Appointment of the Officers, and the Authority of training the Militia according to the discipline prescribed by Congress;

To exercise exclusive Legislation in all Cases whatsoever, over such District (not exceeding ten Miles square) as may, by Cession of particular States, and the acceptance of Congress, become the Seat of the Government of the United States, and to exercise like Authority over all Places purchased by the Consent of the Legislature of the State in which the Same shall be, for the Erection of Forts, Magazines, Arsenals, dock-Yards, and other needful Buildings; And

To make all Laws which shall be necessary and proper for carrying into Execution the foregoing Powers, and all other Powers vested by this Constitution in the Government of the United States, or in any Department or Officer thereof.

Section 9

The Migration or Importation of such Persons as any of the States now existing shall think proper to admit, shall not be prohibited by the Congress prior to the Year one thousand eight hundred and eight, but a tax or duty may be imposed on such Importation, not exceeding ten dollars for each Person.

The privilege of the Writ of Habeas Corpus shall not be suspended, unless when in Cases of Rebellion or Invasion the public Safety may require it.

No Bill of Attainder or ex post facto Law shall be passed.

No capitation, or other direct, Tax shall be laid, unless in Proportion to the Census or Enumeration herein before directed to be taken.

No Tax or Duty shall be laid on Articles exported from any State.

No Preference shall be given by any Regulation of Commerce or Revenue to the Ports of one State over those of another: nor shall Vessels bound to, or from, one State, be obliged to enter, clear, or pay Duties in another.

No Money shall be drawn from the Treasury, but in Consequence of Appropriations made by Law; and a regular Statement and Account of the Receipts and Expenditures of all public Money shall be published from time to time.

No Title of Nobility shall be granted by the United States: And no Person holding any Office of Profit or Trust under them, shall, without the Consent of the Congress, accept of any present, Emolument, Office, or Title, of any kind whatever, from any King, Prince or foreign State.

Section 10

No State shall enter into any Treaty, Alliance, or Confederation; grant Letters of Marque and Reprisal; coin Money; emit Bills of Credit; make any Thing but gold and silver Coin a Tender in Payment of Debts; pass any Bill of Attainder, ex post facto Law, or Law impairing the Obligation of Contracts, or grant any Title of Nobility.

No State shall, without the Consent of the Congress, lay any Imposts or Duties on Imports or Exports, except what may be absolutely necessary for executing its inspection Laws: and the net Produce of all Duties and Imposts, laid by any State on Imports or Exports, shall be for the Use of the Treasury of the United States; and all such Laws shall be subject to the Revision and Control of the Congress.

No State shall, without the Consent of Congress, lay any duty of Tonnage, keep Troops, or Ships of War in time of Peace, enter into any Agreement or Compact with another State, or with a foreign Power, or engage in War, unless actually invaded, or in such imminent Danger as will not admit of delay.

ARTICLE 2

Section 1

The executive Power shall be vested in a President of the United States of America. He shall hold his Office during the Term of four Years, and, together with the Vice-President chosen for the same Term, be elected, as follows:

Each State shall appoint, in such Manner as the Legislature thereof may direct, a Number of Electors, equal to the whole Number of Senators and Representatives to which the State may be entitled in the Congress: but no Senator or Representative, or Person holding an Office of Trust or Profit under the United States, shall be appointed an Elector.

The Electors shall meet in their respective states, and vote by ballot for two persons, of whom one at least shall not lie an inhabitant of the same State with themselves. And they shall make a List of all the Persons voted for, and of the Number of Votes for each; which List they shall sign and certify, and transmit sealed to the Seat of the Government of the United States, directed to the President of the Senate. The President of the Senate shall, in the Presence of the Senate and House of Representatives, open all the Certificates, and the Votes shall then be counted. The Person having the greatest Number of Votes shall be the President, if such Number be a Majority of the whole Number of Electors appointed; and if there be more than one who have such Majority, and have an equal Number of Votes, then the House of Representatives shall immediately choose by Ballot one of them for President; and if no Person have a Majority, then from the five highest on the List the said House shall in like Manner choose the President. But in choosing the President, the Votes shall be taken by States, the Representation from each State having one Vote; a quorum for this Purpose shall consist of a Member or Members from two-thirds of the States, and a Majority of all the States shall be necessary to a Choice. In every Case, after the Choice of the President, the Person having the greatest Number of Votes of the Electors shall be the Vice-President. But if there should remain two or more who have equal Votes, the Senate shall choose from them by Ballot the Vice-President.

The Congress may determine the Time of choosing the Electors, and the Day on which they shall give their Votes; which Day shall be the same throughout the United States.

No person except a natural born Citizen, or a Citizen of the United States, at the time of the Adoption of this Constitution, shall be eligible to the Office of President; neither shall any Person be eligible to that Office who shall not have attained to the Age of thirty-five Years, and been fourteen Years a Resident within the United States.

In Case of the Removal of the President from Office, or of his Death, Resignation, or Inability to discharge the Powers and Duties of the said Office, the same shall devolve on the Vice-President, and the Congress may by Law provide for the Case of Removal, Death, Resignation or Inability, both of the President and Vice-President, declaring what Officer shall then act as President, and such Officer shall act accordingly, until the Disability be removed, or a President shall be elected.

The President shall, at stated Times, receive for his Services, a Compensation, which shall neither be increased nor diminished during the Period for which he shall have been elected, and he shall not receive within that Period any other Emolument from the United States, or any of them.

Before he enter on the Execution of his Office, he shall take the following Oath or Affirmation:

> "I do solemnly swear (or affirm) that I will faithfully execute the Office of President of the United States, and will to the best of my Ability, preserve, protect and defend the Constitution of the United States."

Section 2

The President shall be Commander in Chief of the Army and Navy of the United States, and of the Militia of the several States, when called into the actual Service of the United States; he may require the Opinion, in writing, of the principal Officer in each of the executive Departments, upon any subject relating to the Duties of their respective Offices, and he shall have Power to Grant Reprieves and Pardons for Offenses against the United States, except in Cases of Impeachment.

He shall have Power, by and with the Advice and Consent of the Senate, to make Treaties, provided two thirds of the Senators present concur; and he shall nominate, and by and with the Advice and Consent of the Senate, shall appoint Ambassadors, other public Ministers and Consuls, Judges of the supreme Court, and all other Officers of the United States, whose Appointments are not herein otherwise provided for, and which shall be established by Law: but the Congress may by Law vest the Appointment of such inferior Officers, as they think proper, in the President alone, in the Courts of Law, or in the Heads of Departments.

The President shall have Power to fill up all Vacancies that may happen during the Recess of the Senate, by granting Commissions, which shall expire at the End of their next Session.

Section 3

He shall from time to time give to the Congress Information of the State of the Union, and recommend to their Consideration such Measures as he shall judge necessary and expedient; he may, on extraordinary Occasions, convene both Houses, or either of them, and in Case of Disagreement between them, with Respect to the Time of Adjournment, he may adjourn them to such Time as he shall think proper; he shall receive Ambassadors and other public Ministers; he shall take Care that the Laws be faithfully executed, and shall Commission all the Officers of the United States.

Section 4

The President, Vice-President and all civil Officers of the United States, shall be removed from Office on Impeachment for, and Conviction of, Treason, Bribery, or other high Crimes and Misdemeanors.

ARTICLE 3

Section 1

The judicial Power of the United States shall be vested in one Supreme Court, and in such inferior Courts as the Congress may from time to time ordain and establish. The Judges, both of the supreme and inferior Courts, shall hold their Offices during good Behavior, and shall, at stated Times, receive for their Services a Compensation which shall not be diminished during their Continuance in Office.

Section 2

The judicial Power shall extend to all Cases, in Law and Equity, arising under this Constitution, the Laws of the United States, and Treaties made, or which shall be made, under their Authority; to all Cases affecting Ambassadors, other public Ministers and Consuls; to all Cases of admiralty and maritime Jurisdiction; to Controversies to which the United States shall be a Party; to Controversies between two or more States; between a State and Citizens of another State; between Citizens of different States; between Citizens of the same State claiming Lands under Grants of different States, and between a State, or the Citizens thereof, and foreign States, Citizens or Subjects.

In all Cases affecting Ambassadors, other public Ministers and Consuls, and those in which a State shall be Party, the Supreme Court shall have original Jurisdiction. In all the other Cases before mentioned, the Supreme Court shall have appellate Jurisdiction, both as to Law and Fact, with such Exceptions, and under such Regulations as the Congress shall make.

The Trial of all Crimes, except in Cases of Impeachment, shall be by Jury; and such Trial shall be held in the State where the said Crimes shall have been committed; but when not committed within any State, the Trial shall be at such Place or Places as the Congress may by Law have directed.

Section 3

Treason against the United States shall consist only in levying War against them, or in adhering to their Enemies, giving them Aid and Comfort. No Person shall be convicted of Treason unless on the Testimony of two Witnesses to the same overt Act, or on Confession in open Court.

The Congress shall have power to declare the Punishment of Treason, but no Attainder of Treason shall work Corruption of Blood, or Forfeiture except during the Life of the Person attainted.

ARTICLE 4

Section 1

Full Faith and Credit shall be given in each State to the public Acts, Records, and judicial Proceedings of every other State. And the Congress may by general Laws prescribe the Manner in which such Acts, Records and Proceedings shall be proved, and the Effect thereof.

Section 2

The Citizens of each State shall be entitled to all Privileges and Immunities of Citizens in the several States.

A Person charged in any State with Treason, Felony, or other Crime, who shall flee from Justice, and be found in another State, shall on demand of the executive Authority of the State from which he fled, be delivered up, to be removed to the State having Jurisdiction of the Crime.

No Person held to Service or Labor in one State, under the Laws thereof, escaping into another, shall, in Consequence of any Law or Regulation therein, be discharged from such Service or Labor, But shall be delivered up on Claim of the Party to whom such Service or Labor may be due.

Section 3

New States may be admitted by the Congress into this Union; but no new States shall be formed or erected within the Jurisdiction of any other State; nor any State be formed by the Junction of two or more States, or parts of States, without the Consent of the Legislatures of the States concerned as well as of the Congress.

The Congress shall have Power to dispose of and make all needful Rules and Regulations respecting the Territory or other Property belonging to the United States; and nothing in this Constitution shall be so construed as to Prejudice any Claims of the United States, or of any particular State.

Section 4

The United States shall guarantee to every State in this Union a Republican Form of Government, and shall protect each of them against Invasion; and on Application of the Legislature, or of the Executive (when the Legislature cannot be convened) against domestic Violence.

ARTICLE 5

The Congress, whenever two-thirds of both Houses shall deem it necessary, shall propose Amendments to this Constitution, or, on the Application of the Legislatures of two-thirds of the several States, shall call a Convention for proposing Amendments, which, in either Case, shall be valid to all Intents and Purposes, as part of this Constitution, when ratified by the Legislatures of three-fourths of the several States, or by Conventions in three-fourths thereof, as the one or the other Mode of Ratification may be proposed by the Congress; Provided that no Amendment which may be made prior to the Year One thousand eight hundred and eight shall in any Manner affect the first and fourth Clauses in the Ninth Section of the first Article; and that no State, without its Consent, shall be deprived of its equal Suffrage in the Senate.

ARTICLE 6

All Debts contracted and Engagements entered into, before the Adoption of this Constitution, shall be as valid against the United States under this Constitution, as under the Confederation.

This Constitution, and the Laws of the United States which shall be made in Pursuance thereof; and all Treaties made, or which shall be made, under the Authority of the United States, shall be the supreme Law of the Land; and the Judges in every State shall be bound thereby, any Thing in the Constitution or Laws of any State to the Contrary notwithstanding.

The Senators and Representatives before mentioned, and the Members of the several State Legislatures, and all executive and judicial Officers, both of the United States and of the several States, shall be bound by Oath or Affirmation, to support this Constitution; but no religious Test shall ever be required as a Qualification to any Office or public Trust under the United States.

ARTICLE 7

The ratification of the Conventions of nine states shall be sufficient for the establishment of this Constitution between the states so ratifying the same.

Done in Convention by the Unanimous Consent of the States present the Seventeenth Day of September in the Year of our Lord one thousand seven hundred and Eighty-seven and of the Independence of the United States of America the Twelfth. In witness whereof we have hereunto subscribed our Names.

George Washington—President and deputy from Virginia

New Hampshire—John Langdon, Nicholas Gilman

Massachusetts—Nathaniel Gorham, Rufus King

Connecticut—William Samuel Johnson, Roger Sherman

New York—Alexander Hamilton

New Jersey—William Livingston, David Brearley, William Paterson, Jonathan Dayton

Pennsylvania—Benjamin Franklin, Thomas Mifflin, Robert Morris, George Clymer, Thomas Fitzsimons, Jared Ingersoll, James Wilson, Gouvernour Morris

Delaware—George Read, Gunning Bedford Jr., John Dickinson, Richard Bassett, Jacob Broom

Maryland—James McHenry, Daniel of St Thomas Jenifer, Daniel Carroll

Virginia—John Blair, James Madison Jr.

North Carolina—William Blount, Richard Dobbs Spaight, Hugh Williamson

South Carolina—John Rutledge, Charles Cotesworth Pinckney, Charles Pinckney, Pierce Butler

Georgia—William Few, Abraham Baldwin

Attest—William Jackson, Secretary

AMENDMENT 1

Congress shall make no law respecting an establishment of religion, or prohibiting the free exercise thereof; or abridging the freedom of speech, or of the press; or the right of the people peaceably to assemble, and to petition the Government for a redress of grievances.

AMENDMENT 2

A well regulated Militia, being necessary to the security of free State, the right of the people to keep and bear Arms, shall not be infringed.

AMENDMENT 3

No Soldier shall, in time of peace be quartered in any house, without the consent of the Owner, nor in time of war, but in a manner to be prescribed by law.

AMENDMENT 4

The right of the people to be secure in their persons, houses, papers, and effects, against unreasonable searches and seizures, shall not be violated, and no Warrants shall issue, but upon probable cause, supported by Oath or affirmation, and particularly describing the place to be searched, and the persons or things to be seized.

AMENDMENT 5

No person shall be held to answer for a capital, or otherwise infamous crime, unless on a presentment or indictment of a Grand Jury, except in cases arising in the land or naval forces, or in the Militia, when in actual service in time of War or public danger; nor shall any person be subject for the same offense to be twice put in jeopardy of life or limb; nor shall be compelled in any criminal case to be a witness against himself, nor be deprived of life, liberty, or property, without due process of law; nor shall private property be taken for public use, without just compensation.

AMENDMENT 6

In all criminal prosecutions, the accused shall enjoy the right to a speedy and public trial, by an impartial jury of the State and district wherein the crime shall have been committed, which district shall have been previously ascertained by law, and to be informed of the nature and cause of the accusation; to be confronted with the witnesses against him; to have compulsory process for obtaining witnesses in his favor, and to have the Assistance of Counsel for his defense.

AMENDMENT 7

In Suits at common law, where the value in controversy shall exceed twenty dollars, the right of trial by jury shall be preserved, and no fact tried by a jury, shall be otherwise reexamined in any Court of the United States, than according to the rules of the common law.

AMENDMENT 8

Excessive bail shall not be required, nor excessive fines imposed, nor cruel and unusual punishments inflicted.

AMENDMENT 9

The enumeration in the Constitution, of certain rights, shall not be construed to deny or disparage others retained by the people.

AMENDMENT 10

The powers not delegated to the United States by the Constitution, nor prohibited by it to the States, are reserved to the States respectively, or to the people.

AMENDMENT 11

The Judicial power of the United States shall not be construed to extend to any suit in law or equity, commenced or prosecuted against one of the United States by Citizens of another State, or by Citizens or Subjects of any Foreign State.

AMENDMENT 12

The Electors shall meet in their respective states, and vote by ballot for President and Vice-President, one of whom, at least, shall not be an inhabitant of the same state with themselves; they shall name in their ballots the person voted for as President, and in distinct ballots the person voted for as Vice-President, and they shall make distinct lists of all persons voted for as President, and of all persons voted for as Vice-President and of the number of votes for each, which lists they shall sign and certify, and transmit sealed to the seat of the government of the United States, directed to the President of the Senate.

The President of the Senate shall, in the presence of the Senate and House of Representatives, open all the certificates and the votes shall then be counted.

The person having the greatest Number of votes for President, shall be the President, if such number be a majority of the whole number of Electors appointed; and if no person have such majority, then from the persons having the highest numbers not exceeding three on the list of those voted for as President, the House of Representatives shall choose immediately, by ballot, the President. But in choosing the President, the votes shall be taken by states, the representation from each state having one vote; a quorum for this purpose shall consist of a member or members from two-thirds of the states, and a majority of all the states shall be necessary to a choice. And if the House of Representatives shall not choose a President whenever the right of choice shall devolve upon them, before the fourth day of March next following, then the Vice-President shall act as President, as in the case of the death or other constitutional disability of the President.

The person having the greatest number of votes as Vice-President, shall be the Vice-President, if such number be a majority of the whole number of Electors appointed, and if no person have a majority, then from the two highest numbers on the list, the Senate shall choose the Vice-President; a quorum for the purpose shall consist of two-thirds of the whole number of Senators, and a majority of the whole number shall be necessary to a choice. But no person constitutionally ineligible to the office of President shall be eligible to that of Vice-President of the United States.

AMENDMENT 13

1. Neither slavery nor involuntary servitude, except as a punishment for crime whereof the party shall have been duly convicted, shall exist within the United States, or any place subject to their jurisdiction.
2. Congress shall have power to enforce this article by appropriate legislation.

AMENDMENT 14

1. All persons born or naturalized in the United States, and subject to the jurisdiction thereof, are citizens of the United States and of the State wherein they reside. No State shall make or enforce any law which shall abridge the privileges or immunities of citizens of the United States; nor shall any State deprive any person of life, liberty, or property, without due process of law; nor deny to any person within its jurisdiction the equal protection of the laws.
2. Representatives shall be apportioned among the several States according to their respective numbers, counting the whole number of persons in each State, excluding Indians not taxed. But when the right to vote at any election for the choice of electors for President and Vice-President of the United States, Representatives in Congress, the Executive and Judicial officers of a State, or the members of the Legislature thereof, is denied to any of the male inhabitants of such State, being twenty-one years of age, and citizens of the United States, or in any way abridged, except for participation in rebellion, or other crime, the basis of representation therein shall be reduced in the proportion which the number of such male citizens shall bear to the whole number of male citizens twenty-one years of age in such State.
3. No person shall be a Senator or Representative in Congress, or elector of President and Vice-President, or hold any office, civil or military, under the United States, or under any State, who, having previously taken an oath, as a member of Congress, or as an officer of the United States, or as a member of any State legislature, or as an executive or judicial officer of any State, to support the Constitution of the United States, shall have engaged in insurrection or rebellion against the same, or given aid or comfort to the enemies thereof. But Congress may by a vote of two-thirds of each House, remove such disability.
4. The validity of the public debt of the United States, authorized by law, including debts incurred for payment of pensions and bounties for services in suppressing insurrection or rebellion, shall not be questioned. But neither the United States nor any State shall assume or pay any debt or obligation incurred in aid of insurrection or rebellion against the United States, or any claim for the loss or emancipation of any slave; but all such debts, obligations and claims shall be held illegal and void.
5. The Congress shall have power to enforce, by appropriate legislation, the provisions of this article.

AMENDMENT 15

1. The right of citizens of the United States to vote shall not be denied or abridged by the United States or by any State on account of race, color, or previous condition of servitude.
2. The Congress shall have power to enforce this article by appropriate legislation.

AMENDMENT 16

The Congress shall have power to lay and collect taxes on incomes, from whatever source derived, without apportionment among the several States, and without regard to any census or enumeration.

AMENDMENT 17

The Senate of the United States shall be composed of two Senators from each State, elected by the people thereof, for six years; and each Senator shall have one vote. The electors in each State shall have the qualifications requisite for electors of the most numerous branch of the State legislatures.

When vacancies happen in the representation of any State in the Senate, the executive authority of such State shall issue writs of election to fill such vacancies: Provided, That the legislature of any State may empower the executive thereof to make temporary appointments until the people fill the vacancies by election as the legislature may direct.

This amendment shall not be so construed as to affect the election or term of any Senator chosen before it becomes valid as part of the Constitution.

AMENDMENT 18

1. After one year from the ratification of this article the manufacture, sale, or transportation of intoxicating liquors within, the importation thereof into, or the exportation thereof from the United States and all territory subject to the jurisdiction thereof for beverage purposes is hereby prohibited.
2. The Congress and the several States shall have concurrent power to enforce this article by appropriate legislation.
3. This article shall be inoperative unless it shall have been ratified as an amendment to the Constitution by the legislatures of the several States, as provided in the Constitution, within seven years from the date of the submission hereof to the States by the Congress.

AMENDMENT 19

The right of citizens of the United States to vote shall not be denied or abridged by the United States or by any State on account of sex.

Congress shall have power to enforce this article by appropriate legislation.

AMENDMENT 20

1. The terms of the President and Vice-President shall end at noon on the 20th day of January, and the terms of Senators and Representatives at noon on the 3rd day of January, of the years in which such terms would have ended if this article had not been ratified; and the terms of their successors shall then begin.
2. The Congress shall assemble at least once in every year, and such meeting shall begin at noon on the 3d day of January, unless they shall by law appoint a different day.
3. If, at the time fixed for the beginning of the term of the President, the President elect shall have died, the Vice-President elect shall become President. If a President shall not have been chosen before the time fixed for the beginning of his term, or if the President elect shall have failed to qualify, then the Vice-President elect shall act as President until a President shall have qualified; and the Congress may by law provide for the case wherein neither a President elect nor a Vice-President elect shall have qualified, declaring who shall then act as President, or the manner in which one who is to act shall be selected, and such person shall act accordingly until a President or Vice-President shall have qualified.
4. The Congress may by law provide for the case of the death of any of the persons from whom the House of Representatives may choose a President whenever the right of choice shall have devolved upon them, and for the case of the death of any of the persons from whom the Senate may choose a Vice-President whenever the right of choice shall have devolved upon them.
5. Sections 1 and 2 shall take effect on the 15th day of October following the ratification of this article.
6. This article shall be inoperative unless it shall have been ratified as an amendment to the Constitution by the legislatures of three-fourths of the several States within seven years from the date of its submission.

AMENDMENT 21

1. The eighteenth article of amendment to the Constitution of the United States is hereby repealed.
2. The transportation or importation into any State, Territory, or possession of the United States for delivery or use therein of intoxicating liquors, in violation of the laws thereof, is hereby prohibited.
3. The article shall be inoperative unless it shall have been ratified as an amendment to the Constitution by conventions in the several States, as provided in the Constitution, within seven years from the date of the submission hereof to the States by the Congress.

AMENDMENT 22

1. No person shall be elected to the office of the President more than twice, and no person who has held the office of President, or acted as President, of more than two years of a term to which some other person was elected President shall be elected to the office of the President more than once. But this Article shall not apply to any person holding the office of President, when this Article was proposed by the Congress, and shall not prevent any person who may be holding the office of President, or acting as President, during the term within which this Article becomes operative from holding the office of President or acting as President during the remainder of such term.
2. This article shall be inoperative unless it shall have been ratified as an amendment to the Constitution by the legislatures of three-fourths of the several States within seven years from the date of its submission to the States by the Congress.

AMENDMENT 23

1. The District constituting the seat of Government of the United States shall appoint in such manner as the

Congress may direct: A number of electors of President and Vice-President equal to the whole number of Senators and Representatives in Congress to which the District would be entitled if it were a State, but in no event more than the least populous State; they shall be in addition to those appointed by the States, but they shall be considered, for the purposes of the election of President and Vice-President, to be electors appointed by a State; and they shall meet in the District and perform such duties as provided by the twelfth article of amendment.

2. The Congress shall have power to enforce this article by appropriate legislation.

AMENDMENT 24

1. The right of citizens of the United States to vote in any primary or other election for President or Vice-President, for electors for President or Vice-President, or for Senator or Representative in Congress, shall not be denied or abridged by the United States or any State by reason of failure to pay any poll tax or other tax.
2. The Congress shall have power to enforce this article by appropriate legislation.

AMENDMENT 25

1. In case of the removal of the President from office or of his death or resignation, the Vice-President shall become President.
2. Whenever there is a vacancy in the office of the Vice-President, the President shall nominate a Vice-President who shall take office upon confirmation by a majority vote of both Houses of Congress.
3. Whenever the President transmits to the President pro tempore of the Senate and the Speaker of the House of Representatives his written declaration that he is unable to discharge the powers and duties of his office, and until he transmits to them a written declaration to the contrary, such powers and duties shall be discharged by the Vice-President as Acting President.
4. Whenever the Vice-President and a majority of either the principal officers of the executive departments or of such other body as Congress may by law provide, transmit to the President pro tempore of the Senate and the Speaker of the House of Representatives their written declaration that the President is unable to discharge the powers and duties of his office, the Vice-President shall immediately assume the powers and duties of the office as Acting President.

Thereafter, when the President transmits to the President pro tempore of the Senate and the Speaker of the House of Representatives his written declaration that no inability exists, he shall resume the powers and duties of his office unless the Vice-President and a majority of either the principal officers of the executive department or of such other body as Congress may by law provide, transmit within four days to the President pro tempore of the Senate and the Speaker of the House of Representatives their written declaration that the President is unable to discharge the powers and duties of his office. Thereupon Congress shall decide the issue, assembling within forty-eight hours for that purpose if not in session. If the Congress, within twenty-one days after receipt of the latter written declaration, or, if Congress is not in session, within twenty-one days after Congress is required to assemble, determines by two-thirds vote of both Houses that the President is unable to discharge the powers and duties of his office, the Vice-President shall continue to discharge the same as Acting President; otherwise, the President shall resume the powers and duties of his office.

AMENDMENT 26

1. The right of citizens of the United States, who are eighteen years of age or older, to vote shall not be denied or abridged by the United States or by any State on account of age.
2. The Congress shall have power to enforce this article by appropriate legislation.

AMENDMENT 27

No law, varying the compensation for the services of the Senators and Representatives, shall take effect, until an election of Representatives shall have intervened.

THE ENGINEERING AND HEALTH HISTORY OF LOVE CANAL

by Dr. Rita Meyninger

Former Region II Director of the Federal Emergency Management Administration

A.1 THE ENGINEERING AND HISTORY OF LOVE CANAL

The words **Love Canal** are now a synonym for the problems of land disposal of chemical wastes. But the canal site should also be viewed from another perspective—a continuum of almost one hundred years of history of a series of related and ordinary civil engineering projects. The projects involved classic civil engineering subjects such as soils, hydrology, groundwater flows, and waste disposal. They included ordinary civil engineering construction projects: a canal, streets, highways, sanitary and storm sewers, and a waster distribution system. The projects included span the traditional civil engineering feasibility studies, land use and site planning, and involvement with community planning groups, local politics, law, and economics. Because chemicals placed in a dumpsite closed for almost three decades surfaced in a residential community, civil engineers became involved in remediation of the site, environmental health studies, emergency evacuation, and litigation.

The prime professional responsibility of civil engineers is for public health and public safety. Love Canal was a situation where the public believed its health and safety were endangered and the professional community was not able to adequately respond to public concerns.

An 1890 Planned Model Community

In the 1890s, William T. Love planned a model community for more than half a million people that also included attractive inducements to industrialists looking for suitable plant sites. A seven mile long navigable canal to connect the upper and lower levels of the Niagara River and the production of inexpensive direct current electricity were part of the plan. Work began on excavation for the canal in 1894 but an economic recession in the country, combined with the advent of alternating current, caused the project to collapse by 1910. At that time it was sold at public auction.

A 200 foot by 3,200 foot long segment of the canal was completed before the project was halted. The canal was used as a local swimming hole in the ensuing years. Then the unfinished canal was used as a dumpsite for the disposal of 21,800 tons of various chemical wastes by Hooker Electrochemical Company between 1942 and 1953. The city of Niagara Falls regularly unloaded its municipal refuse into the canal.

A Chemical Waste Landfill and a Public School

In September 1941, the Hooker Electrochemical Company undertook feasibility studies to determine the suitability of using the unfinished canal as a disposal site for wastes from the Niagara Falls manufacturing operation. Although defects in the chain of title prevented the completion of the sale until April 29, 1947, Hooker obtained a license from the Niagara Power and Development Company in 1942 to use the property for the disposal of waste materials from the manufacture of chlorinated hydrocarbons and caustics.

After World War II, the postwar baby boom produced a need for the construction of more schools. Since the area around the canal site was mostly vacant the land prices were low. In early 1951, the Niagara Falls Board of Education completed a site study and drew a map showing a planned school on 99th Street to be built right over the center of the canal. Some lots abutting the canal were purchased by the Board. Letters were sent to other property owners—not eager to sell—advising them that condemnation actions had been instituted to acquire the property for educational purposes.

It was in this climate of possible condemnation that in March of 1952 the executive vice president of Hooker visited the site with school board officials. Hooker prepared a map showing where wastes were deposited, how they were covered, and the results of test borings that had recently been completed. The property was sold to the Board for $1.00 when the Board threatened to take the property by eminent domain. This option gave Hooker the opportunity to transfer title with a deed, the last paragraph of which states in part "that the premises . . . have been filled . . . with waste products resulting from the manufacture of chemicals" and that "no claim, suit, action or demand of any nature whatsoever . . ." could be made against Hooker "for injury to a person or person, including death resulting therefrom . . ."

The school district deputy corporation counsel advised the Board that, in his opinion, they incurred risk and possible liability for injury or damage to people and to property because of the chemical wastes in the canal landfill. He also advised them to engage a chemical engineer as a consultant to assess the situation. Apparently this advice was ignored and the Board voted on May 7, 1953, and the deed was accepted. The deed was recorded on July 6, 1953. The new school building on 99th Street was completed and its doors opened to 500 students in February of 1955.

Construction Projects of Special Significance

There are many indications that the covering on the landfill was removed. Within a month of the transfer of title, the Board approved the removal of 4,000 cubic yards of soil for grading at another school and again in January 1954 they approved removal of another 3,000 cubic yards of fill from the canal site. Following completion of the 99th Street Elementary School on the site in 1955, 10,000 cubic yards of soil was removed from the top of the canal to grade the surrounding area.

The contractor building the school found drums containing waste materials beneath the original area planned for the building and the playground. The architect moved the location of the school eighty feet north of the original site and eliminated plans for a basement.

A sanitary sewer installed between 97th and 99th Streets beneath Wheatfield Avenue in 1957 crossed the canal south of the school. Records in the city engineer's office show no sewers in the canal before this one. A storm sewer was laid under Read Avenue and connected to a catch basin in the canal to allow flow in the direction of 97th Street. This breeched the west wall of the canal. In addition to the sewer lines, the French drain system placed around the school was connected to a storm sewer emptying into the Niagara River. Another catch basin was installed by a homeowner on 97th Street and connected to storm sewers that carried flows into the Black Creek.

In 1968, the **New York State Department of Transportation** built an expressway south of the canal and relocated a local street, Frontier Boulevard. They called upon Hooker Chemical to remove forty truckloads of chemical wastes that were uncovered during construction.

A review of these construction projects lead to the conclusion that whatever integrity the canal dumpsite may have had—with a bottom, sides, and cover of impermeable clay—was breached and the chemical wastes inside the canal could now migrate through trenches into the surrounding area.

A.2 SITE DESCRIPTION

Love Canal is a sixteen acre site in the southeast corner of Niagara Falls. This small area of the city contains an interesting combination of numerous bodies of surface water, dozens of dried up streambeds called **swales**, a high water table, and highly impermeable clay soils. This combination, together with prolonged periods of heavy rainfall and a rising water table, forced the chemical wastes to flow over the top of the clay dump pit and onto adjacent properties in what has been termed the **overflow bathtub**, effect.

The Love Canal site and surrounding environs are flat and the groundwater is near the surface. The general area of the canal is surrounded by creeks and rivers. The Niagara River is about 1,500 feet south of other nearby streams that eventually flow into this river. Berholtz Creek and Black Creek are a few blocks to the north and the Little Niagara River and Cayuga Creek are a few blocks to the west. As the area was studied it was discovered that the flow in these streams could reverse direction. When the Niagara River was high during the spring, the adjacent streams would flow north and at other times in the year they would flow south. This was later found to be a significant factor in explaining why high levels of chemical contaminants were found in stream sediments both above and below storm sewer outfall pipes.

Old aerial photographs of this area show geological formations called swales—which are old stream beds—and these may have served as natural permeable pathways through which the waste chemicals migrated over the years.

From the ground surface to five feet below grade there is a permeable layer of sand and silt. Below this to a depth of twenty-five feet are impermeable lacustrine clays. This is underlain by a highly permeable glacial till to a depth of forty feet. Below the glacial till is limestone bedrock.

There are two groundwater systems: an overburden shallow system and a bedrock aquifer, which serves as the principal aquifer in the Niagara Falls area.

A.3 A RAINY SEASON AND SURFACING OF THE CHEMICALS

In 1976, after six years of unusually heavy rainfall and snow, rain filled the canal and it was prevented by the canal clay bottom and sides from percolating any deeper. The canal overflowed like a bathtub. Reports began in October that chemicals had seeped into the basements of some of the homes on the periphery of the Love Canal property and a thick black residue accumulated in basement sump pumps. Portions of the landfill subsided and drums surfaced at some locations. Ponded surface water, heavily contaminated with chemicals, was found in the school area and the backyards of home immediately adjacent to the canal site.

A.4 ENGINEERING STUDIES AND A HEALTH EMERGENCY

Homes were built around the canal site beginning in the 1950s. With the canal at its center, the neighborhood ultimately consisted of approximately eight hundred private single family homes and two hundred and forty-four low income and senior citizen apartments to the west. All were within 1,500 feet of the canal. Starting in the 1960s, residents repeatedly complained about strong odors and about substances surfacing in yards and on the school playground. Following the long period of heavy rainfall in 1976 there were many signs of surfacing chemicals and the materials reacted with rainwater, turning milky white in color and the fumes causes eyes to tear. Because of the complaints, the city and the county hired a consulting firm, the Calspan Corporation, to investigate the situation in 1976.

The Calspan Corporation conducted a study of the canal area. They found drums just beneath the surface of the canal and high levels of PCBs in the storm sewer system. They also found chemical fumes in the air and oily residues in the sump pumps of a high percentage of the homes at the southern end of the canal. Calspan recommended that the canal be covered with a clay cap, home sump pumps be sealed off and a tile drainage system be installed to control migration of the wastes. None of this work was done.

In 1978, New York State started studies to learn what chemical wastes had migrated from the canal and where they were in the area. Samples were collected and analyzed from nearby storm sewers, stream sediments, drinking water taps in homes, shallow groundwater around the homes, and the leachate from the French drain system installed at the 99th Street Elementary School. Unacceptable levels of toxic vapors associated with eighty chemical compounds were found in the basements of many of the homes that were directly adjacent to the canal and this was called "ring 1." Additional studies showed contamination of some of the homes across the street from the Canal and this are was called "ring 2."

In March, Dr. Robert Whalen, the New York State Health Commissioner, initiated a house to house survey of the ninety-seven families in ring 1. This detailed health survey, a twenty-nine page questionnaire, was reviewed and quickly expanded to include all of the people residing in the two hundred and thirty-nine homes in rings 1 and 2. This study was completed in August and Commissioner Whalen declared a health emergency and immediately ordered the evacuation of pregnant women and children under the age of two. He ordered the closure of the 99th Street Elementary School and recommended that families not eat vegetables from their gardens and that residents limit the amount of time spent in their basements where air samples had been taken. The study demonstrated an increased amount of reproductive problems among women.

A few days after announcement of these results, the state agreed to purchase all two hundred and thirty-nine homes. This area was evacuated, the homes were boarded up, and a fence was placed around the site. The homes have since been demolished.

A.5 CONSTRUCTION REMEDIATION PROGRAM

After the homes were vacated a remedial construction program was started in October of the same year and completed the next year, which was 1979. The objective of the remediation program was to halt the flow of waste materials into the surrounding area. Tests that were done on the material in the canal showed a total organic carbon (TOC) range of 3,000 to 5,000 parts per million (ppm), depending on the rainfall. The **Environmental Protection Agency** brought in a portable treatment plant to pilot test treatment methods on the canal leachate. Results showed that the leachate could be treated if it was prefiltered, pH adjusted, flocculated, and then the clarified leachate fed into carbon beds.

A tile drainage system of eight inch perforated pipe surrounded by gravel was located in the backyards of adjoining properties, totally encircling the canal site and twelve to eighteen feet below the ground surface. A clay cap was placed on the site to prevent more surface water from entering the canal. By 1979, a permanent treatment plant was built on the canal site. The leachate collected flows into a 55,000 gallon well. From the well, the leachate is pumped into a 25,000 gallon holding/settling tank. The supernatant liquid passes through activated carbon columns in series where chlorinated hydrocarbons, other organic, and some inorganic contaminants are removed. The treated leachate, now at a TOC of less than 400 ppm, is discharged into the Niagara Falls sanitary sewer system.

A.6 A FEDERAL HEALTH EMERGENCY DECLARATION

In January 1980, the **U.S. Department of Justice** contracted for a limited cytogenetic study of twenty-six Love Canal residents for use in the preparation of a civil lawsuit against the Hooker Chemical Company for improper hazardous waste disposal at four Niagara Falls sites. The results of the study were released on May 19, 1980 and the study disclosed that there were excessive chromosomal abnormalities in the residents tested (Heath et al., 1984).

An emergency meeting at the White House followed this announcement. It included White House staff members and officials from the **Center for Disease Control**, the **Environmental Protection Agency**, and the **Federal Emergency Management Administration**. It was declared that the Environmental Protection Agency would undertake a comprehensive environmental testing program to determine what chemicals had migrated from the landfill and were in the air, soil, and water in the neighborhood and in the homes. The **Federal Emergency Management Administration** was directed to move the residents into temporary living quarters in order to allow the homes to be monitored continuously for chemical contamination. The **Center for Disease Control** was to undertake a complete health study of the residents to determine possible health effects from living near the chemical waste landfill. The objective of this coordinated federal effort was to correlate the health effects observed in the residents with the chemicals and their exposure pathways and concentrations and then determine whether or not living at the site constituted a substantial risk. "Is the site habitable" was the question to be answered. What happened at the White House meeting was the first confrontation with the dilemma, which is central to making decisions about hazardous chemical waste landfills (U.S. Environmental Protection Agency, May 21, 1980).

On May 21, 1980, President Carter declared, "that the damage resulting from the adverse impacts of chemical wastes . . . is of such severity and magnitude as to warrant an emergency declaration under Public Law 93-288." This notice in the Federal Register also stated "I hereby appoint Ms. Rita Meyninger of the Federal Emergency Management Administration to act as the Federal Coordinating Officer of this declared emergency" (*Federal Register*, June 4, 1980, p. 37733).

A.7 THE EMERGENCY RESPONSE PROGRAMS

From August through October 1980 the Environmental Protection Agency undertook the largest environmental testing program ever performed in the U.S. It involved a general contractor and eighteen subcontractors who collected samples and performed analytical laboratory work at a cost of $5.4 million. The study included a major hydroelectric investigation and the collection and analysis of more than 6,000 field samples of water, soil, sediment, and biota. Approximately 150,000 individual measurements of environmental contamination levels were compiled (Office of Technology Assessment, 1983).

Simultaneously, the Center for Disease Control engaged medical organizations in the Niagara Falls-Buffalo area to aid in the design of a comprehensive medical testing program. The program was divided into two phases, the first was to be a physical examination, medical history, and laboratory examination available to all residents living at the canal, as of June 1978. The second phase of the tests was suppose to be a random sample of the resident population and it would focus on specific biological systems known to be affected by chemicals at the canal. The tests included chromosome studies, psychological studies, and nerve conduction studies. As there was little knowledge of specific effects on human exposure to toxic wastes, a broad range of possibilities was included in the test protocols. Control groups were selected from similar socio-economic populations, one within Niagara Falls and one from a "clean" area of western New York State with no heavy industry or known dumpsites. Over $300,000 was spent on this planning effort. The projected $3 million for the cost of the program was never allocated and the testing was never done.

In October of 1980, funds were provided to buy out the houses and to allow the residents in the Love Canal neighborhood to permanently leave the area. Of the 800 families involved, approximately 570 had been relocated by May of 1982. Not all of the residents wanted to leave. There were about 550 single family homes and as of September of 1983 150 of these were still occupied. The **Love Canal Revitalization Agency**, using federal and state funds, bought 400 of the 550 homes in order to resell them to new owners under favorable financial arrangements when it is decided the area is habitable.

A.8 HABITABILITY OF THE LOVE CANAL HOMES

It was not until May of 1982, eighteen months after the Environmental Protection Agency completed the sample collection phase in the Love Canal Emergency Declaration Area that the results were published (U.S. Environmental Protection Agency, May 1982). The long delay was caused by many questions being raised about the validity of the testing protocol and the meaning of the test results.

A review by the **National Bureau of Standards** resulted in criticism both about the lack of control area data and questions about the significance of the ninety percent of the laboratory analysis results, which were characterized as "not detected" or "trace." The Bureau stated "Unless measured values, including 'none detected,' are accompanied by estimates of uncertainty, they are incomplete and of limited usefulness for further interpretation and for drawing conclusions". For these reasons, performance ". . . in the low parts-per-billion range' has not been demonstrated to our satisfaction in the documentation." They believe that such measurements could actually represent concentrations of one part per million (U.S. Department of Commerce, May 1981).

The **Department of Health and Human Services** was assigned the task of answering the question of "habitability" based on the results of the environmental testing program. First they declared the area was safe for resettlement, next they withdrew this conclusion, and finally in July of 1982 they reversed themselves again (The Risks of Living Near Love Canal, August 1982). One of the questions they formulated was "based on the available data, can you conclude that the area is not habitable?" Since the National Bureau of Standards review had questioned the validity of the "available data," federal officials and the eleven consultants they hired could not make the assessment that the data showed the area "not habitable." Ergo, it was habitable (U.S. Finds Love Canal Neighborhood is Habitable, July 15, 1982). The habitability recommendation included the requirements that "obvious contamination be cleaned up and that optimal containment methods be installed and maintained," i.e. the underground wall being planned to encircle the site to prevent migration of the chemical contaminants into the surrounding neighborhood.

A.9 CONSTRUCTION OF THE CONTAINMENT WALL ABORTED

The first part of a remediation program was completed when the leachate collection system and treatment facility was installed in 1978. After that a number of alternatives were evaluated with the objective of preventing the movement of the contaminants offsite through utility systems such as storm and sanitary sewers and of reducing the flow of shallow groundwater from the site (Camp Dresser and McKee, Inc., 1982). It was decided to construct a slurry wall containment system around the entire site. The system planned was a fifteen foot deep two foot thick reinforced concrete wall extending into the soft clay, which underlies the site. Before construction started, a borehole investigation was done in 1983 to explore the subsurface chemical conditions along the alignment of the proposed cutoff wall. The results of this investigation showed substantially higher levels of some chemicals than were reported at the same location in the Environmental Protection Agency testing program done in 1980 (E.C. Jordan Company, October 1983). The Agency reported that there had been "significant migration of chemicals" beyond the site of the proposed containment wall (*New York Times*, September 27, 1983).

Key Terms

- Center for Disease Control
- Department of Health and Human Services
- Environmental Protection Agency
- Federal Emergency Management Administration
- Love Canal
- Love Canal Revitalization Agency
- National Bureau of Standards
- New York State Department of Transportation
- Niagara Power and Development Company
- overflow bathtub
- swales
- U.S. Department of Justice

References

Camp Dresser and McKee, Inc. 1980. CH2M Hill, *Love Canal Remedial Action Program Environmental Information Document, Project Objectives*. Newark, New Jersey, pp. 2–1.

E.C. Jordan Company. October 1983. *Love Canal Remedial Project Borehole Investigation Volume I: Text*.

Federal Register. June 4, 1980. *Notices*, 45 (109) Washington, DC: Government Printing Office, p. 37733.

Heath et al. March 16, 1984. Cytogentic findings in persons living near the Love Canal. *The Journal of the American Medical Association*, 251(11): 1437–1440.

Jack McGraw Interview. September 1994. Acting Administrator, Office of Solid Waste and Emergency Response, U.S. Environmental Protection Agency, Washington, DC.

New York Times. September 27, 1983. Love Canal chemicals leak; Resettlement plans set back, p. A1.

Office of Technology Assessment. June 1983. *Habitability of the Love Canal Area Technical Memorandum*. Washington, DC: Congress of the United States.

The risks of living near Love Canal. August 1982. *Science*, 217: 808–810.

U.S. Department of Commerce. May 1982. National Bureau of Standards, *Review Material Provided by the EPA on the Analysis for Organic Chemical in the EPA Love Canal Monitoring Study*, Washington, DC.

U.S. Environmental Protection Agency Fact Sheet. May 21, 1980. *Love Canal Environmental and Health Data*. Washington, DC: Government Printing Office.

U.S. Environmental Protection Agency. May 1982. *Environmental Monitoring at Love Canal*, Vol. 1, EPA-600/4-82-030a, Washington, DC.

U.S. Finds Love Canal neighborhood is habitable. July 15, 1982. *New York Times*, pp. A1 and B5.

AMERICAN SOCIETY OF CIVIL ENGINEERS CODE OF ETHICS[1]

FUNDAMENTAL PRINCIPLES[2]

Engineers uphold and advance the integrity, honor and dignity of the engineering profession by:

1. using their knowledge and skill for the enhancement of human welfare and the environment;
2. being honest and impartial and serving with fidelity the public, their employers and clients;
3. striving to increase the competence and prestige of the engineering profession; and
4. supporting the professional and technical societies of their disciplines.

FUNDAMENTAL CANONS

1. Engineers shall hold paramount the safety, health and welfare of the public and shall strive to comply with the principles of sustainable development[3] in the performance of their professional duties.
2. Engineers shall perform services only in areas of their competence.
3. Engineers shall issue public statements only in an objective and truthful manner.
4. Engineers shall act in professional matters for each employer or client as faithful agents or trustees, and shall avoid conflicts of interest.
5. Engineers shall build their professional reputation on the merit of their services and shall not compete unfairly with others.
6. Engineers shall act in such a manner as to uphold and enhance the honor, integrity, and dignity of the engineering profession and shall act with zero-tolerance for bribery, fraud, and corruption.
7. Engineers shall continue their professional development throughout their careers, and shall provide opportunities for the professional development of those engineers under their supervision.

GUIDELINES TO PRACTICE UNDER THE FUNDAMENTAL CANONS OF ETHICS

CANON 1

Engineers shall hold paramount the safety, health and welfare of the public and shall strive to comply with the principles of sustainable development in the performance of their professional duties.

1. Engineers shall recognize that the lives, safety, health and welfare of the general public are dependent upon engineering judgments, decisions and practices incorporated into structures, machines, products, processes and devices.

[1]The Society's Code of Ethics was adopted on September 2, 1914 and was most recently amended on July 23, 2006. Pursuant to the Society's Bylaws, it is the duty of every Society member to report promptly to the Committee on Professional Conduct any observed violation of the Code of Ethics.

[2]In April 1975, the ASCE Board of Direction adopted the fundamental principles of the Code of Ethics of Engineers as accepted by the Accreditation Board for Engineering and Technology, Inc. (ABET).

[3]In November 1996, the ASCE Board of Direction adopted the following definition of Sustainable Development: "Sustainable Development is the challenge of meeting human needs for natural resources, industrial products, energy, food, transportation, shelter, and effective waste management while conserving and protecting environmental quality and the natural resource base essential for future development."

2. Engineers shall approve or seal only those design documents, reviewed or prepared by them, which are determined to be safe for public health and welfare in conformity with accepted engineering standards.
3. Engineers whose professional judgment is overruled under circumstances where the safety, health and welfare of the public are endangered, or the principles of sustainable development ignored, shall inform their clients or employers of the possible consequences.
4. Engineers who have knowledge or reason to believe that another person or firm may be in violation of any of the provisions of Canon 1 shall present such information to the proper authority in writing and shall cooperate with the proper authority in furnishing such further information or assistance as may be required.
5. Engineers should seek opportunities to be of constructive service in civic affairs and work for the advancement of the safety, health and well being of their communities, and the protection of the environment through the practice of sustainable development.
6. Engineers should be committed to improving the environment by adherence to the principles of sustainable development so as to enhance the quality of life of the general public.

CANON 2

Engineers shall perform services only in areas of their competence.

1. Engineers shall undertake to perform engineering assignments only when qualified by education or experience in the technical field of engineering involved.
2. Engineers may accept an assignment requiring education or experience outside of their own fields of competence, provided their services are restricted to those phases of the project in which they are qualified. All other phases of such project shall be performed by qualified associates, consultants, or employees.
3. Engineers shall not affix their signatures or seals to any engineering plan or document dealing with subject matter in which they lack competence by virtue of education or experience or to any such plan or document not reviewed or prepared under their supervisory control.

CANON 3

Engineers shall issue public statements only in an objective and truthful manner.

1. Engineers should endeavor to extend the public knowledge of engineering and sustainable development, and shall not participate in the dissemination of untrue, unfair or exaggerated statements regarding engineering.
2. Engineers shall be objective and truthful in professional reports, statements, or testimony. They shall include all relevant and pertinent information in such reports, statements, or testimony.
3. Engineers, when serving as expert witnesses, shall express an engineering opinion only when it is founded upon adequate knowledge of the facts, upon a background of technical competence, and upon honest conviction.
4. Engineers shall issue no statements, criticisms, or arguments on engineering matters which are inspired or paid for by interested parties, unless they indicate on whose behalf the statements are made.
5. Engineers shall be dignified and modest in explaining their work and merit, and will avoid any act tending to promote their own interests at the expense of the integrity, honor and dignity of the profession.

CANON 4

Engineers shall act in professional matters for each employer or client as faithful agents or trustees, and shall avoid conflicts of interest.

1. Engineers shall avoid all known or potential conflicts of interest with their employers or clients and shall promptly inform their employers or clients of any business association, interests, or circumstances which could influence their judgment or the quality of their services.
2. Engineers shall not accept compensation from more than one party for services on the same project, or for services pertaining to the same project, unless the circumstances are fully disclosed to and agreed to, by all interested parties.
3. Engineers shall not solicit or accept gratuities, directly or indirectly, from contractors, their agents, or other parties dealing with their clients or employers in connection with work for which they are responsible.
4. Engineers in public service as members, advisors, or employees of a governmental body or department shall not participate in considerations or actions with respect to services solicited or provided by them or their organization in private or public engineering practice.
5. Engineers shall advise their employers or clients when, as a result of their studies, they believe a project will not be successful.
6. Engineers shall not use confidential information coming to them in the course of their assignments as a means of making personal profit if such action is adverse to the interests of their clients, employers or the public.
7. Engineers shall not accept professional employment outside of their regular work or interest without the knowledge of their employers.

CANON 5

Engineers shall build their professional reputation on the merit of their services and shall not compete unfairly with others.

1. Engineers shall not give, solicit or receive either directly or indirectly, any political contribution, gratuity, or unlawful consideration in order to secure work,

exclusive of securing salaried positions through employment agencies.

2. Engineers should negotiate contracts for professional services fairly and on the basis of demonstrated competence and qualifications for the type of professional service required.
3. Engineers may request, propose or accept professional commissions on a contingent basis only under circumstances in which their professional judgments would not be compromised.
4. Engineers shall not falsify or permit misrepresentation of their academic or professional qualifications or experience.
5. Engineers shall give proper credit for engineering work to those to whom credit is due, and shall recognize the proprietary interests of others. Whenever possible, they shall name the person or persons who may be responsible for designs, inventions, writings or other accomplishments.
6. Engineers may advertise professional services in a way that does not contain misleading language or is in any other manner derogatory to the dignity of the profession. Examples of permissible advertising are as follows:
   - Professional cards in recognized, dignified publications, and listings in rosters or directories published by responsible organizations, provided that the cards or listings are consistent in size and content and are in a section of the publication regularly devoted to such professional cards.
   - Brochures which factually describe experience, facilities, personnel and capacity to render service, providing they are not misleading with respect to the engineer's participation in projects described.
   - Display advertising in recognized dignified business and professional publications, providing it is factual and is not misleading with respect to the engineer's extent of participation in projects described.
   - A statement of the engineers' names or the name of the firm and statement of the type of service posted on projects for which they render services.
   - Preparation or authorization of descriptive articles for the lay or technical press, which are factual and dignified. Such articles shall not imply anything more than direct participation in the project described.
   - Permission by engineers for their names to be used in commercial advertisements, such as may be published by contractors, material suppliers, etc., only by means of a modest, dignified notation acknowledging the engineers' participation in the project described. Such permission shall not include public endorsement of proprietary products.
   - Engineers shall not maliciously or falsely, directly or indirectly, injure the professional reputation, prospects, practice or employment of another engineer or indiscriminately criticize another's work.
   - Engineers shall not use equipment, supplies, laboratory or office facilities of their employers to carry on outside private practice without the consent of their employers.

CANON 6

Engineers shall act in such a manner as to uphold and enhance the honor, integrity, and dignity of the engineering profession and shall act with zero tolerance for bribery, fraud, and corruption.

1. Engineers shall not knowingly engage in business or professional practices of a fraudulent, dishonest or unethical nature.
2. Engineers shall be scrupulously honest in their control and spending of monies, and promote effective use of resources through open, honest and impartial service with fidelity to the public, employers, associates and clients.
3. Engineers shall act with zero-tolerance for bribery, fraud, and corruption in all engineering or construction activities in which they are engaged.
4. Engineers should be especially vigilant to maintain appropriate ethical behavior where payments of gratuities or bribes are institutionalized practices.
5. Engineers should strive for transparency in the procurement and execution of projects. Transparency includes disclosure of names, addresses, purposes, and fees or commissions paid for all agents facilitating projects.
6. Engineers should encourage the use of certifications specifying zero tolerance for bribery, fraud, and corruption in all contracts.

CANON 7

Engineers shall continue their professional development throughout their careers, and shall provide opportunities for the professional development of those engineers under their supervision.

1. Engineers should keep current in their specialty fields by engaging in professional practice, participating in continuing education courses, reading in the technical literature, and attending professional meetings and seminars.
2. Engineers should encourage their engineering employees to become registered at the earliest possible date.
3. Engineers should encourage engineering employees to attend and present papers at professional and technical society meetings.
4. Engineers shall uphold the principle of mutually satisfying relationships between employers and employees with respect to terms of employment including professional grade descriptions, salary ranges, and fringe benefits.

CODES OF ETHICS OF ENGINEERS

AMERICAN SOCIETY OF AGRICULTURAL AND BIOLOGICAL ENGINEERS CODE OF ETHICS

THE FUNDAMENTAL PRINCIPLES

Engineers uphold and advance the integrity, honor, and dignity of the engineering profession by:

I. using their knowledge and skill for the enhancement of human welfare;

II. being honest and impartial, and serving with fidelity the public, their employers, and clients;

III. striving to increase the competence and prestige of the engineering profession; and

IV. supporting the professional and technical societies of the disciplines.

THE FUNDAMENTAL CANNONS

1. Engineers shall hold paramount the safety, health, and welfare of the public in the performance of their professional duties.
2. Engineers shall perform services only in the areas of their competence.
3. Engineers shall issue public statement only in an objective and truthful manner.
4. Engineers shall act in professional matters for each employer or client as faithful agents or trustees, and shall avoid conflicts of interest.
5. Engineers shall build their professional reputation on the merit of their services and shall not compete unfairly with others.
6. Engineers shall act in such a manner as to uphold and enhance the honor, integrity, and dignity of the profession.
7. Engineers shall continue their professional development throughout their careers and shall provide opportunities for the professional development of those engineers under their supervision.

AMERICAN INSTITUTE OF CHEMICAL ENGINEERS CODE OF ETHICS

(Revised January 17, 2003)

Members of the American Institute of Chemical Engineers shall uphold and advance the integrity, honor and dignity of the engineering profession by: being honest and impartial and serving with fidelity their employers, their clients, and the public; striving to increase the competence and prestige of the engineering profession; and using their knowledge and skill for the enhancement of human welfare. To achieve these goals, members shall:

- Hold paramount the safety, health and welfare of the public and protect the environment in performance of their professional duties.
- Formally advise their employers or clients (and consider further disclosure, if warranted) if they perceive that a consequence of their duties will adversely affect the present or future health or safety of their colleagues or the public.
- Accept responsibility for their actions, seek and heed critical review of their work and offer objective criticism of the work of others.
- Issue statements or present information only in an objective and truthful manner.

- Act in professional matters for each employer or client as faithful agents or trustees, avoiding conflicts of interest and never breaching confidentiality.
- Treat fairly and respectfully all colleagues and co-workers, recognizing their unique contributions and capabilities.
- Perform professional services only in areas of their competence.
- Build their professional reputations on the merits of their services.
- Continue their professional development throughout their careers, and provide opportunities for the professional development of those under their supervision.
- Never tolerate harassment.
- Conduct themselves in a fair, honorable and respectful manner.

AMERICAN SOCEITY OF MECHANICAL ENGINEERS CODE OF ETHICS

THE FUNDAMENTAL PRINCIPLES

Engineers uphold and advance the integrity, honor and dignity of the engineering profession by:

I. using their knowledge and skill for the enhancement of human welfare;
II. being honest and impartial, and serving with fidelity their clients (including their employers) and the public; and
III. striving to increase the competence and prestige of the engineering profession.

THE FUNDAMENTAL CANONS

1. Engineers shall hold paramount the safety, health and welfare of the public in the performance of their professional duties.
2. Engineers shall perform services only in the areas of their competence; they shall build their professional reputation on the merit of their services and shall not compete unfairly with others.
3. Engineers shall continue their professional development throughout their careers and shall provide opportunities for the professional and ethical development of those engineers under their supervision.
4. Engineers shall act in professional matters for each employer or client as faithful agents or trustees, and shall avoid conflicts of interest or the appearance of conflicts of interest.
5. Engineers shall respect the proprietary information and intellectual property rights of others, including charitable organizations and professional societies in the engineering field.
6. Engineers shall associate only with reputable persons or organizations.
7. Engineers shall issue public statements only in an objective and truthful manner and shall avoid any conduct which brings discredit upon the profession.
8. Engineers shall consider environmental impact and sustainable development in the performance of their professional duties.
9. Engineers shall not seek ethical sanction against another engineer unless there is good reason to do so under the relevant codes, policies and procedures governing that engineer's ethical conduct.
10. Engineers who are members of the Society shall endeavor to abide by the Constitution, By-Laws and Policies of the Society, and they shall disclose knowledge of any matter involving another member's alleged violation of this Code of Ethics or the Society's Conflicts of Interest Policy in a prompt, complete and truthful manner to the chair of the Committee on Ethical Standards and Review.

INSTITUTE OF ELECTRONIC AND ELECTRICAL ENGINEERS CODE OF ETHICS

As per IEEE Bylaw I-104.14, membership in IEEE in any grade shall carry the obligation to abide by the IEEE Code of Ethics (IEEE Policy 7.8) as stated below.

We, the members of the IEEE, in recognition of the importance of our technologies in affecting the quality of life throughout the world, and in accepting a personal obligation to our profession, its members and the communities we serve, do hereby commit ourselves to the highest ethical and professional conduct and agree:

1. to accept responsibility in making decisions consistent with the safety, health and welfare of the public, and to disclose promptly factors that might endanger the public or the environment;
2. to avoid real or perceived conflicts of interest whenever possible, and to disclose them to affected parties when they do exist;
3. to be honest and realistic in stating claims or estimates based on available data;
4. to reject bribery in all its forms;
5. to improve the understanding of technology, its appropriate application, and potential consequences;

6. to maintain and improve our technical competence and to undertake technological tasks for others only if qualified by training or experience, or after full disclosure of pertinent limitations;
7. to seek, accept, and offer honest criticism of technical work, to acknowledge and correct errors, and to credit properly the contributions of others;
8. to treat fairly all persons regardless of such factors as race, religion, gender, disability, age, or national origin;
9. to avoid injuring others, their property, reputation, or employment by false or malicious action;
10. to assist colleagues and co-workers in their professional development and to support them in following this code of ethics.

INSTITUTE OF INDUSTRIAL ENGINEERS CODE OF ETHICS

IIE endorses the Canon of Ethics provided by the Accreditation Board for Engineering and Technology.

THE FUNDAMENTAL PRINCIPLES

Engineers uphold and advance the integrity, honor and dignity of the engineering profession by:

1. Using their knowledge and skill for the enhancement of human welfare;
2. Being honest and impartial, and serving with fidelity the public, their employers and clients;
3. Striving to increase the competence and prestige of the engineering profession; and
4. Supporting the professional and technical societies of their disciplines.

THE FUNDAMENTAL CANONS

1. Engineers shall hold paramount the safety, health and welfare of the public in the performance of their professional duties.
2. Engineers shall perform services only in the areas of their competence.
3. Engineers shall issue public statements only in an objective and truthful manner.
4. Engineers shall act in professional matters for each employer or client as faithful agents or trustees, and shall avoid conflicts of interest.
5. Engineers shall build their professional reputation on the merit of their services and shall not compete unfairly with others.
6. Engineers shall associate only with reputable persons or organizations.
7. Engineers shall continue their professional development throughout their careers and shall provide opportunities for the professional development of those engineers under their supervision.

Visit the Accreditation Board for Engineering and Technology (ABET) site to learn more about the Code of Ethics and the accreditation process.

Approved by the IEEE Board of Directors, *February 2006*.

Engineers Joint Documents Committee
Design and Construction Related Documents
Instructions and License Agreement

Instructions

Before you use any EJCDC document:

1. Read the License Agreement. You agree to it and are bound by its terms when you use the EJCDC document.

2. Make sure that you have the correct version for your word processing software.

How to Use:

1. While EJCDC has expended considerable effort to make the software translations exact, it can be that a few document controls (e.g., bold, underline) did not carry over.

2. Similarly, your software may change the font specification if the font is not available in your system. It will choose a font that is close in appearance. In this event, the pagination may not match the control set.

3. If you modify the document, you must follow the instructions in the License Agreement about notification.

4. Also note the instruction in the License Agreement about the EJCDC copyright.

License Agreement

You should carefully read the following terms and conditions before using this document. Commencement of use of this document indicates your acceptance of these terms and conditions. If you do not agree to them, you should promptly return the materials to the vendor, and your money will be refunded.

The Engineers Joint Contract Documents Committee ("EJCDC") provides **EJCDC Design and Construction Related Documents** and licenses their use worldwide. You assume sole responsibility for the selection of specific documents or portions thereof to achieve your intended results, and for the installation, use, and results obtained from **EJCDC Design and Construction Related Documents**.

You acknowledge that you understand that the text of the contract documents of **EJCDC Design and Construction Related Documents** has important legal consequences and that consultation with an attorney is recommended with respect to use or modification of the text. You further acknowledge that EJCDC documents are protected by the copyright laws of the United States.

License:
You have a limited nonexclusive license to:

1. Use **EJCDC Design and Construction Related Documents** on any number of machines owned, leased or rented by your company or organization.

2. Use **EJCDC Design and Construction Related Documents** in printed form for bona fide contract documents.

3. Copy **EJCDC Design and Construction Related Documents** into any machine readable or printed form for backup or modification purposes in support of your use of **EJCDC Design and Construction Related Documents**.

You agree that you will:

1. Reproduce and include EJCDC's copyright notice on any printed or machine-readable copy, modification, or portion merged into another document or program. All proprietary rights in **EJCDC Design and Construction Related Documents** are and shall remain the property of EJCDC.

2. Not represent that any of the contract documents you generate from **EJCDC Design and Construction Related Documents** are EJCDC documents unless (i) the document text is used without alteration or (ii) all additions and changes to, and deletions from, the text are clearly shown.

You may not use, copy, modify, or transfer EJCDC Design and Construction Related Documents, or any copy, modification or merged portion, in whole or in part, except as expressly provided for in this license. Reproduction of EJCDC Design and Construction Related Documents in printed or machine-readable format for resale or educational purposes is expressly prohibited.

If you transfer possession of any copy, modification or merged portion of EJCDC Design and Construction Related Documents to another party, your license is automatically terminated.

Term:
The license is effective until terminated. You may terminate it at any time by destroying **EJCDC Design and Construction Related Documents** altogether with all

copies, modifications and merged portions in any form. It will also terminate upon conditions set forth elsewhere in this Agreement or if you fail to comply with any term or condition of this Agreement. You agree upon such termination to destroy **EJCDC Design and Construction Related Documents** along with all copies, modifications and merged portions in any form.

Limited Warranty:
EJCDC warrants the CDs and diskettes on which **EJCDC Design and Construction Related Documents** is furnished to be free from defects in materials and workmanship under normal use for a period of ninety (90) days from the date of delivery to you as evidenced by a copy of your receipt.

There is no other warranty of any kind, either expressed or implied, including, but not limited to the implied warranties of merchantability and fitness for a particular purpose. Some states do not allow the exclusion of implied warranties, so the above exclusion may not apply to you. This warranty gives you specific legal rights and you may also have other rights which vary from state to state.

EJCDC does not warrant that the functions contained in **EJCDC Design and Construction Related Documents** will meet your requirements or that the operation of **EJCDC Design and Construction Related Documents** will be uninterrupted or error free.

Limitations of Remedies:
EJCDC's entire liability and your exclusive remedy shall be:

1. the replacement of any document not meeting EJCDC's "Limited Warranty" which is returned to EJCDC's selling agent with a copy of your receipt, or

2. if EJCDC's selling agent is unable to deliver a replacement CD or diskette which is free of defects in materials and workmanship, you may terminate this Agreement by returning EJCDC Document and your money will be refunded.

In no event will EJCDC be liable to you for any damages, including any lost profits, lost savings or other incidental or consequential damages arising out of the use or inability to use **EJCDC Design and Construction Related Documents** even if EJCDC has been advised of the possibility of such damages, or for any claim by any other party.

Some states do not allow the limitation or exclusion of liability for incidental or consequential damages, so the above limitation or exclusion may not apply to you.

General:
You may not sublicense, assign, or transfer this license except as expressly provided in this Agreement. Any attempt otherwise to sublicense, assign, or transfer any of the rights, duties, or obligations hereunder is void.

This Agreement shall be governed by the laws of the State of Virginia. Should you have any questions concerning this Agreement, you may contact EJCDC by writing to:

Arthur Schwartz, Esq.
General Counsel
National Society of Professional Engineers
1420 King Street
Alexandria, VA 22314

Phone: (703) 684-2845
Fax: (703) 836-4875
e-mail: aschwartz@nspe.org

You acknowledge that you have read this agreement, understand it and agree to be bound by its terms and conditions. You further agree that it is the complete and exclusive statement of the agreement between us which supersedes any proposal or prior agreement, oral or written, and any other communications between us relating to the subject matter of this agreement.

This document has important legal consequences; consultation with an attorney is encouraged with respect to its use or modification. This document should be adapted to the particular circumstances of the contemplated Project and the controlling Laws and Regulations.

SUGGESTED FORM OF AGREEMENT BETWEEN OWNER AND CONTRACTOR FOR CONSTRUCTION CONTRACT (STIPULATED PRICE)

Prepared by

ENGINEERS JOINT CONTRACT DOCUMENTS COMMITTEE

and

Issued and Published Jointly by

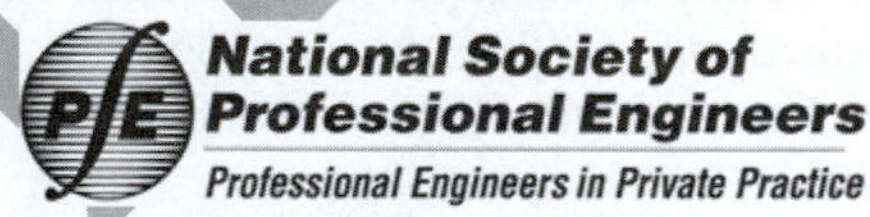

AMERICAN COUNCIL OF ENGINEERING COMPANIES

ASSOCIATED GENERAL CONTRACTORS OF AMERICA

AMERICAN SOCIETY OF CIVIL ENGINEERS

PROFESSIONAL ENGINEERS IN PRIVATE PRACTICE
A Practice Division of the
NATIONAL SOCIETY OF PROFESSIONAL ENGINEERS

Endorsed by

CONSTRUCTION SPECIFICATIONS INSTITUTE

This Suggested Form of Agreement has been prepared for use with the Standard General Conditions of the Construction Contract (EJCDC C-700, 2007 Edition). Their provisions are interrelated, and a change in one may necessitate a change in the other. The language contained in the Suggested Instructions to Bidders (EJCDC C-200, 2007 Edition) is also carefully interrelated with the language of this Agreement. Their usage is discussed in the Narrative Guide to the 2007 EJCDC Construction Documents (EJCDC C-001, 2007 Edition).

INTRODUCTION

This Suggested Form of Agreement between Owner and Contractor for Construction Contract (Stipulated Price) ("Agreement") has been prepared for use with the Suggested Instructions to Bidders for Construction Contracts ("Instructions to Bidders") (EJCDC C-200, 2007 Edition); the Suggested Bid Form for Construction Contracts ("Bid Form") (EJCDC C-410, 2007 Edition); and the Standard General Conditions of the Construction Contract ("General Conditions") (EJCDC C-700, 2007 Edition). Their provisions are interrelated, and a change in one may necessitate a change in the others. See also the Guide to the Preparation of Supplementary Conditions (EJCDC C-800, 2007 Edition), and the Narrative Guide to the 2007 EJCDC Construction Documents (EJCDC C-001, 2007 Edition).

This Agreement form assumes use of a Project Manual that contains the following documentary information for a construction project:

- Bidding Requirements, which include the advertisement or invitation to bid, the Instructions to Bidders, and the Bid Form that is suggested or prescribed, all of which provide information and guidance for all Bidders; and

- Contract Documents, which include the Agreement, performance and payment bonds, the General Conditions, the Supplementary Conditions, the Drawings, and the Specifications.

The Bidding Requirements are not Contract Documents because much of their substance pertains to the relationships prior to the award of the Contract and has little effect or impact thereafter, and because many contracts are awarded without going through the bidding process. In some cases, however, the actual Bid may be attached as an exhibit to the Agreement to avoid extensive rekeying. (The definitions of terms used in this Agreement, including "Bidding Documents," "Bidding Requirements," and "Contract Documents," are set forth Article 1 of the General Conditions.)

Suggested provisions are accompanied by "Notes to User" to assist in preparing the Agreement. The provisions have been coordinated with the other forms produced by EJCDC. Much of the language should be usable on most projects, but modifications and additional provisions will often be necessary. When modifying the suggested language or writing additional provisions, the user must check the other documents thoroughly for conflicts and coordination of terms and make appropriate revisions in all affected documents.

All parties involved in construction projects benefit significantly from a standardized approach in the location of subject matter throughout the documents. Experience confirms the danger of addressing the same subject matter in more than one location: doing so frequently leads to confusion and unanticipated legal consequences. When preparing documents for a construction project, careful attention should be given to the guidance provided in the Uniform Location of Subject Matter (EJCDC N-122).

EJCDC has designated Section 00520 for this Agreement. If this convention is used, the first page of the Agreement would be numbered 00520-1. If CSI's MasterFormat 04™ is being used for the Project Manual, consult MasterFormat 04 for the appropriate section number and number the pages accordingly.

For brevity, paragraphs of the Instructions to Bidders are referenced with the prefix "I," those of the Bid Form are referenced with the prefix "BF," and those of this Agreement are referenced with the prefix "A."

NOTE: EJCDC publications may be purchased from any of the organizations listed on the page immediately following the cover page of this document.

SUGGESTED FORM OF AGREEMENT BETWEEN OWNER AND CONTRACTOR FOR CONSTRUCTION CONTRACT (STIPULATED PRICE)

THIS AGREEMENT is by and between ______________________________ ("Owner") and ______________________________ ("Contractor").

Owner and Contractor hereby agree as follows:

ARTICLE 1 – WORK

1.01 Contractor shall complete all Work as specified or indicated in the Contract Documents. The Work is generally described as follows:

ARTICLE 2 – THE PROJECT

2.01 The Project for which the Work under the Contract Documents may be the whole or only a part is generally described as follows:

ARTICLE 3 – ENGINEER

3.01 The Project has been designed by _____ (Engineer), which is to act as Owner's representative, assume all duties and responsibilities, and have the rights and authority assigned to Engineer in the Contract Documents in connection with the completion of the Work in accordance with the Contract Documents.

ARTICLE 4 – CONTRACT TIMES

4.01 *Time of the Essence*

A. All time limits for Milestones, if any, Substantial Completion, and completion and readiness for final payment as stated in the Contract Documents are of the essence of the Contract.

4.02 *Dates for Substantial Completion and Final Payment*

A. The Work will be substantially completed on or before _____, and completed and ready for final payment in accordance with Paragraph 14.07 of the General Conditions on or before _____.

[or]

4.02 *Days to Achieve Substantial Completion and Final Payment*

A. The Work will be substantially completed within _____ days after the date when the Contract Times commence to run as provided in Paragraph 2.03 of the General Conditions, and completed and ready for final payment in accordance with Paragraph 14.07 of the General Conditions within _____ days after the date when the Contract Times commence to run.

4.03 *Liquidated Damages*

A. Contractor and Owner recognize that time is of the essence as stated in Paragraph 4.01 above and that Owner will suffer financial loss if the Work is not completed within the times specified in Paragraph 4.02 above, plus any extensions thereof allowed in accordance with Article 12 of the General Conditions. The parties also recognize the delays, expense, and difficulties involved in proving in a legal or arbitration proceeding the actual loss suffered by Owner if the Work is not completed on time. Accordingly, instead of requiring any such proof, Owner and Contractor agree that as liquidated damages for delay (but not as a penalty), Contractor shall pay Owner $_____ for each day that expires after the time specified in Paragraph 4.02 above for Substantial Completion until the Work is substantially complete. After Substantial Completion, if Contractor shall neglect, refuse, or fail to complete the remaining Work within the Contract Time or any proper extension thereof granted by Owner, Contractor shall pay Owner $_____ for each day that expires after the time specified in Paragraph 4.02 above for completion and readiness for final payment until the Work is completed and ready for final payment.

NOTE TO USER

If failure to reach a Milestone on time is of such consequence that the assessment of liquidated damages for failure to reach one or more Milestones on time is to be provided, appropriate amending or supplementing language should be inserted here.

ARTICLE 5 – CONTRACT PRICE

5.01 Owner shall pay Contractor for completion of the Work in accordance with the Contract Documents an amount in current funds equal to the sum of the amounts determined pursuant to Paragraphs 5.01.A, 5.01.B, and 5.01.C below:

A. For all Work other than Unit Price Work, a lump sum of: $______________________

All specific cash allowances are included in the above price in accordance with Paragraph 11.02 of the General Conditions.

B. For all Unit Price Work, an amount equal to the sum of the established unit price for each separately identified item of Unit Price Work times the actual quantity of that item:

UNIT PRICE WORK

Item No.	Description	Unit	Estimated Quantity	Bid Unit Price	Bid Price

Total of all Bid Prices (Unit Price Work) $

The Bid prices for Unit Price Work set forth as of the Effective Date of the Agreement are based on estimated quantities. As provided in Paragraph 11.03 of the General Conditions, estimated quantities are not guaranteed, and determinations of actual quantities and classifications are to be made by Engineer as provided in Paragraph 9.07 of the General Conditions.

C. For all Work, at the prices stated in Contractor's Bid, attached hereto as an exhibit.

NOTES TO USER

1. *If adjustment prices for variations from stipulated Base Bid quantities have been agreed to, insert appropriate provisions.*

2. *Depending upon the particular project bid form used, use 5.01.A alone, 5.01.A and 5.01.B together, 5.01.B alone, or 5.01.C alone, deleting those not used and renumbering accordingly. If 5.01.C is used, Contractor's Bid is attached as an exhibit and listed as a Contract Document in A-9.*

ARTICLE 6 – PAYMENT PROCEDURES

6.01 *Submittal and Processing of Payments*

A. Contractor shall submit Applications for Payment in accordance with Article 14 of the General Conditions. Applications for Payment will be processed by Engineer as provided in the General Conditions.

6.02 *Progress Payments; Retainage*

A. Owner shall make progress payments on account of the Contract Price on the basis of Contractor's Applications for Payment on or about the _____ day of each month during performance of the Work as provided in Paragraph 6.02.A.1 below. All such payments will be measured by the schedule of values established as provided in Paragraph 2.07.A of the General Conditions (and in the case of Unit Price Work based on the number of units completed) or, in the event there is no schedule of values, as provided in the General Requirements.

1. Prior to Substantial Completion, progress payments will be made in an amount equal to the percentage indicated below but, in each case, less the aggregate of payments previously made and less such amounts as Engineer may determine or Owner may withhold, including but not limited to liquidated damages, in accordance with Paragraph 14.02 of the General Conditions.

   a. _____ percent of Work completed (with the balance being retainage). If the Work has been 50 percent completed as determined by Engineer, and if the character and progress of the Work have been satisfactory to Owner and Engineer, then as long as the character and progress of the Work remain satisfactory to Owner and Engineer, there will be no additional retainage; and

   b. _____ percent of cost of materials and equipment not incorporated in the Work (with the balance being retainage).

B. Upon Substantial Completion, Owner shall pay an amount sufficient to increase total payments to Contractor to _____ percent of the Work completed, less such amounts as Engineer shall determine in accordance with Paragraph 14.02.B.5 of the General Conditions and less _____percent of Engineer's estimate of the value of Work to be completed or corrected as shown on the tentative list of items to be completed or corrected attached to the certificate of Substantial Completion.

NOTE TO USER

Typical values used in Paragraph 6.02.B are 100 percent and 200 percent respectively.

6.03 *Final Payment*

A. Upon final completion and acceptance of the Work in accordance with Paragraph 14.07 of the General Conditions, Owner shall pay the remainder of the Contract Price as recommended by Engineer as provided in said Paragraph 14.07.

ARTICLE 7 – INTEREST

7.01 All moneys not paid when due as provided in Article 14 of the General Conditions shall bear interest at the rate of _____ percent per annum.

ARTICLE 8 – CONTRACTOR'S REPRESENTATIONS

8.01 In order to induce Owner to enter into this Agreement, Contractor makes the following representations:

A. Contractor has examined and carefully studied the Contract Documents and the other related data identified in the Bidding Documents.

B. Contractor has visited the Site and become familiar with and is satisfied as to the general, local, and Site conditions that may affect cost, progress, and performance of the Work.

C. Contractor is familiar with and is satisfied as to all federal, state, and local Laws and Regulations that may affect cost, progress, and performance of the Work.

D. Contractor has carefully studied all: (1) reports of explorations and tests of subsurface conditions at or contiguous to the Site and all drawings of physical conditions relating to existing surface or subsurface structures at the Site (except Underground Facilities), if any, that have been identified in Paragraph SC-4.02 of the Supplementary Conditions as containing reliable "technical data," and (2) reports and drawings of Hazardous Environmental Conditions, if any, at the Site that have been identified in Paragraph SC-4.06 of the Supplementary Conditions as containing reliable "technical data."

NOTE TO USER

Modify the above paragraph if there are no such reports or drawings.

E. Contractor has considered the information known to Contractor; information commonly known to contractors doing business in the locality of the Site; information and observations obtained from visits to the Site; the Contract Documents; and the Site-related reports and drawings identified in the Contract Documents, with respect to the effect of such information, observations, and documents on (1) the cost, progress, and performance of the Work; (2) the means, methods, techniques, sequences, and procedures of construction to be employed by Contractor, including any specific means, methods, techniques, sequences, and procedures of construction expressly required by the Contract Documents; and (3) Contractor's safety precautions and programs.

NOTE TO USER

If the Contract Documents do not identify any Site-related reports and drawings, modify this paragraph accordingly.

F. Based on the information and observations referred to in Paragraph 8.01.E above, Contractor does not consider that further examinations, investigations, explorations, tests, studies, or data are necessary for the performance of the Work at the Contract Price, within the Contract Times, and in accordance with the other terms and conditions of the Contract Documents.

G. Contractor is aware of the general nature of work to be performed by Owner and others at the Site that relates to the Work as indicated in the Contract Documents.

H. Contractor has given Engineer written notice of all conflicts, errors, ambiguities, or discrepancies that Contractor has discovered in the Contract Documents, and the written resolution thereof by Engineer is acceptable to Contractor.

I. The Contract Documents are generally sufficient to indicate and convey understanding of all terms and conditions for performance and furnishing of the Work.

ARTICLE 9 – CONTRACT DOCUMENTS

9.01 *Contents*

A. The Contract Documents consist of the following:

1. This Agreement (pages 1 to __, inclusive).

2. Performance bond (pages _____ to _____, inclusive).

3. Payment bond (pages _____ to _____, inclusive).

4. Other bonds (pages _____ to _____, inclusive).

   a. _____ (pages _____ to _____, inclusive).

   b. _____ (pages _____ to _____, inclusive).

   c. _____ (pages _____ to _____, inclusive).

5. General Conditions (pages _____ to _____, inclusive).

6. Supplementary Conditions (pages _____ to _____, inclusive).

7. Specifications as listed in the table of contents of the Project Manual.

8. Drawings consisting of _____ sheets with each sheet bearing the following general title: _____ [or] the Drawings listed on attached sheet index.

9. Addenda (numbers _____ to _____, inclusive).

10. Exhibits to this Agreement (enumerated as follows):

    a. Contractor's Bid (pages _____ to _____, inclusive).

    b. Documentation submitted by Contractor prior to Notice of Award (pages _____ to _____, inclusive).

    c. *[List other required attachments (if any), such as documents required by funding or lending agencies].*

11. The following which may be delivered or issued on or after the Effective Date of the Agreement and are not attached hereto:

    a. Notice to Proceed (pages _____ to _____, inclusive).

    b. Work Change Directives.

    c. Change Orders.

NOTE TO USER

If any of the items listed are not to be included as Contract Documents, remove such item from the list and renumber the remaining items.

B. The documents listed in Paragraph 9.01.A are attached to this Agreement (except as expressly noted otherwise above).

C. There are no Contract Documents other than those listed above in this Article 9.

D. The Contract Documents may only be amended, modified, or supplemented as provided in Paragraph 3.04 of the General Conditions.

ARTICLE 10 – MISCELLANEOUS

10.01 *Terms*

A. Terms used in this Agreement will have the meanings stated in the General Conditions and the Supplementary Conditions.

10.02 *Assignment of Contract*

A. No assignment by a party hereto of any rights under or interests in the Contract will be binding on another party hereto without the written consent of the party sought to be bound; and, specifically but without limitation, moneys that may become due and moneys that are due may not be assigned without such consent (except to the extent that the effect of this restriction may be limited by law), and unless specifically stated to the contrary in any written consent to an assignment, no assignment will release or discharge the assignor from any duty or responsibility under the Contract Documents.

10.03 *Successors and Assigns*

A. Owner and Contractor each binds itself, its partners, successors, assigns, and legal representatives to the other party hereto, its partners, successors, assigns, and legal representatives in respect to all covenants, agreements, and obligations contained in the Contract Documents.

10.04 *Severability*

A. Any provision or part of the Contract Documents held to be void or unenforceable under any Law or Regulation shall be deemed stricken, and all remaining provisions shall continue to be valid and binding upon Owner and Contractor, who agree that the Contract Documents shall be reformed to replace such stricken provision or part thereof with a valid and enforceable provision that comes as close as possible to expressing the intention of the stricken provision.

10.05 *Contractor's Certifications*

A. Contractor certifies that it has not engaged in corrupt, fraudulent, collusive, or coercive practices in competing for or in executing the Contract. For the purposes of this Paragraph 10.05:

1. "corrupt practice" means the offering, giving, receiving, or soliciting of any thing of value likely to influence the action of a public official in the bidding process or in the Contract execution;

2. "fraudulent practice" means an intentional misrepresentation of facts made (a) to influence the bidding process or the execution of the Contract to the detriment of Owner, (b) to establish Bid or Contract prices at artificial non-competitive levels, or (c) to deprive Owner of the benefits of free and open competition;

3. "collusive practice" means a scheme or arrangement between two or more Bidders, with or without the knowledge of Owner, a purpose of which is to establish Bid prices at artificial, non-competitive levels; and

4. "coercive practice" means harming or threatening to harm, directly or indirectly, persons or their property to influence their participation in the bidding process or affect the execution of the Contract.

10.06 *Other Provisions*

NOTES TO USER

1. *If Owner intends to assign a procurement contract (for goods and services) to the Contractor, see Notes to User at Article 23 of Suggested Instructions to Bidders for Procurement Contracts (EJCDC P-200, 2000 Edition) for provisions to be inserted in this Article.*

2. *Insert other provisions here if applicable.*

IN WITNESS WHEREOF, Owner and Contractor have signed this Agreement. Counterparts have been delivered to Owner and Contractor. All portions of the Contract Documents have been signed or have been identified by Owner and Contractor or on their behalf.

NOTE TO USER
See I-21 and correlate procedures for format and signing of the documents.

This Agreement will be effective on _____ (which is the Effective Date of the Agreement).

NOTE TO USER
The Effective Date of the Agreement and the dates of any Construction Performance Bond (EJCDC C-610) and Construction Payment Bond (EJCDC C-615) should be the same, if possible. In no case may the date of any bonds be earlier then the Effective Date of the Agreement.

OWNER:

____

By: ______________________________

Title: ______________________________

Attest: ______________________________

Title: ______________________________

Address for giving notices:

____

____

____

(If Owner is a corporation, attach evidence of authority to sign. If Owner is a public body, attach evidence of authority to sign and resolution or other documents authorizing execution of this Agreement.)

CONTRACTOR

____

By: ______________________________

Title: ______________________________

(If Contractor is a corporation, a partnership, or a joint venture, attach evidence of authority to sign.)

Attest: ______________________________

Title: ______________________________

Address for giving notices:

____

____

____

License No.: ______________________________
(Where applicable)

NOTE TO USER: Use in those states or other jurisdictions where applicable or required.

Agent for service of process:

____

Engineers Joint Documents Committee
Design and Construction Related Documents
Instructions and License Agreement

Instructions

Before you use any EJCDC document:

1. Read the License Agreement. You agree to it and are bound by its terms when you use the EJCDC document.

2. Make sure that you have the correct version for your word processing software.

How to Use:

1. While EJCDC has expended considerable effort to make the software translations exact, it can be that a few document controls (e.g., bold, underline) did not carry over.

2. Similarly, your software may change the font specification if the font is not available in your system. It will choose a font that is close in appearance. In this event, the pagination may not match the control set.

3. If you modify the document, you must follow the instructions in the License Agreement about notification.

4. Also note the instruction in the License Agreement about the EJCDC copyright.

License Agreement

You should carefully read the following terms and conditions before using this document. Commencement of use of this document indicates your acceptance of these terms and conditions. If you do not agree to them, you should promptly return the materials to the vendor, and your money will be refunded.

The Engineers Joint Contract Documents Committee ("EJCDC") provides **EJCDC Design and Construction Related Documents** and licenses their use worldwide. You assume sole responsibility for the selection of specific documents or portions thereof to achieve your intended results, and for the installation, use, and results obtained from **EJCDC Design and Construction Related Documents**.

You acknowledge that you understand that the text of the contract documents of **EJCDC Design and Construction Related Documents** has important legal consequences and that consultation with an attorney is recommended with respect to use or modification of the text. You further acknowledge that EJCDC documents are protected by the copyright laws of the United States.

License:
You have a limited nonexclusive license to:

1. Use **EJCDC Design and Construction Related Documents** on any number of machines owned, leased or rented by your company or organization.

2. Use **EJCDC Design and Construction Related Documents** in printed form for bona fide contract documents.

3. Copy **EJCDC Design and Construction Related Documents** into any machine readable or printed form for backup or modification purposes in support of your use of **EJCDC Design and Construction Related Documents**.

You agree that you will:

1. Reproduce and include EJCDC's copyright notice on any printed or machine-readable copy, modification, or portion merged into another document or program. All proprietary rights in **EJCDC Design and Construction Related Documents** are and shall remain the property of EJCDC.

2. Not represent that any of the contract documents you generate from **EJCDC Design and Construction Related Documents** are EJCDC documents unless (i) the document text is used without alteration or (ii) all additions and changes to, and deletions from, the text are clearly shown.

You may not use, copy, modify, or transfer EJCDC Design and Construction Related Documents, or any copy, modification or merged portion, in whole or in part, except as expressly provided for in this license. Reproduction of EJCDC Design and Construction Related Documents in printed or machine-readable format for resale or educational purposes is expressly prohibited.

If you transfer possession of any copy, modification or merged portion of EJCDC Design and Construction Related Documents to another party, your license is automatically terminated.

Term:
The license is effective until terminated. You may terminate it at any time by destroying **EJCDC Design and Construction Related Documents** altogether with all copies, modifications and merged portions in any form. It will also terminate upon conditions set forth elsewhere in this Agreement or if you fail to comply with any term or condition of this Agreement. You agree upon such termination to destroy **EJCDC Design and Construction Related Documents** along with all copies, modifications and merged portions in any form.

Limited Warranty:
EJCDC warrants the CDs and diskettes on which **EJCDC Design and Construction Related Documents** is furnished to be free from defects in materials and workmanship under normal use for a period of ninety (90) days from the date of delivery to you as evidenced by a copy of your receipt.

There is no other warranty of any kind, either expressed or implied, including, but not limited to the implied warranties of merchantability and fitness for a particular purpose. Some states do not allow the exclusion of implied warranties, so the above exclusion may not apply to you. This warranty gives you specific legal rights and you may also have other rights which vary from state to state.

EJCDC does not warrant that the functions contained in **EJCDC Design and Construction Related Documents** will meet your requirements or that the operation of **EJCDC Design and Construction Related Documents** will be uninterrupted or error free.

Limitations of Remedies:
EJCDC's entire liability and your exclusive remedy shall be:

1. the replacement of any document not meeting EJCDC's "Limited Warranty" which is returned to EJCDC's selling agent with a copy of your receipt, or

2. if EJCDC's selling agent is unable to deliver a replacement CD or diskette which is free of defects in materials and workmanship, you may terminate this Agreement by returning EJCDC Document and your money will be refunded.

In no event will EJCDC be liable to you for any damages, including any lost profits, lost savings or other incidental or consequential damages arising out of the use or inability to use **EJCDC Design and Construction Related Documents** even if EJCDC has been advised of the possibility of such damages, or for any claim by any other party.

Some states do not allow the limitation or exclusion of liability for incidental or consequential damages, so the above limitation or exclusion may not apply to you.

General:
You may not sublicense, assign, or transfer this license except as expressly provided in this Agreement. Any attempt otherwise to sublicense, assign, or transfer any of the rights, duties, or obligations hereunder is void.

This Agreement shall be governed by the laws of the State of Virginia. Should you have any questions concerning this Agreement, you may contact EJCDC by writing to:

Arthur Schwartz, Esq.
General Counsel
National Society of Professional Engineers
1420 King Street
Alexandria, VA 22314

Phone: (703) 684-2845
Fax: (703) 836-4875
e-mail: aschwartz@nspe.org

You acknowledge that you have read this agreement, understand it and agree to be bound by its terms and conditions. You further agree that it is the complete and exclusive statement of the agreement between us which supersedes any proposal or prior agreement, oral or written, and any other communications between us relating to the subject matter of this agreement.

This document has important legal consequences; consultation with an attorney is encouraged with respect to its use or modification. This document should be adapted to the particular circumstances of the contemplated Project and the Controlling Laws and Regulations.

AGREEMENT BETWEEN OWNER AND ENGINEER FOR PROFESSIONAL SERVICES

Prepared by

ENGINEERS JOINT CONTRACT DOCUMENTS COMMITTEE

and

Issued and Published Jointly by

AMERICAN COUNCIL OF ENGINEERING COMPANIES

ASSOCIATED GENERAL CONTRACTORS OF AMERICA

AMERICAN SOCIETY OF CIVIL ENGINEERS

PROFESSIONAL ENGINEERS IN PRIVATE PRACTICE
A Practice Division of the
NATIONAL SOCIETY OF PROFESSIONAL ENGINEERS

This Agreement has been prepared for use with the Standard General Conditions of the Construction Contract (EJCDC C-700, 2007 Edition). Their provisions are interrelated, and a change in one may necessitate a change in the other. For guidance on the completion and use of this Agreement, see EJCDC User's Guide to the Owner-Engineer Agreement, EJCDC E-001, 2009 Edition.

1420 King Street, Alexandria, VA 22314-2794
(703) 684-2882
www.nspe.org

American Council of Engineering Companies
1015 15th Street N.W., Washington, DC 20005
(202) 347-7474
www.acec.org

American Society of Civil Engineers
1801 Alexander Bell Drive, Reston, VA 20191-4400
(800) 548-2723
www.asce.org

Associated General Contractors of America
2300 Wilson Boulevard, Suite 400, Arlington, VA 22201-3308
(703) 548-3118
www.agc.org

TABLE OF CONTENTS

Page

ARTICLE 1 – SERVICES OF ENGINEER 1
1.01 Scope 1

ARTICLE 2 – OWNER'S RESPONSIBILITIES 1
2.01 General 1

ARTICLE 3 – SCHEDULE FOR RENDERING SERVICES 2
3.01 Commencement 2
3.02 Time for Completion 2

ARTICLE 4 – INVOICES AND PAYMENTS 2
4.01 Invoices 2
4.02 Payments 2

ARTICLE 5 – OPINIONS OF COST 3
5.01 Opinions of Probable Construction Cost 3
5.02 Designing to Construction Cost Limit 3
5.03 Opinions of Total Project Costs 3

ARTICLE 6 – GENERAL CONSIDERATIONS 3
6.01 Standards of Performance 3
6.02 Design Without Construction Phase Services 5
6.03 Use of Documents 5
6.04 Insurance 6
6.05 Suspension and Termination 7
6.06 Controlling Law 8
6.07 Successors, Assigns, and Beneficiaries 9
6.08 Dispute Resolution 9
6.09 Environmental Condition of Site 9
6.10 Indemnification and Mutual Waiver 10
6.11 Miscellaneous Provisions 11

ARTICLE 7 – DEFINITIONS 11
7.01 Defined Terms 11

ARTICLE 8 – EXHIBITS AND SPECIAL PROVISIONS 15
8.01 Exhibits Included 15
8.02 Total Agreement 15
8.03 Designated Representatives 15
8.04 Engineer's Certifications 16

AGREEMENT
BETWEEN OWNER AND ENGINEER
FOR
PROFESSIONAL SERVICES

THIS IS AN AGREEMENT effective as of _________, ____________ ("Effective Date") between

____ ("Owner") and

____ ("Engineer").

Owner's Project, of which Engineer's services under this Agreement are a part, is generally identified as follows:

____ ("Project").

Engineer's services under this Agreement are generally identified as follows:

____

____

Owner and Engineer further agree as follows:

ARTICLE 1 – SERVICES OF ENGINEER

1.01 *Scope*

A. Engineer shall provide, or cause to be provided, the services set forth herein and in Exhibit A.

ARTICLE 2 – OWNER'S RESPONSIBILITIES

2.01 *General*

A. Owner shall have the responsibilities set forth herein and in Exhibit B.

B. Owner shall pay Engineer as set forth in Exhibit C.

C. Owner shall be responsible for, and Engineer may rely upon, the accuracy and completeness of all requirements, programs, instructions, reports, data, and other information furnished by Owner to Engineer pursuant to this Agreement. Engineer may use such requirements, programs, instructions, reports, data, and information in performing or furnishing services under this Agreement.

ARTICLE 3 – SCHEDULE FOR RENDERING SERVICES

3.01 *Commencement*

A. Engineer is authorized to begin rendering services as of the Effective Date.

3.02 *Time for Completion*

A. Engineer shall complete its obligations within a reasonable time. Specific periods of time for rendering services are set forth or specific dates by which services are to be completed are provided in Exhibit A, and are hereby agreed to be reasonable.

B. If, through no fault of Engineer, such periods of time or dates are changed, or the orderly and continuous progress of Engineer's services is impaired, or Engineer's services are delayed or suspended, then the time for completion of Engineer's services, and the rates and amounts of Engineer's compensation, shall be adjusted equitably.

C. If Owner authorizes changes in the scope, extent, or character of the Project, then the time for completion of Engineer's services, and the rates and amounts of Engineer's compensation, shall be adjusted equitably.

D. Owner shall make decisions and carry out its other responsibilities in a timely manner so as not to delay the Engineer's performance of its services.

E. If Engineer fails, through its own fault, to complete the performance required in this Agreement within the time set forth, as duly adjusted, then Owner shall be entitled, as its sole remedy, to the recovery of direct damages, if any, resulting from such failure.

ARTICLE 4 – INVOICES AND PAYMENTS

4.01 *Invoices*

A. *Preparation and Submittal of Invoices*: Engineer shall prepare invoices in accordance with its standard invoicing practices and the terms of Exhibit C. Engineer shall submit its invoices to Owner on a monthly basis. Invoices are due and payable within 30 days of receipt.

4.02 *Payments*

A. *Application to Interest and Principal*: Payment will be credited first to any interest owed to Engineer and then to principal.

B. *Failure to Pay*: If Owner fails to make any payment due Engineer for services and expenses within 30 days after receipt of Engineer's invoice, then:

1. amounts due Engineer will be increased at the rate of 1.0% per month (or the maximum rate of interest permitted by law, if less) from said thirtieth day; and

2. Engineer may, after giving seven days written notice to Owner, suspend services under this Agreement until Owner has paid in full all amounts due for services, expenses, and other related charges. Owner waives any and all claims against Engineer for any such suspension.

C. *Disputed Invoices:* If Owner contests an invoice, Owner shall promptly advise Engineer of the specific basis for doing so, may withhold only that portion so contested, and must pay the undisputed portion.

D. *Legislative Actions:* If after the Effective Date any governmental entity takes a legislative action that imposes taxes, fees, or charges on Engineer's services or compensation under this Agreement, then the Engineer may invoice such new taxes, fees, or charges as a Reimbursable Expense to which a factor of 1.0 shall be applied. Owner shall reimburse Engineer for the cost of such invoiced new taxes, fees, and charges; such reimbursement shall be in addition to the compensation to which Engineer is entitled under the terms of Exhibit C.

ARTICLE 5 – OPINIONS OF COST

5.01 *Opinions of Probable Construction Cost*

A. Engineer's opinions of probable Construction Cost are to be made on the basis of Engineer's experience and qualifications and represent Engineer's best judgment as an experienced and qualified professional generally familiar with the construction industry. However, because Engineer has no control over the cost of labor, materials, equipment, or services furnished by others, or over contractors' methods of determining prices, or over competitive bidding or market conditions, Engineer cannot and does not guarantee that proposals, bids, or actual Construction Cost will not vary from opinions of probable Construction Cost prepared by Engineer. If Owner requires greater assurance as to probable Construction Cost, Owner must employ an independent cost estimator as provided in Exhibit B.

5.02 *Designing to Construction Cost Limit*

A. If a Construction Cost limit is established between Owner and Engineer, such Construction Cost limit and a statement of Engineer's rights and responsibilities with respect thereto will be specifically set forth in Exhibit F, "Construction Cost Limit," to this Agreement.

5.03 *Opinions of Total Project Costs*

A. The services, if any, of Engineer with respect to Total Project Costs shall be limited to assisting the Owner in collating the various cost categories which comprise Total Project Costs. Engineer assumes no responsibility for the accuracy of any opinions of Total Project Costs.

ARTICLE 6 – GENERAL CONSIDERATIONS

6.01 *Standards of Performance*

A. *Standard of Care:* The standard of care for all professional engineering and related services performed or furnished by Engineer under this Agreement will be the care and skill ordinarily used by members of the subject profession practicing under similar circumstances at the same time and in the same locality. Engineer makes no warranties, express or implied, under this Agreement or otherwise, in connection with Engineer's services.

B. *Technical Accuracy:* Owner shall not be responsible for discovering deficiencies in the technical accuracy of Engineer's services. Engineer shall correct deficiencies in technical accuracy without additional compensation, unless such corrective action is directly attributable to deficiencies in Owner-furnished information.

C. *Consultants:* Engineer may employ such Consultants as Engineer deems necessary to assist in the performance or furnishing of the services, subject to reasonable, timely, and substantive objections by Owner.

D. *Reliance on Others:* Subject to the standard of care set forth in Paragraph 6.01.A, Engineer and its Consultants may use or rely upon design elements and information ordinarily or customarily furnished by others, including, but not limited to, specialty contractors, manufacturers, suppliers, and the publishers of technical standards.

E. *Compliance with Laws and Regulations, and Policies and Procedures*:

1. Engineer and Owner shall comply with applicable Laws and regulations.

2. Prior to the Effective Date, Owner provided to Engineer in writing any and all policies and procedures of Owner applicable to Engineer's performance of services under this Agreement. provided to Engineer in writing. Engineer shall comply with such policies and procedures, subject to the standard of care set forth in Paragraph 6.01.A, and to the extent compliance is not inconsistent with professional practice requirements.

3. This Agreement is based on Laws and Regulations and Owner-provided written policies and procedures as of the Effective Date. Changes after the Effective Date to these Laws and Regulations, or to Owner-provided written policies and procedures, may be the basis for modifications to Owner's responsibilities or to Engineer's scope of services, times of performance, or compensation.

F. Engineer shall not be required to sign any documents, no matter by whom requested, that would result in the Engineer having to certify, guarantee, or warrant the existence of conditions whose existence the Engineer cannot ascertain. Owner agrees not to make resolution of any dispute with the Engineer or payment of any amount due to the Engineer in any way contingent upon the Engineer signing any such documents.

G. The general conditions for any construction contract documents prepared hereunder are to be the "Standard General Conditions of the Construction Contract" as prepared by the Engineers Joint Contract Documents Committee (EJCDC C-700, 2007 Edition) unless both parties mutually agree to use other general conditions by specific reference in Exhibit J.

H. Engineer shall not at any time supervise, direct, control, or have authority over any contractor work, nor shall Engineer have authority over or be responsible for the means, methods, techniques, sequences, or procedures of construction selected or used by any contractor, or the safety precautions and programs incident thereto, for security or safety at the Site, nor for any failure of a contractor to comply with Laws and Regulations applicable to such contractor's furnishing and performing of its work.

I. Engineer neither guarantees the performance of any Contractor nor assumes responsibility for any Contractor's failure to furnish and perform the Work in accordance with the Contract Documents.

J. Engineer shall not provide or have any responsibility for surety bonding or insurance-related advice, recommendations, counseling, or research, or enforcement of construction insurance or surety bonding requirements.

K. Engineer shall not be responsible for the acts or omissions of any Contractor, Subcontractor, or Supplier, or of any of their agents or employees or of any other persons (except Engineer's own agents, employees, and Consultants) at the Site or otherwise furnishing or performing any Work; or for any decision made regarding the Contract Documents, or any application, interpretation, or clarification, of the Contract Documents, other than those made by Engineer.

L. While at the Site, Engineer's employees and representatives shall comply with the specific applicable requirements of Contractor's and Owner's safety programs of which Engineer has been informed in writing.

6.02 *Design Without Construction Phase Services*

A. Engineer shall be responsible only for those Construction Phase services expressly required of Engineer in Exhibit A, Paragraph A1.05. With the exception of such expressly required services, Engineer shall have no design, Shop Drawing review, or other obligations during construction and Owner assumes all responsibility for the application and interpretation of the Contract Documents, review and response to Contractor claims, contract administration, processing Change Orders, revisions to the Contract Documents during construction, construction surety bonding and insurance requirements, construction observation and review, review of payment applications, and all other necessary Construction Phase engineering and professional services. Owner waives all claims against the Engineer that may be connected in any way to Construction Phase engineering or professional services except for those services that are expressly required of Engineer in Exhibit A, Paragraph A1.05.

6.03 *Use of Documents*

A. All Documents are instruments of service in respect to this Project, and Engineer shall retain an ownership and property interest therein (including the copyright and the right of reuse at the discretion of the Engineer) whether or not the Project is completed. Owner shall not rely in any way on any Document unless it is in printed form, signed or sealed by the Engineer or one of its Consultants.

B. Either party to this Agreement may rely that data or information set forth on paper (also known as hard copies) that the party receives from the other party by mail, hand delivery, or facsimile,are the items that the other party intended to send. Files in electronic media format of text, data, graphics, or other types that are furnished by one party to the other are furnished only for convenience, not reliance by the receiving party. Any conclusion or information obtained or derived from such electronic files will be at the user's sole risk. If there is a discrepancy between the electronic files and the hard copies, the hard copies govern. If the parties agree to other electronic transmittal procedures, such are set forth in Exhibit J.

C. Because data stored in electronic media format can deteriorate or be modified inadvertently or otherwise without authorization of the data's creator, the party receiving electronic files agrees that it will perform acceptance tests or procedures within 60 days, after which the receiving party shall be deemed to have accepted the data thus transferred. Any transmittal errors detected within the 60-day acceptance period will be corrected by the party delivering the electronic files.

D. When transferring documents in electronic media format, the transferring party makes no representations as to long-term compatibility, usability, or readability of such documents resulting from the use of software application packages, operating systems, or computer hardware differing from those used by the documents' creator.

E. Owner may make and retain copies of Documents for information and reference in connection with use on the Project by Owner. Engineer grants Owner a limited license to use the Documents on the Project, extensions of the Project, and for related uses of the Owner, subject to receipt by Engineer of full payment for all services relating to preparation of the Documents and subject to the following limitations: (1) Owner acknowledges that such Documents are not intended or represented to be suitable for use on the Project unless completed by Engineer, or for use or reuse by Owner or others on extensions of the Project, on any other project, or for any other use or purpose, without written verification or adaptation by Engineer; (2) any such use or reuse, or any modification of the Documents, without written verification, completion, or adaptation by Engineer, as appropriate for the specific purpose intended, will be at Owner's sole risk and without liability or legal exposure to Engineer or to its officers, directors, members, partners, agents, employees, and Consultants; (3) Owner shall indemnify and hold harmless Engineer and its officers, directors, members, partners, agents, employees, and Consultants from all claims, damages, losses, and expenses, including attorneys' fees, arising out of or resulting from any use, reuse, or modification of the Documents without written verification, completion, or adaptation by Engineer; and (4) such limited license to Owner shall not create any rights in third parties.

F. If Engineer at Owner's request verifies the suitability of the Documents, completes them, or adapts them for extensions of the Project or for any other purpose, then Owner shall compensate Engineer at rates or in an amount to be agreed upon by Owner and Engineer.

6.04 *Insurance*

A. Engineer shall procure and maintain insurance as set forth in Exhibit G, "Insurance." Engineer shall cause Owner to be listed as an additional insured on any applicable general liability insurance policy carried by Engineer.

B. Owner shall procure and maintain insurance as set forth in Exhibit G, "Insurance." Owner shall cause Engineer and its Consultants to be listed as additional insureds on any general liability policies and as loss payees on any property insurance policies carried by Owner which are applicable to the Project.

C. Owner shall require Contractor to purchase and maintain policies of insurance covering workers' compensation, general liability, property damage (other than to the Work itself), motor vehicle damage and injuries, and other insurance necessary to protect Owner's and Engineer's interests in the Project. Owner shall require Contractor to cause Engineer and its Consultants to be listed as additional insureds with respect to such liability and other insurance purchased and maintained by Contractor for the Project.

D. Owner and Engineer shall each deliver to the other certificates of insurance evidencing the coverages indicated in Exhibit G. Such certificates shall be furnished prior to commencement of Engineer's services and at renewals thereafter during the life of the Agreement.

E. All policies of property insurance relating to the Project shall contain provisions to the effect that Engineer's and its Consultants' interests are covered and that in the event of payment of any loss or damage the insurers will have no rights of recovery against Engineer or its Consultants, or any insureds, additional insureds, or loss payees thereunder.

F. All policies of insurance shall contain a provision or endorsement that the coverage afforded will not be canceled or reduced in limits by endorsement, and thatrenewal will not be refused, until at least 30 days prior written notice has been given to Owner and Engineer and to each other additional insured (if any) to which a certificate of insurance has been issued.

G. At any time, Owner may request that Engineer or its Consultants, at Owner's sole expense, provide additional insurance coverage, increased limits, or revised deductibles that are more protective than those specified in Exhibit G. If so requested by Owner, and if commercially available, Engineer shall obtain and shall require its Consultants to obtain such additional insurance coverage, different limits, or revised deductibles for such periods of time as requested by Owner, and Exhibit G will be supplemented to incorporate these requirements.

6.05 *Suspension and Termination*

A. *Suspension*:

1. By Owner: Owner may suspend the Project for up to 90 days upon seven days written notice to Engineer.

2. By Engineer: Engineer may, after giving seven days written notice to Owner, suspend services under this Agreement if Engineer's performance has been substantially delayed through no fault of Engineer.

B. *Termination*: The obligation to provide further services under this Agreement may be terminated:

1. For cause,

a. By either party upon 30 days written notice in the event of substantial failure by the other party to perform in accordance with the terms hereof through no fault of the terminating party.

b. By Engineer:

1) upon seven days written notice if Owner demands that Engineer furnish or perform services contrary to Engineer's responsibilities as a licensed professional; or

2) upon seven days written notice if the Engineer's services for the Project are delayed or suspended for more than 90 days for reasons beyond Engineer's control.

3) Engineer shall have no liability to Owner on account of such termination.

c. Notwithstanding the foregoing, this Agreement will not terminate under Paragraph 6.05.B.1.a if the party receiving such notice begins, within seven days of receipt of such notice, to correct its substantial failure to perform and proceeds diligently to cure such failure within no more than 30 days of receipt thereof; provided, however, that if and to the extent such substantial failure cannot be reasonably cured within such 30 day period, and if such party has diligently attempted to cure the same and thereafter continues diligently to cure the same, then the cure period provided for herein shall extend up to, but in no case more than, 60 days after the date of receipt of the notice.

2. For convenience,

a. By Owner effective upon Engineer's receipt of notice from Owner.

C. *Effective Date of Termination*: The terminating party under Paragraph 6.05.B may set the effective date of termination at a time up to 30 days later than otherwise provided to allow Engineer to demobilize personnel and equipment from the Site, to complete tasks whose value would otherwise be lost, to prepare notes as to the status of completed and uncompleted tasks, and to assemble Project materials in orderly files.

D. *Payments Upon Termination*:

1. In the event of any termination under Paragraph 6.05, Engineer will be entitled to invoice Owner and to receive full payment for all services performed or furnished in accordance with this Agreement and all Reimbursable Expenses incurred through the effective date of termination. Upon making such payment, Owner shall have the limited right to the use of Documents, at Owner's sole risk, subject to the provisions of Paragraph 6.03.E.

2. In the event of termination by Owner for convenience or by Engineer for cause, Engineer shall be entitled, in addition to invoicing for those items identified in Paragraph 6.05.D.1, to invoice Owner and to payment of a reasonable amount for services and expenses directly attributable to termination, both before and after the effective date of termination, such as reassignment of personnel, costs of terminating contracts with Engineer's Consultants, and other related close-out costs, using methods and rates for Additional Services as set forth in Exhibit C.

6.06 *Controlling Law*

A. This Agreement is to be governed by the law of the state or jurisdiction in which the Project is located.

6.07 *Successors, Assigns, and Beneficiaries*

A. Owner and Engineer are hereby bound and the successors, executors, administrators, and legal representatives of Owner and Engineer (and to the extent permitted by Paragraph 6.07.B the assigns of Owner and Engineer) are hereby bound to the other party to this Agreement and to the successors, executors, administrators and legal representatives (and said assigns) of such other party, in respect of all covenants, agreements, and obligations of this Agreement.

B. Neither Owner nor Engineer may assign, sublet, or transfer any rights under or interest (including, but without limitation, moneys that are due or may become due) in this Agreement without the written consent of the other, except to the extent that any assignment, subletting, or transfer is mandated or restricted by law. Unless specifically stated to the contrary in any written consent to an assignment, no assignment will release or discharge the assignor from any duty or responsibility under this Agreement.

C. Unless expressly provided otherwise in this Agreement:

1. Nothing in this Agreement shall be construed to create, impose, or give rise to any duty owed by Owner or Engineer to any Contractor, Subcontractor, Supplier, other individual or entity, or to any surety for or employee of any of them.

2. All duties and responsibilities undertaken pursuant to this Agreement will be for the sole and exclusive benefit of Owner and Engineer and not for the benefit of any other party.

3. Owner agrees that the substance of the provisions of this Paragraph 6.07.C shall appear in the Contract Documents.

6.08 *Dispute Resolution*

A. Owner and Engineer agree to negotiate all disputes between them in good faith for a period of 30 days from the date of notice prior to invoking the procedures of Exhibit H or other provisions of this Agreement, or exercising their rights under law.

B. If the parties fail to resolve a dispute through negotiation under Paragraph 6.08.A, then either or both may invoke the procedures of Exhibit H. If Exhibit H is not included, or if no dispute resolution method is specified in Exhibit H, then the parties may exercise their rights under law.

6.09 *Environmental Condition of Site*

A. Owner has disclosed to Engineer in writing the existence of all known and suspected Asbestos, PCBs, Petroleum, Hazardous Waste, Radioactive Material, hazardous substances, and other Constituents of Concern located at or near the Site, including type, quantity, and location.

B. Owner represents to Engineer that to the best of its knowledge no Constituents of Concern, other than those disclosed in writing to Engineer, exist at the Site.

C. If Engineer encounters or learns of an undisclosed Constituent of Concern at the Site, then Engineer shall notify (1) Owner and (2) appropriate governmental officials if Engineer reasonably concludes that doing so is required by applicable Laws or Regulations.

D. It is acknowledged by both parties that Engineer's scope of services does not include any services related to Constituents of Concern. If Engineer or any other party encounters an undisclosed Constituent of Concern, or if investigative or remedial action, or other professional services, are necessary with respect to disclosed or undisclosed Constituents of Concern, then Engineer may, at its option and without liability for consequential or any other damages, suspend performance of services on the portion of the Project affected thereby until Owner: (1) retains appropriate specialist consultants or contractors to identify and, as appropriate, abate, remediate, or remove the Constituents of Concern; and (2) warrants that the Site is in full compliance with applicable Laws and Regulations.

E. If the presence at the Site of undisclosed Constituents of Concern adversely affects the performance of Engineer's services under this Agreement, then the Engineer shall have the option of (1) accepting an equitable adjustment in its compensation or in the time of completion, or both; or (2) terminating this Agreement for cause on 30 days notice.

F. Owner acknowledges that Engineer is performing professional services for Owner and that Engineer is not and shall not be required to become an "owner" "arranger," "operator," "generator," or "transporter" of hazardous substances, as defined in the Comprehensive Environmental Response, Compensation, and Liability Act (CERCLA), as amended, which are or may be encountered at or near the Site in connection with Engineer's activities under this Agreement.

6.10 *Indemnification and Mutual Waiver*

A. *Indemnification by Engineer*: To the fullest extent permitted by law, Engineer shall indemnify and hold harmless Owner, and Owner's officers, directors, members, partners, agents, consultants, and employees from reasonable claims, costs, losses, and damages arising out of or relating to the Project, provided that any such claim, cost, loss, or damage is attributable to bodily injury, sickness, disease, or death, or to injury to or destruction of tangible property (other than the Work itself), including the loss of use resulting therefrom, but only to the extent caused by any negligent act or omission of Engineer or Engineer's officers, directors, members, partners, agents, employees, or Consultants. **This indemnification provision is subject to and limited by the provisions, if any, agreed to by Owner and Engineer in Exhibit I, "Limitations of Liability."**

B. *Indemnification by Owner*: Owner shall indemnify and hold harmless Engineer and its officers, directors, members, partners, agents, employees, and Consultants as required by Laws and Regulations and to the extent (if any) required in Exhibit I, Limitations of Liability.

C. *Environmental Indemnification*: To the fullest extent permitted by law, Owner shall indemnify and hold harmless Engineer and its officers, directors, members, partners, agents, employees, and Consultants from and against any and all claims, costs, losses, and damages (including but not limited to all fees and charges of engineers, architects, attorneys and other professionals, and all court, arbitration, or other dispute resolution costs) caused by, arising out of, relating to, or resulting from a Constituent of Concern at, on, or under the Site, provided that (1) any such claim, cost, loss, or damage is attributable to bodily injury, sickness, disease, or death, or to injury to or destruction of tangible property (other than the Work itself), including the loss of use resulting therefrom, and (2) nothing in this paragraph shall obligate Owner to indemnify any individual or entity from and against the consequences of that individual's or entity's own negligence or willful misconduct.

D. *Percentage Share of Negligence*: To the fullest extent permitted by law, a party's total liability to the other party and anyone claiming by, through, or under the other party for any cost, loss, or damages caused in part by the negligence of the party and in part by the negligence of the other party or any other negligent entity or individual, shall not exceed the percentage share that the party's negligence bears to the total negligence of Owner, Engineer, and all other negligent entities and individuals.

E. *Mutual Waiver*: To the fullest extent permitted by law, Owner and Engineer waive against each other, and the other's employees, officers, directors, members, agents, insurers, partners, and consultants, any and all claims for or entitlement to special, incidental, indirect, or consequential damages arising out of, resulting from, or in any way related to the Project.

6.11 *Miscellaneous Provisions*

A. *Notices*: Any notice required under this Agreement will be in writing, addressed to the appropriate party at its address on the signature page and given personally, by facsimile, by registered or certified mail postage prepaid, or by a commercial courier service. All notices shall be effective upon the date of receipt.

B. *Survival*: All express representations, waivers, indemnifications, and limitations of liability included in this Agreement will survive its completion or termination for any reason.

C. *Severability*: Any provision or part of the Agreement held to be void or unenforceable under any Laws or Regulations shall be deemed stricken, and all remaining provisions shall continue to be valid and binding upon Owner and Engineer, which agree that the Agreement shall be reformed to replace such stricken provision or part thereof with a valid and enforceable provision that comes as close as possible to expressing the intention of the stricken provision.

D. *Waiver*: A party's non-enforcement of any provision shall not constitute a waiver of that provision, nor shall it affect the enforceability of that provision or of the remainder of this Agreement.

E. *Accrual of Claims:* To the fullest extent permitted by law, all causes of action arising under this Agreement shall be deemed to have accrued, and all statutory periods of limitation shall commence, no later than the date of Substantial Completion.

ARTICLE 7 – DEFINITIONS

7.01 *Defined Terms*

A. Wherever used in this Agreement (including the Exhibits hereto) terms (including the singular and plural forms) printed with initial capital letters have the meanings indicated in the text above, in the exhibits, or in the following provisions:

1. *Additional Services* – The services to be performed for or furnished to Owner by Engineer in accordance with Part 2 of Exhibit A of this Agreement.

2. *Agreement* – This written contract for professional services between Owner and Engineer, including all exhibits identified in Paragraph 8.01 and any duly executed amendments.

3. *Asbestos* – Any material that contains more than one percent asbestos and is friable or is releasing asbestos fibers into the air above current action levels established by the United States Occupational Safety and Health Administration.

4. *Basic Services* – The services to be performed for or furnished to Owner by Engineer in accordance with Part 1 of Exhibit A of this Agreement.

5. *Construction Contract* – The entire and integrated written agreement between Owner and Contractor concerning the Work.

6. *Construction Cost* – The cost to Owner of those portions of the entire Project designed or specified by Engineer. Construction Cost does not include costs of services of Engineer or other design professionals and consultants; cost of land or rights-of-way, or compensation for damages to properties; Owner's costs for legal, accounting, insurance counseling or auditing services; interest or financing charges incurred in connection with the Project; or the cost of other services to be provided by others to Owner pursuant to Exhibit B of this Agreement. Construction Cost is one of the items comprising Total Project Costs.

7. *Constituent of Concern* – Any substance, product, waste, or other material of any nature whatsoever (including, but not limited to, Asbestos, Petroleum, Radioactive Material, and PCBs) which is or becomes listed, regulated, or addressed pursuant to (a) the Comprehensive Environmental Response, Compensation and Liability Act, 42 U.S.C. §§9601 et seq. ("CERCLA"); (b) the Hazardous Materials Transportation Act, 49 U.S.C. §§1801 et seq.; (c) the Resource Conservation and Recovery Act, 42 U.S.C. §§6901 et seq. ("RCRA"); (d) the Toxic Substances Control Act, 15 U.S.C. §§2601 et seq.; (e) the Clean Water Act, 33 U.S.C. §§1251 et seq.; (f) the Clean Air Act, 42 U.S.C. §§7401 et seq.; and (g) any other federal, state, or local statute, law, rule, regulation, ordinance, resolution, code, order, or decree regulating, relating to, or imposing liability or standards of conduct concerning, any hazardous, toxic, or dangerous waste, substance, or material.

8. *Consultants* – Individuals or entities having a contract with Engineer to furnish services with respect to this Project as Engineer's independent professional associates and consultants; subcontractors; or vendors.

9. *Contract Documents* – Those items so designated in the Construction Contract, including the Drawings, Specifications, construction agreement, and general and supplementary conditions. Only printed or hard copies of the items listed in the Construction Contract are Contract Documents. Approved Shop Drawings, other Contractor submittals, and the reports and drawings of subsurface and physical conditions are not Contract Documents.

10. *Contractor* – The entity or individual with which Owner has entered into a Construction Contract.

11. *Documents* – Data, reports, Drawings, Specifications, Record Drawings, and other deliverables, whether in printed or electronic media format, provided or furnished in appropriate phases by Engineer to Owner pursuant to this Agreement.

12. *Drawings* – That part of the Contract Documents prepared or approved by Engineer which graphically shows the scope, extent, and character of the Work to be performed by Contractor. Shop Drawings are not Drawings as so defined.

13. *Effective Date* – The date indicated in this Agreement on which it becomes effective, but if no such date is indicated, the date on which this Agreement is signed and delivered by the last of the parties to sign and deliver.

14. *Engineer* – The individual or entity named as such in this Agreement.

15. *Hazardous Waste* – The term Hazardous Waste shall have the meaning provided in Section 1004 of the Solid Waste Disposal Act (42 USC Section 6903) as amended from time to time.

16. *Laws and Regulations; Laws or Regulations* – Any and all applicable laws, rules, regulations, ordinances, codes, and orders of any and all governmental bodies, agencies, authorities, and courts having jurisdiction.

17. *Owner* – The individual or entity with which Engineer has entered into this Agreement and for which the Engineer's services are to be performed. Unless indicated otherwise, this is the same individual or entity that will enter into any Construction Contracts concerning the Project.

18. *PCBs* – Polychlorinated biphenyls.

19. *Petroleum* – Petroleum, including crude oil or any fraction thereof which is liquid at standard conditions of temperature and pressure (60 degrees Fahrenheit and 14.7 pounds per square inch absolute), such as oil, petroleum, fuel oil, oil sludge, oil refuse, gasoline, kerosene, and oil mixed with other non-hazardous waste and crude oils.

20. *Project* – The total construction of which the Work to be performed under the Contract Documents may be the whole, or a part.

21. *Radioactive Material* – Source, special nuclear, or byproduct material as defined by the Atomic Energy Act of 1954 (42 USC Section 2011 et seq.) as amended from time to time.

22. *Record Drawings* – Drawings depicting the completed Project, prepared by Engineer as an Additional Service and based solely on Contractor's record copy of all Drawings, Specifications, addenda, change orders, work change directives, field orders, and written interpretations and clarifications, as delivered to Engineer and annotated by Contractor to show changes made during construction.

23. *Reimbursable Expenses* – The expenses incurred directly by Engineer in connection with the performing or furnishing of Basic and Additional Services for the Project.

24. *Resident Project Representative* – The authorized representative of Engineer assigned to assist Engineer at the Site during the Construction Phase. As used herein, the term Resident Project Representative or "RPR" includes any assistants or field staff of Resident Project Representative agreed to by Owner. The duties and responsibilities of the Resident Project Representative, if any, are as set forth in Exhibit D.

25. *Samples* – Physical examples of materials, equipment, or workmanship that are representative of some portion of the Work and which establish the standards by which such portion of the Work will be judged.

26. *Shop Drawings* – All drawings, diagrams, illustrations, schedules, and other data or information which are specifically prepared or assembled by or for Contractor and submitted by Contractor to illustrate some portion of the Work.

27. *Site* – Lands or areas to be indicated in the Contract Documents as being furnished by Owner upon which the Work is to be performed, including rights-of-way and easements for access thereto, and such other lands furnished by Owner which are designated for the use of Contractor.

28. *Specifications* – That part of the Contract Documents consisting of written technical descriptions of materials, equipment, systems, standards, and workmanship as applied to the Work and certain administrative details applicable thereto.

29. *Subcontractor* – An individual or entity having a direct contract with Contractor or with any other Subcontractor for the performance of a part of the Work at the Site.

30. *Substantial Completion* – The time at which the Work (or a specified part thereof) has progressed to the point where, in the opinion of Engineer, the Work (or a specified part thereof) is sufficiently complete, in accordance with the Contract Documents, so that the Work (or a specified part thereof) can be utilized for the purposes for which it is intended. The terms "substantially complete" and "substantially completed" as applied to all or part of the Work refer to Substantial Completion thereof.

31. *Supplier* – A manufacturer, fabricator, supplier, distributor, materialman, or vendor having a direct contract with Contractor or with any Subcontractor to furnish materials or equipment to be incorporated in the Work by Contractor or Subcontractor.

32. *Total Project Costs* – The sum of the Construction Cost, allowances for contingencies, and the total costs of services of Engineer or other design professionals and consultants, together with such other Project-related costs that Owner furnishes for inclusion, including but not limited to cost of land, rights-of-way, compensation for damages to properties, Owner's costs for legal, accounting, insurance counseling and auditing services, interest and financing charges incurred in connection with the Project, and the cost of other services to be provided by others to Owner pursuant to Exhibit B of this Agreement.

33. *Work* – The entire construction or the various separately identifiable parts thereof required to be provided under the Contract Documents. Work includes and is the result of performing or providing all labor, services, and documentation necessary to produce such

construction, and furnishing, installing, and incorporating all materials and equipment into such construction, all as required by the Contract Documents.

ARTICLE 8 – EXHIBITS AND SPECIAL PROVISIONS

8.01 *Exhibits Included:*

A. Exhibit A, Engineer's Services.

B. Exhibit B, Owner's Responsibilities.

C. Exhibit C, Payments to Engineer for Services and Reimbursable Expenses.

D. Exhibit D, Duties, Responsibilities and Limitations of Authority of Resident Project Representative.

E. Exhibit E, Notice of Acceptability of Work.

F. Exhibit F, Construction Cost Limit.

G. Exhibit G, Insurance.

H. Exhibit H, Dispute Resolution.

I. Exhibit I, Limitations of Liability.

J. Exhibit J, Special Provisions.

K. Exhibit K, Amendment to Owner-Engineer Agreement.

[NOTE TO USER: If an exhibit is not included, indicate "not included" after the listed exhibit item]

8.02 *Total Agreement:*

A. This Agreement, (together with the exhibits identified above) constitutes the entire agreement between Owner and Engineer and supersedes all prior written or oral understandings. This Agreement may only be amended, supplemented, modified, or canceled by a duly executed written instrument based on the format of Exhibit K to this Agreement.

8.03 *Designated Representatives:*

A. With the execution of this Agreement, Engineer and Owner shall designate specific individuals to act as Engineer's and Owner's representatives with respect to the services to be performed or furnished by Engineer and responsibilities of Owner under this Agreement. Such an individual shall have authority to transmit instructions, receive information, and render decisions relative to the Project on behalf of the respective party whom the individual represents.

8.04 *Engineer's Certifications:*

A. Engineer certifies that it has not engaged in corrupt, fraudulent, or coercive practices in competing for or in executing the Agreement. For the purposes of this Paragraph 8.04:

1. "corrupt practice" means the offering, giving, receiving, or soliciting of any thing of value likely to influence the action of a public official in the selection process or in the Agreement execution;

2. "fraudulent practice" means an intentional misrepresentation of facts made (a) to influence the selection process or the execution of the Agreement to the detriment of Owner, or (b) to deprive Owner of the benefits of free and open competition;

3. "coercive practice" means harming or threatening to harm, directly or indirectly, persons or their property to influence their participation in the selection process or affect the execution of the Agreement.

IN WITNESS WHEREOF, the parties hereto have executed this Agreement, the Effective Date of which is indicated on page 1.

Owner:	Engineer:
______________________	______________________
By: ______________________	By: ______________________
Title: ______________________	Title: ______________________
Date Signed: ______________________	Date Signed: ______________________
	Engineer License or Firm's Certificate No. ________
	State of: ______________________
Address for giving notices:	Address for giving notices:
______________________	______________________
______________________	______________________
______________________	______________________
Designated Representative (Paragraph 8.03.A):	Designated Representative (Paragraph 8.03.A):
______________________	______________________
Title: ______________________	Title: ______________________
Phone Number: ______________________	Phone Number: ______________________

Facsimile Number:	______________	Facsimile Number:	______________
E-Mail Address:	______________	E-Mail Address:	______________

This is **EXHIBIT A**, consisting of _____ pages, referred to in and part of the **Agreement between Owner and Engineer for Professional Services** dated _____, _____.

Engineer's Services

Article 1 of the Agreement is supplemented to include the following agreement of the parties.

Engineer shall provide Basic and Additional Services as set forth below.

PART 1 – BASIC SERVICES

A1.01 *Study and Report Phase*

A. Engineer shall:

1. Consult with Owner to define and clarify Owner's requirements for the Project and available data.

2. Advise Owner of any need for Owner to provide data or services of the types described in Exhibit B which are not part of Engineer's Basic Services.

3. Identify, consult with, and analyze requirements of governmental authorities having jurisdiction to approve the portions of the Project designed or specified by Engineer, including but not limited to mitigating measures identified in the environmental assessment.

4. Identify and evaluate [*insert specific number or list here*] alternate solutions available to Owner and, after consultation with Owner, recommend to Owner those solutions which in Engineer's judgment meet Owner's requirements for the Project.

5. Prepare a report (the "Report") which will, as appropriate, contain schematic layouts, sketches, and conceptual design criteria with appropriate exhibits to indicate the agreed-to requirements, considerations involved, and those alternate solutions available to Owner which Engineer recommends. For each recommended solution Engineer will provide the following, which will be separately itemized: opinion of probable Construction Cost; proposed allowances for contingencies; the estimated total costs of design, professional, and related services to be provided by Engineer and its Consultants; and, on the basis of information furnished by Owner, a summary of allowances for other items and services included within the definition of Total Project Costs.

6. Perform or provide the following additional Study and Report Phase tasks or deliverables: [*here list any such tasks or deliverables*]

7. Furnish ___ review copies of the Report and any other deliverables to Owner within ___ calendar days of the Effective Date and review it with Owner. Within ___ calendar days of receipt, Owner shall submit to Engineer any comments regarding the Report and any other deliverables.

8. Revise the Report and any other deliverables in response to Owner's comments, as appropriate, and furnish ___ copies of the revised Report and any other deliverables to the Owner within ___ calendar days of receipt of Owner's comments.

B. Engineer's services under the Study and Report Phase will be considered complete on the date when the revised Report and any other deliverables have been delivered to Owner.

A1.02 *Preliminary Design Phase*

A. After acceptance by Owner of the Report and any other deliverables, selection by Owner of a recommended solution and indication of any specific modifications or changes in the scope, extent, character, or design requirements of the Project desired by Owner, and upon written authorization from Owner, Engineer shall:

1. Prepare Preliminary Design Phase documents consisting of final design criteria, preliminary drawings, outline specifications, and written descriptions of the Project.

2. Provide necessary field surveys and topographic and utility mapping for design purposes. Utility mapping will be based upon information obtained from utility owners.

3. Advise Owner if additional reports, data, information, or services of the types described in Exhibit B are necessary and assist Owner in obtaining such reports, data, information, or services.

4. Based on the information contained in the Preliminary Design Phase documents, prepare a revised opinion of probable Construction Cost, and assist Owner in collating the various cost categories which comprise Total Project Costs.

5. Perform or provide the following additional Preliminary Design Phase tasks or deliverables: [*here list any such tasks or deliverables*]

6. Furnish ___ review copies of the Preliminary Design Phase documents and any other deliverables to Owner within ___ calendar days of authorization to proceed with this phase, and review them with Owner. Within ___ calendar days of receipt, Owner shall submit to Engineer any comments regarding the Preliminary Design Phase documents and any other deliverables.

7. Revise the Preliminary Design Phase documents and any other deliverables in response to Owner's comments, as appropriate, and furnish to Owner ___ copies of the revised Preliminary Design Phase documents, revised opinion of probable Construction Cost, and any other deliverables within ___ calendar days after receipt of Owner's comments.

B. Engineer's services under the Preliminary Design Phase will be considered complete on the date when the revised Preliminary Design Phase documents, revised opinion of probable Construction Cost, and any other deliverables have been delivered to Owner.

A1.03 *Final Design Phase*

A. After acceptance by Owner of the Preliminary Design Phase documents, revised opinion of probable Construction Cost as determined in the Preliminary Design Phase, and any other deliverables subject to any Owner-directed modifications or changes in the scope, extent, character, or design requirements of or for the Project, and upon written authorization from Owner, Engineer shall:

1. Prepare final Drawings and Specifications indicating the scope, extent, and character of the Work to be performed and furnished by Contractor.

2. Provide technical criteria, written descriptions, and design data for Owner's use in filing applications for permits from or approvals of governmental authorities having jurisdiction to review or approve the final design of the Project; assist Owner in consultations with such authorities; and revise the Drawings and Specifications in response to directives from such authorities.

3. Advise Owner of any adjustments to the opinion of probable Construction Cost known to Engineer.

4. Perform or provide the following additional Final Design Phase tasks or deliverables: *[here list any such tasks or deliverables]*

5. Prepare and furnish bidding documents for review by Owner, its legal counsel, and other advisors, and assist Owner in the preparation of other related documents. Within ___ days of receipt, Owner shall submit to Engineer any comments and, subject to the provisions of Paragraph 6.01.G, instructions for revisions.

6. Revise the bidding documents in accordance with comments and instructions from the Owner, as appropriate, and submit ___ final copies of the bidding documents, a revised opinion of probable Construction Cost, and any other deliverables to Owner within ___ calendar days after receipt of Owner's comments and instructions.

B. Engineer's services under the Final Design Phase will be considered complete on the date when the submittals required by Paragraph A1.03.A.6 have been delivered to Owner.

C. In the event that the Work designed or specified by Engineer is to be performed or furnished under more than one prime contract, or if Engineer's services are to be separately sequenced with the work of one or more prime Contractors (such as in the case of fast-tracking), Owner and Engineer shall, prior to commencement of the Final Design Phase, develop a schedule for performance of Engineer's services during the Final Design, Bidding or Negotiating, Construction, and Post-Construction Phases in order to sequence and coordinate properly such services as are applicable to the work under such separate prime contracts. This schedule is to be prepared and included in or become an amendment to Exhibit A whether or not the work under such contracts is to proceed concurrently.

D. The number of prime contracts for Work designed or specified by Engineer upon which the Engineer's compensation has been established under this Agreement is _____. If more prime contracts are awarded, Engineer shall be entitled to an equitable increase in its compensation under this Agreement.

A1.04 *Bidding or Negotiating Phase*

A. After acceptance by Owner of the bidding documents and the most recent opinion of probable Construction Cost as determined in the Final Design Phase, and upon written authorization by Owner to proceed, Engineer shall:

1. Assist Owner in advertising for and obtaining bids or proposals for the Work and, where applicable, maintain a record of prospective bidders to whom Bidding Documents have been issued, attend pre-bid conferences, if any, and receive and process contractor deposits or charges for the bidding documents.

2. Issue addenda as appropriate to clarify, correct, or change the bidding documents.

3. Provide information or assistance needed by Owner in the course of any negotiations with prospective contractors.

4. Consult with Owner as to the acceptability of subcontractors, suppliers, and other individuals and entities proposed by prospective contractors for those portions of the Work as to which such acceptability is required by the bidding documents.

5. If bidding documents require, the Engineer shall evaluate and determine the acceptability of "or equals" and substitute materials and equipment proposed by bidders, but subject to the provisions of paragraph A2.02.A.2 of this Exhibit A.

6. Attend the Bid opening, prepare Bid tabulation sheets, and assist Owner in evaluating Bids or proposals and in assembling and awarding contracts for the Work.

7. Perform or provide the following additional Bidding or Negotiating Phase tasks or deliverables: [*here list any such tasks or deliverables*]

B. The Bidding or Negotiating Phase will be considered complete upon commencement of the Construction Phase or upon cessation of negotiations with prospective contractors (except as may be required if Exhibit F is a part of this Agreement).

A1.05 *Construction Phase*

A. Upon successful completion of the Bidding and Negotiating Phase, and upon written authorization from Owner, Engineer shall:

1. *General Administration of Construction Contract:* Consult with Owner and act as Owner's representative as provided in the Construction Contract. The extent and limitations of the duties, responsibilities, and authority of Engineer as assigned in the Construction Contract shall not be modified, except as Engineer may otherwise agree in writing. All of Owner's instructions to Contractor will be issued through Engineer, which shall have authority to act

on behalf of Owner in dealings with Contractor to the extent provided in this Agreement and the Construction Contract except as otherwise provided in writing.

2. *Resident Project Representative (RPR):* Provide the services of an RPR at the Site to assist the Engineer and to provide more extensive observation of Contractor's work. Duties, responsibilities, and authority of the RPR are as set forth in Exhibit D. The furnishing of such RPR's services will not limit, extend, or modify Engineer's responsibilities or authority except as expressly set forth in Exhibit D. [If *Engineer will not be providing the services of an RPR, then delete this Paragraph 2 by inserting the word "DELETED" after the paragraph title, and do not include Exhibit D.*]

3. *Selecting Independent Testing Laboratory:* Assist Owner in the selection of an independent testing laboratory to perform the services identified in Exhibit B, Paragraph B2.01.0.

4. *Pre-Construction Conference:* Participate in a Pre-Construction Conference prior to commencement of Work at the Site.

5. *Schedules:* Receive, review, and determine the acceptability of any and all schedules that Contractor is required to submit to Engineer, including the Progress Schedule, Schedule of Submittals, and Schedule of Values.

6. *Baselines and Benchmarks:* As appropriate, establish baselines and benchmarks for locating the Work which in Engineer's judgment are necessary to enable Contractor to proceed.

7. *Visits to Site and Observation of Construction:* In connection with observations of Contractor's Work while it is in progress:

   a. Make visits to the Site at intervals appropriate to the various stages of construction, as Engineer deems necessary, to observe as an experienced and qualified design professional the progress of Contractor's executed Work. Such visits and observations by Engineer, and the Resident Project Representative, if any, are not intended to be exhaustive or to extend to every aspect of Contractor's Work in progress or to involve detailed inspections of Contractor's Work in progress beyond the responsibilities specifically assigned to Engineer in this Agreement and the Contract Documents, but rather are to be limited to spot checking, selective sampling, and similar methods of general observation of the Work based on Engineer's exercise of professional judgment, as assisted by the Resident Project Representative, if any. Based on information obtained during such visits and observations, Engineer will determine in general if the Work is proceeding in accordance with the Contract Documents, and Engineer shall keep Owner informed of the progress of the Work.

   b. The purpose of Engineer's visits to, and representation by the Resident Project Representative, if any, at the Site, will be to enable Engineer to better carry out the duties and responsibilities assigned to and undertaken by Engineer during the Construction Phase, and, in addition, by the exercise of Engineer's efforts as an experienced and qualified design professional, to provide for Owner a greater degree of confidence that the completed Work will conform in general to the Contract Documents and that Contractor has implemented and maintained the integrity of the design concept of the completed Project as a functioning whole as indicated in the Contract Documents. Engineer shall not, during such visits or as a result of such observations of Contractor's Work in progress, supervise, direct, or have control over Contractor's Work, nor shall Engineer have authority over or responsibility for the means, methods, techniques, sequences, or procedures of construction selected or used by Contractor, for security or safety at the Site, for safety precautions and programs incident to Contractor's Work, nor for any failure of Contractor to comply with Laws and Regulations applicable to Contractor's furnishing and performing the Work. Accordingly, Engineer neither guarantees the performance of any Contractor nor assumes responsibility for any Contractor's failure to furnish or perform the Work in accordance with the Contract Documents.

8. *Defective Work:* Reject Work if, on the basis of Engineer's observations, Engineer believes that such Work (a) is defective under the standards set forth in the Contract Documents, (b) will not produce a completed Project that conforms to the Contract Documents, or (c) will imperil the integrity of the design concept of the completed Project as a functioning whole as indicated by the Contract Documents.

9. *Clarifications and Interpretations; Field Orders:* Issue necessary clarifications and interpretations of the Contract Documents as appropriate to the orderly completion of Contractor's work. Such clarifications and interpretations will be consistent with the intent of and reasonably inferable from the Contract Documents. Subject to any limitations in the Contract Documents, Engineer may issue field orders authorizing minor variations in the Work from the requirements of the Contract Documents.

10. *Change Orders and Work Change Directives:* Recommend change orders and work change directives to Owner, as appropriate, and prepare change orders and work change directives as required.

11. *Shop Drawings and Samples:* Review and approve or take other appropriate action in respect to Shop Drawings and Samples and other data which Contractor is required to submit, but only for conformance with the information given in the Contract Documents and compatibility with the design concept of the completed Project as a functioning whole as indicated by the Contract Documents. Such reviews and approvals or other action will not extend to means, methods, techniques, sequences, or procedures of construction or to safety precautions and programs incident thereto. Engineer shall meet any Contractor's submittal schedule that Engineer has accepted.

12. *Substitutes and "or-equal":* Evaluate and determine the acceptability of substitute or "or-equal" materials and equipment proposed by Contractor, but subject to the provisions of Paragraph A2.02.A.2 of this Exhibit A.

13. *Inspections and Tests:* Require such special inspections or tests of Contractor's work as deemed reasonably necessary, and receive and review all certificates of inspections, tests, and approvals required by Laws and Regulations or the Contract Documents. Engineer's review of such certificates will be for the purpose of determining that the results certified

indicate compliance with the Contract Documents and will not constitute an independent evaluation that the content or procedures of such inspections, tests, or approvals comply with the requirements of the Contract Documents. Engineer shall be entitled to rely on the results of such tests.

14. *Disagreements between Owner and Contractor*: Render formal written decisions on all duly submitted issues relating to the acceptability of Contractor's work or the interpretation of the requirements of the Contract Documents pertaining to the execution, performance, or progress of Contractor's Work; review each duly submitted Claim by Owner or Contractor, and in writing either deny such Claim in whole or in part, approve such Claim, or decline to resolve such Claim if Engineer in its discretion concludes that to do so would be inappropriate. In rendering such decisions, Engineer shall be fair and not show partiality to Owner or Contractor and shall not be liable in connection with any decision rendered in good faith in such capacity.

15. *Applications for Payment:* Based on Engineer's observations as an experienced and qualified design professional and on review of Applications for Payment and accompanying supporting documentation:

    a. Determine the amounts that Engineer recommends Contractor be paid. Such recommendations of payment will be in writing and will constitute Engineer's representation to Owner, based on such observations and review, that, to the best of Engineer's knowledge, information and belief, Contractor's Work has progressed to the point indicated, the Work is generally in accordance with the Contract Documents (subject to an evaluation of the Work as a functioning whole prior to or upon Substantial Completion, to the results of any subsequent tests called for in the Contract Documents, and to any other qualifications stated in the recommendation), and the conditions precedent to Contractor's being entitled to such payment appear to have been fulfilled in so far as it is Engineer's responsibility to observe Contractor's Work. In the case of unit price work, Engineer's recommendations of payment will include final determinations of quantities and classifications of Contractor's Work (subject to any subsequent adjustments allowed by the Contract Documents).

    b. By recommending any payment, Engineer shall not thereby be deemed to have represented that observations made by Engineer to check the quality or quantity of Contractor's Work as it is performed and furnished have been exhaustive, extended to every aspect of Contractor's Work in progress, or involved detailed inspections of the Work beyond the responsibilities specifically assigned to Engineer in this Agreement and the Contract Documents. Neither Engineer's review of Contractor's Work for the purposes of recommending payments nor Engineer's recommendation of any payment including final payment will impose on Engineer responsibility to supervise, direct, or control Contractor's Work in progress or for the means, methods, techniques, sequences, or procedures of construction or safety precautions or programs incident thereto, or Contractor's compliance with Laws and Regulations applicable to Contractor's furnishing and performing the Work. It will also not impose responsibility on Engineer to make any examination to ascertain how or for what purposes Contractor has used the moneys paid on account of the Contract Price, or to determine that title to any portion of the Work in progress, materials, or equipment has passed to Owner free and clear of any liens, claims, security interests, or encumbrances, or that there may not be other matters at issue between Owner and Contractor that might affect the amount that should be paid.

16. *Contractor's Completion Documents:* Receive, review, and transmit to Owner maintenance and operating instructions, schedules, guarantees, bonds, certificates or other evidence of insurance required by the Contract Documents, certificates of inspection, tests and approvals, Shop Drawings, Samples and other data approved as provided under Paragraph A1.05.A.11, and transmit the annotated record documents which are to be assembled by Contractor in accordance with the Contract Documents to obtain final payment. The extent of such review by Engineer will be limited as provided in Paragraph A1.05.A.11.

17. *Substantial Completion:* Promptly after notice from Contractor that Contractor considers the entire Work ready for its intended use, in company with Owner and Contractor, visit the Project to determine if the Work is substantially complete. If after considering any objections of Owner, Engineer considers the Work substantially complete, Engineer shall deliver a certificate of Substantial Completion to Owner and Contractor.

18. *Additional Tasks:* Perform or provide the following additional Construction Phase tasks or deliverables: [*here list any such tasks or deliverables*].

19. *Final Notice of Acceptability of the Work:* Conduct a final visit to the Project to determine if the completed Work of Contractor is acceptable so that Engineer may recommend, in writing, final payment to Contractor. Accompanying the recommendation for final payment, Engineer shall also provide a notice in the form attached hereto as Exhibit E (the "Notice of Acceptability of Work") that the Work is acceptable (subject to the provisions of Paragraph A1.05.A.15.b) to the best of Engineer's knowledge, information, and belief and based on the extent of the services provided by Engineer under this Agreement.

B. *Duration of Construction Phase:* The Construction Phase will commence with the execution of the first Construction Contract for the Project or any part thereof and will terminate upon written recommendation by Engineer for final payment to Contractors. If the Project involves more than one prime contract as indicated in Paragraph A1.03.C, then Construction Phase services may be rendered at different times in respect to the separate contracts. Subject to the provisions of Article 3, Engineer shall be entitled to an equitable increase in compensation if Construction Phase services (including Resident Project Representative services, if any) are required after the original date for completion and readiness for final payment of Contractor as set forth in the Construction Contract.

C. *Limitation of Responsibilities:* Engineer shall not be responsible for the acts or omissions of any Contractor, Subcontractor or Supplier, or other individuals or entities performing or furnishing any of the Work, for safety or security at the Site, or for safety precautions and programs incident to Contractor's Work, during the Construction Phase or otherwise. Engineer shall not be responsible for the failure of any Contractor to perform or furnish the Work in accordance with the Contract Documents.

A1.06 *Post-Construction Phase*

A. Upon written authorization from Ownerduring the Post-Construction Phase Engineer shall:

1. Together with Owner, visit the Project to observe any apparent defects in the Work, assist Owner in consultations and discussions with Contractor concerning correction of any such defects, and make recommendations as to replacement or correction of defective Work, if any.

2. Together with Owner or Owner's representative, visit the Project within one month before the end of the correction period to ascertain whether any portion of the Work is subject to correction.

3. Perform or provide the following additional Post-Construction Phase tasks or deliverables: *[Here list any such tasks or deliverables]*

B. The Post-Construction Phase services may commence during the Construction Phase and, if not otherwise modified in this Exhibit A, will terminate twelve months after the commencement of the Construction Contract's correction period.

PART 2 – ADDITIONAL SERVICES

A2.01 *Additional Services Requiring Owner's Written Authorization*

A. If authorized in writing by Owner, Engineer shall furnish or obtain from others Additional Services of the types listed below.

1. Preparation of applications and supporting documents (in addition to those furnished under Basic Services) for private or governmental grants, loans, or advances in connection with the Project; preparation or review of environmental assessments and impact statements; review and evaluation of the effects on the design requirements for the Project of any such statements and documents prepared by others; and assistance in obtaining approvals of authorities having jurisdiction over the anticipated environmental impact of the Project.

2. Services to make measured drawings of or to investigate existing conditions or facilities, or to verify the accuracy of drawings or other information furnished by Owner or others.

3. Services resulting from significant changes in the scope, extent, or character of the portions of the Project designed or specified by Engineer or its design requirements including, but not limited to, changes in size, complexity, Owner's schedule, character of construction, or method of financing; and revising previously accepted studies, reports, Drawings, Specifications, or Contract Documents when such revisions are required by changes in Laws and Regulations enacted subsequent to the Effective Date or are due to any other causes beyond Engineer's control.

4. Services resulting from Owner's request to evaluate additional Study and Report Phase alternative solutions beyond those identified in Paragraph A1.01.A.4.

5. Services required as a result of Owner's providing incomplete or incorrect Project information to Engineer.

6. Providing renderings or models for Owner's use.

7. Undertaking investigations and studies including, but not limited to, detailed consideration of operations, maintenance, and overhead expenses; the preparation of financial feasibility and cash flow studies, rate schedules, and appraisals; assistance in obtaining financing for the Project; evaluating processes available for licensing, and assisting Owner in obtaining process licensing; detailed quantity surveys of materials, equipment, and labor; and audits or inventories required in connection with construction performed by Owner.

8. Furnishing services of Consultants for other than Basic Services.

9. Services attributable to more prime construction contracts than specified in Paragraph A1.03.D.

10. Services during out-of-town travel required of Engineer other than for visits to the Site or Owner's office.

11. Preparing for, coordinating with, participating in and responding to structured independent review processes, including, but not limited to, construction management, cost estimating, project peer review, value engineering, and constructibility review requested by Owner; and performing or furnishing services required to revise studies, reports, Drawings, Specifications, or other Bidding Documents as a result of such review processes.

12. Preparing additional Bidding Documents or Contract Documents for alternate bids or prices requested by Owner for the Work or a portion thereof.

13. Assistance in connection with Bid protests, rebidding, or renegotiating contracts for construction, materials, equipment, or services, except when such assistance is required by Exhibit F.

14. Providing construction surveys and staking to enable Contractor to perform its work other than as required under Paragraph A1.05.A.6, and any type of property surveys or related engineering services needed for the transfer of interests in real property; and providing other special field surveys.

15. Providing Construction Phase services beyond the original date for completion and readiness for final payment of Contractor.

16. Providing assistance in responding to the presence of any Constituent of Concern at the Site, in compliance with current Laws and Regulations.

17. Preparing Record Drawings showing appropriate record information based on Project annotated record documents received from Contractor, and furnishing such Record Drawings to Owner.

18. Preparation of operation and maintenance manuals.

19. Preparing to serve or serving as a consultant or witness for Owner in any litigation, arbitration, or other dispute resolution process related to the Project.

20. Providing more extensive services required to enable Engineer to issue notices or certifications requested by Owner.

21. Assistance in connection with the adjusting of Project equipment and systems.

22. Assistance to Owner in training Owner's staff to operate and maintain Project equipment and systems.

23. Assistance to Owner in developing procedures for (a) control of the operation and maintenance of Project equipment and systems, and (b) related record-keeping.

24. Overtime work requiring higher than regular rates.

25. Other services performed or furnished by Engineer not otherwise provided for in this Agreement.

A2.02 *Additional Services Not Requiring Owner's Written Authorization*

A. Engineer shall advise Owner in advance that Engineer is will immediately commence to perform or furnish the Additional Services of the types listed below. For such Additional Services, Engineer need not request or obtain specific advance written authorization from Owner. Engineer shall cease performing or furnishing such Additional Services upon receipt of written notice from Owner.

1. Services in connection with work change directives and change orders to reflect changes requested by Owner.

2. Services in making revisions to Drawings and Specifications occasioned by the acceptance of substitute materials or equipment other than "or-equal" items; services after the award of the Construction Contract in evaluating and determining the acceptability of a proposed "or equal" or substitution which is found to be inappropriate for the Project; evaluation and determination of an excessive number of proposed "or equals" or substitutions, whether proposed before or after award of the Construction Contract.

3. Services resulting from significant delays, changes, or price increases occurring as a direct or indirect result of materials, equipment, or energy shortages.

4. Additional or extended services during construction made necessary by (1) emergencies or acts of God endangering the Work (advance notice not required), (2) the presence at the Site of any Constituent of Concern or items of historical or cultural significance, (3) Work damaged by fire or other cause during construction, (4) a significant amount of defective, neglected, or delayed work by Contractor, (5) acceleration of the progress schedule involving services beyond normal working hours, or (6) default by Contractor.

5. Services (other than Basic Services during the Post-Construction Phase) in connection with any partial utilization of any part of the Work by Owner prior to Substantial Completion.

6. Evaluating an unreasonable claim or an excessive number of claims submitted by Contractor or others in connection with the Work.

7. Services during the Construction Phase rendered after the original date for completion of the Work referred to in A1.05.B.

8. Reviewing a Shop Drawing more than three times, as a result of repeated inadequate submissions by Contractor.

9. While at the Site, compliance by Engineer and its staff with those terms of Owner's or Contractor's safety program provided to Engineer subsequent to the Effective Date that exceed those normally required of engineering personnel by federal, state, or local safety authorities for similar construction sites.

This is **EXHIBIT B**, consisting of _____ pages, referred to in and part of the **Agreement between Owner and Engineer for Professional Services** dated _____, _____.

Owner's Responsibilities

Article 2 of the Agreement is supplemented to include the following agreement of the parties.

B2.01 In addition to other responsibilities of Owner as set forth in this Agreement, Owner shall at its expense:

A. Provide Engineer with all criteria and full information as to Owner's requirements for the Project, including design objectives and constraints, space, capacity and performance requirements, flexibility, and expandability, and any budgetary limitations; and furnish copies of all design and construction standards which Owner will require to be included in the Drawings and Specifications; and furnish copies of Owner's standard forms, conditions, and related documents for Engineer to include in the Bidding Documents, when applicable.

B. Furnish to Engineer any other available information pertinent to the Project including reports and data relative to previous designs, or investigation at or adjacent to the Site.

C. Following Engineer's assessment of initially-available Project information and data and upon Engineer's request, furnish or otherwise make available such additional Project related information and data as is reasonably required to enable Engineer to complete its Basic and Additional Services. Such additional information or data would generally include the following:

1. Property descriptions.

2. Zoning, deed, and other land use restrictions.

3. Property, boundary, easement, right-of-way, and other special surveys or data, including establishing relevant reference points.

4. Explorations and tests of subsurface conditions at or contiguous to the Site, drawings of physical conditions relating to existing surface or subsurface structures at the Site, or hydrographic surveys, with appropriate professional interpretation thereof.

5. Environmental assessments, audits, investigations, and impact statements, and other relevant environmental or cultural studies as to the Project, the Site, and adjacent areas.

6. Data or consultations as required for the Project but not otherwise identified in the Agreement or the Exhibits thereto.

D. Give prompt written notice to Engineer whenever Owner observes or otherwise becomes aware of the presence at the Site of any Constituent of Concern, or of any other development that affects the scope or time of performance of Engineer's services, or any defect or nonconformance in Engineer's services, the Work, or in the performance of any Contractor.

E. Authorize Engineer to provide Additional Services as set forth in Part 2 of Exhibit A of the Agreement as required.

F. Arrange for safe access to and make all provisions for Engineer to enter upon public and private property as required for Engineer to perform services under the Agreement.

G. Examine all alternate solutions, studies, reports, sketches, Drawings, Specifications, proposals, and other documents presented by Engineer (including obtaining advice of an attorney, insurance counselor, and other advisors or consultants as Owner deems appropriate with respect to such examination) and render in writing timely decisions pertaining thereto.

H. Provide reviews, approvals, and permits from all governmental authorities having jurisdiction to approve all phases of the Project designed or specified by Engineer and such reviews, approvals, and consents from others as may be necessary for completion of each phase of the Project.

I. Recognizing and acknowledging that Engineer's services and expertise do not include the following services, provide, as required for the Project:

   1. Accounting, bond and financial advisory, independent cost estimating, and insurance counseling services.

   2. Legal services with regard to issues pertaining to the Project as Owner requires, Contractor raises, or Engineer reasonably requests.

   3. Such auditing services as Owner requires to ascertain how or for what purpose Contractor has used the moneys paid.

J. Place and pay for advertisement for Bids in appropriate publications.

K. Advise Engineer of the identity and scope of services of any independent consultants employed by Owner to perform or furnish services in regard to the Project, including, but not limited to, cost estimating, project peer review, value engineering, and constructibility review.

L. Furnish to Engineer data as to Owner's anticipated costs for services to be provided by others (including, but not limited to, accounting, bond and financial, independent cost estimating, insurance counseling, and legal advice) for Owner so that Engineer may assist Owner in collating the various cost categories which comprise Total Project Costs.

M. If Owner designates a construction manager or an individual or entity other than, or in addition to, Engineer to represent Owner at the Site, define and set forth as an attachment to this Exhibit B the duties, responsibilities, and limitations of authority of such other party and the relation thereof to the duties, responsibilities, and authority of Engineer.

N. If more than one prime contract is to be awarded for the Work designed or specified by Engineer, designate a person or entity to have authority and responsibility for coordinating the activities among the various prime Contractors, and define and set forth the duties, responsibilities, and limitations of authority of such individual or entity and the relation thereof to the duties, responsibilities, and authority of Engineer as an attachment to this Exhibit B that is to be mutually agreed upon and made a part of this Agreement before such services begin.

O. Attend the pre-bid conference, bid opening, pre-construction conferences, construction progress and other job related meetings, and Substantial Completion and final payment visits to the Project.

P. Provide the services of an independent testing laboratory to perform all inspections, tests, and approvals of samples, materials, and equipment required by the Contract Documents, or to evaluate the performance of materials, equipment, and facilities of Owner, prior to their incorporation into the Work with appropriate professional interpretation thereof.

Q. Provide Engineer with the findings and reports generated by the entities providing services to Owner pursuant to this paragraph.

R. Inform Engineer in writing of any specific requirements of safety or security programs that are applicable to Engineer, as a visitor to the Site.

S. Perform or provide the following additional services: *[Here list any such additional services].*

COMPENSATION DECISION GUIDE FOR USE WITH EXHIBIT C TO EJCDC E-500, 2008 EDITION

1. **Compensation for Basic Services (<u>not</u> including Resident Project Representative) (as described in Exhibit A, Part I)**

Decision Question: Which method of compensation is to be used?

	Lump Sum	**Standard Hourly Rates**	**Percentage of Construction Costs**	**Direct Labor Costs Times a Factor**	**Direct Labor Costs Plus Overhead Plus a Fixed Fee**	**Salary Costs Times a Factor**
Use This Base Compensation Packet	Packet BC-1	Packet BC-2	Packet BC-3	Packet BC-4	Packet BC-5	Packet BC-6
Include This Appendix	N/A	Appendices 1 and 2	N/A	Appendix 1	Appendix 1	Appendix 1

2. **Compensation for Resident Project Representative (as described in Exhibit A, Paragraph A1.05.A.2, and in Exhibit D)**

Decision Question: Which method of compensation is to be used?

	Lump Sum	**Standard Hourly Rates**	**Percentage of Construction Costs**	**Direct Labor Costs Times a Factor**	**Salary Costs Times a Factor**
Use This RPR Compensation Packet	Packet RPR-1	Packet RPR-2	Packet RPR-3	Packet RPR-4	Packet RPR-5
Include This Appendix	N/A	Appendices 1 and 2	N/A	Appendix 1	Appendix 1

3. **Compensation for Additional Services (as described in Exhibit A, Part 2)**

Decision Question: Which method of compensation is to be used?

	Standard Hourly Rates	**Direct Labor Costs Times a Factor**	**Salary Costs Times a Factor**
Use This Additional Services Compensation Packet	Packet AS-1	Packet AS-2	Packet AS-3
Include This Appendix	Appendices 1 and 2	Appendix 1	Appendix 1

Example: If Basic Services (other than RPR) will be compensated using Lump Sum; RPR services using Direct Labor Times a Factor; and Additional Services using Standard Hourly Rates; then use Packet BC-1; Packet RPR-4; Packet AS-1; and Appendices 1 and 2 to form Exhibit C.

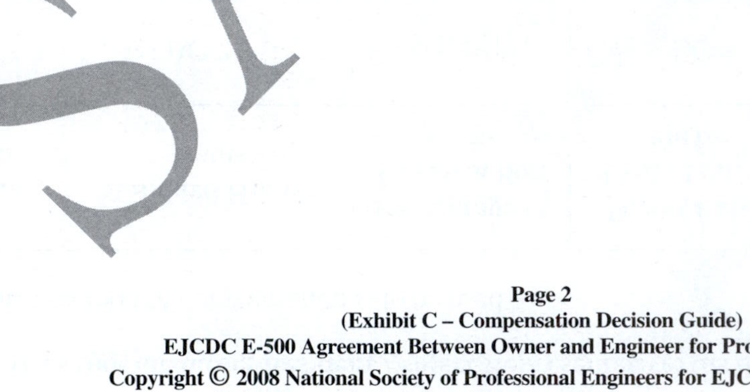

SUGGESTED FORMAT
(for use with E-500, 2008 Edition)

This is **EXHIBIT C**, consisting of _____ pages, referred to in and part of the Agreement between Owner and Engineer for Professional Services dated _____, _____.

Payments to Engineer for Services and Reimbursable Expenses
COMPENSATION PACKET BC-1: Basic Services – Lump Sum

Article 2 of the Agreement is supplemented to include the following agreement of the parties:

ARTICLE 2 – OWNER'S RESPONSIBILITIES

C2.01 *Compensation for Basic Services (other than Resident Project Representative) – Lump Sum Method of Payment*

A. Owner shall pay Engineer for Basic Services set forth in Exhibit A, except for services of Engineer's Resident Project Representative, if any, as follows:

1. A Lump Sum amount of $_____ based on the following estimated distribution of compensation:

   a. Study and Report Phase $_____

   b. Preliminary Design Phase $_____

   c. Final Design Phase $_____

   d. Bidding and Negotiating Phase $_____

   e. Construction Phase $_____

   f. Post-Construction Phase $_____

2. Engineer may alter the distribution of compensation between individual phases noted herein to be consistent with services actually rendered, but shall not exceed the total Lump Sum amount unless approved in writing by the Owner.

3. The Lump Sum includes compensation for Engineer's services and services of Engineer's Consultants, if any. Appropriate amounts have been incorporated in the Lump Sum to account for labor, overhead, profit, and Reimbursable Expenses.

4. The portion of the Lump Sum amount billed for Engineer's services will be based upon Engineer's estimate of the percentage of the total services actually completed during the billing period.

B. *Period of Service:* The compensation amount stipulated in Compensation Packet BC-1 is conditioned on a period of service not exceeding ____ months. If such period of service is extended, the compensation amount for Engineer's services shall be appropriately adjusted.

C. *Estimated Compensation Amounts:*

1. Engineer's estimate of the amounts that will become payable for specified services are only estimates for planning purposes, are not binding on the parties, and are not the minimum or maximum amounts payable to Engineer under the Agreement.

2. When estimated compensation amounts have been stated herein and it subsequently becomes apparent to Engineer that the total compensation amount thus estimated will be exceeded, Engineer shall give Owner written notice thereof, allowing Owner to consider its options, including suspension or termination of Engineer's services for Owner's convenience. Upon notice, Owner and Engineer promptly shall review the matter of services remaining to be performed and compensation for such services. Owner shall either exercise its right to suspend or terminate Engineer's services for Owner's convenience, agree to such compensation exceeding said estimated amount, or agree to a reduction in the remaining services to be rendered by Engineer, so that total compensation for such services will not exceed said estimated amount when such services are completed. If Owner decides not to suspend the Engineer's services during the negotiations and Engineer exceeds the estimated amount before Owner and Engineer have agreed to an increase in the compensation due Engineer or a reduction in the remaining services, then Engineer shall be paid for all services rendered hereunder.

D. To the extent necessary to verify Engineer's charges and upon Owner's timely request, Engineer shall make copies of such records available to Owner at cost.

This is **EXHIBIT C**, consisting of _____ pages, referred to in and part of the **Agreement between Owner and Engineer for Professional Services** dated _____, _____.

Payments to Engineer for Services and Reimbursable Expenses
COMPENSATION PACKET BC-3: Basic Services – Percentage of Construction Cost

Article 2 of the Agreement is supplemented to include the following agreement of the parties:

ARTICLE 2 – OWNER'S RESPONSIBILITIES

C2.01 *Compensation for Basic Services (other than Resident Project Representative) – Percentage of Construction Cost Method of Payment*

A. Owner shall pay Engineer for Basic Services set forth in Exhibit A, except for services of Engineer's Resident Project Representative, if any, as follows:

1. *General:* An amount equal to _____ percent of the Construction Cost. This amount includes compensation for Engineer's Services and services of Engineer's Consultants, if any. The percentage of Construction Cost noted herein accounts for labor, overhead, profit, and Reimbursable Expenses.

2. As a basis for payment to Engineer, Construction Cost will be based on one or more of the following determinations with precedence in the order listed for Work designed or specified by Engineer:

   a. For Work designed or specified and incorporated in the completed Project, the actual final price of the Construction Contract(s), as duly adjusted by change orders.

   b. For Work designed or specified but not constructed, the lowest bona fide Bid received from a qualified bidder for such Work; or, if the Work is not bid, the lowest bona fide negotiated proposal for such Work.

   c. For Work designed or specified but not constructed upon which no such Bid or proposal is received, Engineer's most recent opinion of probable Construction Cost.

   d. Labor furnished by Owner for the Project will be included in the Construction Cost at current market rates including a reasonable allowance for overhead and profit. Materials and equipment furnished by Owner will be included at current market prices.

   e. For purposes of determining Construction Cost under this provision, no deduction is to be made from Construction Contract pricing on account of any penalty, liquidated damages, or other amounts withheld from payments to Contractor(s).

3. *Progress Payments:*

a. The portion of the amounts billed for Engineer's services which is on account of the Percentage of Construction Cost will be based upon Engineer's estimate of the percentage of the total services actually completed during the billing period.

b. Upon conclusion of each phase of Basic Services, Owner shall pay such additional amount, if any, as may be necessary to bring total compensation paid during such phase on account of the percentage of Construction Cost to the following estimated percentages of total compensation payable on account of the percentage of Construction Cost for all phases of Basic Services:

Study and Report Phase	______%
Preliminary Design Phase	______%
Final Design Phase	______%
Bidding or Negotiating Phase	______%
Construction Phase	______%
	100%

c. Engineer may alter the distribution of compensation between individual phases of the work noted herein to be consistent with services actually rendered, but shall not exceed the total estimated compensation amount unless approved in writing by Owner.

This is **EXHIBIT C**, consisting of _____ pages, referred to in and part of the **Agreement between Owner and Engineer for Professional Services** dated _____, _____.

Payments to Engineer for Services and Reimbursable Expenses
COMPENSATION PACKET BC-4: Basic Services – Direct Labor Costs Times a Factor

Article 2 of the Agreement is supplemented to include the following agreement of the parties:

ARTICLE 2 – OWNER'S RESPONSIBILITIES

C2.01 *Compensation for Basic Services (other than Resident Project Representative) – Direct Labor Costs Times a Factor Method of Payment*

A. Owner shall pay Engineer for Basic Services set forth in Exhibit A, except for services of Engineer's Resident Project Representative, if any, as follows:

1. An amount equal to Engineer's Direct Labor Costs times a factor of _____ for the services of Engineer's personnel engaged on the Project, plus Reimbursable Expenses, estimated to be $_________________, and Engineer's Consultant's charges, if any, estimated to be $_________________.

2. Engineer's Reimbursable Expenses Schedule is attached to this Exhibit C as Appendix 1.

3. The total compensation for services under Paragraph C2.01 is estimated to be $_________________ based on the following distribution of compensation:

a. Study and Report Phase $_____________________

b. Preliminary Design Phase $_____________________

c. Final Design Phase $_____________________

d. Bidding or Negotiating Phase $_____________________

e. Construction Phase $_____________________

f. Post-Construction Phase $_____________________

4. Engineer may alter the distribution of compensation between individual phases of the work noted herein to be consistent with services actually rendered, but shall not exceed the total estimated compensation amount unless approved in writing by Owner. See C2.03.C.2 below.

5. The total estimated compensation for Engineer's services included in the breakdown by phases as noted in Paragraph C2.01.A.3, incorporates all labor, overhead, profit, Reimbursable Expenses, and Engineer's Consultant's charges.

6. The portion of the amounts billed for Engineer's services which are related to services rendered on a Direct Labor Costs times a Factor basis will be billed based on the applicable Direct Labor Costs for the cumulative hours charged to the Project by Engineer's principals and employees multiplied by the above-designated factor, plus Reimbursable Expenses and Engineer's Consultant's charges incurred during the billing period.

7. Direct Labor Costs means salaries and wages paid to employees but does not include payroll-related costs or benefits.

8. The Direct Labor Costs and the factor applied to Direct Labor Costs will be adjusted annually (as of ___) to reflect equitable changes to the compensation payable to Engineer.

C2.02 *Compensation for Reimbursable Expenses*

A. Owner shall pay Engineer for all Reimbursable Expenses at the rates set forth in Appendix 1 to this Exhibit C.

B. Reimbursable Expenses include the following categories: transportation and subsistence t incidental thereto; providing and maintaining field office facilities including furnishings and utilities; toll telephone calls and mobile phone charges; reproduction of reports, Drawings, Specifications, Bidding Documents, and similar Project-related items in addition to those required under Exhibit A. In addition, if authorized in advance by Owner, Reimbursable Expenses will also include expenses incurred for the use of highly specialized equipment.

C. The amounts payable to Engineer for Reimbursable Expenses will be the Project-related internal expenses actually incurred or allocated by Engineer, plus all invoiced external Reimbursable Expenses allocable to the Project, the latter multiplied by a factor of ______.

D. The Reimbursable Expenses Schedule will be adjusted annually (as of ___) to reflect equitable changes in the compensation payable to Engineer.

C2.03 *Other Provisions Concerning Payment*

A. Whenever Engineer is entitled to compensation for the charges of Engineer's Consultants, those charges shall be the amounts billed by Engineer's Consultants to Engineer times a factor of _____.

B. *Factors:* The external Reimbursable Expenses and Engineer's Consultant's factors include Engineer's overhead and profit associated with Engineer's responsibility for the administration of such services and costs.

C. *Estimated Compensation Amounts:*

1. Engineer's estimate of the amounts that will become payable for specified services are only estimates for planning purposes, are not binding on the parties, and are not the minimum or maximum amounts payable to Engineer under the Agreement.

2. When estimated compensation amounts have been stated herein and it subsequently becomes apparent to Engineer that the total compensation amount thus estimated will be exceeded, Engineer shall give Owner written notice thereof, allowing Owner to consider its options, including suspension or termination of Engineer's services for Owner's convenience. Upon notice, Owner and Engineer promptly shall review the matter of services remaining to be performed and compensation for such services. Owner shall either exercise its right to suspend or terminate Engineer's services for Owner's convenience, agree to such compensation exceeding said estimated amount, or agree to a reduction in the remaining services to be rendered by Engineer,so that total compensation for such services will not exceed said estimated amount when such services are completed. If Owner decides not to suspend Engineer's services during negotiations and Engineer exceeds the estimated amount before Owner and Engineer have agreed to an increase in the compensation due Engineer or a reduction in the remaining services, then Engineer shall be paid for all services rendered hereunder.

3. To the extent necessary to verify Engineer's charges and upon Owner's timely request, Engineer shall make copies of such records available to Owner at cost.

This is **EXHIBIT C**, consisting of _____ pages, referred to in and part of the **Agreement between Owner and Engineer for Professional Services** dated _____, _____.

Payments to Engineer for Services and Reimbursable Expenses
COMPENSATION PACKET BC-5: Basic Services – Direct Labor Costs Plus Overhead Plus a Fixed Fee

Article 2 of the Agreement is supplemented to include the following agreement of the parties:

ARTICLE 2 – OWNER'S RESPONSIBILITIES

C2.01 *Compensation for Basic Services (other than Resident Project Representative) – Direct Labor Costs Plus Overhead Plus a Fixed Fee Method of Payment*

A. Owner shall pay Engineer for Basic Services set forth in Exhibit A, except for services of Engineer's Resident Project Representative, if any, as follows:

1. An amount equal to Engineer's Direct Labor Costs plus overhead for the services of Engineer's personnel engaged directly on the Project, plus Reimbursable Expenses estimated to be $__________________, plus Engineer's Consultant's charges, if any, estimated to be $__________________, plus a fixed fee of $__________________.

2. Engineer's Reimbursable Expenses Schedule is attached to this Exhibit C as Appendix 1.

3. The total compensation for services under Paragraph C2.01 is estimated to be $__________________ based on the following estimated distribution of compensation:

   a. Study and Report Phase $______________________

   b. Preliminary Design Phase $______________________

   c. Final Design Phase $______________________

   d. Bidding or Negotiating Phase $______________________

   e. Construction Phase $______________________

   f. Post-Construction Phase $______________________

4. Engineer may alter the distribution of compensation between individual phases of the work noted herein to be consistent with services actually rendered, but shall not exceed the total estimated compensation amount unless approved in writing by Owner. See Paragraph C2.03.C.2 below.

5. The total estimated compensation for Engineer's services, included in the breakdown by phases as noted in Paragraph C2.01.A.3, incorporates all labor, overhead, fixed fees, Reimbursable Expenses, and Engineer's Consultant's charges.

6. The portion of the amounts billed for Engineer's services will be based on the applicable Direct Labor Costs for the cumulative hours charged to the Project during the billing period by Engineer's employees plus overhead, Reimbursable Expenses, Engineer's Consultant's charges, and the proportionate portion of the fixed fee.

7. Direct Labor Costs means salaries and wages paid to employees but does not include payroll-related costs or benefits.

8. Overhead includes the cost of customary and statutory benefits including, but not limited to, social security contributions, unemployment, excise and payroll taxes, workers' compensation, health and retirement benefits, bonuses, sick leave, vacation, and holiday pay applicable thereto; the cost of general and administrative overhead which includes salaries and wages of employees engaged in business operations not directly chargeable to projects, plus non-Project operating costs, including but not limited to, business taxes, legal, rent, utilities, office supplies, insurance, and other operating costs. Overhead shall be computed as a percentage of Direct Labor Costs. Fixed fee is the lump sum amount paid to Engineer by Owner as margin or profit and will only be adjusted by an amendment to this agreement.

9. Direct Labor Costs and Overhead applied to Direct Labor Costs will be adjusted annually (as of ___________) to reflect equitable changes in the compensation payable to Engineer.

C2.02 *Compensation for Reimbursable Expenses*

A. Owner shall pay Engineer for all Reimbursable Expenses at the rates set forth in Appendix 1 to this Exhibit C.

B. Reimbursable Expenses include the following categories: transportation and subsistence incidental thereto; providing and maintaining field office facilities including furnishings and utilities; toll telephone calls and mobile phone charges, reproduction of reports, Drawings, Specifications, Bidding Documents, and similar Project-related items in addition to those required under Exhibit A. In addition, if authorized in advance by Owner, Reimbursable Expenses will also include expenses incurred for the use of highly specialized equipment.

C. The amounts payable to Engineer for Reimbursable Expenses will be the Project-related internal expenses actually incurred or allocated by Engineer, plus all invoiced external Reimbursable Expenses allocable to the Project, the latter multiplied by a factor of _____.

D. The Reimbursable Expenses Schedule will be adjusted annually (as of ___________) to reflect equitable changes in the compensation payable to Engineer.

C2.03 *Other Provisions Concerning Payment*

A. Whenever Engineer is entitled to compensation for the charges of Engineer's Consultants, those charges shall be the amounts billed by Engineer's Consultants to Engineer times a factor of _____.

B. *Factors:* The external Reimbursable Expenses and Engineer's Consultant's factors include Engineer's overhead and profit associated with Engineer's responsibility for the administration of such services and costs.

C. *Estimated Compensation Amounts:*

1. Engineer's estimate of the amounts that will become payable for specified services are only estimates for planning purposes, are not binding on the parties, and are not the minimum or maximum amounts payable to Engineer under the Agreement.

2. When estimated compensation amounts have been stated herein and it subsequently becomes apparent to Engineer that the total compensation amount thus estimated will be exceeded, Engineer shall give Owner written notice thereof, allowing Owner to consider its options, including suspension or termination of Engineer's services for Owner's convenience. Upon notice, Owner and Engineer promptly shall review the matter of services remaining to be performed and compensation for such services. Owner shall either exercise its right to suspend or terminate Engineer's services for Owner's convenience, agree to such compensation exceeding said estimated amount, or agree to a reduction in the remaining services to be rendered by Engineer, so that total compensation for such services will not exceed said estimated amount when such services are completed. If Owner decides not to suspend Engineer's services during negotiations and Engineer exceeds the estimated amount before Owner and Engineer have agreed to an increase in the compensation due Engineer or a reduction in the remaining services, then Engineer shall be paid for all services rendered hereunder.

D. To the extent necessary to verify Engineer's charges and upon Owner's timely request, Engineer shall make copies of such records available to Owner at cost.

This is **EXHIBIT C**, consisting of _____ pages, referred to in and part of the **Agreement between Owner and Engineer for Professional Services** dated _____, _____.

Payments to Engineer for Services and Reimbursable Expenses
COMPENSATION PACKET BC-6: Basic Services – Salary Costs Times a Factor

Article 2 of the Agreement is supplemented to include the following agreement of the parties:

ARTICLE 2 – OWNER'S RESPONSIBILITIES

C2.01 *Compensation for Basic Services (other than Resident Project Representative) – Salary Costs Times a Factor Method of Payment*

A. Owner shall pay Engineer for Basic Services set forth in Exhibit A, except for services of Engineer's Resident Project Representative, if any, as follows:

1. An amount equal to Engineer's Salary Costs times a factor of _____ for all Basic Services by principals and employees engaged directly on the Project, plus Reimbursable Expenses, estimated to be $____________________, and Engineer's Consultant's charges, if any, estimated to be $____________________.

2. Engineer's Reimbursable Expenses Schedule is attached to this Exhibit C as Appendix 1.

3. The total compensation for services under Paragraph C2.01 is estimated to be $____________________ based on the following assumed distribution of compensation:

   a. Study and Report Phase $____________________

   b. Preliminary Design Phase $____________________

   c. Final Design Phase $____________________

   d. Bidding or Negotiating Phase $____________________

   e. Construction Phase $____________________

   f. Post-Construction Phase $____________________

4. Engineer may alter the distribution of compensation between individual phases of the work noted herein to be consistent with services actually rendered, but shall not exceed the total estimated compensation amount unless approved in writing by Owner. See also Paragraph C2.03.C.2 below.

5. The total compensation for Engineer's services, included in the breakdown by phases as noted in Paragraph C2.01.A.3, incorporates all labor, overhead, profit, Reimbursable Expenses, and Engineer's Consultant's charges.

6. The portion of the amounts billed for Engineer's services will be based on the applicable Salary Costs for the cumulative hours charged to the Project incurred during the billing period by Engineer's principals and employees multiplied by the above designated factor, plus Reimbursable Expenses and Engineer's Consultant's charges.

7. Salary Costs means salaries and wages paid to Engineer's employees plus the cost of customary and statutory benefits including, but not limited to, social security contributions, unemployment, excise and payroll taxes, workers' compensation, health and retirement benefits, bonuses, sick leave, vacation, and holiday pay applicable thereto.

8. The Salary Costs and the factor applied to Salary Costs will be adjusted annually (as of ___) to reflect equitable changes in the compensation payable to Engineer.

C2.02 *Compensation for Reimbursable Expenses*

A. Owner shall pay Engineer for all Reimbursable Expenses at the rates set forth in Appendix 1 to this Exhibit C.

B. Reimbursable Expenses include the following categories: transportation and subsistence incidental thereto; providing and maintaining field office facilities including furnishings and utilities; toll telephone calls and mobile phone charges; reproduction of reports, Drawings, Specifications, Bidding Documents, and similar Project-related items in addition to those required under Exhibit A. In addition, if authorized in advance by Owner, Reimbursable Expenses will also include expenses incurred for the use of highly specialized equipment.

C. The amounts payable to Engineer for Reimbursable Expenses will be the Project-related internal expenses actually incurred or allocated by Engineer, plus all invoiced external Reimbursable Expenses allocable to the Project, the latter multiplied by a factor of______.

D. The Reimbursable Expenses Schedule will be adjusted annually (as of ___) to reflect equitable changes in the compensation payable to Engineer.

C2.03 *Other Provisions Concerning Payment*

A. Whenever Engineer is entitled to compensation for the charges of Engineer's Consultants, those charges shall be the amounts billed by Engineer's Consultants to Engineer times a factor of _____.

B. *Factors:* The external Reimbursable Expenses and Engineer's Consultant's factors include Engineer's overhead and profit associated with Engineer's responsibility for the administration of such services and costs.

C. *Estimated Compensation Amounts:*

1. Engineer's estimate of the amounts that will become payable for specified services are only estimates for planning purposes, are not binding on the parties, and are not the minimum or maximum amounts payable to Engineer under the Agreement.

2. When estimated compensation amounts have been stated herein and it subsequently becomes apparent to Engineer that the total compensation amount thus estimated will be exceeded, Engineer shall give Owner written notice thereof, allowing Owner to consider its options, including suspension or termination of Engineer's services for Owner's convenience. Upon notice, Owner and Engineer promptly shall review the matter of services remaining to be performed and compensation for such services. Owner shall either exercise its right to suspend or terminate Engineer's services for Owner's convenience, agree to such compensation exceeding said estimated amount, or agree to a reduction in the remaining services to be rendered by Engineer, so that total compensation for such services will not exceed said estimated amount when such services are completed. If Owner decides not to suspend Engineer's services during negotiations and Engineer exceeds the estimated amount before Owner and Engineer have agreed to an increase in the compensation due Engineer or a reduction in the remaining services, then Engineer shall be paid for all services rendered hereunder.

D. To the extent necessary to verify Engineer's charges and upon Owner's timely request, Engineer shall make copies of such records available to Owner at cost.

COMPENSATION PACKET RPR-1:
Resident Project Representative – Lump Sum

Article 2 of the Agreement is supplemented to include the following agreement of the parties:

C2.04 *Compensation for Resident Project Representative Basic Services – Lump Sum Method of Payment*

A. Owner shall pay Engineer for Resident Project Representative Basic Services as follows:

1. *Resident Project Representative Services*: For services of Engineer's Resident Project Representative, if any, under Paragraph A1.05 of Exhibit A, the Lump Sum amount of $_________________. The Lump Sum includes compensation for the Resident Project Representative's services, and for the services of any direct assistants to the Resident Project Representative. Appropriate amounts have been incorporated in the Lump Sum to account for labor, overhead, profit, and Reimbursable Expenses related to the Resident Project Representative's Services.

2. *Resident Project Representative Schedule*: The Lump Sum amount set forth in Paragraph C2.04.A.1 above is based on full-time RPR services on an eight-hour workday Monday through Friday over a ___ day construction schedule. Modifications to the schedule shall entitle Engineer to an equitable adjustment of compensation for RPR services.

COMPENSATION PACKET RPR-2:
Resident Project Representative – Standard Hourly Rates

Article 2 of the Agreement is supplemented to include the following agreement of the parties:

C2.04 *Compensation for Resident Project Representative Basic Services – Standard Hourly Rates Method of Payment*

A. *Owner shall pay Engineer for Resident Project Representative Basic Services as follows:*

1. *Resident Project Representative Services:* For services of Engineer's Resident Project Representative under Paragraph A1.05A of Exhibit A, an amount equal to the cumulative hours charged to the Project by each class of Engineer's personnel times Standard Hourly Rates for each applicable billing class for all Resident Project Representative services performed on the Project, plus related Reimbursable Expenses and Engineer's Consultant's charges, if any. The total compensation under this Paragraph is estimated to be $_____ based upon full-time RPR services on an eight-hour workday, Monday through Friday, over a ___ day construction schedule.

B. *Compensation for Reimbursable Expenses:*

1. For those Reimbursable Expenses that are not accounted for in the compensation for Basic Services under Paragraph C2.01, and are directly related to the provision of Resident Project Representative or Post-Construction Basic Services, Owner shall pay Engineer at the rates set forth in Appendix 1 to this Exhibit C.

2. Reimbursable Expenses include the following categories: transportation and subsistence incidental thereto; ; providing and maintaining field office facilities including furnishings and utilities; subsistence and transportation of Resident Project Representative and assistants; toll telephone calls and mobile phone charges; reproduction of reports, Drawings, Specifications, Bidding Documents, and similar Project-related items in addition to those required under Exhibit A. In addition, if authorized in advance by Owner, Reimbursable Expenses will also include expenses incurred for the use of highly specialized equipment.

3. The amounts payable to Engineer for Reimbursable Expenses, if any, will be those internal expenses related to the Resident Project Representative Basic Services that are actually incurred or allocated by Engineer, plus all invoiced external Reimbursable Expenses allocable to such services, the latter multiplied by a factor of_____.

4. The Reimbursable Expenses Schedule will be adjusted annually (as of ___) to reflect equitable changes in the compensation payable to Engineer.

C. *Other Provisions Concerning Payment Under this Paragraph C2.04:*

1. Whenever Engineer is entitled to compensation for the charges of Engineer's Consultants, those charges shall be the amounts billed by Engineer's Consultants to Engineer times a factor of______.

2. *Factors*: The external Reimbursable Expenses and Engineer's Consultant's factors include Engineer's overhead and profit associated with Engineer's responsibility for the administration of such services and costs.

3. *Estimated Compensation Amounts*:

   a. Engineer's estimate of the amounts that will become payable for specified services are only estimates for planning purposes, are not binding on the parties, and are not the minimum or maximum amounts payable to Engineer under the Agreement.

   b. When estimated compensation amounts have been stated herein and it subsequently becomes apparent to Engineer that the total compensation amount thus estimated will be exceeded, Engineer shall give Owner written notice thereof, allowing Owner to consider its options, including suspension or termination of Engineer's services for Owner's convenience. Upon notice Owner and Engineer promptly shall review the matter of services remaining to be performed and compensation for such services. Owner shall either exercise its right to suspend or terminate Engineer's services for Owner's convenience, agree to such compensation exceeding said estimated amount, or agree to a reduction in the remaining services to be rendered by Engineer, so that total compensation for such services will not exceed said estimated amount when such services are completed. If Owner decides not to suspend Engineer's services during negotiations and Engineer exceeds the estimated amount before Owner and Engineer have agreed to an increase in the compensation due Engineer or a reduction in the remaining services, then Engineer shall be paid for all services rendered hereunder.

4. To the extent necessary to verify Engineer's charges and upon Owner's timely request, Engineer shall make copies of such records available to Owner at cost.

COMPENSATION PACKET RPR-3:
Resident Project Representative – Percentage of Construction Cost

Article 2 of the Agreement is supplmented to include the following agreement of the parties:

C2.04 *Compensation for Resident Project Representative Basic Services – Percentage of Construction Cost Method of Payment*

A. Owner shall pay Engineer for:

1. *Resident Project Representative Services:* For services of Engineer's Resident Project Representative under Paragraph A1.05 of Exhibit A of the Agreement, an amount equal to _____ percent of the Construction Cost. This amount includes compensation for Resident Project Representative's services, and those of any assistants to the Resident Project Representative. The percentage of Construction Cost noted herein accounts for labor, overhead, profit, and Reimbursable Expenses. The total compensation under this Paragraph is estimated to be $_________, based upon full-time RPR services on an eight-hour workday, Monday through Friday, over a ___ day construction schedule.

2. As a basis for payment to Engineer, Construction Cost will be based on one or more of the following determinations with precedence in the order listed for Work designed or specified by Engineer.

   a. For Work designed or specified and incorporated in the completed Project, the actual final price of the Construction Contract(s), as duly adjusted by change orders.

   b. For Work designed or specified but not constructed, the lowest bona fide Bid received from a qualified bidder for such Work; or, if the Work is not Bid, the lowest bona fide negotiated proposal for such Work.

   c. For Work designed or specified but not constructed upon which no such Bid or proposal is received, Engineer's most recent opinion of probable Construction Cost.

   d. Labor furnished by Owner for the Project will be included in the Construction Cost at current market rates including a reasonable allowance for overhead and profit. Materials and equipment furnished by Owner will be included at current market prices.

   e. For purposes of determining Construction Cost under this provision, no deduction is to be made from Construction Contract price on account of any penalty, liquidated damages, or other amounts withheld from payments to Contractor(s).

COMPENSATION PACKET RPR-4:
Resident Project Representative – Direct Labor Times a Factor

Article 2 of the Agreement is supplmented to include the following agreement of the parties:

C2.04 *Compensation for Resident Project Representative Basic Services – Direct Labor Costs Times a Factor Method of Payment*

A. Owner shall pay Engineer for:

1. *Resident Project Representative Services:* For services of Engineer's Resident Project Representative under Paragraph A1.05.A.2 of Exhibit A of the Agreement, an amount equal to Engineer's Direct Labor Costs times a factor of _____ for the services of Engineer's personnel engaged directly in resident Project representation, plus related Reimbursable Expenses and Engineer's Consultant's charges, if any. The total compensation under this paragraph is estimated to be $______________, based upon full-time RPR services on an eight-hour workday, Monday through Friday, over a ___ day construction schedule.

B. *Compensation for Reimbursable Expenses:*

1. For those Reimbursable Expenses that are not accounted for in the compensation for Basic Services under Paragraph C2.01, and are directly related to the provision of Resident Project Representative or Post-Construction Basic Services, Owner shall pay Engineer at the rates set forth in Appendix 1 to this Exhibit C.

2. Reimbursable Expenses include the following categories: transportation and subsistence incidental thereto; ; providing and maintaining field office facilities including furnishings and utilities; subsistence and transportation of Resident Project Representative and assistants; toll telephone calls and mobile phone charges; reproduction of reports, Drawings, Specifications, Bidding Documents, and similar Project-related items in addition to those required under Exhibit A. In addition, if authorized in advance by Owner, Reimbursable Expenses will also include expenses incurred for computer time and the use of other highly specialized equipment.

3. The amounts payable to Engineer for Reimbursable Expenses, if any, will be those internal expenses related to the Resident Project Representative Basic Services that are actually incurred or allocated by Engineer, plus all invoiced external Reimbursable Expenses allocable to such services, the latter multiplied by a factor of_____.

4. The Reimbursable Expenses Schedule will be adjusted annually (as of ________) to reflect equitable changes in the compensation payable to Engineer.

C. *Other Provisions Concerning Payment Under this Paragraph C2.04:*

1. Whenever Engineer is entitled to compensation for the charges of Engineer's Consultants, those charges shall be the amounts billed by Engineer's Consultants to Engineer times a factor of _____.

2. *Factors:* The external Reimbursable Expenses and Engineer's Consultant's factors include Engineer's overhead and profit associated with Engineer's responsibility for the administration of such services and costs.

3. *Estimated Compensation Amounts:*

   a. Engineer's estimate of the amounts that will become payable for specified services are only estimates for planning purposes, are not binding on the parties, and are not the minimum or maximum amounts payable to Engineer under the Agreement.

   b. When estimated compensation amounts have been stated herein and it subsequently becomes apparent to Engineer that the total compensation amount thus estimated will be exceeded, Engineer shall give Owner written notice thereof, allowing Owner to consider its options, including suspension or termination of Engineer's services for Owner's convenience. Upon notice, Owner and Engineer promptly shall review the matter of services remaining to be performed and compensation for such services. Owner shall either exercise its right to suspend or terminate Engineer's services for Owner's convenience, agree to such compensation exceeding said estimated amount, or agree to a reduction in the remaining services to be rendered by Engineer, so that total compensation for such services will not exceed said estimated amount when such services are completed. If Owner decides not to suspend Engineer's services during negotiations and Engineer exceeds the estimated amount before Owner and Engineer have agreed to an increase in the compensation due Engineer or a reduction in the remaining services, then Engineer shall be paid for all services rendered hereunder.

4. To the extent necessary to verify Engineer's charges and upon Owner's timely request, Engineer shall make copies of such records available to Owner at cost.

COMPENSATION PACKET RPR-5:
Resident Project Representative – Salary Costs Times a Factor

Article 2 of the Agreement is supplmented to include the following agreement of the parties:

C2.04 *Compensation for Resident Project Representative Basic Services – Salary Costs Times a Factor Method of Payment*

A. Owner shall pay Engineer for:

1. *Resident Project Representative Services:* For services of Engineer's Resident Project Representative, if any, under Paragraph A1.05.A.2 of Exhibit A, an amount equal to the Engineer's Salary Costs times a factor of _____ for services of Engineer's personnel engaged directly in resident Project representation, plus related Reimbursable Expenses and Engineer's Consultant's charges, if any. The total compensation under this paragraph is estimated to be $________________, based upon RPR services on an eight-hour workday, Monday through Friday, over a ____ day construction schedule.

B. *Compensation for Reimbursable Expenses:*

1. For those Reimbursable Expenses that are not accounted for in the compensation for Basic Services under Paragraph C2.01 and are directly related to the provision of Resident Project Representative or Post-Construction Basic Services, Owner shall pay Engineer at the rates set forth in Appendix 1 to this Exhibit C.

2. Reimbursable Expenses include the following categories: transportation and subsistence incidental thereto; ; providing and maintaining field office facilities including furnishings and utilities; subsistence and transportation of Resident Project Representative and assistants; toll telephone calls and mobile phone charges; reproduction of reports, Drawings, Specifications, Bidding Documents, and similar Project-related items in addition to those required under Exhibit A. In addition, if authorized in advance by Owner, Reimbursable Expenses will also include expenses incurred for the use of highly specialized equipment.

3. The amounts payable to Engineer for Reimbursable Expenses, if any, will be those internal expenses related to the Resident Project Representative or Basic Services that are actually incurred or allocated by Engineer, plus all invoiced external Reimbursable Expenses allocable to such services, the latter multiplied by a factor of_____.

4. The Reimbursable Expenses Schedule will be adjusted annually (as of _____) to reflect equitable changes in the compensation payable to Engineer.

C. *Other Provisions Concerning Payment Under this Paragraph C2.04:*

5. Whenever Engineer is entitled to compensation for the charges of Engineer's Consultants, those charges shall be the amounts billed by Engineer's Consultants to Engineer times a factor of _____.

6. *Factors:* The external Reimbursable Expenses and Engineer's Consultant's factors include Engineer's overhead and profit associated with Engineer's responsibility for the administration of such services and costs.

7. *Estimated Compensation Amounts:*

   a. Engineer's estimate of the amounts that will become payable for specified services are only estimates for planning purposes, are not binding on the parties, and are not the minimum or maximum amounts payable to Engineer under the Agreement.

   b. When estimated compensation amounts have been stated herein and it subsequently becomes apparent to Engineer that the total compensation amount thus estimated will be exceeded, Engineer shall give Owner written notice thereof, allowing Owner to consider its options, including suspension or termination of Engineer's services for Owner's convenience. Upon notice, Owner and Engineer promptly shall review the matter of services remaining to be performed and compensation for such services. Owner shall either exercise its right to suspend or terminate Engineer's services for Owner's convenience, agree to such compensation exceeding said estimated amount, or agree to a reduction in the remaining services to be rendered by Engineer, so that total compensation for such services will not exceed said estimated amount when such services are completed. If Owner decides not to suspend Engineer's services during the negotiations and Engineer exceeds the estimated amount before Owner and Engineer have agreed to an increase in the compensation due Engineer or a reduction in the remaining services, then Engineer shall be paid for all services rendered hereunder.

8. To the extent necessary to verify Engineer's charges and upon Owner's timely request, Engineer shall make copies of such records available to Owner at cost.

COMPENSATION PACKET AS-1:
Additional Services – Standard Hourly Rates

Article 2 of the Agreement is supplmented to include the following agreement of the parties:

C2.05 Compensation for Additional Services – Standard Hourly Rates Method of Payment

A. Owner shall pay Engineer for Additional Services, if any, as follows:

1. *General*: For services of Engineer's personnel engaged directly on the Project pursuant to Paragraph A2.01 or A2.02 of Exhibit A, except for services as a consultant or witness under Paragraph A2.01.A.20, (which if needed shall be separately negotiated based on the nature of the required consultation or testimony) an amount equal to the cumulative hours charged to the Project by each class of Engineer's personnel times Standard Hourly Rates for each applicable billing class for all Additional Services performed on the Project, plus related Reimbursable Expenses and Engineer's Consultant's charges, if any.

B. *Compensation For Reimbursable Expenses:*

1. For those Reimbursable Expenses that are not accounted for in the compensation for Basic Services under Paragraph C2.01 and are directly related to the provision of Additional Services, Owner shall pay Engineer at the rates set forth in Appendix 1 to this Exhibit C.

2. Reimbursable Expenses include the following categories: transportation and subsistence incidental thereto; providing and maintaining field office facilities including furnishings and utilities; toll telephone calls and mobile phone charges; reproduction of reports, Drawings, Specifications, Bidding Documents, and similar Project-related items in addition to those required under Exhibit A. In addition, if authorized in advance by Owner, Reimbursable Expenses will also include expenses incurred for the use of highly specialized equipment.

3. The amounts payable to Engineer for Reimbursable Expenses, if any,will be the Additional Services-related internal expenses actually incurred or allocated by Engineer, plus all invoiced external Reimbursable Expenses allocable to such Additional Services, the latter multiplied by a factor of _____.

4. The Reimbursable Expenses Schedule will be adjusted annually (as of ___) to reflect equitable changes in the compensation payable to Engineer.

C. *Other Provisions Concerning Payment For Additional Services:*

1. Whenever Engineer is entitled to compensation for the charges of Engineer's Consultants, those charges shall be the amounts billed by Engineer's Consultants to Engineer times a factor of ______.

2. *Factors:* The external Reimbursable Expenses and Engineer's Consultant's Factors include Engineer's overhead and profit associated with Engineer's responsibility for the administration of such services and costs.

3. To the extent necessary to verify Engineer's charges and upon Owner's timely request, Engineer shall make copies of such records available to Owner at cost.

COMPENSATION PACKET AS-2:
Additional Services – Direct Labor Costs Times a Factor

Article 2 of the Agreement is supplmented to include the following agreement of the parties:

C2.05 *Compensation for Additional Services – Direct Labor Costs Times a Factor Method of Payment*

A. Owner shall pay Engineer for Additional Services as follows:

1. *General:* For services of Engineer's personnel engaged directly on the Project pursuant to Paragraph A2.01 or A2.02 of Exhibit A of the Agreement, except for services as a consultant or witness under Paragraph A2.01.A.20, (which if needed shall be separately negotiated based on the nature of the required consultation or testimony) an amount equal to Engineer's Direct Labor Costs times a factor of _____, plus related Reimbursable Expenses and Engineer's Consultant's charges, if any.

B. *Compensation for Reimbursable Expenses:*

1. For those Reimbursable Expenses that are not accounted for in the compensation for Basic Services under Paragraph C2.01 and are directly related to the provision of Additional Services, Owner shall pay Engineer at the rates set forth in Appendix 1 to this Exhibit C.

2. Reimbursable Expenses include the following categories: transportation and subsistence incidental thereto; providing and maintaining field office facilities including furnishings and utilities; toll telephone calls and mobile phone charges; reproduction of reports, Drawings, Specifications, Bidding Documents, and similar Project-related items in addition to those required under Exhibit A. In addition, if authorized in advance by Owner, Reimbursable Expenses will also include expenses incurred for and the use of highly specialized equipment.

3. The amounts payable to Engineer for Reimbursable Expenses, if any,will be the Additional Services-related internal expenses actually incurred or allocated by Engineer, plus all invoiced external Reimbursable Expenses allocable to such Additional Services, the latter multiplied by a factor of ______.

4. The Reimbursable Expenses Schedule will be adjusted annually (as of ___) to reflect equitable changes in the compensation payable to Engineer.

C. *Other Provisions Concerning Payment for Additional Services:*

1. Whenever Engineer is entitled to compensation for the charges of Engineer's Consultants, those charges shall be the amounts billed by Engineer's Consultants to Engineer times a factor of _____.

2. *Factors:* The external Reimbursable Expenses and Engineer's Consultant's factors include Engineer's overhead and profit associated with Engineer's responsibility for the administration of such services and costs.

3. To the extent necessary to verify Engineer's charges and upon Owner's timely request, Engineer shall make copies of such records available to Owner at cost.

COMPENSATION PACKET AS-3:
Additional Services – Salary Costs Times a Factor

Article 2 of the Agreement is supplmented to include the following agreement of the parties:

C2.05 *Compensation for Additional Services – Salary Costs Times a Factor Method of Payment*

A. Owner shall pay Engineer for Additional Services as follows:

1. *General:* For services of Engineer's personnel engaged directly on the Project pursuant to Paragraph A2.01 or A2.02 of Exhibit A, except for services as a consultant or witness under Paragraph A2.01.A.20, (which if needed shall be separately negotiated based on the nature of the required consutlation or testimony) an amount equal to the cumulative hours charged to the Project by each Engineer's personnel times the Engineer's applicable Salary Costs times a factor of _____, plus related Reimbursable Expenses and Engineer's Consultant's charges, if any.

B. *Compensation for Reimbursable Expenses:*

1. For those Reimbursable Expenses that are not accounted for in the compensation for Basic Services under Paragraph C2.01 and are directly related to the provision of Additional Services, Owner shall pay Engineer at the rates set forth in Appendix 1 to this Exhibit C.

2. Reimbursable Expenses include the following categories: transportation and subsistence incidental thereto; providing and maintaining field office facilities including furnishings and utilities; toll telephone calls and mobile phone charges; reproduction of reports, Drawings, Specifications, Bidding Documents, and similar Project-related items in addition to those required under Exhibit A. In addition, if authorized in advance by Owner, Reimbursable Expenses will also include expenses incurred for and the use of highly specialized equipment.

3. The amounts payable to Engineer for Reimbursable Expenses, if any,will be the Additional Services-related internal expenses actually incurred or allocated by Engineer, plus all invoiced external Reimbursable Expenses allocable to Additional Services, the latter multiplied by a factor of _____.

4. The Reimbursable Expenses Schedule will be adjusted annually (as of _____) to reflect equitable changes in the compensation payable to Engineer.

C. *Other Provisions Concerning Payment for Additional Services:*

1. Whenever Engineer is entitled to compensation for the charges of Engineer's Consultants, those charges shall be the amounts billed by Engineer's Consultants to Engineer times a factor of _____.

2. *Factors:* The external Reimbursable Expenses and Engineer's Consultant's factors include Engineer's overhead and profit associated with Engineer's responsibility for the administration of such services and costs.

3. To the extent necessary to verify Engineer's charges and upon Owner's timely request, Engineer shall make copies of such records available to Owner at cost.

This is **Appendix 1 to EXHIBIT C**, consisting of _____ pages, referred to in and part of the **Agreement between Owner and Engineer for Professional Services** dated _____, _____.

Reimbursable Expenses Schedule

Current agreements for engineering services stipulate that the Reimbursable Expenses are subject to review and adjustment per Exhibit C. Reimbursable expenses for services performed on the date of the Agreement are:

Fax	$_____/page
8"x11" Copies/Impressions	_____/page
Blue Print Copies	_____/sq. ft.
Reproducible Copies (Mylar)	_____/sq. ft.
Reproducible Copies (Paper)	_____/sq. ft.
Mileage (auto)	_____/mile
Field Truck Daily Charge	_____/day
Mileage (Field Truck)	_____/mile
Field Survey Equipment	_____/day
Confined Space Equipment	_____/day plus expenses
Resident Project Representative Equipment	_____/month
Specialized Software	_____/hour
CAD Charge	_____/hour
CAE Terminal Charge	_____/hour
Video Equipment Charge	_____/day, $_____/week, or $_____/month
Electrical Meters Charge	_____/week, or $_____/month
Flow Meter Charge	_____/week, or $_____/month
Rain Gauge	_____/week, or $_____/month
Sampler Charge	_____/week, or $_____/month
Dissolved Oxygen Tester Charge	_____/week
Fluorometer	_____/week
Laboratory Pilot Testing Charge	_____/week, or $_____/month
Soil Gas Kit	_____/day
Submersible Pump	_____/day
Water Level Meter	_____/day, or $_____/month
Soil Sampling	_____/sample
Groundwater Sampling	_____/sample
Health and Safety Level D	_____/day
Health and Safety Level C	_____/day
Electronic Media Charge	_____/hour
Long Distance Phone Calls	at cost
Mobile Phone	_____/day
Meals and Lodging	at cost

[Note to User: Customize this Schedule to reflect anticipated reimbursable expenses on this specific Project]

This is **Appendix 2 to EXHIBIT C**, consisting of _____ pages, referred to in and part of the **Agreement between Owner and Engineer for Professional Services** dated _____, _____.

Standard Hourly Rates Schedule

A. *Standard Hourly Rates:*

1. Standard Hourly Rates are set forth in this Appendix 2 to this Exhibit C and include salaries and wages paid to personnel in each billing class plus the cost of customary and statutory benefits, general and administrative overhead, non-project operating costs, and operating margin or profit.

2. The Standard Hourly Rates apply only as specified in Article C2.

B. *Schedule:*

Hourly rates for services performed on or after the date of the Agreement are:

Billing Class VIII	$ _____/hour
Billing Class VII	_____/hour
Billing Class VI	_____/hour
Billing Class V	_____/hour
Billing Class IV	_____/hour
Billing Class III	_____/hour
Billing Class II	_____/hour
Billing Class I	_____/hour
Support Staff	_____/hour

This is **EXHIBIT D**, consisting of _____ pages, referred to in and part of the **Agreement between Owner and Engineer for Professional Services** dated _____, _____.

[Note to User: Delete this Exhibit D if Engineer will not be providing Resident Project Representative Services under Paragraph A1.05.A.2]

Duties, Responsibilities, and Limitations of Authority of Resident Project Representative

Article 1 of the Agreement is supplemented to include the following agreement of the parties:

D1.01 *Resident Project Representative*

C. Engineer shall furnish a Resident Project Representative ("RPR") to assist Engineer in observing progress and quality of the Work. The RPR may provide full time representation or may provide representation to a lesser degree.

D. Through RPR's observations of Contractor's work in progress and field checks of materials and equipment, Engineer shall endeavor to provide further protection for Owner against defects and deficiencies in the Work. However, Engineer shall not, during such RPR field checks or as a result of such RPR observations of Contractor's work in progress, supervise, direct, or have control over Contractor's Work, nor shall Engineer (including the RPR) have authority over or responsibility for the means, methods, techniques, sequences, or procedures of construction selected or used by any contractor, for security or safety at the Site, for safety precautions and programs incident to any contractor's work in progress, or for any failure of a contractor to comply with Laws and Regulations applicable to such contractor's performing and furnishing of its work. The Engineer (including RPR) neither guarantee the performances of any contractor nor assumes responsibility for Contractor's failure to furnish and perform the Work in accordance with the Contract Documents. In addition, the specific terms set forth in Paragraph A1.05 of Exhibit A of the Agreement are applicable.

E. The duties and responsibilities of the RPR are as follows:

1. *General:* RPR is Engineer's representative at the Site, will act as directed by and under the supervision of Engineer, and will confer with Engineer regarding RPR's actions. RPR's dealings in matters pertaining to the Contractor's work in progress shall in general be with Engineer and Contractor. RPR's dealings with Subcontractors shall only be through or with the full knowledge and approval of Contractor. RPR shall generally communicate with Owner only with the knowledge of and under the direction of Engineer.

2. *Schedules:* Review the progress schedule, schedule of Shop Drawing and Sample submittals, and schedule of values prepared by Contractor and consult with Engineer concerning acceptability.

3. *Conferences and Meetings:* Attend meetings with Contractor, such as preconstruction conferences, progress meetings, job conferences and other project-related meetings, and prepare and circulate copies of minutes thereof.

4. *Liaison:*

   a. Serve as Engineer's liaison with Contractor. Working principally through Contractor's authorized representative or designee, assist in providing information regarding the intent of the Contract Documents.

   b. Assist Engineer in serving as Owner's liaison with Contractor when Contractor's operations affect Owner's on-Site operations.

   c. Assist in obtaining from Owner additional details or information, when required for proper execution of the Work.

5. *Interpretation of Contract Documents:* Report to Engineer when clarifications and interpretations of the Contract Documents are needed and transmit to Contractor clarifications and interpretations as issued by Engineer.

6. *Shop Drawings and Samples:*

   a. Record date of receipt of Samples and approved Shop Drawings.

   b. Receive Samples which are furnished at the Site by Contractor, and notify Engineer of availability of Samples for examination.

   c. Advise Engineer and Contractor of the commencement of any portion of the Work requiring a Shop Drawing or Sample submittal for which RPR believes that the submittal has not been approved by Engineer.

7. M*odifications:* Consider and evaluate Contractor's suggestions for modifications in Drawings or Specifications and report such suggestions, together with RPR's recommendations, to Engineer. Transmit to Contractor in writing decisions as issued by Engineer.

8. *Review of Work and Rejection of Defective Work:*

   a. Conduct on-Site observations of Contractor's work in progress to assist Engineer in determining if the Work is in general proceeding in accordance with the Contract Documents.

   b. Report to Engineer whenever RPR believes that any part of Contractor's work in progress will not produce a completed Project that conforms generally to the Contract Documents or will imperil the integrity of the design concept of the completed Project as a functioning whole as indicated in the Contract Documents, or has been damaged, or does not meet the requirements of any inspection, test or approval required to be made; and advise Engineer of that part of work in progress that RPR believes should be corrected or rejected or should be uncovered for observation, or requires special testing, inspection, or approval.

9. *Inspections, Tests, and System Start-ups:*

a. Consult with Engineer in advance of scheduled inspections, tests, and systems start-ups.

b. Verify that tests, equipment, and systems start-ups and operating and maintenance training are conducted in the presence of appropriate Owner's personnel, and that Contractor maintains adequate records thereof.

c. Observe, record, and report to Engineer appropriate details relative to the test procedures and systems start-ups.

d. Accompany visiting inspectors representing public or other agencies having jurisdiction over the Project, record the results of these inspections, and report to Engineer.

10. *Records:*

a. Maintain at the Site orderly files for correspondence, reports of job conferences, reproductions of original Contract Documents including all change orders, field orders, work change directives, addenda, additional Drawings issued subsequent to the execution of the Construction Contract, Engineer's clarifications and interpretations of the Contract Documents, progress reports, Shop Drawing and Sample submittals received from and delivered to Contractor, and other Project-related documents.

b. Prepare a daily report or keep a diary or log book, recording Contractor's hours on the Site, weather conditions, data relative to questions of change orders, field orders, work change directives, or changed conditions, Site visitors, daily activities, decisions, observations in general, and specific observations in more detail as in the case of observing test procedures; and send copies to Engineer.

c. Record names, addresses, fax numbers, e-mail addresses, web site locations, and telephone numbers of all Contractors, Subcontractors, and major Suppliers of materials and equipment.

d. Maintain records for use in preparing Project documentation.

e. Upon completion of the Work, furnish original set of all RPR Project documentation to Engineer.

11. *Reports:*

a. Furnish to Engineer periodic reports as required of progress of the Work and of Contractor's compliance with the progress schedule and schedule of Shop Drawing and Sample submittals.

b. Draft and recommend to Engineer proposed change orders, work change directives, and field orders. Obtain backup material from Contractor.

c. Furnish to Engineer and Owner copies of all inspection, test, and system start-up reports.

d. Immediately notify Engineer of the occurrence of any Site accidents, emergencies, acts of God endangering the Work, damage to property by fire or other causes, or the discovery of any Constituent of Concern.

12. *Payment Requests:* Review applications for payment with Contractor for compliance with the established procedure for their submission and forward with recommendations to Engineer, noting particularly the relationship of the payment requested to the schedule of values, Work completed, and materials and equipment delivered at the Site but not incorporated in the Work.

13. *Certificates, Operation and Maintenance Manuals:* During the course of the Work, verify that materials and equipment certificates, operation and maintenance manuals and other data required by the Contract Documents to be assembled and furnished by Contractor are applicable to the items actually installed and in accordance with the Contract Documents, and have these documents delivered to Engineer for review and forwarding to Owner prior to payment for that part of the Work.

14. *Completion*:

a. Participate in visits to the Project to determine Substantial Completion, assist in the determination of Substantial Completion and the preparation of lists of items to be completed or corrected.

b. Participate in a final visit to the Project in the company of Engineer, Owner, and Contractor, and prepare a final list of items to be completed and deficiencies to be remedied.

c. Observe whether all items on the final list have been completed or corrected and make recommendations to Engineer concerning acceptance and issuance of the Notice of Acceptability of the Work (Exhibit E).

F. Resident Project Representative shall not:

1. Authorize any deviation from the Contract Documents or substitution of materials or equipment (including "or-equal" items).

2. Exceed limitations of Engineer's authority as set forth in this Agreement.

3. Undertake any of the responsibilities of Contractor, Subcontractors or Suppliers.

4. Advise on, issue directions relative to, or assume control over any aspect of the means, methods, techniques, sequences or procedures of Contractor's work.

5. Advise on, issue directions regarding, or assume control over security or safety practices, precautions, and programs in connection with the activities or operations of Owner or Contractor.

6. Participate in specialized field or laboratory tests or inspections conducted off-site by others except as specifically authorized by Engineer.

7. Accept shop drawing or sample submittals from anyone other than Contractor.

8. Authorize Owner to occupy the Project in whole or in part.

This is **EXHIBIT E**, consisting of _____ pages, referred to in and part of the **Agreement between Owner and Engineer for Professional Services** dated _____, _____.

NOTICE OF ACCEPTABILITY OF WORK

PROJECT:

OWNER:

CONTRACTOR:

OWNER'S CONSTRUCTION CONTRACT IDENTIFICATION:

EFFECTIVE DATE OF THE CONSTRUCTION CONTRACT:

ENGINEER:

NOTICE DATE:

To: ____________________
Owner

And To: ____________________
Contractor

From: ____________________
Engineer

The Engineer hereby gives notice to the above Owner and Contractor that the completed Work furnished and performed by Contractor under the above Contract is acceptable, expressly subject to the provisions of the related Contract Documents, the Agreement between Owner and Engineer for Professional Services dated _____, _____, and the terms and conditions set forth in this Notice.

By: ______________________________

Title: ______________________________

Dated: ______________________________

CONDITIONS OF NOTICE OF ACCEPTABILITY OF WORK

The Notice of Acceptability of Work ("Notice") is expressly made subject to the following terms and conditions to which all those who receive said Notice and rely thereon agree:

1. This Notice is given with the skill and care ordinarily used by members of the engineering profession practicing under similar conditions at the same time and in the same locality.

2. This Notice reflects and is an expression of the professional judgment of Engineer.

3. This Notice is given as to the best of Engineer's knowledge, information, and belief as of the Notice Date.

4. This Notice is based entirely on and expressly limited by the scope of services Engineer has been employed by Owner to perform or furnish during construction of the Project (including observation of the Contractor's work) under Engineer's Agreement with Owner and under the Construction Contract referred to in this Notice, and applies only to facts that are within Engineer's knowledge or could reasonably have been ascertained by Engineer as a result of carrying out the responsibilities specifically assigned to Engineer under such Agreement and Construction Contract.

5. This Notice is not a guarantee or warranty of Contractor's performance under the Construction Contract referred to in this Notice, nor an assumption of responsibility for any failure of Contractor to furnish and perform the Work thereunder in accordance with the Contract Documents.

This is **EXHIBIT F**, consisting of _____ pages, referred to in and part of the **Agreement between Owner and Engineer for Professional Services** dated _____, _____.

Construction Cost Limit

Paragraph 5.02 of the Agreement is supplemented to include the following agreement of the parties:

F5.02 *Designing to Construction Cost Limit*

A. Owner and Engineer hereby agree to a Construction Cost limit in the amount of $_____.

B. A bidding or negotiating contingency of _____ percent will be added to any Construction Cost limit established.

C. The acceptance by Owner at any time during Basic Services of a revised opinion of probable Construction Cost in excess of the then established Construction Cost limit will constitute a corresponding increase in the Construction Cost limit.

D. Engineer will be permitted to determine what types and quality of materials, equipment and component systems are to be included in the Drawings and Specifications. Engineer may make reasonable adjustments in the scope, extent, and character of the Project to the extent consistent with the Project requirements and sound engineering practices, to bring the Project within the Construction Cost limit.

E. If the Bidding or Negotiating Phase has not commenced within three months after completion of the Final Design Phase, or if industry-wide prices are changed because of unusual or unanticipated events affecting the general level of prices or times of delivery in the construction industry, the established Construction Cost limit will not be binding on Engineer. In such cases, Owner shall consent to an adjustment in the Construction Cost limit commensurate with any applicable change in the general level of prices in the construction industry between the date of completion of the Final Design Phase and the date on which proposals or Bids are sought.

F. If the lowest bona fide proposal or Bid exceeds the established Construction Cost limit, Owner shall (1) give written approval to increase such Construction Cost limit, or (2) authorize negotiating or rebidding the Project within a reasonable time, or (3) cooperate in revising the Project's scope, extent, or character to the extent consistent with the Project's requirements and with sound engineering practices. In the case of (3), Engineer shall modify the Contract Documents as necessary to bring the Construction Cost within the Construction Cost Limit. Owner shall pay Engineer's cost to provide such modification services, including the costs of the services of its Consultants, all overhead expenses reasonably related thereto, and Reimbursable Expenses, but without profit to Engineer on account of such services. The providing of such services will be the limit of Engineer's responsibility in this regard and, having done so, Engineer shall be entitled to payment for services and expenses in accordance with this Agreement and will not otherwise be liable for damages attributable to the lowest bona fide proposal or bid exceeding the established Construction Cost limit.

This is **EXHIBIT G**, consisting of _____ pages, referred to in and part of the **Agreement between Owner and Engineer for Professional Services** dated _____, _____.

Insurance

Paragraph 6.04 of the Agreement is supplemented to include the following agreement of the parties.

G6.04 *Insurance*

A. The limits of liability for the insurance required by Paragraph 6.04.A and 6.04.B of the Agreement are as follows:

1. By Engineer:

a. Workers' Compensation: Statutory

b. Employer's Liability --

1) Each Accident: $_______________
2) Disease, Policy Limit: $_______________
3) Disease, Each Employee: $_______________

c. General Liability --

1) Each Occurrence (Bodily Injury and Property Damage): $_______________
2) General Aggregate: $_______________

d. Excess or Umbrella Liability --

1) Each Occurrence: $_______________
2) General Aggregate: $_______________

e. Automobile Liability --Combined Single Limit (Bodily Injury and Property Damage):

Each Accident $_______________

f. Professional Liability –

1) Each Claim Made $_______________
2) Annual Aggregate $_______________

g. Other (specify): $_______________

2. By Owner:

   a. Workers' Compensation: Statutory

   b. Employer's Liability --

      1) Each Accident $_______________
      2) Disease, Policy Limit $_______________
      3) Disease, Each Employee $_______________

   c. General Liability --

      1) General Aggregate: $_______________
      2) Each Occurrence (Bodily Injury and Property Damage): $_______________

   d. Excess Umbrella Liability --

      1) Each Occurrence: $_______________
      2) General Aggregate: $_______________

   e. Automobile Liability --Combined Single Limit (Bodily Injury and Property Damage):

      Each Accident:
      $_______________

   f. Other (specify): $_______________

B. *Additional Insureds:*

1. The following persons or entities are to be listed on Owner's general liability policies of insurance as additional insureds, and on any applicable property insurance policy as loss payees, as provided in Paragraph 6.04.B:

a. ____
Engineer

b. ____
Engineer's Consultant

c. ____
Engineer's Consultant

2. During the term of this Agreement the Engineer shall notify Owner of any other Consultant to be listed as an additional insured on Owner's general liability and property policies of insurance.

3. The Owner shall be listed on Engineer's general liability policy as provided in Paragraph 6.04.A.

This is **EXHIBIT H**, consisting of _____ pages, referred to in and part of the **Agreement between Owner and Engineer for Professional Services** dated _____, _____.

Dispute Resolution

Paragraph 6.08 of the Agreement is amended and supplemented to include the following agreement of the parties:

[***NOTE TO USER: Select one of the two alternatives provided***]

H6.08 *Dispute Resolution*

A. *Mediation*: Owner and Engineer agree that they shall first submit any and all unsettled claims, counterclaims, disputes, and other matters in question between them arising out of or relating to this Agreement or the breach thereof ("Disputes") to mediation by *[insert name of mediator, or mediation service]*. Owner and Engineer agree to participate in the mediation process in good faith. The process shall be conducted on a confidential basis, and shall be completed within 120 days. If such mediation is unsuccessful in resolving a Dispute, then (1) the parties may mutually agree to a dispute resolution of their choice, or (2) either party may seek to have the Dispute resolved by a court of competent jurisdiction.

[or]

A. *Arbitration:* All Disputes between Owner and Engineer shall be settled by arbitration in accordance with the [*here insert the name of a specified arbitration service or organization*] rules effective at the Effective Date, subject to the conditions stated below. This agreement to arbitrate and any other agreement or consent to arbitrate entered into in accordance with this Paragraph H6.08.A will be specifically enforceable under prevailing law of any court having jurisdiction.

1. Notice of the demand for arbitration must be filed in writing with the other party to the Agreement and with the [*specified arbitration service or organization*]. The demand must be made within a reasonable time after the Dispute has arisen. In no event may the demand for arbitration be made after the date when institution of legal or equitable proceedings based on such Dispute would be barred by the applicable statute of limitations.

2. All demands for arbitration and all answering statements thereto which include any monetary claims must contain a statement that the total sum or value in controversy as alleged by the party making such demand or answering statement is not more than $_____ (exclusive of interest and costs). The arbitrators will not have jurisdiction, power, or authority to consider, or make findings (except in denial of their own jurisdiction) concerning any Dispute if the amount in controversy in such Dispute is more than $_____ (exclusive of interest and costs), or to render a monetary award in response thereto against any party which totals more than $_____ (exclusive of interest and costs). Disputes that

are not subject to arbitration under this paragraph may be resolved in any court of competent jurisdiction.

3. The award rendered by the arbitrators shall be in writing, and shall include: (i) a precise breakdown of the award; and (ii) a written explanation of the award specifically citing the Agreement provisions deemed applicable and relied on in making the award.

4. The award rendered by the arbitrators will be consistent with the Agreement of the parties and final, and judgment may be entered upon it in any court having jurisdiction thereof, and will not be subject to appeal or modification.

5. If a Dispute in question between Owner and Engineer involves the work of a Contractor, Subcontractor, or consultants to the Owner or Engineer (each a "Joinable Party"), and such Joinable Party has agreed contractually or otherwise to participate in a consolidated arbitration concerning this Project, then either Owner or Engineer may join such Joinable Party as a party to the arbitration between Owner and Engineer hereunder. Nothing in this Paragraph H6.08.A.5 nor in the provision of such contract consenting to joinder shall create any claim, right, or cause of action in favor of the Joinable Party and against Owner or Engineer that does not otherwise exist.

This is **EXHIBIT I**, consisting of _____ pages, referred to in and part of the **Agreement between Owner and Engineer for Professional Services** dated _____, _____.

Limitations of Liability

Paragraph 6.10 of the Agreement is supplemented to include the following agreement of the parties:

A. *Limitation of Engineer's Liability*

[***NOTE TO USER: Select one of the three alternatives listed below for I6.10 A.1***]

1. *Engineer's Liability Limited to Amount of Engineer's Compensation:* To the fullest extent permitted by law, and notwithstanding any other provision of this Agreement, the total liability, in the aggregate, of Engineer and Engineer's officers, directors, members, partners, agents, employees, and Consultants, to Owner and anyone claiming by, through, or under Owner for any and all claims, losses, costs, or damages whatsoever arising out of, resulting from, or in any way related to the Project or the Agreement from any cause or causes, including but not limited to the negligence, professional errors or omissions, strict liability, breach of contract, indemnity obligations, or warranty express or implied of Engineer or Engineer's officers, directors, members, partners, agents, employees, or Consultants shall not exceed the total compensation received by Engineer under this Agreement.

[or]

1. *Engineer's Liability Limited to Amount of Insurance Proceeds:* Engineer shall procure and maintain insurance as required by and set forth in Exhibit G to this Agreement. Notwithstanding any other provision of this Agreement, and to the fullest extent permitted by law, the total liability, in the aggregate, of Engineer and Engineer's officers, directors, members, partners, agents, employees, and Consultants to Owner and anyone claiming by, through, or under Owner for any and all claims, losses, costs, or damages whatsoever arising out of, resulting from, or in any way related to the Project or the Agreement from any cause or causes, including but not limited to the negligence, professional errors or omissions, strict liability, breach of contract, indemnity obligations, or warranty express or implied, of Engineer or Engineer's officers, directors, members, partners, agents, employees,or Consultantss (hereafter "Owner's Claims"), shall not exceed the total insurance proceeds paid on behalf of or to Engineer by Engineer's insurers in settlement or satisfaction of Owner's Claims under the terms and conditions of Engineer's insurance policies applicable thereto (excluding fees, costs and expenses of investigation, claims adjustment, defense, and appeal). If no such insurance coverage is provided with respect to Owner's Claims, then the total liability, in the aggregate, of Engineer and Engineer's officers, directors, members, partners, agents, employees, and Consultants to Owner and anyone claiming by, through, or under Owner for any and all such uninsured Owner's Claims shall not exceed $____________ [*or*]

1. *Engineer's Liability Limited to the Amount of $____________:* Notwithstanding any other provision of this Agreement, and to the fullest extent permitted by law, the total liability, in the aggregate, of Engineer and Engineer's officers, directors, members, partners, agents, employees, and Consultants, to Owner and anyone claiming by, through, or under Owner for any and all claims, losses, costs, or damages whatsoever arising out of, resulting from, or in any way related to the Project or the Agreement from any cause or causes, including but not limited to the negligence, professional errors or omissions, strict liability, breach of contract, indemnity obligations, or warranty express or implied of Engineer or Engineer's officers, directors, members, partners, agents, employees, or Consultants shall not exceed the total amount of $____________.

[NOTE TO USER: If appropriate and desired, include I6.10.A.2 below as a supplement to Paragraph 6.10, which contains a mutual waiver of damages applicable to the benefit of both Owner and Engineer]

2. *Exclusion of Special, Incidental, Indirect, and Consequential Damages:* To the fullest extent permitted by law, and notwithstanding any other provision in the Agreement, consistent with the terms of Paragraph 6.10. the Engineer and Engineer's officers, directors, members, partners, agents, Consultants, and employees shall not be liable to Owner or anyone claiming by, through, or under Owner for any special, incidental, indirect, or consequential damages whatsoever arising out of, resulting from, or in any way related to the Project or the Agreement from any cause or causes, including but not limited to any such damages caused by the negligence, professional errors or omissions, strict liability, breach of contract, indemnity obligations, or warrantyexpress or implied of Engineer or Engineer's officers, directors, members, partners, agents, employees, or Consultants, and including but not limited to:

    [NOTE TO USER: list here particular types of damages that may be of special concern because of the nature of the project or specific circumstances, e.g., cost of replacement power, loss of use of equipment or of the facility, loss of profits or revenue, loss of financing, regulatory fines, etc. If the parties prefer to leave the language general, then end the sentence after the word "employees"]

    [NOTE TO USER: the above exclusion of consequential and other damages can be converted to a limitation on the amount of such damages, following the format of Paragraph I6.10.A.1 above, by providing that "Engineer's total liability for such damages shall not exceed $_______."]

[NOTE TO USER: If appropriate and desired, include I6.10.A.3 below]

3. *Agreement Not to Claim for Cost of Certain Change Orders:* Owner recognizes and expects that certain Change Orders may be required to be issued as the result in whole or part of imprecision, incompleteness, errors, omissions, ambiguities, or inconsistencies in

the Drawings, Specifications, and other design documentation furnished by Engineer or in the other professional services performed or furnished by Engineer under this Agreement ("Covered Change Orders"). Accordingly, Owner agrees not to sue or to make any claim directly or indirectly against Engineer on the basis of professional negligence, breach of contract, or otherwise with respect to the costs of approved Covered Change Orders unless the costs of such approved Covered Change Orders exceed _____% of Construction Cost, and then only for an amount in excess of such percentage. Any responsibility of Engineer for the costs of Covered Change Orders in excess of such percentage will be determined on the basis of applicable contractual obligations and professional liability standards. For purposes of this paragraph, the cost of Covered Change Orders will not include any costs that Owner would have incurred if the Covered Change Order work had been included originally without any imprecision, incompleteness, error, omission, ambiguity, or inconsistency in the Contract Documents and without any other error or omission of Engineer related thereto. Nothing in this provision creates a presumption that, or changes the professional liability standard for determining if, Engineer is liable for the cost of Covered Change Orders in excess of the percentage of Construction Cost stated above or for any other Change Order. Wherever used in this paragraph, the term Engineer includes Engineer's officers, directors, members, partners, agents, employees, and Consultants.

[NOTE TO USER: The parties may wish to consider the additional limitation contained in the following sentence.]

Owner further agrees not to sue or to make any claim directly or indirectly against Engineer with respect to any Covered Change Order not in excess of such percentage stated above, and Owner agrees to hold Engineer harmless from and against any suit or claim made by the Contractor relating to any such Covered Change Order.]

[NOTE TO USER: Many professional service agreements contain mutual indemnifications. If the parties elect to provide a mutual counterpart to the indemnification of Owner by Engineer in Paragraph 6.10.A, then supplement Paragraph 6.10.B by including the following indemnification of Engineer by Owner as Paragraph I6.10.B.]

B. *Indemnification by Owner*: To the fullest extent permitted by law, Owner shall indemnify and hold harmless Engineer and its officers, directors, members, partners, agents, employees, and Consultants from and against any and all claims, costs, losses, and damages (including but not limited to all fees and charges of engineers, architects, attorneys, and other professionals, and all court, arbitration, or other dispute resolution costs) arising out of or relating to the Project, provided that any such claim, cost, loss, or damage is attributable to bodily injury, sickness, disease, or death or to injury to or destruction of tangible property (other than the Work itself), including the loss of use resulting therefrom, but only to the extent caused by any negligent act or omission of Owner or Owner's officers, directors, members, partners, agents, employees, consultants, or others retained by or under contract to the Owner with respect to this Agreement or to the Project.

This is **EXHIBIT J**, consisting of _____ pages, referred to in and part of the **Agreement between Owner and Engineer for Professional Services** dated _____, _____.

Special Provisions

Paragraph(s) ___ of the Agreement is/are amended to include the following agreement(s) of the parties:

This is **EXHIBIT K**, consisting of _____ pages, referred to in and part of the **Agreement between Owner and Engineer for Professional Services** dated _____, _____.

AMENDMENT TO OWNER-ENGINEER AGREEMENT
Amendment No. _____

1. *Background Data:*

   a. Effective Date of Owner-Engineer Agreement: ______________________

   b. Owner: ______________________

   c. Engineer: ______________________

   d. Project: ______________________

2. *Description of Modifications:*

[NOTE TO USER: Include the following paragraphs that are appropriate and delete those not applicable to this amendment. Refer to paragraph numbers used in the Agreement or a previous amendment for clarity with respect to the modifications to be made. Use paragraph numbers in this document for ease of reference herein and in future correspondence or amendments.]

   a. Engineer shall perform or furnish the following Additional Services:

   b. The Scope of Services currently authorized to be performed by Engineer in accordance with the Agreement and previous amendments, if any, is modified as follows:

   c. The responsibilities of Owner are modified as follows:

   d. For the Additional Services or the modifications to services set forth above, Owner shall pay Engineer the following additional or modified compensation:

   e. The schedule for rendering services is modified as follows:

   f. Other portions of the Agreement (including previous amendments, if any) are modified as follows:

[List other Attachments, if any]

5. Agreement Summary (Reference only)

a. Original Agreement amount: $________________
b. Net change for prior amendments: $________________
c. This amendment amount: $________________
d. Adjusted Agreement amount: $________________

The foregoing Agreement Summary is for reference only and does not alter the terms of the Agreement, including those set forth in Exhibit C.

Owner and Engineer hereby agree to modify the above-referenced Agreement as set forth in this Amendment. All provisions of the Agreement not modified by this or previous Amendments remain in effect. The Effective Date of this Amendment is ________________.

OWNER:	ENGINEER:
________________	________________
By: ________________	By: ________________
Title: ________________	Title: ________________
Date Signed: ________________	Date Signed: ________________

Engineers Joint Documents Committee
Design and Construction Related Documents
Instructions and License Agreement

Instructions

Before you use any EJCDC document:

1. Read the License Agreement. You agree to it and are bound by its terms when you use the EJCDC document.

2. Make sure that you have the correct version for your word processing software.

How to Use:

1. While EJCDC has expended considerable effort to make the software translations exact, it can be that a few document controls (e.g., bold, underline) did not carry over.

2. Similarly, your software may change the font specification if the font is not available in your system. It will choose a font that is close in appearance. In this event, the pagination may not match the control set.

3. If you modify the document, you must follow the instructions in the License Agreement about notification.

4. Also note the instruction in the License Agreement about the EJCDC copyright.

License Agreement

You should carefully read the following terms and conditions before using this document. Commencement of use of this document indicates your acceptance of these terms and conditions. If you do not agree to them, you should promptly return the materials to the vendor, and your money will be refunded.

The Engineers Joint Contract Documents Committee ("EJCDC") provides **EJCDC Design and Construction Related Documents** and licenses their use worldwide. You assume sole responsibility for the selection of specific documents or portions thereof to achieve your intended results, and for the installation, use, and results obtained from **EJCDC Design and Construction Related Documents**.

You acknowledge that you understand that the text of the contract documents of **EJCDC Design and Construction Related Documents** has important legal consequences and that consultation with an attorney is recommended with respect to use or modification of the text. You further acknowledge that EJCDC documents are protected by the copyright laws of the United States.

License:
You have a limited nonexclusive license to:

1. Use **EJCDC Design and Construction Related Documents** on any number of machines owned, leased or rented by your company or organization.

2. Use **EJCDC Design and Construction Related Documents** in printed form for bona fide contract documents.

3. Copy **EJCDC Design and Construction Related Documents** into any machine readable or printed form for backup or modification purposes in support of your use of **EJCDC Design and Construction Related Documents**.

You agree that you will:

1. Reproduce and include EJCDC's copyright notice on any printed or machine-readable copy, modification, or portion merged into another document or program. All proprietary rights in **EJCDC Design and Construction Related Documents** are and shall remain the property of EJCDC.

2. Not represent that any of the contract documents you generate from **EJCDC Design and Construction Related Documents** are EJCDC documents unless (i) the document text is used without alteration or (ii) all additions and changes to, and deletions from, the text are clearly shown.

You may not use, copy, modify, or transfer EJCDC Design and Construction Related Documents, or any copy, modification or merged portion, in whole or in part, except as expressly provided for in this license. Reproduction of EJCDC Design and Construction Related Documents in printed or machine-readable format for resale or educational purposes is expressly prohibited.

If you transfer possession of any copy, modification or merged portion of EJCDC Design and Construction Related Documents to another party, your license is automatically terminated.

Term:
The license is effective until terminated. You may terminate it at any time by destroying **EJCDC Design and Construction Related Documents** altogether with all copies, modifications and merged portions in any form. It will also terminate upon conditions set forth elsewhere in this Agreement or if you fail to comply with any term or condition of this Agreement. You agree upon

such termination to destroy **EJCDC Design and Construction Related Documents** along with all copies, modifications and merged portions in any form.

Limited Warranty:
EJCDC warrants the CDs and diskettes on which **EJCDC Design and Construction Related Documents** is furnished to be free from defects in materials and workmanship under normal use for a period of ninety (90) days from the date of delivery to you as evidenced by a copy of your receipt.

There is no other warranty of any kind, either expressed or implied, including, but not limited to the implied warranties of merchantability and fitness for a particular purpose. Some states do not allow the exclusion of implied warranties, so the above exclusion may not apply to you. This warranty gives you specific legal rights and you may also have other rights which vary from state to state.

EJCDC does not warrant that the functions contained in **EJCDC Design and Construction Related Documents** will meet your requirements or that the operation of **EJCDC Design and Construction Related Documents** will be uninterrupted or error free.

Limitations of Remedies:
EJCDC's entire liability and your exclusive remedy shall be:

1. the replacement of any document not meeting EJCDC's "Limited Warranty" which is returned to EJCDC's selling agent with a copy of your receipt, or

2. if EJCDC's selling agent is unable to deliver a replacement CD or diskette which is free of defects in materials and workmanship, you may terminate this Agreement by returning EJCDC Document and your money will be refunded.

In no event will EJCDC be liable to you for any damages, including any lost profits, lost savings or other incidental or consequential damages arising out of the use or inability to use **EJCDC Design and Construction Related Documents** even if EJCDC has been advised of the possibility of such damages, or for any claim by any other party.

Some states do not allow the limitation or exclusion of liability for incidental or consequential damages, so the above limitation or exclusion may not apply to you.

General:
You may not sublicense, assign, or transfer this license except as expressly provided in this Agreement. Any attempt otherwise to sublicense, assign, or transfer any of the rights, duties, or obligations hereunder is void.

This Agreement shall be governed by the laws of the State of Virginia. Should you have any questions concerning this Agreement, you may contact EJCDC by writing to:

Arthur Schwartz, Esq.
General Counsel
National Society of Professional Engineers
1420 King Street
Alexandria, VA 22314

Phone: (703) 684-2845
Fax: (703) 836-4875
e-mail: aschwartz@nspe.org

You acknowledge that you have read this agreement, understand it and agree to be bound by its terms and conditions. You further agree that it is the complete and exclusive statement of the agreement between us which supersedes any proposal or prior agreement, oral or written, and any other communications between us relating to the subject matter of this agreement.

This document has important legal consequences; consultation with an attorney is encouraged with respect to its use or modification. This document should be adapted to the particular circumstances of the contemplated Project and the Controlling Law.

SHORT FORM OF AGREEMENT BETWEEN OWNER AND ENGINEER FOR PROFESSIONAL SERVICES

Prepared by

ENGINEERS JOINT CONTRACT DOCUMENTS COMMITTEE

and

Issued and Published Jointly by

PROFESSIONAL ENGINEERS IN PRIVATE PRACTICE
a practice division of the
NATIONAL SOCIETY OF PROFESSIONAL ENGINEERS

AMERICAN COUNCIL OF ENGINEERING COMPANIES

AMERICAN SOCIETY OF CIVIL ENGINEERS

This Agreement has been prepared for use with the Standard General Conditions of the Construction Contract (No. C-700, 2002 Edition) of the Engineers Joint Contract Documents Committee. Their provisions are interrelated, and a change in one may necessitate a change in the other.

SPECIAL NOTE ON USE OF THIS FORM:
This abbreviated Agreement form is intended for use only for professional services of limited scope and complexity. It does not address the full range of issues of importance on most projects. In most cases, Owner and Engineer will be better served by the Standard Form of Agreement Between Owner and Engineer for Professional Services (No. E-500, 2002 Edition), or one of the several special purpose EJCDC professional services agreement forms.

American Council of Engineering Companies
1015 15th Street N.W., Washington, DC 20005

American Society of Civil Engineers
1801 Alexander Bell Drive, Reston, VA 20191-4400

SHORT FORM OF AGREEMENT BETWEEN OWNER AND ENGINEER FOR PROFESSIONAL SERVICES

THIS IS AN AGREEMENT effective as of ______________________________ ("Effective Date") between

____ ("Owner")

and ____ ("Engineer")

Engineer agrees to provide the services described below to Owner for ________________________ ("Project").

Description of Engineer's Services: ____

____

____

____

____

Owner and Engineer further agree as follows:

1.01 Basic Agreement

A. Engineer shall provide, or cause to be provided, the services set forth in this Agreement, and Owner shall pay Engineer for such Services as set forth in Paragraph 9.01.

2.01 Payment Procedures

A. *Preparation of Invoices*. Engineer will prepare a monthly invoice in accordance with Engineer's standard invoicing practices and submit the invoice to Owner.

B. *Payment of Invoices*. Invoices are due and payable within 30 days of receipt. If Owner fails to make any payment due Engineer for services and expenses within 30 days after receipt of Engineer's invoice, the amounts due Engineer will be increased at the rate of 1.0% per month (or the maximum rate of interest permitted by law, if less) from said thirtieth day. In addition, Engineer may, without liability, after giving seven days written notice to Owner, suspend services under this Agreement until Engineer has been paid in full all amounts due for services, expenses, and other related charges. Payments will be credited first to interest and then to principal.

3.01 Additional Services

A. If authorized by Owner, or if required because of changes in the Project, Engineer shall furnish services in addition to those set forth above.

B. Owner shall pay Engineer for such additional services as follows: For additional services of Engineer's employees engaged directly on the Project an amount equal to the cumulative hours charged to the Project by each class of Engineer's employees times standard hourly rates for each applicable billing class; plus reimbursable expenses and Engineer's consultants' charges, if any.

4.01 Termination

A. The obligation to provide further services under this Agreement may be terminated:

1. For cause,

a. By either party upon 30 days written notice in the event of substantial failure by the other party to perform in accordance with the Agreement's terms through no fault of the terminating party.

b. By Engineer:

1) upon seven days written notice if Engineer believes that Engineer is being requested by Owner to furnish or perform services contrary to Engineer's

responsibilities as a licensed professional; or

2) upon seven days written notice if the Engineer's services for the Project are delayed or suspended for more than 90 days for reasons beyond Engineer's control.

3) Engineer shall have no liability to Owner on account of such termination.

c. Notwithstanding the foregoing, this Agreement will not terminate as a result of a substantial failure under paragraph 4.01.A.1.a if the party receiving such notice begins, within seven days of receipt of such notice, to correct its failure and proceeds diligently to cure such failure within no more than 30 days of receipt of notice; provided, however, that if and to the extent such substantial failure cannot be reasonably cured within such 30 day period, and if such party has diligently attempted to cure the same and thereafter continues diligently to cure the same, then the cure period provided for herein shall extend up to, but in no case more than, 60 days after the date of receipt of the notice.

2. For convenience, by Owner effective upon the receipt of notice by Engineer.

B. The terminating party under paragraphs 4.01.A.1 or 4.01.A.2 may set the effective date of termination at a time up to 30 days later than otherwise provided to allow Engineer to demobilize personnel and equipment from the Project site, to complete tasks whose value would otherwise be lost, to prepare notes as to the status of completed and uncompleted tasks, and to assemble Project materials in orderly files.

5.01 Controlling Law

A. This Agreement is to be governed by the law of the state in which the Project is located.

6.01 Successors, Assigns, and Beneficiaries

A. Owner and Engineer each is hereby bound and the partners, successors, executors, administrators, and legal representatives of Owner and Engineer (and to the extent permitted by paragraph 6.01.B the assigns of Owner and Engineer) are hereby bound to the other party to this Agreement and to the partners, successors, executors, administrators, and legal representatives (and said assigns) of such other party, in respect of all covenants, agreements, and obligations of this Agreement.

B. Neither Owner nor Engineer may assign, sublet, or transfer any rights under or interest (including, but without limitation, moneys that are due or may become due) in this Agreement without the written consent of the other, except to the extent that any assignment, subletting, or transfer is mandated or restricted by law. Unless specifically stated to the contrary in any written consent to an assignment, no assignment will release or discharge the assignor from any duty or responsibility under this Agreement.

7.01 General Considerations

A. The standard of care for all professional engineering and related services performed or furnished by Engineer under this Agreement will be the care and skill ordinarily used by members of the subject profession practicing under similar circumstances at the same time and in the same locality. Engineer makes no warranties, express or implied, under this Agreement or otherwise, in connection with Engineer's services. Engineer and its consultants may use or rely upon the design services of others, including, but not limited to, contractors, manufacturers, and suppliers.

B. Engineer shall not at any time supervise, direct, or have control over any contractor's work, nor shall Engineer have authority over or responsibility for the means, methods, techniques, sequences, or procedures of construction selected or used by any contractor, for safety precautions and programs incident to a contractor's work progress, nor for any failure of any contractor to comply with laws and regulations applicable to contractor's work.

C. Engineer neither guarantees the performance of any contractor nor assumes responsibility for any contractor's failure to furnish and perform its work in accordance with the contract between Owner and such contractor.

D. Engineer shall not be responsible for the acts or omissions of any contractor, subcontractor, or supplier, or of any contractor's agents or employees or any other persons (except Engineer's own employees) at the Project site or otherwise furnishing or performing any of construction work; or for any decision made on interpretations or clarifications of the construction contract given by Owner without consultation and advice of Engineer.

E. The general conditions for any construction contract documents prepared hereunder are to be the "Standard General Conditions of the Construction Contract" as prepared by the Engineers Joint Contract Documents Committee (No. C-700, 2002 Edition).

F. All design documents prepared or furnished by Engineer are instruments of service, and Engineer retains an ownership and property interest (including the copyright and the right of reuse) in such documents, whether or not the Project is completed.

G. To the fullest extent permitted by law, Owner and Engineer (1) waive against each other, and the other's employees, officers, directors, agents, insurers, partners, and consultants, any and all claims for or entitlement to special, incidental, indirect, or consequential damages arising out of, resulting from, or in any way related to the Project, and (2) agree that Engineer's total liability to Owner under this Agreement shall be limited to $50,000 or the total amount of compensation received by Engineer, whichever is greater.

H. The parties acknowledge that Engineer's scope of services does not include any services related to a Hazardous Environmental Condition (the presence of asbestos, PCBs, petroleum, hazardous substances or waste, and radioactive materials). If Engineer or any other party encounters a Hazardous Environmental Condition, Engineer may, at its option and without liability for consequential or any other damages, suspend performance of services on the portion of the Project affected thereby until Owner: (i) retains appropriate specialist consultants or contractors to identify and, as appropriate, abate, remediate, or remove the Hazardous Environmental Condition; and (ii) warrants that the Site is in full compliance with applicable Laws and Regulations.

8.01 Total Agreement

A. This Agreement (consisting of pages 1 to 4 inclusive together with any expressly incorporated appendix), constitutes the entire agreement between Owner and Engineer and supersedes all prior written or oral understandings. This Agreement may only be amended, supplemented, modified, or canceled by a duly executed written instrument.

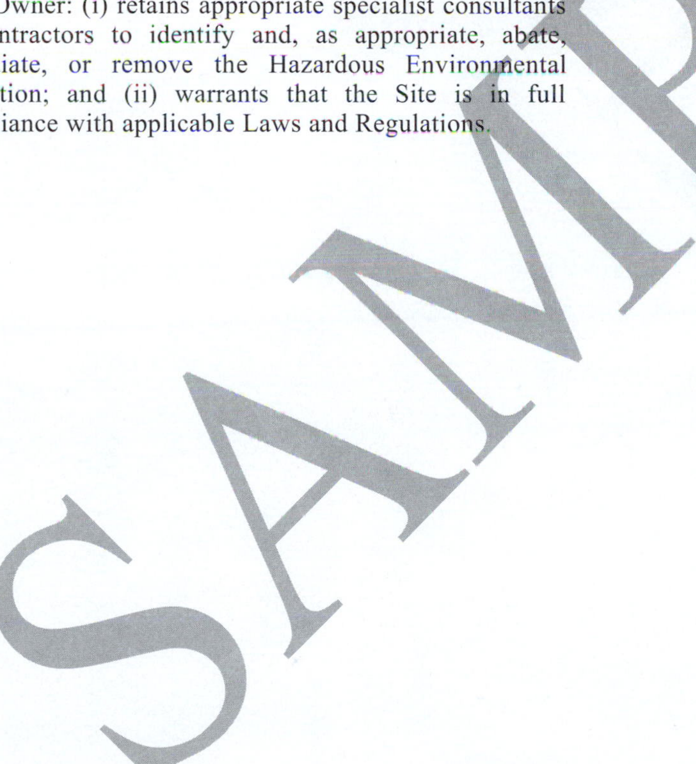

INSTRUCTIONS TO USERS FOR COMPLETION OF AGREEMENT:

1. Select and retain as Page 4 of 4 one of the four method of payment pages that follow.
2. Remove and discard this page and the three unused method of payment pages.

9.01 Payment (Lump Sum Basis)

A. Using the procedures set forth in paragraph 2.01, Owner shall pay Engineer as follows:

1. A Lump Sum amount of $ ________________

B. The Engineer's compensation is conditioned on the time to complete construction not exceeding ________ months. Should the time to complete construction be extended beyond this period, total compensation to Engineer shall be appropriately adjusted.

IN WITNESS WHEREOF, the parties hereto have executed this Agreement, the Effective Date of which is indicated on page 1.

OWNER:

By: ________________________________

Title: ________________________________

Date Signed: ________________________________

Address for giving notices:

____

____

____

ENGINEER:

By: ________________________________

Title: ________________________________

Date Signed: ________________________________

License or Certificate No. and State ________________

____

Address for giving notices:

____

____

____

9.01 Payment (Hourly Rates Plus Reimbursable Expenses)

A. Using the procedures set forth in paragraph 2.01, Owner shall pay Engineer as follows:

1. An amount equal to the cumulative hours charged to the Project by each class of Engineer's employees times standard hourly rates for each applicable billing class for all services performed on the Project, plus reimbursable expenses and Engineer's consultants' charges, if any.

2. Engineer's Standard Hourly Rates are attached as Appendix 1.

3. The total compensation for services and reimbursable expenses is estimated to be $ ____________________

B. The Engineer's compensation is conditioned on the time to complete construction not exceeding ________ months. Should the time to complete construction be extended beyond this period, total compensation to Engineer shall be appropriately adjusted.

IN WITNESS WHEREOF, the parties hereto have executed this Agreement, the Effective Date of which is indicated on page 1.

OWNER:

By: ______________________________

Title: ______________________________

Date Signed: ______________________________

Address for giving notices:

____

____

____

ENGINEER:

By: ______________________________

Title: ______________________________

Date Signed: ______________________________

License or Certificate No. and State ____________________

____

Address for giving notices:

____

____

____

9.01 Payment (Percentage of Construction Cost)

A. Using the procedures set forth in paragraph 2.01, Owner shall pay Engineer as follows:

1. An amount equal to ____ Percent of the cost to construct the work designed or specified by the Engineer ("Construction Cost"). This amount includes compensation for Engineer's Services and services of Engineer's consultants, if any. The percentage of Construction Cost noted herein accounts for labor, overhead, profit, and reimbursable expenses.

2. As a basis for payment to Engineer, Construction Cost will be based on one or more of the following determinations with precedence in the order listed:

a. For work designed or specified by Engineer and incorporated in the completed Project, the actual final cost of the work performed by Contractor.

b. For work designed or specified by Engineer but not constructed, the lowest bona fide bid received from a qualified bidder for such work; or, if the work is not bid, the lowest bona fide negotiated proposal or contractor's estimate for such work.

c. For work designed or specified but not constructed, upon which no bid, proposal, or estimate is received, Engineer's most recent opinion of probable Construction Cost.

B. The Engineer's compensation is conditioned on the time to complete construction not exceeding ______ months. Should the time to complete construction be extended beyond this period, total compensation to Engineer shall be appropriately adjusted.

IN WITNESS WHEREOF, the parties hereto have executed this Agreement, the Effective Date of which is indicated on page 1.

OWNER:	ENGINEER:
By: ______________________	By: ______________________
Title: ______________________	Title: ______________________
Date Signed: ______________________	Date Signed: ______________________
	License or Certificate No. and State ______________________
Address for giving notices:	Address for giving notices:
______________________	______________________
______________________	______________________
______________________	______________________

9.01 Payment (Direct Labor Costs Times Factor; Plus Reimbursables)

A. Using the procedures set forth in paragraph 2.01, Owner shall pay Engineer as follows:

1. An amount equal to Engineer's Direct Labor Costs times a Factor of ____________ for services of Engineer's Employees engaged on the Project, plus reimbursable expenses, and Engineer's consultants' charges, if any.

2. The total compensation for services and reimbursable expenses is estimated to be $ ____________________

B. The Engineer's compensation is conditioned on the time to complete construction not exceeding _____ months. Should the time to complete construction be extended beyond this period, total compensation to Engineer shall be appropriately adjusted.

IN WITNESS WHEREOF, the parties hereto have executed this Agreement, the Effective Date of which is indicated on page 1.

OWNER:	ENGINEER:
By: ______________________________	By: ______________________________
Title: ______________________________	Title: ______________________________
Date Signed: ______________________________	Date Signed: ______________________________
	License or Certificate No. and State ____________________
	______________________________
Address for giving notices:	Address for giving notices:
______________________________	______________________________
______________________________	______________________________
______________________________	______________________________

Engineers Joint Documents Committee
Design and Construction Related Documents
Instructions and License Agreement

Instructions

Before you use any EJCDC document:

1. Read the License Agreement. You agree to it and are bound by its terms when you use the EJCDC document.

2. Make sure that you have the correct version for your word processing software.

How to Use:

1. While EJCDC has expended considerable effort to make the software translations exact, it can be that a few document controls (e.g., bold, underline) did not carry over.

2. Similarly, your software may change the font specification if the font is not available in your system. It will choose a font that is close in appearance. In this event, the pagination may not match the control set.

3. If you modify the document, you must follow the instructions in the License Agreement about notification.

4. Also note the instruction in the License Agreement about the EJCDC copyright.

License Agreement

You should carefully read the following terms and conditions before using this document. Commencement of use of this document indicates your acceptance of these terms and conditions. If you do not agree to them, you should promptly return the materials to the vendor, and your money will be refunded.

The Engineers Joint Contract Documents Committee ("EJCDC") provides **EJCDC Design and Construction Related Documents** and licenses their use worldwide. You assume sole responsibility for the selection of specific documents or portions thereof to achieve your intended results, and for the installation, use, and results obtained from **EJCDC Design and Construction Related Documents**.

You acknowledge that you understand that the text of the contract documents of **EJCDC Design and Construction Related Documents** has important legal consequences and that consultation with an attorney is recommended with respect to use or modification of the text. You further acknowledge that EJCDC documents are protected by the copyright laws of the United States.

License:
You have a limited nonexclusive license to:

1. Use **EJCDC Design and Construction Related Documents** on any number of machines owned, leased or rented by your company or organization.

2. Use **EJCDC Design and Construction Related Documents** in printed form for bona fide contract documents.

3. Copy **EJCDC Design and Construction Related Documents** into any machine readable or printed form for backup or modification purposes in support of your use of **EJCDC Design and Construction Related Documents**.

You agree that you will:

1. Reproduce and include EJCDC's copyright notice on any printed or machine-readable copy, modification, or portion merged into another document or program. All proprietary rights in **EJCDC Design and Construction Related Documents** are and shall remain the property of EJCDC.

2. Not represent that any of the contract documents you generate from **EJCDC Design and Construction Related Documents** are EJCDC documents unless (i) the document text is used without alteration or (ii) all additions and changes to, and deletions from, the text are clearly shown.

You may not use, copy, modify, or transfer EJCDC Design and Construction Related Documents, or any copy, modification or merged portion, in whole or in part, except as expressly provided for in this license. Reproduction of EJCDC Design and Construction Related Documents in printed or machine-readable format for resale or educational purposes is expressly prohibited.

If you transfer possession of any copy, modification or merged portion of EJCDC Design and Construction Related Documents to another party, your license is automatically terminated.

Term:
The license is effective until terminated. You may terminate it at any time by destroying **EJCDC Design and Construction Related Documents** altogether with all copies, modifications and merged portions in any form. It will also terminate upon conditions set forth elsewhere in this Agreement or if you fail to comply with any term or condition of this Agreement. You agree upon such termination to destroy **EJCDC Design and Construction**

Related Documents along with all copies, modifications and merged portions in any form.

Limited Warranty:

EJCDC warrants the CDs and diskettes on which **EJCDC Design and Construction Related Documents** is furnished to be free from defects in materials and workmanship under normal use for a period of ninety (90) days from the date of delivery to you as evidenced by a copy of your receipt.

There is no other warranty of any kind, either expressed or implied, including, but not limited to the implied warranties of merchantability and fitness for a particular purpose. Some states do not allow the exclusion of implied warranties, so the above exclusion may not apply to you. This warranty gives you specific legal rights and you may also have other rights which vary from state to state.

EJCDC does not warrant that the functions contained in **EJCDC Design and Construction Related Documents** will meet your requirements or that the operation of **EJCDC Design and Construction Related Documents** will be uninterrupted or error free.

Limitations of Remedies:

EJCDC's entire liability and your exclusive remedy shall be:

1. the replacement of any document not meeting EJCDC's "Limited Warranty" which is returned to EJCDC's selling agent with a copy of your receipt, or

2. if EJCDC's selling agent is unable to deliver a replacement CD or diskette which is free of defects in materials and workmanship, you may terminate this Agreement by returning EJCDC Document and your money will be refunded.

In no event will EJCDC be liable to you for any damages, including any lost profits, lost savings or other incidental or consequential damages arising out of the use or inability to use **EJCDC Design and Construction Related Documents** even if EJCDC has been advised of the possibility of such damages, or for any claim by any other party.

Some states do not allow the limitation or exclusion of liability for incidental or consequential damages, so the above limitation or exclusion may not apply to you.

General:

You may not sublicense, assign, or transfer this license except as expressly provided in this Agreement. Any attempt otherwise to sublicense, assign, or transfer any of the rights, duties, or obligations hereunder is void.

This Agreement shall be governed by the laws of the State of Virginia. Should you have any questions concerning this Agreement, you may contact EJCDC by writing to:

Arthur Schwartz, Esq.
General Counsel
National Society of Professional Engineers
1420 King Street
Alexandria, VA 22314

Phone: (703) 684-2845
Fax: (703) 836-4875
e-mail: aschwartz@nspe.org

You acknowledge that you have read this agreement, understand it and agree to be bound by its terms and conditions. You further agree that it is the complete and exclusive statement of the agreement between us which supersedes any proposal or prior agreement, oral or written, and any other communications between us relating to the subject matter of this agreement.

This document has important legal consequences; consultation with an attorney is encouraged with respect to its use or modification. This document should be adapted to the particular circumstances of the contemplated Project and the controlling Laws and Regulations.

STANDARD GENERAL CONDITIONS OF THE CONSTRUCTION CONTRACT

Prepared by

ENGINEERS JOINT CONTRACT DOCUMENTS COMMITTEE

and

Issued and Published Jointly by

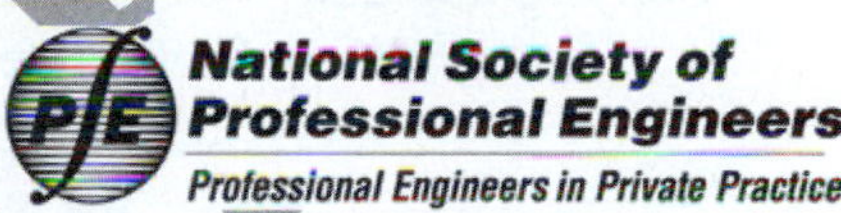

AMERICAN COUNCIL OF ENGINEERING COMPANIES

ASSOCIATED GENERAL CONTRACTORS OF AMERICA

AMERICAN SOCIETY OF CIVIL ENGINEERS

PROFESSIONAL ENGINEERS IN PRIVATE PRACTICE
A Practice Division of the
NATIONAL SOCIETY OF PROFESSIONAL ENGINEERS

Endorsed by

CONSTRUCTION SPECIFICATIONS INSTITUTE

These General Conditions have been prepared for use with the Suggested Forms of Agreement Between Owner and Contractor (EJCDC C-520 or C-525, 2007 Editions). Their provisions are interrelated and a change in one may necessitate a change in the other. Comments concerning their usage are contained in the Narrative Guide to the EJCDC Construction Documents (EJCDC C-001, 2007 Edition). For guidance in the preparation of Supplementary Conditions, see Guide to the Preparation of Supplementary Conditions (EJCDC C-800, 2007 Edition).

1420 King Street, Alexandria, VA 22314-2794
(703) 684-2882
www.nspe.org

American Council of Engineering Companies
1015 15th Street N.W., Washington, DC 20005
(202) 347-7474
www.acec.org

American Society of Civil Engineers
1801 Alexander Bell Drive, Reston, VA 20191-4400
(800) 548-2723
www.asce.org

Associated General Contractors of America
2300 Wilson Boulevard, Suite 400, Arlington, VA 22201-3308
(703) 548-3118
www.agc.org

STANDARD GENERAL CONDITIONS OF THE CONSTRUCTION CONTRACT

TABLE OF CONTENTS

Page

Article 1 – Definitions and Terminology 1
1.01 Defined Terms 1
1.02 Terminology 5

Article 2 – Preliminary Matters 6
2.01 Delivery of Bonds and Evidence of Insurance 6
2.02 Copies of Documents 6
2.03 Commencement of Contract Times; Notice to Proceed 6
2.04 Starting the Work 7
2.05 Before Starting Construction 7
2.06 Preconstruction Conference; Designation of Authorized Representatives 7
2.07 Initial Acceptance of Schedules 7

Article 3 – Contract Documents: Intent, Amending, Reuse 8
3.01 Intent 8
3.02 Reference Standards 8
3.03 Reporting and Resolving Discrepancies 8
3.04 Amending and Supplementing Contract Documents 9
3.05 Reuse of Documents 10
3.06 Electronic Data 10

Article 4 – Availability of Lands; Subsurface and Physical Conditions; Hazardous Environmental Conditions; Reference Points 10
4.01 Availability of Lands 10
4.02 Subsurface and Physical Conditions 11
4.03 Differing Subsurface or Physical Conditions 11
4.04 Underground Facilities 13
4.05 Reference Points 14
4.06 Hazardous Environmental Condition at Site 14

Article 5 – Bonds and Insurance 16
5.01 Performance, Payment, and Other Bonds 16
5.02 Licensed Sureties and Insurers 16
5.03 Certificates of Insurance 16
5.04 Contractor's Insurance 17
5.05 Owner's Liability Insurance 18
5.06 Property Insurance 18
5.07 Waiver of Rights 20
5.08 Receipt and Application of Insurance Proceeds 21

5.09 Acceptance of Bonds and Insurance; Option to Replace 21
5.10 Partial Utilization, Acknowledgment of Property Insurer 21

Article 6 – Contractor's Responsibilities 22
6.01 Supervision and Superintendence 22
6.02 Labor; Working Hours 22
6.03 Services, Materials, and Equipment 22
6.04 Progress Schedule 23
6.05 Substitutes and "Or-Equals" 23
6.06 Concerning Subcontractors, Suppliers, and Others 25
6.07 Patent Fees and Royalties 27
6.08 Permits 27
6.09 Laws and Regulations 27
6.10 Taxes 28
6.11 Use of Site and Other Areas 28
6.12 Record Documents 29
6.13 Safety and Protection 29
6.14 Safety Representative 30
6.15 Hazard Communication Programs 30
6.16 Emergencies 30
6.17 Shop Drawings and Samples 30
6.18 Continuing the Work 32
6.19 Contractor's General Warranty and Guarantee 32
6.20 Indemnification 33
6.21 Delegation of Professional Design Services 34

Article 7 – Other Work at the Site 35
7.01 Related Work at Site 35
7.02 Coordination 35
7.03 Legal Relationships 36

Article 8 – Owner's Responsibilities 36
8.01 Communications to Contractor 36
8.02 Replacement of Engineer 36
8.03 Furnish Data 36
8.04 Pay When Due 36
8.05 Lands and Easements; Reports and Tests 36
8.06 Insurance 36
8.07 Change Orders 36
8.08 Inspections, Tests, and Approvals 37
8.09 Limitations on Owner's Responsibilities 37
8.10 Undisclosed Hazardous Environmental Condition 37
8.11 Evidence of Financial Arrangements 37
8.12 Compliance with Safety Program 37

Article 9 – Engineer's Status During Construction 37
9.01 Owner's Representative 37
9.02 Visits to Site 37

9.03 Project Representative 38
9.04 Authorized Variations in Work 38
9.05 Rejecting Defective Work 38
9.06 Shop Drawings, Change Orders and Payments 38
9.07 Determinations for Unit Price Work 39
9.08 Decisions on Requirements of Contract Documents and Acceptability of Work 39
9.09 Limitations on Engineer's Authority and Responsibilities 39
9.10 Compliance with Safety Program 40

Article 10 – Changes in the Work; Claims 40
10.01 Authorized Changes in the Work 40
10.02 Unauthorized Changes in the Work 40
10.03 Execution of Change Orders 41
10.04 Notification to Surety 41
10.05 Claims 41

Article 11 – Cost of the Work; Allowances; Unit Price Work 42
11.01 Cost of the Work 42
11.02 Allowances 45
11.03 Unit Price Work 45

Article 12 – Change of Contract Price; Change of Contract Times 46
12.01 Change of Contract Price 46
12.02 Change of Contract Times 47
12.03 Delays 47

Article 13 – Tests and Inspections; Correction, Removal or Acceptance of Defective Work 48
13.01 Notice of Defects 48
13.02 Access to Work 48
13.03 Tests and Inspections 48
13.04 Uncovering Work 49
13.05 Owner May Stop the Work 50
13.06 Correction or Removal of Defective Work 50
13.07 Correction Period 50
13.08 Acceptance of Defective Work 51
13.09 Owner May Correct Defective Work 51

Article 14 – Payments to Contractor and Completion 52
14.01 Schedule of Values 52
14.02 Progress Payments 52
14.03 Contractor's Warranty of Title 55
14.04 Substantial Completion 55
14.05 Partial Utilization 56
14.06 Final Inspection 56
14.07 Final Payment 57
14.08 Final Completion Delayed 58
14.09 Waiver of Claims 58

Article 15 – Suspension of Work and Termination 58
15.01 Owner May Suspend Work 58
15.02 Owner May Terminate for Cause 58
15.03 Owner May Terminate For Convenience 60
15.04 Contractor May Stop Work or Terminate 60

Article 16 – Dispute Resolution 61
16.01 Methods and Procedures 61

Article 17 – Miscellaneous 61
17.01 Giving Notice 61
17.02 Computation of Times 61
17.03 Cumulative Remedies 62
17.04 Survival of Obligations 62
17.05 Controlling Law 62
17.06 Headings 62

ARTICLE 1 – DEFINITIONS AND TERMINOLOGY

1.01 *Defined Terms*

A. Wherever used in the Bidding Requirements or Contract Documents and printed with initial capital letters, the terms listed below will have the meanings indicated which are applicable to both the singular and plural thereof. In addition to terms specifically defined, terms with initial capital letters in the Contract Documents include references to identified articles and paragraphs, and the titles of other documents or forms.

1. *Addenda*—Written or graphic instruments issued prior to the opening of Bids which clarify, correct, or change the Bidding Requirements or the proposed Contract Documents.

2. *Agreement*—The written instrument which is evidence of the agreement between Owner and Contractor covering the Work.

3. *Application for Payment*—The form acceptable to Engineer which is to be used by Contractor during the course of the Work in requesting progress or final payments and which is to be accompanied by such supporting documentation as is required by the Contract Documents.

4. *Asbestos*—Any material that contains more than one percent asbestos and is friable or is releasing asbestos fibers into the air above current action levels established by the United States Occupational Safety and Health Administration.

5. *Bid*—The offer or proposal of a Bidder submitted on the prescribed form setting forth the prices for the Work to be performed.

6. *Bidder*—The individual or entity who submits a Bid directly to Owner.

7. *Bidding Documents*—The Bidding Requirements and the proposed Contract Documents (including all Addenda).

8. *Bidding Requirements*—The advertisement or invitation to bid, Instructions to Bidders, Bid security of acceptable form, if any, and the Bid Form with any supplements.

9. *Change Order*—A document recommended by Engineer which is signed by Contractor and Owner and authorizes an addition, deletion, or revision in the Work or an adjustment in the Contract Price or the Contract Times, issued on or after the Effective Date of the Agreement.

10. *Claim*—A demand or assertion by Owner or Contractor seeking an adjustment of Contract Price or Contract Times, or both, or other relief with respect to the terms of the Contract. A demand for money or services by a third party is not a Claim.

11. *Contract*—The entire and integrated written agreement between the Owner and Contractor concerning the Work. The Contract supersedes prior negotiations, representations, or agreements, whether written or oral.

12. *Contract Documents*—Those items so designated in the Agreement. Only printed or hard copies of the items listed in the Agreement are Contract Documents. Approved Shop Drawings, other Contractor submittals, and the reports and drawings of subsurface and physical conditions are not Contract Documents.

13. *Contract Price*—The moneys payable by Owner to Contractor for completion of the Work in accordance with the Contract Documents as stated in the Agreement (subject to the provisions of Paragraph 11.03 in the case of Unit Price Work).

14. *Contract Times*—The number of days or the dates stated in the Agreement to: (i) achieve Milestones, if any; (ii) achieve Substantial Completion; and (iii) complete the Work so that it is ready for final payment as evidenced by Engineer's written recommendation of final payment.

15. *Contractor*—The individual or entity with whom Owner has entered into the Agreement.

16. *Cost of the Work*—See Paragraph 11.01 for definition.

17. *Drawings*—That part of the Contract Documents prepared or approved by Engineer which graphically shows the scope, extent, and character of the Work to be performed by Contractor. Shop Drawings and other Contractor submittals are not Drawings as so defined.

18. *Effective Date of the Agreement*—The date indicated in the Agreement on which it becomes effective, but if no such date is indicated, it means the date on which the Agreement is signed and delivered by the last of the two parties to sign and deliver.

19. *Engineer*—The individual or entity named as such in the Agreement.

20. *Field Order*—A written order issued by Engineer which requires minor changes in the Work but which does not involve a change in the Contract Price or the Contract Times.

21. *General Requirements*—Sections of Division 1 of the Specifications.

22. *Hazardous Environmental Condition*—The presence at the Site of Asbestos, PCBs, Petroleum, Hazardous Waste, or Radioactive Material in such quantities or circumstances that may present a substantial danger to persons or property exposed thereto.

23. *Hazardous Waste*—The term Hazardous Waste shall have the meaning provided in Section 1004 of the Solid Waste Disposal Act (42 USC Section 6903) as amended from time to time.

24. *Laws and Regulations; Laws or Regulations*—Any and all applicable laws, rules, regulations, ordinances, codes, and orders of any and all governmental bodies, agencies, authorities, and courts having jurisdiction.

25. *Liens*—Charges, security interests, or encumbrances upon Project funds, real property, or personal property.

26. *Milestone*—A principal event specified in the Contract Documents relating to an intermediate completion date or time prior to Substantial Completion of all the Work.

27. *Notice of Award*—The written notice by Owner to the Successful Bidder stating that upon timely compliance by the Successful Bidder with the conditions precedent listed therein, Owner will sign and deliver the Agreement.

28. *Notice to Proceed*—A written notice given by Owner to Contractor fixing the date on which the Contract Times will commence to run and on which Contractor shall start to perform the Work under the Contract Documents.

29. *Owner*—The individual or entity with whom Contractor has entered into the Agreement and for whom the Work is to be performed.

30. *PCBs*—Polychlorinated biphenyls.

31. *Petroleum*—Petroleum, including crude oil or any fraction thereof which is liquid at standard conditions of temperature and pressure (60 degrees Fahrenheit and 14.7 pounds per square inch absolute), such as oil, petroleum, fuel oil, oil sludge, oil refuse, gasoline, kerosene, and oil mixed with other non-Hazardous Waste and crude oils.

32. *Progress Schedule*—A schedule, prepared and maintained by Contractor, describing the sequence and duration of the activities comprising the Contractor's plan to accomplish the Work within the Contract Times.

33. *Project*—The total construction of which the Work to be performed under the Contract Documents may be the whole, or a part.

34. *Project Manual*—The bound documentary information prepared for bidding and constructing the Work. A listing of the contents of the Project Manual, which may be bound in one or more volumes, is contained in the table(s) of contents.

35. *Radioactive Material*—Source, special nuclear, or byproduct material as defined by the Atomic Energy Act of 1954 (42 USC Section 2011 et seq.) as amended from time to time.

36. *Resident Project Representative*—The authorized representative of Engineer who may be assigned to the Site or any part thereof.

37. *Samples*—Physical examples of materials, equipment, or workmanship that are representative of some portion of the Work and which establish the standards by which such portion of the Work will be judged.

38. *Schedule of Submittals*—A schedule, prepared and maintained by Contractor, of required submittals and the time requirements to support scheduled performance of related construction activities.

39. *Schedule of Values*—A schedule, prepared and maintained by Contractor, allocating portions of the Contract Price to various portions of the Work and used as the basis for reviewing Contractor's Applications for Payment.

40. *Shop Drawings*—All drawings, diagrams, illustrations, schedules, and other data or information which are specifically prepared or assembled by or for Contractor and submitted by Contractor to illustrate some portion of the Work.

41. *Site*—Lands or areas indicated in the Contract Documents as being furnished by Owner upon which the Work is to be performed, including rights-of-way and easements for access thereto, and such other lands furnished by Owner which are designated for the use of Contractor.

42. *Specifications*—That part of the Contract Documents consisting of written requirements for materials, equipment, systems, standards and workmanship as applied to the Work, and certain administrative requirements and procedural matters applicable thereto.

43. *Subcontractor*—An individual or entity having a direct contract with Contractor or with any other Subcontractor for the performance of a part of the Work at the Site.

44. *Substantial Completion*—The time at which the Work (or a specified part thereof) has progressed to the point where, in the opinion of Engineer, the Work (or a specified part thereof) is sufficiently complete, in accordance with the Contract Documents, so that the Work (or a specified part thereof) can be utilized for the purposes for which it is intended. The terms "substantially complete" and "substantially completed" as applied to all or part of the Work refer to Substantial Completion thereof.

45. *Successful Bidder*—The Bidder submitting a responsive Bid to whom Owner makes an award.

46. *Supplementary Conditions*—That part of the Contract Documents which amends or supplements these General Conditions.

47. *Supplier*—A manufacturer, fabricator, supplier, distributor, materialman, or vendor having a direct contract with Contractor or with any Subcontractor to furnish materials or equipment to be incorporated in the Work by Contractor or Subcontractor.

48. *Underground Facilities*—All underground pipelines, conduits, ducts, cables, wires, manholes, vaults, tanks, tunnels, or other such facilities or attachments, and any encasements containing such facilities, including those that convey electricity, gases, steam, liquid petroleum products, telephone or other communications, cable television, water, wastewater, storm water, other liquids or chemicals, or traffic or other control systems.

49. *Unit Price Work*—Work to be paid for on the basis of unit prices.

50. *Work*—The entire construction or the various separately identifiable parts thereof required to be provided under the Contract Documents. Work includes and is the result of performing or providing all labor, services, and documentation necessary to produce such construction, and furnishing, installing, and incorporating all materials and equipment into such construction, all as required by the Contract Documents.

51. *Work Change Directive*—A written statement to Contractor issued on or after the Effective Date of the Agreement and signed by Owner and recommended by Engineer ordering an

addition, deletion, or revision in the Work, or responding to differing or unforeseen subsurface or physical conditions under which the Work is to be performed or to emergencies. A Work Change Directive will not change the Contract Price or the Contract Times but is evidence that the parties expect that the change ordered or documented by a Work Change Directive will be incorporated in a subsequently issued Change Order following negotiations by the parties as to its effect, if any, on the Contract Price or Contract Times.

1.02 *Terminology*

A. The words and terms discussed in Paragraph 1.02.B through F are not defined but, when used in the Bidding Requirements or Contract Documents, have the indicated meaning.

B. *Intent of Certain Terms or Adjectives:*

1. The Contract Documents include the terms "as allowed," "as approved," "as ordered," "as directed" or terms of like effect or import to authorize an exercise of professional judgment by Engineer. In addition, the adjectives "reasonable," "suitable," "acceptable," "proper," "satisfactory," or adjectives of like effect or import are used to describe an action or determination of Engineer as to the Work. It is intended that such exercise of professional judgment, action, or determination will be solely to evaluate, in general, the Work for compliance with the information in the Contract Documents and with the design concept of the Project as a functioning whole as shown or indicated in the Contract Documents (unless there is a specific statement indicating otherwise). The use of any such term or adjective is not intended to and shall not be effective to assign to Engineer any duty or authority to supervise or direct the performance of the Work, or any duty or authority to undertake responsibility contrary to the provisions of Paragraph 9.09 or any other provision of the Contract Documents.

C. *Day:*

1. The word "day" means a calendar day of 24 hours measured from midnight to the next midnight.

D. *Defective:*

1. The word "defective," when modifying the word "Work," refers to Work that is unsatisfactory, faulty, or deficient in that it:

a. does not conform to the Contract Documents; or

b. does not meet the requirements of any applicable inspection, reference standard, test, or approval referred to in the Contract Documents; or

c. has been damaged prior to Engineer's recommendation of final payment (unless responsibility for the protection thereof has been assumed by Owner at Substantial Completion in accordance with Paragraph 14.04 or 14.05).

E. *Furnish, Install, Perform, Provide:*

1. The word "furnish," when used in connection with services, materials, or equipment, shall mean to supply and deliver said services, materials, or equipment to the Site (or some other specified location) ready for use or installation and in usable or operable condition.

2. The word "install," when used in connection with services, materials, or equipment, shall mean to put into use or place in final position said services, materials, or equipment complete and ready for intended use.

3. The words "perform" or "provide," when used in connection with services, materials, or equipment, shall mean to furnish and install said services, materials, or equipment complete and ready for intended use.

4. When "furnish," "install," "perform," or "provide" is not used in connection with services, materials, or equipment in a context clearly requiring an obligation of Contractor, "provide" is implied.

F. Unless stated otherwise in the Contract Documents, words or phrases that have a well-known technical or construction industry or trade meaning are used in the Contract Documents in accordance with such recognized meaning.

ARTICLE 2 – PRELIMINARY MATTERS

2.01 *Delivery of Bonds and Evidence of Insurance*

A. When Contractor delivers the executed counterparts of the Agreement to Owner, Contractor shall also deliver to Owner such bonds as Contractor may be required to furnish.

B. *Evidence of Insurance:* Before any Work at the Site is started, Contractor and Owner shall each deliver to the other, with copies to each additional insured identified in the Supplementary Conditions, certificates of insurance (and other evidence of insurance which either of them or any additional insured may reasonably request) which Contractor and Owner respectively are required to purchase and maintain in accordance with Article 5.

2.02 *Copies of Documents*

A. Owner shall furnish to Contractor up to ten printed or hard copies of the Drawings and Project Manual. Additional copies will be furnished upon request at the cost of reproduction.

2.03 *Commencement of Contract Times; Notice to Proceed*

A. The Contract Times will commence to run on the thirtieth day after the Effective Date of the Agreement or, if a Notice to Proceed is given, on the day indicated in the Notice to Proceed. A Notice to Proceed may be given at any time within 30 days after the Effective Date of the Agreement. In no event will the Contract Times commence to run later than the sixtieth day after the day of Bid opening or the thirtieth day after the Effective Date of the Agreement, whichever date is earlier.

2.04 *Starting the Work*

A. Contractor shall start to perform the Work on the date when the Contract Times commence to run. No Work shall be done at the Site prior to the date on which the Contract Times commence to run.

2.05 *Before Starting Construction*

A. *Preliminary Schedules:* Within 10 days after the Effective Date of the Agreement (unless otherwise specified in the General Requirements), Contractor shall submit to Engineer for timely review:

1. a preliminary Progress Schedule indicating the times (numbers of days or dates) for starting and completing the various stages of the Work, including any Milestones specified in the Contract Documents;

2. a preliminary Schedule of Submittals; and

3. a preliminary Schedule of Values for all of the Work which includes quantities and prices of items which when added together equal the Contract Price and subdivides the Work into component parts in sufficient detail to serve as the basis for progress payments during performance of the Work. Such prices will include an appropriate amount of overhead and profit applicable to each item of Work.

2.06 *Preconstruction Conference; Designation of Authorized Representatives*

A. Before any Work at the Site is started, a conference attended by Owner, Contractor, Engineer, and others as appropriate will be held to establish a working understanding among the parties as to the Work and to discuss the schedules referred to in Paragraph 2.05.A, procedures for handling Shop Drawings and other submittals, processing Applications for Payment, and maintaining required records.

B. At this conference Owner and Contractor each shall designate, in writing, a specific individual to act as its authorized representative with respect to the services and responsibilities under the Contract. Such individuals shall have the authority to transmit instructions, receive information, render decisions relative to the Contract, and otherwise act on behalf of each respective party.

2.07 *Initial Acceptance of Schedules*

A. At least 10 days before submission of the first Application for Payment a conference attended by Contractor, Engineer, and others as appropriate will be held to review for acceptability to Engineer as provided below the schedules submitted in accordance with Paragraph 2.05.A. Contractor shall have an additional 10 days to make corrections and adjustments and to complete and resubmit the schedules. No progress payment shall be made to Contractor until acceptable schedules are submitted to Engineer.

1. The Progress Schedule will be acceptable to Engineer if it provides an orderly progression of the Work to completion within the Contract Times. Such acceptance will not impose on Engineer responsibility for the Progress Schedule, for sequencing, scheduling, or progress of

the Work, nor interfere with or relieve Contractor from Contractor's full responsibility therefor.

2. Contractor's Schedule of Submittals will be acceptable to Engineer if it provides a workable arrangement for reviewing and processing the required submittals.

3. Contractor's Schedule of Values will be acceptable to Engineer as to form and substance if it provides a reasonable allocation of the Contract Price to component parts of the Work.

ARTICLE 3 – CONTRACT DOCUMENTS: INTENT, AMENDING, REUSE

3.01 *Intent*

A. The Contract Documents are complementary; what is required by one is as binding as if required by all.

B. It is the intent of the Contract Documents to describe a functionally complete project (or part thereof) to be constructed in accordance with the Contract Documents. Any labor, documentation, services, materials, or equipment that reasonably may be inferred from the Contract Documents or from prevailing custom or trade usage as being required to produce the indicated result will be provided whether or not specifically called for, at no additional cost to Owner.

C. Clarifications and interpretations of the Contract Documents shall be issued by Engineer as provided in Article 9.

3.02 *Reference Standards*

A. Standards, Specifications, Codes, Laws, and Regulations

1. Reference to standards, specifications, manuals, or codes of any technical society, organization, or association, or to Laws or Regulations, whether such reference be specific or by implication, shall mean the standard, specification, manual, code, or Laws or Regulations in effect at the time of opening of Bids (or on the Effective Date of the Agreement if there were no Bids), except as may be otherwise specifically stated in the Contract Documents.

2. No provision of any such standard, specification, manual, or code, or any instruction of a Supplier, shall be effective to change the duties or responsibilities of Owner, Contractor, or Engineer, or any of their subcontractors, consultants, agents, or employees, from those set forth in the Contract Documents. No such provision or instruction shall be effective to assign to Owner, Engineer, or any of their officers, directors, members, partners, employees, agents, consultants, or subcontractors, any duty or authority to supervise or direct the performance of the Work or any duty or authority to undertake responsibility inconsistent with the provisions of the Contract Documents.

3.03 *Reporting and Resolving Discrepancies*

A. *Reporting Discrepancies:*

1. *Contractor's Review of Contract Documents Before Starting Work*: Before undertaking each part of the Work, Contractor shall carefully study and compare the Contract Documents and check and verify pertinent figures therein and all applicable field measurements. Contractor shall promptly report in writing to Engineer any conflict, error, ambiguity, or discrepancy which Contractor discovers, or has actual knowledge of, and shall obtain a written interpretation or clarification from Engineer before proceeding with any Work affected thereby.

2. *Contractor's Review of Contract Documents During Performance of Work*: If, during the performance of the Work, Contractor discovers any conflict, error, ambiguity, or discrepancy within the Contract Documents, or between the Contract Documents and (a) any applicable Law or Regulation , (b) any standard, specification, manual, or code, or (c) any instruction of any Supplier, then Contractor shall promptly report it to Engineer in writing. Contractor shall not proceed with the Work affected thereby (except in an emergency as required by Paragraph 6.16.A) until an amendment or supplement to the Contract Documents has been issued by one of the methods indicated in Paragraph 3.04.

3. Contractor shall not be liable to Owner or Engineer for failure to report any conflict, error, ambiguity, or discrepancy in the Contract Documents unless Contractor had actual knowledge thereof.

B. *Resolving Discrepancies:*

1. Except as may be otherwise specifically stated in the Contract Documents, the provisions of the Contract Documents shall take precedence in resolving any conflict, error, ambiguity, or discrepancy between the provisions of the Contract Documents and:

a. the provisions of any standard, specification, manual, or code, or the instruction of any Supplier (whether or not specifically incorporated by reference in the Contract Documents); or

b. the provisions of any Laws or Regulations applicable to the performance of the Work (unless such an interpretation of the provisions of the Contract Documents would result in violation of such Law or Regulation).

3.04 *Amending and Supplementing Contract Documents*

A. The Contract Documents may be amended to provide for additions, deletions, and revisions in the Work or to modify the terms and conditions thereof by either a Change Order or a Work Change Directive.

B. The requirements of the Contract Documents may be supplemented, and minor variations and deviations in the Work may be authorized, by one or more of the following ways:

1. A Field Order;

2. Engineer's approval of a Shop Drawing or Sample (subject to the provisions of Paragraph 6.17.D.3); or

3. Engineer's written interpretation or clarification.

3.05 *Reuse of Documents*

A. Contractor and any Subcontractor or Supplier shall not:

1. have or acquire any title to or ownership rights in any of the Drawings, Specifications, or other documents (or copies of any thereof) prepared by or bearing the seal of Engineer or its consultants, including electronic media editions; or

2. reuse any such Drawings, Specifications, other documents, or copies thereof on extensions of the Project or any other project without written consent of Owner and Engineer and specific written verification or adaptation by Engineer.

B. The prohibitions of this Paragraph 3.05 will survive final payment, or termination of the Contract. Nothing herein shall preclude Contractor from retaining copies of the Contract Documents for record purposes.

3.06 *Electronic Data*

A. Unless otherwise stated in the Supplementary Conditions, the data furnished by Owner or Engineer to Contractor, or by Contractor to Owner or Engineer, that may be relied upon are limited to the printed copies (also known as hard copies). Files in electronic media format of text, data, graphics, or other types are furnished only for the convenience of the receiving party. Any conclusion or information obtained or derived from such electronic files will be at the user's sole risk. If there is a discrepancy between the electronic files and the hard copies, the hard copies govern.

B. Because data stored in electronic media format can deteriorate or be modified inadvertently or otherwise without authorization of the data's creator, the party receiving electronic files agrees that it will perform acceptance tests or procedures within 60 days, after which the receiving party shall be deemed to have accepted the data thus transferred. Any errors detected within the 60-day acceptance period will be corrected by the transferring party.

C. When transferring documents in electronic media format, the transferring party makes no representations as to long term compatibility, usability, or readability of documents resulting from the use of software application packages, operating systems, or computer hardware differing from those used by the data's creator.

ARTICLE 4 – AVAILABILITY OF LANDS; SUBSURFACE AND PHYSICAL CONDITIONS; HAZARDOUS ENVIRONMENTAL CONDITIONS; REFERENCE POINTS

4.01 *Availability of Lands*

A. Owner shall furnish the Site. Owner shall notify Contractor of any encumbrances or restrictions not of general application but specifically related to use of the Site with which Contractor must comply in performing the Work. Owner will obtain in a timely manner and pay for easements for permanent structures or permanent changes in existing facilities. If Contractor and Owner are unable to agree on entitlement to or on the amount or extent, if any, of any adjustment in the

Contract Price or Contract Times, or both, as a result of any delay in Owner's furnishing the Site or a part thereof, Contractor may make a Claim therefor as provided in Paragraph 10.05.

B. Upon reasonable written request, Owner shall furnish Contractor with a current statement of record legal title and legal description of the lands upon which the Work is to be performed and Owner's interest therein as necessary for giving notice of or filing a mechanic's or construction lien against such lands in accordance with applicable Laws and Regulations.

C. Contractor shall provide for all additional lands and access thereto that may be required for temporary construction facilities or storage of materials and equipment.

4.02 *Subsurface and Physical Conditions*

A. *Reports and Drawings:* The Supplementary Conditions identify:

1. those reports known to Owner of explorations and tests of subsurface conditions at or contiguous to the Site; and

2. those drawings known to Owner of physical conditions relating to existing surface or subsurface structures at the Site (except Underground Facilities).

B. *Limited Reliance by Contractor on Technical Data Authorized:* Contractor may rely upon the accuracy of the "technical data" contained in such reports and drawings, but such reports and drawings are not Contract Documents. Such "technical data" is identified in the Supplementary Conditions. Except for such reliance on such "technical data," Contractor may not rely upon or make any claim against Owner or Engineer, or any of their officers, directors, members, partners, employees, agents, consultants, or subcontractors with respect to:

1. the completeness of such reports and drawings for Contractor's purposes, including, but not limited to, any aspects of the means, methods, techniques, sequences, and procedures of construction to be employed by Contractor, and safety precautions and programs incident thereto; or

2. other data, interpretations, opinions, and information contained in such reports or shown or indicated in such drawings; or

3. any Contractor interpretation of or conclusion drawn from any "technical data" or any such other data, interpretations, opinions, or information.

4.03 *Differing Subsurface or Physical Conditions*

A. *Notice:* If Contractor believes that any subsurface or physical condition that is uncovered or revealed either:

1. is of such a nature as to establish that any "technical data" on which Contractor is entitled to rely as provided in Paragraph 4.02 is materially inaccurate; or

2. is of such a nature as to require a change in the Contract Documents; or

3. differs materially from that shown or indicated in the Contract Documents; or

4. is of an unusual nature, and differs materially from conditions ordinarily encountered and generally recognized as inherent in work of the character provided for in the Contract Documents;

then Contractor shall, promptly after becoming aware thereof and before further disturbing the subsurface or physical conditions or performing any Work in connection therewith (except in an emergency as required by Paragraph 6.16.A), notify Owner and Engineer in writing about such condition. Contractor shall not further disturb such condition or perform any Work in connection therewith (except as aforesaid) until receipt of written order to do so.

B. *Engineer's Review*: After receipt of written notice as required by Paragraph 4.03.A, Engineer will promptly review the pertinent condition, determine the necessity of Owner's obtaining additional exploration or tests with respect thereto, and advise Owner in writing (with a copy to Contractor) of Engineer's findings and conclusions.

C. *Possible Price and Times Adjustments:*

1. The Contract Price or the Contract Times, or both, will be equitably adjusted to the extent that the existence of such differing subsurface or physical condition causes an increase or decrease in Contractor's cost of, or time required for, performance of the Work; subject, however, to the following:

   a. such condition must meet any one or more of the categories described in Paragraph 4.03.A; and

   b. with respect to Work that is paid for on a unit price basis, any adjustment in Contract Price will be subject to the provisions of Paragraphs 9.07 and 11.03.

2. Contractor shall not be entitled to any adjustment in the Contract Price or Contract Times if:

   a. Contractor knew of the existence of such conditions at the time Contractor made a final commitment to Owner with respect to Contract Price and Contract Times by the submission of a Bid or becoming bound under a negotiated contract; or

   b. the existence of such condition could reasonably have been discovered or revealed as a result of any examination, investigation, exploration, test, or study of the Site and contiguous areas required by the Bidding Requirements or Contract Documents to be conducted by or for Contractor prior to Contractor's making such final commitment; or

   c. Contractor failed to give the written notice as required by Paragraph 4.03.A.

3. If Owner and Contractor are unable to agree on entitlement to or on the amount or extent, if any, of any adjustment in the Contract Price or Contract Times, or both, a Claim may be made therefor as provided in Paragraph 10.05. However, neither Owner or Engineer, or any of their officers, directors, members, partners, employees, agents, consultants, or subcontractors shall be liable to Contractor for any claims, costs, losses, or damages (including but not limited to all fees and charges of engineers, architects, attorneys, and other

professionals and all court or arbitration or other dispute resolution costs) sustained by Contractor on or in connection with any other project or anticipated project.

4.04 *Underground Facilities*

A. *Shown or Indicated:* The information and data shown or indicated in the Contract Documents with respect to existing Underground Facilities at or contiguous to the Site is based on information and data furnished to Owner or Engineer by the owners of such Underground Facilities, including Owner, or by others. Unless it is otherwise expressly provided in the Supplementary Conditions:

1. Owner and Engineer shall not be responsible for the accuracy or completeness of any such information or data provided by others; and

2. the cost of all of the following will be included in the Contract Price, and Contractor shall have full responsibility for:

    a. reviewing and checking all such information and data;

    b. locating all Underground Facilities shown or indicated in the Contract Documents;

    c. coordination of the Work with the owners of such Underground Facilities, including Owner, during construction; and

    d. the safety and protection of all such Underground Facilities and repairing any damage thereto resulting from the Work.

B. *Not Shown or Indicated:*

1. If an Underground Facility is uncovered or revealed at or contiguous to the Site which was not shown or indicated, or not shown or indicated with reasonable accuracy in the Contract Documents, Contractor shall, promptly after becoming aware thereof and before further disturbing conditions affected thereby or performing any Work in connection therewith (except in an emergency as required by Paragraph 6.16.A), identify the owner of such Underground Facility and give written notice to that owner and to Owner and Engineer. Engineer will promptly review the Underground Facility and determine the extent, if any, to which a change is required in the Contract Documents to reflect and document the consequences of the existence or location of the Underground Facility. During such time, Contractor shall be responsible for the safety and protection of such Underground Facility.

2. If Engineer concludes that a change in the Contract Documents is required, a Work Change Directive or a Change Order will be issued to reflect and document such consequences. An equitable adjustment shall be made in the Contract Price or Contract Times, or both, to the extent that they are attributable to the existence or location of any Underground Facility that was not shown or indicated or not shown or indicated with reasonable accuracy in the Contract Documents and that Contractor did not know of and could not reasonably have been expected to be aware of or to have anticipated. If Owner and Contractor are unable to agree on entitlement to or on the amount or extent, if any, of any such adjustment in Contract Price

or Contract Times, Owner or Contractor may make a Claim therefor as provided in Paragraph 10.05.

4.05 *Reference Points*

A. Owner shall provide engineering surveys to establish reference points for construction which in Engineer's judgment are necessary to enable Contractor to proceed with the Work. Contractor shall be responsible for laying out the Work, shall protect and preserve the established reference points and property monuments, and shall make no changes or relocations without the prior written approval of Owner. Contractor shall report to Engineer whenever any reference point or property monument is lost or destroyed or requires relocation because of necessary changes in grades or locations, and shall be responsible for the accurate replacement or relocation of such reference points or property monuments by professionally qualified personnel.

4.06 *Hazardous Environmental Condition at Site*

A. *Reports and Drawings:* The Supplementary Conditions identify those reports and drawings known to Owner relating to Hazardous Environmental Conditions that have been identified at the Site.

B. *Limited Reliance by Contractor on Technical Data Authorized:* Contractor may rely upon the accuracy of the "technical data" contained in such reports and drawings, but such reports and drawings are not Contract Documents. Such "technical data" is identified in the Supplementary Conditions. Except for such reliance on such "technical data," Contractor may not rely upon or make any claim against Owner or Engineer, or any of their officers, directors, members, partners, employees, agents, consultants, or subcontractors with respect to:

1. the completeness of such reports and drawings for Contractor's purposes, including, but not limited to, any aspects of the means, methods, techniques, sequences and procedures of construction to be employed by Contractor and safety precautions and programs incident thereto; or

2. other data, interpretations, opinions and information contained in such reports or shown or indicated in such drawings; or

3. any Contractor interpretation of or conclusion drawn from any "technical data" or any such other data, interpretations, opinions or information.

C. Contractor shall not be responsible for any Hazardous Environmental Condition uncovered or revealed at the Site which was not shown or indicated in Drawings or Specifications or identified in the Contract Documents to be within the scope of the Work. Contractor shall be responsible for a Hazardous Environmental Condition created with any materials brought to the Site by Contractor, Subcontractors, Suppliers, or anyone else for whom Contractor is responsible.

D. If Contractor encounters a Hazardous Environmental Condition or if Contractor or anyone for whom Contractor is responsible creates a Hazardous Environmental Condition, Contractor shall immediately: (i) secure or otherwise isolate such condition; (ii) stop all Work in connection with such condition and in any area affected thereby (except in an emergency as required by

Paragraph 6.16.A); and (iii) notify Owner and Engineer (and promptly thereafter confirm such notice in writing). Owner shall promptly consult with Engineer concerning the necessity for Owner to retain a qualified expert to evaluate such condition or take corrective action, if any. Promptly after consulting with Engineer, Owner shall take such actions as are necessary to permit Owner to timely obtain required permits and provide Contractor the written notice required by Paragraph 4.06.E.

E. Contractor shall not be required to resume Work in connection with such condition or in any affected area until after Owner has obtained any required permits related thereto and delivered written notice to Contractor: (i) specifying that such condition and any affected area is or has been rendered safe for the resumption of Work; or (ii) specifying any special conditions under which such Work may be resumed safely. If Owner and Contractor cannot agree as to entitlement to or on the amount or extent, if any, of any adjustment in Contract Price or Contract Times, or both, as a result of such Work stoppage or such special conditions under which Work is agreed to be resumed by Contractor, either party may make a Claim therefor as provided in Paragraph 10.05.

F. If after receipt of such written notice Contractor does not agree to resume such Work based on a reasonable belief it is unsafe, or does not agree to resume such Work under such special conditions, then Owner may order the portion of the Work that is in the area affected by such condition to be deleted from the Work. If Owner and Contractor cannot agree as to entitlement to or on the amount or extent, if any, of an adjustment in Contract Price or Contract Times as a result of deleting such portion of the Work, then either party may make a Claim therefor as provided in Paragraph 10.05. Owner may have such deleted portion of the Work performed by Owner's own forces or others in accordance with Article 7.

G. To the fullest extent permitted by Laws and Regulations, Owner shall indemnify and hold harmless Contractor, Subcontractors, and Engineer, and the officers, directors, members, partners, employees, agents, consultants, and subcontractors of each and any of them from and against all claims, costs, losses, and damages (including but not limited to all fees and charges of engineers, architects, attorneys, and other professionals and all court or arbitration or other dispute resolution costs) arising out of or relating to a Hazardous Environmental Condition, provided that such Hazardous Environmental Condition: (i) was not shown or indicated in the Drawings or Specifications or identified in the Contract Documents to be included within the scope of the Work, and (ii) was not created by Contractor or by anyone for whom Contractor is responsible. Nothing in this Paragraph 4.06.G shall obligate Owner to indemnify any individual or entity from and against the consequences of that individual's or entity's own negligence.

H. To the fullest extent permitted by Laws and Regulations, Contractor shall indemnify and hold harmless Owner and Engineer, and the officers, directors, members, partners, employees, agents, consultants, and subcontractors of each and any of them from and against all claims, costs, losses, and damages (including but not limited to all fees and charges of engineers, architects, attorneys, and other professionals and all court or arbitration or other dispute resolution costs) arising out of or relating to a Hazardous Environmental Condition created by Contractor or by anyone for whom Contractor is responsible. Nothing in this Paragraph 4.06.H shall obligate Contractor to indemnify any individual or entity from and against the consequences of that individual's or entity's own negligence.

I. The provisions of Paragraphs 4.02, 4.03, and 4.04 do not apply to a Hazardous Environmental Condition uncovered or revealed at the Site.

ARTICLE 5 – BONDS AND INSURANCE

5.01 *Performance, Payment, and Other Bonds*

A. Contractor shall furnish performance and payment bonds, each in an amount at least equal to the Contract Price as security for the faithful performance and payment of all of Contractor's obligations under the Contract Documents. These bonds shall remain in effect until one year after the date when final payment becomes due or until completion of the correction period specified in Paragraph 13.07, whichever is later, except as provided otherwise by Laws or Regulations or by the Contract Documents. Contractor shall also furnish such other bonds as are required by the Contract Documents.

B. All bonds shall be in the form prescribed by the Contract Documents except as provided otherwise by Laws or Regulations, and shall be executed by such sureties as are named in the list of "Companies Holding Certificates of Authority as Acceptable Sureties on Federal Bonds and as Acceptable Reinsuring Companies" as published in Circular 570 (amended) by the Financial Management Service, Surety Bond Branch, U.S. Department of the Treasury. All bonds signed by an agent or attorney-in-fact must be accompanied by a certified copy of that individual's authority to bind the surety. The evidence of authority shall show that it is effective on the date the agent or attorney-in-fact signed each bond.

C. If the surety on any bond furnished by Contractor is declared bankrupt or becomes insolvent or its right to do business is terminated in any state where any part of the Project is located or it ceases to meet the requirements of Paragraph 5.01.B, Contractor shall promptly notify Owner and Engineer and shall, within 20 days after the event giving rise to such notification, provide another bond and surety, both of which shall comply with the requirements of Paragraphs 5.01.B and 5.02.

5.02 *Licensed Sureties and Insurers*

A. All bonds and insurance required by the Contract Documents to be purchased and maintained by Owner or Contractor shall be obtained from surety or insurance companies that are duly licensed or authorized in the jurisdiction in which the Project is located to issue bonds or insurance policies for the limits and coverages so required. Such surety and insurance companies shall also meet such additional requirements and qualifications as may be provided in the Supplementary Conditions.

5.03 *Certificates of Insurance*

A. Contractor shall deliver to Owner, with copies to each additional insured and loss payee identified in the Supplementary Conditions, certificates of insurance (and other evidence of insurance requested by Owner or any other additional insured) which Contractor is required to purchase and maintain.

B. Owner shall deliver to Contractor, with copies to each additional insured and loss payee identified in the Supplementary Conditions, certificates of insurance (and other evidence of insurance requested by Contractor or any other additional insured) which Owner is required to purchase and maintain.

C. Failure of Owner to demand such certificates or other evidence of Contractor's full compliance with these insurance requirements or failure of Owner to identify a deficiency in compliance from the evidence provided shall not be construed as a waiver of Contractor's obligation to maintain such insurance.

D. Owner does not represent that insurance coverage and limits established in this Contract necessarily will be adequate to protect Contractor.

E. The insurance and insurance limits required herein shall not be deemed as a limitation on Contractor's liability under the indemnities granted to Owner in the Contract Documents.

5.04 *Contractor's Insurance*

A. Contractor shall purchase and maintain such insurance as is appropriate for the Work being performed and as will provide protection from claims set forth below which may arise out of or result from Contractor's performance of the Work and Contractor's other obligations under the Contract Documents, whether it is to be performed by Contractor, any Subcontractor or Supplier, or by anyone directly or indirectly employed by any of them to perform any of the Work, or by anyone for whose acts any of them may be liable:

1. claims under workers' compensation, disability benefits, and other similar employee benefit acts;

2. claims for damages because of bodily injury, occupational sickness or disease, or death of Contractor's employees;

3. claims for damages because of bodily injury, sickness or disease, or death of any person other than Contractor's employees;

4. claims for damages insured by reasonably available personal injury liability coverage which are sustained:

   a. by any person as a result of an offense directly or indirectly related to the employment of such person by Contractor, or

   b. by any other person for any other reason;

5. claims for damages, other than to the Work itself, because of injury to or destruction of tangible property wherever located, including loss of use resulting therefrom; and

6. claims for damages because of bodily injury or death of any person or property damage arising out of the ownership, maintenance or use of any motor vehicle.

B. The policies of insurance required by this Paragraph 5.04 shall:

1. with respect to insurance required by Paragraphs 5.04.A.3 through 5.04.A.6 inclusive, be written on an occurrence basis, include as additional insureds (subject to any customary exclusion regarding professional liability) Owner and Engineer, and any other individuals or entities identified in the Supplementary Conditions, all of whom shall be listed as additional insureds, and include coverage for the respective officers, directors, members, partners, employees, agents, consultants, and subcontractors of each and any of all such additional insureds, and the insurance afforded to these additional insureds shall provide primary coverage for all claims covered thereby;

2. include at least the specific coverages and be written for not less than the limits of liability provided in the Supplementary Conditions or required by Laws or Regulations, whichever is greater;

3. include contractual liability insurance covering Contractor's indemnity obligations under Paragraphs 6.11 and 6.20;

4. contain a provision or endorsement that the coverage afforded will not be canceled, materially changed or renewal refused until at least 30 days prior written notice has been given to Owner and Contractor and to each other additional insured identified in the Supplementary Conditions to whom a certificate of insurance has been issued (and the certificates of insurance furnished by the Contractor pursuant to Paragraph 5.03 will so provide);

5. remain in effect at least until final payment and at all times thereafter when Contractor may be correcting, removing, or replacing defective Work in accordance with Paragraph 13.07; and

6. include completed operations coverage:

   a. Such insurance shall remain in effect for two years after final payment.

   b. Contractor shall furnish Owner and each other additional insured identified in the Supplementary Conditions, to whom a certificate of insurance has been issued, evidence satisfactory to Owner and any such additional insured of continuation of such insurance at final payment and one year thereafter.

5.05 *Owner's Liability Insurance*

A. In addition to the insurance required to be provided by Contractor under Paragraph 5.04, Owner, at Owner's option, may purchase and maintain at Owner's expense Owner's own liability insurance as will protect Owner against claims which may arise from operations under the Contract Documents.

5.06 *Property Insurance*

A. Unless otherwise provided in the Supplementary Conditions, Owner shall purchase and maintain property insurance upon the Work at the Site in the amount of the full replacement cost thereof (subject to such deductible amounts as may be provided in the Supplementary Conditions or required by Laws and Regulations). This insurance shall:

1. include the interests of Owner, Contractor, Subcontractors, and Engineer, and any other individuals or entities identified in the Supplementary Conditions, and the officers, directors, members, partners, employees, agents, consultants, and subcontractors of each and any of them, each of whom is deemed to have an insurable interest and shall be listed as a loss payee;

2. be written on a Builder's Risk "all-risk" policy form that shall at least include insurance for physical loss or damage to the Work, temporary buildings, falsework, and materials and equipment in transit, and shall insure against at least the following perils or causes of loss: fire, lightning, extended coverage, theft, vandalism and malicious mischief, earthquake, collapse, debris removal, demolition occasioned by enforcement of Laws and Regulations, water damage (other than that caused by flood), and such other perils or causes of loss as may be specifically required by the Supplementary Conditions.

3. include expenses incurred in the repair or replacement of any insured property (including but not limited to fees and charges of engineers and architects);

4. cover materials and equipment stored at the Site or at another location that was agreed to in writing by Owner prior to being incorporated in the Work, provided that such materials and equipment have been included in an Application for Payment recommended by Engineer;

5. allow for partial utilization of the Work by Owner;

6. include testing and startup; and

7. be maintained in effect until final payment is made unless otherwise agreed to in writing by Owner, Contractor, and Engineer with 30 days written notice to each other loss payee to whom a certificate of insurance has been issued.

B. Owner shall purchase and maintain such equipment breakdown insurance or additional property insurance as may be required by the Supplementary Conditions or Laws and Regulations which will include the interests of Owner, Contractor, Subcontractors, and Engineer, and any other individuals or entities identified in the Supplementary Conditions, and the officers, directors, members, partners, employees, agents, consultants and subcontractors of each and any of them, each of whom is deemed to have an insurable interest and shall be listed as a loss payee.

C. All the policies of insurance (and the certificates or other evidence thereof) required to be purchased and maintained in accordance with this Paragraph 5.06 will contain a provision or endorsement that the coverage afforded will not be canceled or materially changed or renewal refused until at least 30 days prior written notice has been given to Owner and Contractor and to each other loss payee to whom a certificate of insurance has been issued and will contain waiver provisions in accordance with Paragraph 5.07.

D. Owner shall not be responsible for purchasing and maintaining any property insurance specified in this Paragraph 5.06 to protect the interests of Contractor, Subcontractors, or others in the Work to the extent of any deductible amounts that are identified in the Supplementary Conditions. The risk of loss within such identified deductible amount will be borne by Contractor, Subcontractors, or others suffering any such loss, and if any of them wishes property

E. If Contractor requests in writing that other special insurance be included in the property insurance policies provided under this Paragraph 5.06, Owner shall, if possible, include such insurance, and the cost thereof will be charged to Contractor by appropriate Change Order. Prior to commencement of the Work at the Site, Owner shall in writing advise Contractor whether or not such other insurance has been procured by Owner.

5.07 *Waiver of Rights*

A. Owner and Contractor intend that all policies purchased in accordance with Paragraph 5.06 will protect Owner, Contractor, Subcontractors, and Engineer, and all other individuals or entities identified in the Supplementary Conditions as loss payees (and the officers, directors, members, partners, employees, agents, consultants, and subcontractors of each and any of them) in such policies and will provide primary coverage for all losses and damages caused by the perils or causes of loss covered thereby. All such policies shall contain provisions to the effect that in the event of payment of any loss or damage the insurers will have no rights of recovery against any of the insureds or loss payees thereunder. Owner and Contractor waive all rights against each other and their respective officers, directors, members, partners, employees, agents, consultants and subcontractors of each and any of them for all losses and damages caused by, arising out of or resulting from any of the perils or causes of loss covered by such policies and any other property insurance applicable to the Work; and, in addition, waive all such rights against Subcontractors and Engineer, and all other individuals or entities identified in the Supplementary Conditions as loss payees (and the officers, directors, members, partners, employees, agents, consultants, and subcontractors of each and any of them) under such policies for losses and damages so caused. None of the above waivers shall extend to the rights that any party making such waiver may have to the proceeds of insurance held by Owner as trustee or otherwise payable under any policy so issued.

B. Owner waives all rights against Contractor, Subcontractors, and Engineer, and the officers, directors, members, partners, employees, agents, consultants and subcontractors of each and any of them for:

1. loss due to business interruption, loss of use, or other consequential loss extending beyond direct physical loss or damage to Owner's property or the Work caused by, arising out of, or resulting from fire or other perils whether or not insured by Owner; and

2. loss or damage to the completed Project or part thereof caused by, arising out of, or resulting from fire or other insured peril or cause of loss covered by any property insurance maintained on the completed Project or part thereof by Owner during partial utilization pursuant to Paragraph 14.05, after Substantial Completion pursuant to Paragraph 14.04, or after final payment pursuant to Paragraph 14.07.

C. Any insurance policy maintained by Owner covering any loss, damage or consequential loss referred to in Paragraph 5.07.B shall contain provisions to the effect that in the event of payment of any such loss, damage, or consequential loss, the insurers will have no rights of recovery against Contractor, Subcontractors, or Engineer, and the officers, directors, members, partners, employees, agents, consultants and subcontractors of each and any of them.

5.08 *Receipt and Application of Insurance Proceeds*

A. Any insured loss under the policies of insurance required by Paragraph 5.06 will be adjusted with Owner and made payable to Owner as fiduciary for the loss payees, as their interests may appear, subject to the requirements of any applicable mortgage clause and of Paragraph 5.08.B. Owner shall deposit in a separate account any money so received and shall distribute it in accordance with such agreement as the parties in interest may reach. If no other special agreement is reached, the damaged Work shall be repaired or replaced, the moneys so received applied on account thereof, and the Work and the cost thereof covered by an appropriate Change Order.

B. Owner as fiduciary shall have power to adjust and settle any loss with the insurers unless one of the parties in interest shall object in writing within 15 days after the occurrence of loss to Owner's exercise of this power. If such objection be made, Owner as fiduciary shall make settlement with the insurers in accordance with such agreement as the parties in interest may reach. If no such agreement among the parties in interest is reached, Owner as fiduciary shall adjust and settle the loss with the insurers and, if required in writing by any party in interest, Owner as fiduciary shall give bond for the proper performance of such duties.

5.09 *Acceptance of Bonds and Insurance; Option to Replace*

A. If either Owner or Contractor has any objection to the coverage afforded by or other provisions of the bonds or insurance required to be purchased and maintained by the other party in accordance with Article 5 on the basis of non-conformance with the Contract Documents, the objecting party shall so notify the other party in writing within 10 days after receipt of the certificates (or other evidence requested) required by Paragraph 2.01.B. Owner and Contractor shall each provide to the other such additional information in respect of insurance provided as the other may reasonably request. If either party does not purchase or maintain all of the bonds and insurance required of such party by the Contract Documents, such party shall notify the other party in writing of such failure to purchase prior to the start of the Work, or of such failure to maintain prior to any change in the required coverage. Without prejudice to any other right or remedy, the other party may elect to obtain equivalent bonds or insurance to protect such other party's interests at the expense of the party who was required to provide such coverage, and a Change Order shall be issued to adjust the Contract Price accordingly.

5.10 *Partial Utilization, Acknowledgment of Property Insurer*

A. If Owner finds it necessary to occupy or use a portion or portions of the Work prior to Substantial Completion of all the Work as provided in Paragraph 14.05, no such use or occupancy shall commence before the insurers providing the property insurance pursuant to Paragraph 5.06 have acknowledged notice thereof and in writing effected any changes in coverage necessitated thereby. The insurers providing the property insurance shall consent by endorsement on the policy or policies, but the property insurance shall not be canceled or permitted to lapse on account of any such partial use or occupancy.

ARTICLE 6 – CONTRACTOR'S RESPONSIBILITIES

6.01 *Supervision and Superintendence*

A. Contractor shall supervise, inspect, and direct the Work competently and efficiently, devoting such attention thereto and applying such skills and expertise as may be necessary to perform the Work in accordance with the Contract Documents. Contractor shall be solely responsible for the means, methods, techniques, sequences, and procedures of construction. Contractor shall not be responsible for the negligence of Owner or Engineer in the design or specification of a specific means, method, technique, sequence, or procedure of construction which is shown or indicated in and expressly required by the Contract Documents.

B. At all times during the progress of the Work, Contractor shall assign a competent resident superintendent who shall not be replaced without written notice to Owner and Engineer except under extraordinary circumstances.

6.02 *Labor; Working Hours*

A. Contractor shall provide competent, suitably qualified personnel to survey and lay out the Work and perform construction as required by the Contract Documents. Contractor shall at all times maintain good discipline and order at the Site.

B. Except as otherwise required for the safety or protection of persons or the Work or property at the Site or adjacent thereto, and except as otherwise stated in the Contract Documents, all Work at the Site shall be performed during regular working hours. Contractor will not permit the performance of Work on a Saturday, Sunday, or any legal holiday without Owner's written consent (which will not be unreasonably withheld) given after prior written notice to Engineer.

6.03 *Services, Materials, and Equipment*

A. Unless otherwise specified in the Contract Documents, Contractor shall provide and assume full responsibility for all services, materials, equipment, labor, transportation, construction equipment and machinery, tools, appliances, fuel, power, light, heat, telephone, water, sanitary facilities, temporary facilities, and all other facilities and incidentals necessary for the performance, testing, start-up, and completion of the Work.

B. All materials and equipment incorporated into the Work shall be as specified or, if not specified, shall be of good quality and new, except as otherwise provided in the Contract Documents. All special warranties and guarantees required by the Specifications shall expressly run to the benefit of Owner. If required by Engineer, Contractor shall furnish satisfactory evidence (including reports of required tests) as to the source, kind, and quality of materials and equipment.

C. All materials and equipment shall be stored, applied, installed, connected, erected, protected, used, cleaned, and conditioned in accordance with instructions of the applicable Supplier, except as otherwise may be provided in the Contract Documents.

6.04 *Progress Schedule*

A. Contractor shall adhere to the Progress Schedule established in accordance with Paragraph 2.07 as it may be adjusted from time to time as provided below.

1. Contractor shall submit to Engineer for acceptance (to the extent indicated in Paragraph 2.07) proposed adjustments in the Progress Schedule that will not result in changing the Contract Times. Such adjustments will comply with any provisions of the General Requirements applicable thereto.

2. Proposed adjustments in the Progress Schedule that will change the Contract Times shall be submitted in accordance with the requirements of Article 12. Adjustments in Contract Times may only be made by a Change Order.

6.05 *Substitutes and "Or-Equals"*

A. Whenever an item of material or equipment is specified or described in the Contract Documents by using the name of a proprietary item or the name of a particular Supplier, the specification or description is intended to establish the type, function, appearance, and quality required. Unless the specification or description contains or is followed by words reading that no like, equivalent, or "or-equal" item or no substitution is permitted, other items of material or equipment or material or equipment of other Suppliers may be submitted to Engineer for review under the circumstances described below.

1. *"Or-Equal" Items:* If in Engineer's sole discretion an item of material or equipment proposed by Contractor is functionally equal to that named and sufficiently similar so that no change in related Work will be required, it may be considered by Engineer as an "or-equal" item, in which case review and approval of the proposed item may, in Engineer's sole discretion, be accomplished without compliance with some or all of the requirements for approval of proposed substitute items. For the purposes of this Paragraph 6.05.A.1, a proposed item of material or equipment will be considered functionally equal to an item so named if:

a. in the exercise of reasonable judgment Engineer determines that:

1) it is at least equal in materials of construction, quality, durability, appearance, strength, and design characteristics;

2) it will reliably perform at least equally well the function and achieve the results imposed by the design concept of the completed Project as a functioning whole; and

3) it has a proven record of performance and availability of responsive service.

b. Contractor certifies that, if approved and incorporated into the Work:

1) there will be no increase in cost to the Owner or increase in Contract Times; and

2) it will conform substantially to the detailed requirements of the item named in the Contract Documents.

2. *Substitute Items:*

a. If in Engineer's sole discretion an item of material or equipment proposed by Contractor does not qualify as an "or-equal" item under Paragraph 6.05.A.1, it will be considered a proposed substitute item.

b. Contractor shall submit sufficient information as provided below to allow Engineer to determine if the item of material or equipment proposed is essentially equivalent to that named and an acceptable substitute therefor. Requests for review of proposed substitute items of material or equipment will not be accepted by Engineer from anyone other than Contractor.

c. The requirements for review by Engineer will be as set forth in Paragraph 6.05.A.2.d, as supplemented by the General Requirements, and as Engineer may decide is appropriate under the circumstances.

d. Contractor shall make written application to Engineer for review of a proposed substitute item of material or equipment that Contractor seeks to furnish or use. The application:

1) shall certify that the proposed substitute item will:

a) perform adequately the functions and achieve the results called for by the general design,

b) be similar in substance to that specified, and

c) be suited to the same use as that specified;

2) will state:

a) the extent, if any, to which the use of the proposed substitute item will prejudice Contractor's achievement of Substantial Completion on time,

b) whether use of the proposed substitute item in the Work will require a change in any of the Contract Documents (or in the provisions of any other direct contract with Owner for other work on the Project) to adapt the design to the proposed substitute item, and

c) whether incorporation or use of the proposed substitute item in connection with the Work is subject to payment of any license fee or royalty;

3) will identify:

a) all variations of the proposed substitute item from that specified, and

b) available engineering, sales, maintenance, repair, and replacement services; and

4) shall contain an itemized estimate of all costs or credits that will result directly or indirectly from use of such substitute item, including costs of redesign and claims of other contractors affected by any resulting change.

B. *Substitute Construction Methods or Procedures:* If a specific means, method, technique, sequence, or procedure of construction is expressly required by the Contract Documents, Contractor may furnish or utilize a substitute means, method, technique, sequence, or procedure of construction approved by Engineer. Contractor shall submit sufficient information to allow Engineer, in Engineer's sole discretion, to determine that the substitute proposed is equivalent to that expressly called for by the Contract Documents. The requirements for review by Engineer will be similar to those provided in Paragraph 6.05.A.2.

C. *Engineer's Evaluation:* Engineer will be allowed a reasonable time within which to evaluate each proposal or submittal made pursuant to Paragraphs 6.05.A and 6.05.B. Engineer may require Contractor to furnish additional data about the proposed substitute item. Engineer will be the sole judge of acceptability. No "or equal" or substitute will be ordered, installed or utilized until Engineer's review is complete, which will be evidenced by a Change Order in the case of a substitute and an approved Shop Drawing for an "or equal." Engineer will advise Contractor in writing of any negative determination.

D. *Special Guarantee:* Owner may require Contractor to furnish at Contractor's expense a special performance guarantee or other surety with respect to any substitute.

E. *Engineer's Cost Reimbursement*: Engineer will record Engineer's costs in evaluating a substitute proposed or submitted by Contractor pursuant to Paragraphs 6.05.A.2 and 6.05.B. Whether or not Engineer approves a substitute so proposed or submitted by Contractor, Contractor shall reimburse Owner for the reasonable charges of Engineer for evaluating each such proposed substitute. Contractor shall also reimburse Owner for the reasonable charges of Engineer for making changes in the Contract Documents (or in the provisions of any other direct contract with Owner) resulting from the acceptance of each proposed substitute.

F. *Contractor's Expense*: Contractor shall provide all data in support of any proposed substitute or "or-equal" at Contractor's expense.

6.06 *Concerning Subcontractors, Suppliers, and Others*

A. Contractor shall not employ any Subcontractor, Supplier, or other individual or entity (including those acceptable to Owner as indicated in Paragraph 6.06.B), whether initially or as a replacement, against whom Owner may have reasonable objection. Contractor shall not be required to employ any Subcontractor, Supplier, or other individual or entity to furnish or perform any of the Work against whom Contractor has reasonable objection.

B. If the Supplementary Conditions require the identity of certain Subcontractors, Suppliers, or other individuals or entities to be submitted to Owner in advance for acceptance by Owner by a specified date prior to the Effective Date of the Agreement, and if Contractor has submitted a list thereof in accordance with the Supplementary Conditions, Owner's acceptance (either in writing or by failing to make written objection thereto by the date indicated for acceptance or objection in the Bidding Documents or the Contract Documents) of any such Subcontractor, Supplier, or

C. Contractor shall be fully responsible to Owner and Engineer for all acts and omissions of the Subcontractors, Suppliers, and other individuals or entities performing or furnishing any of the Work just as Contractor is responsible for Contractor's own acts and omissions. Nothing in the Contract Documents:

1. shall create for the benefit of any such Subcontractor, Supplier, or other individual or entity any contractual relationship between Owner or Engineer and any such Subcontractor, Supplier or other individual or entity; nor

2. shall create any obligation on the part of Owner or Engineer to pay or to see to the payment of any moneys due any such Subcontractor, Supplier, or other individual or entity except as may otherwise be required by Laws and Regulations.

D. Contractor shall be solely responsible for scheduling and coordinating the Work of Subcontractors, Suppliers, and other individuals or entities performing or furnishing any of the Work under a direct or indirect contract with Contractor.

E. Contractor shall require all Subcontractors, Suppliers, and such other individuals or entities performing or furnishing any of the Work to communicate with Engineer through Contractor.

F. The divisions and sections of the Specifications and the identifications of any Drawings shall not control Contractor in dividing the Work among Subcontractors or Suppliers or delineating the Work to be performed by any specific trade.

G. All Work performed for Contractor by a Subcontractor or Supplier will be pursuant to an appropriate agreement between Contractor and the Subcontractor or Supplier which specifically binds the Subcontractor or Supplier to the applicable terms and conditions of the Contract Documents for the benefit of Owner and Engineer. Whenever any such agreement is with a Subcontractor or Supplier who is listed as a loss payee on the property insurance provided in Paragraph 5.06, the agreement between the Contractor and the Subcontractor or Supplier will contain provisions whereby the Subcontractor or Supplier waives all rights against Owner, Contractor, Engineer, and all other individuals or entities identified in the Supplementary Conditions to be listed as insureds or loss payees (and the officers, directors, members, partners, employees, agents, consultants, and subcontractors of each and any of them) for all losses and damages caused by, arising out of, relating to, or resulting from any of the perils or causes of loss covered by such policies and any other property insurance applicable to the Work. If the insurers on any such policies require separate waiver forms to be signed by any Subcontractor or Supplier, Contractor will obtain the same.

6.07 *Patent Fees and Royalties*

A. Contractor shall pay all license fees and royalties and assume all costs incident to the use in the performance of the Work or the incorporation in the Work of any invention, design, process, product, or device which is the subject of patent rights or copyrights held by others. If a particular invention, design, process, product, or device is specified in the Contract Documents for use in the performance of the Work and if, to the actual knowledge of Owner or Engineer, its use is subject to patent rights or copyrights calling for the payment of any license fee or royalty to others, the existence of such rights shall be disclosed by Owner in the Contract Documents.

B. To the fullest extent permitted by Laws and Regulations, Owner shall indemnify and hold harmless Contractor, and its officers, directors, members, partners, employees, agents, consultants, and subcontractors from and against all claims, costs, losses, and damages (including but not limited to all fees and charges of engineers, architects, attorneys, and other professionals, and all court or arbitration or other dispute resolution costs) arising out of or relating to any infringement of patent rights or copyrights incident to the use in the performance of the Work or resulting from the incorporation in the Work of any invention, design, process, product, or device specified in the Contract Documents, but not identified as being subject to payment of any license fee or royalty to others required by patent rights or copyrights.

C. To the fullest extent permitted by Laws and Regulations, Contractor shall indemnify and hold harmless Owner and Engineer, and the officers, directors, members, partners, employees, agents, consultants and subcontractors of each and any of them from and against all claims, costs, losses, and damages (including but not limited to all fees and charges of engineers, architects, attorneys, and other professionals and all court or arbitration or other dispute resolution costs) arising out of or relating to any infringement of patent rights or copyrights incident to the use in the performance of the Work or resulting from the incorporation in the Work of any invention, design, process, product, or device not specified in the Contract Documents.

6.08 *Permits*

A. Unless otherwise provided in the Supplementary Conditions, Contractor shall obtain and pay for all construction permits and licenses. Owner shall assist Contractor, when necessary, in obtaining such permits and licenses. Contractor shall pay all governmental charges and inspection fees necessary for the prosecution of the Work which are applicable at the time of opening of Bids, or, if there are no Bids, on the Effective Date of the Agreement. Owner shall pay all charges of utility owners for connections for providing permanent service to the Work.

6.09 *Laws and Regulations*

A. Contractor shall give all notices required by and shall comply with all Laws and Regulations applicable to the performance of the Work. Except where otherwise expressly required by applicable Laws and Regulations, neither Owner nor Engineer shall be responsible for monitoring Contractor's compliance with any Laws or Regulations.

B. If Contractor performs any Work knowing or having reason to know that it is contrary to Laws or Regulations, Contractor shall bear all claims, costs, losses, and damages (including but not limited to all fees and charges of engineers, architects, attorneys, and other professionals and all court or arbitration or other dispute resolution costs) arising out of or relating to such Work. However, it shall not be Contractor's responsibility to make certain that the Specifications and Drawings are in accordance with Laws and Regulations, but this shall not relieve Contractor of Contractor's obligations under Paragraph 3.03.

C. Changes in Laws or Regulations not known at the time of opening of Bids (or, on the Effective Date of the Agreement if there were no Bids) having an effect on the cost or time of performance of the Work shall be the subject of an adjustment in Contract Price or Contract Times. If Owner and Contractor are unable to agree on entitlement to or on the amount or extent, if any, of any such adjustment, a Claim may be made therefor as provided in Paragraph 10.05.

6.10 *Taxes*

A. Contractor shall pay all sales, consumer, use, and other similar taxes required to be paid by Contractor in accordance with the Laws and Regulations of the place of the Project which are applicable during the performance of the Work.

6.11 *Use of Site and Other Areas*

A. *Limitation on Use of Site and Other Areas:*

1. Contractor shall confine construction equipment, the storage of materials and equipment, and the operations of workers to the Site and other areas permitted by Laws and Regulations, and shall not unreasonably encumber the Site and other areas with construction equipment or other materials or equipment. Contractor shall assume full responsibility for any damage to any such land or area, or to the owner or occupant thereof, or of any adjacent land or areas resulting from the performance of the Work.

2. Should any claim be made by any such owner or occupant because of the performance of the Work, Contractor shall promptly settle with such other party by negotiation or otherwise resolve the claim by arbitration or other dispute resolution proceeding or at law.

3. To the fullest extent permitted by Laws and Regulations, Contractor shall indemnify and hold harmless Owner and Engineer, and the officers, directors, members, partners, employees, agents, consultants and subcontractors of each and any of them from and against all claims, costs, losses, and damages (including but not limited to all fees and charges of engineers, architects, attorneys, and other professionals and all court or arbitration or other dispute resolution costs) arising out of or relating to any claim or action, legal or equitable, brought by any such owner or occupant against Owner, Engineer, or any other party indemnified hereunder to the extent caused by or based upon Contractor's performance of the Work.

B. *Removal of Debris During Performance of the Work:* During the progress of the Work Contractor shall keep the Site and other areas free from accumulations of waste materials, rubbish, and other debris. Removal and disposal of such waste materials, rubbish, and other debris shall conform to applicable Laws and Regulations.

C. *Cleaning:* Prior to Substantial Completion of the Work Contractor shall clean the Site and the Work and make it ready for utilization by Owner. At the completion of the Work Contractor

D. *Loading Structures:* Contractor shall not load nor permit any part of any structure to be loaded in any manner that will endanger the structure, nor shall Contractor subject any part of the Work or adjacent property to stresses or pressures that will endanger it.

6.12 *Record Documents*

A. Contractor shall maintain in a safe place at the Site one record copy of all Drawings, Specifications, Addenda, Change Orders, Work Change Directives, Field Orders, and written interpretations and clarifications in good order and annotated to show changes made during construction. These record documents together with all approved Samples and a counterpart of all approved Shop Drawings will be available to Engineer for reference. Upon completion of the Work, these record documents, Samples, and Shop Drawings will be delivered to Engineer for Owner.

6.13 *Safety and Protection*

A. Contractor shall be solely responsible for initiating, maintaining and supervising all safety precautions and programs in connection with the Work. Such responsibility does not relieve Subcontractors of their responsibility for the safety of persons or property in the performance of their work, nor for compliance with applicable safety Laws and Regulations. Contractor shall take all necessary precautions for the safety of, and shall provide the necessary protection to prevent damage, injury or loss to:

1. all persons on the Site or who may be affected by the Work;

2. all the Work and materials and equipment to be incorporated therein, whether in storage on or off the Site; and

3. other property at the Site or adjacent thereto, including trees, shrubs, lawns, walks, pavements, roadways, structures, utilities, and Underground Facilities not designated for removal, relocation, or replacement in the course of construction.

B. Contractor shall comply with all applicable Laws and Regulations relating to the safety of persons or property, or to the protection of persons or property from damage, injury, or loss; and shall erect and maintain all necessary safeguards for such safety and protection. Contractor shall notify owners of adjacent property and of Underground Facilities and other utility owners when prosecution of the Work may affect them, and shall cooperate with them in the protection, removal, relocation, and replacement of their property.

C. Contractor shall comply with the applicable requirements of Owner's safety programs, if any. The Supplementary Conditions identify any Owner's safety programs that are applicable to the Work.

D. Contractor shall inform Owner and Engineer of the specific requirements of Contractor's safety program with which Owner's and Engineer's employees and representatives must comply while at the Site.

E. All damage, injury, or loss to any property referred to in Paragraph 6.13.A.2 or 6.13.A.3 caused, directly or indirectly, in whole or in part, by Contractor, any Subcontractor, Supplier, or any other individual or entity directly or indirectly employed by any of them to perform any of the Work, or anyone for whose acts any of them may be liable, shall be remedied by Contractor (except damage or loss attributable to the fault of Drawings or Specifications or to the acts or omissions of Owner or Engineer or anyone employed by any of them, or anyone for whose acts any of them may be liable, and not attributable, directly or indirectly, in whole or in part, to the fault or negligence of Contractor or any Subcontractor, Supplier, or other individual or entity directly or indirectly employed by any of them).

F. Contractor's duties and responsibilities for safety and for protection of the Work shall continue until such time as all the Work is completed and Engineer has issued a notice to Owner and Contractor in accordance with Paragraph 14.07.B that the Work is acceptable (except as otherwise expressly provided in connection with Substantial Completion).

6.14 *Safety Representative*

A. Contractor shall designate a qualified and experienced safety representative at the Site whose duties and responsibilities shall be the prevention of accidents and the maintaining and supervising of safety precautions and programs.

6.15 *Hazard Communication Programs*

A. Contractor shall be responsible for coordinating any exchange of material safety data sheets or other hazard communication information required to be made available to or exchanged between or among employers at the Site in accordance with Laws or Regulations.

6.16 *Emergencies*

A. In emergencies affecting the safety or protection of persons or the Work or property at the Site or adjacent thereto, Contractor is obligated to act to prevent threatened damage, injury, or loss. Contractor shall give Engineer prompt written notice if Contractor believes that any significant changes in the Work or variations from the Contract Documents have been caused thereby or are required as a result thereof. If Engineer determines that a change in the Contract Documents is required because of the action taken by Contractor in response to such an emergency, a Work Change Directive or Change Order will be issued.

6.17 *Shop Drawings and Samples*

A. Contractor shall submit Shop Drawings and Samples to Engineer for review and approval in accordance with the accepted Schedule of Submittals (as required by Paragraph 2.07). Each submittal will be identified as Engineer may require.

1. *Shop Drawings:*

   a. Submit number of copies specified in the General Requirements.

   b. Data shown on the Shop Drawings will be complete with respect to quantities, dimensions, specified performance and design criteria, materials, and similar data to show Engineer the services, materials, and equipment Contractor proposes to provide and to enable Engineer to review the information for the limited purposes required by Paragraph 6.17.D.

2. *Samples:*

   a. Submit number of Samples specified in the Specifications.

   b. Clearly identify each Sample as to material, Supplier, pertinent data such as catalog numbers, the use for which intended and other data as Engineer may require to enable Engineer to review the submittal for the limited purposes required by Paragraph 6.17.D.

B. Where a Shop Drawing or Sample is required by the Contract Documents or the Schedule of Submittals, any related Work performed prior to Engineer's review and approval of the pertinent submittal will be at the sole expense and responsibility of Contractor.

C. *Submittal Procedures:*

1. Before submitting each Shop Drawing or Sample, Contractor shall have:

   a. reviewed and coordinated each Shop Drawing or Sample with other Shop Drawings and Samples and with the requirements of the Work and the Contract Documents;

   b. determined and verified all field measurements, quantities, dimensions, specified performance and design criteria, installation requirements, materials, catalog numbers, and similar information with respect thereto;

   c. determined and verified the suitability of all materials offered with respect to the indicated application, fabrication, shipping, handling, storage, assembly, and installation pertaining to the performance of the Work; and

   d. determined and verified all information relative to Contractor's responsibilities for means, methods, techniques, sequences, and procedures of construction, and safety precautions and programs incident thereto.

2. Each submittal shall bear a stamp or specific written certification that Contractor has satisfied Contractor's obligations under the Contract Documents with respect to Contractor's review and approval of that submittal.

3. With each submittal, Contractor shall give Engineer specific written notice of any variations that the Shop Drawing or Sample may have from the requirements of the Contract Documents. This notice shall be both a written communication separate from the Shop

Drawings or Sample submittal; and, in addition, by a specific notation made on each Shop Drawing or Sample submitted to Engineer for review and approval of each such variation.

D. *Engineer's Review:*

1. Engineer will provide timely review of Shop Drawings and Samples in accordance with the Schedule of Submittals acceptable to Engineer. Engineer's review and approval will be only to determine if the items covered by the submittals will, after installation or incorporation in the Work, conform to the information given in the Contract Documents and be compatible with the design concept of the completed Project as a functioning whole as indicated by the Contract Documents.

2. Engineer's review and approval will not extend to means, methods, techniques, sequences, or procedures of construction (except where a particular means, method, technique, sequence, or procedure of construction is specifically and expressly called for by the Contract Documents) or to safety precautions or programs incident thereto. The review and approval of a separate item as such will not indicate approval of the assembly in which the item functions.

3. Engineer's review and approval shall not relieve Contractor from responsibility for any variation from the requirements of the Contract Documents unless Contractor has complied with the requirements of Paragraph 6.17.C.3 and Engineer has given written approval of each such variation by specific written notation thereof incorporated in or accompanying the Shop Drawing or Sample. Engineer's review and approval shall not relieve Contractor from responsibility for complying with the requirements of Paragraph 6.17.C.1.

E. *Resubmittal Procedures:*

1. Contractor shall make corrections required by Engineer and shall return the required number of corrected copies of Shop Drawings and submit, as required, new Samples for review and approval. Contractor shall direct specific attention in writing to revisions other than the corrections called for by Engineer on previous submittals.

6.18 *Continuing the Work*

A. Contractor shall carry on the Work and adhere to the Progress Schedule during all disputes or disagreements with Owner. No Work shall be delayed or postponed pending resolution of any disputes or disagreements, except as permitted by Paragraph 15.04 or as Owner and Contractor may otherwise agree in writing.

6.19 *Contractor's General Warranty and Guarantee*

A. Contractor warrants and guarantees to Owner that all Work will be in accordance with the Contract Documents and will not be defective. Engineer and its officers, directors, members, partners, employees, agents, consultants, and subcontractors shall be entitled to rely on representation of Contractor's warranty and guarantee.

B. Contractor's warranty and guarantee hereunder excludes defects or damage caused by:

1. abuse, modification, or improper maintenance or operation by persons other than Contractor, Subcontractors, Suppliers, or any other individual or entity for whom Contractor is responsible; or

2. normal wear and tear under normal usage.

C. Contractor's obligation to perform and complete the Work in accordance with the Contract Documents shall be absolute. None of the following will constitute an acceptance of Work that is not in accordance with the Contract Documents or a release of Contractor's obligation to perform the Work in accordance with the Contract Documents:

1. observations by Engineer;

2. recommendation by Engineer or payment by Owner of any progress or final payment;

3. the issuance of a certificate of Substantial Completion by Engineer or any payment related thereto by Owner;

4. use or occupancy of the Work or any part thereof by Owner;

5. any review and approval of a Shop Drawing or Sample submittal or the issuance of a notice of acceptability by Engineer;

6. any inspection, test, or approval by others; or

7. any correction of defective Work by Owner.

6.20 *Indemnification*

A. To the fullest extent permitted by Laws and Regulations, Contractor shall indemnify and hold harmless Owner and Engineer, and the officers, directors, members, partners, employees, agents, consultants and subcontractors of each and any of them from and against all claims, costs, losses, and damages (including but not limited to all fees and charges of engineers, architects, attorneys, and other professionals and all court or arbitration or other dispute resolution costs) arising out of or relating to the performance of the Work, provided that any such claim, cost, loss, or damage is attributable to bodily injury, sickness, disease, or death, or to injury to or destruction of tangible property (other than the Work itself), including the loss of use resulting therefrom but only to the extent caused by any negligent act or omission of Contractor, any Subcontractor, any Supplier, or any individual or entity directly or indirectly employed by any of them to perform any of the Work or anyone for whose acts any of them may be liable .

B. In any and all claims against Owner or Engineer or any of their officers, directors, members, partners, employees, agents, consultants, or subcontractors by any employee (or the survivor or personal representative of such employee) of Contractor, any Subcontractor, any Supplier, or any individual or entity directly or indirectly employed by any of them to perform any of the Work, or anyone for whose acts any of them may be liable, the indemnification obligation under Paragraph 6.20.A shall not be limited in any way by any limitation on the amount or type of damages, compensation, or benefits payable by or for Contractor or any such Subcontractor,

Supplier, or other individual or entity under workers' compensation acts, disability benefit acts, or other employee benefit acts.

C. The indemnification obligations of Contractor under Paragraph 6.20.A shall not extend to the liability of Engineer and Engineer's officers, directors, members, partners, employees, agents, consultants and subcontractors arising out of:

1. the preparation or approval of, or the failure to prepare or approve maps, Drawings, opinions, reports, surveys, Change Orders, designs, or Specifications; or

2. giving directions or instructions, or failing to give them, if that is the primary cause of the injury or damage.

6.21 *Delegation of Professional Design Services*

A. Contractor will not be required to provide professional design services unless such services are specifically required by the Contract Documents for a portion of the Work or unless such services are required to carry out Contractor's responsibilities for construction means, methods, techniques, sequences and procedures. Contractor shall not be required to provide professional services in violation of applicable law.

B. If professional design services or certifications by a design professional related to systems, materials or equipment are specifically required of Contractor by the Contract Documents, Owner and Engineer will specify all performance and design criteria that such services must satisfy. Contractor shall cause such services or certifications to be provided by a properly licensed professional, whose signature and seal shall appear on all drawings, calculations, specifications, certifications, Shop Drawings and other submittals prepared by such professional. Shop Drawings and other submittals related to the Work designed or certified by such professional, if prepared by others, shall bear such professional's written approval when submitted to Engineer.

C. Owner and Engineer shall be entitled to rely upon the adequacy, accuracy and completeness of the services, certifications or approvals performed by such design professionals, provided Owner and Engineer have specified to Contractor all performance and design criteria that such services must satisfy.

D. Pursuant to this Paragraph 6.21, Engineer's review and approval of design calculations and design drawings will be only for the limited purpose of checking for conformance with performance and design criteria given and the design concept expressed in the Contract Documents. Engineer's review and approval of Shop Drawings and other submittals (except design calculations and design drawings) will be only for the purpose stated in Paragraph 6.17.D.1.

E. Contractor shall not be responsible for the adequacy of the performance or design criteria required by the Contract Documents.

ARTICLE 7 – OTHER WORK AT THE SITE

7.01 *Related Work at Site*

A. Owner may perform other work related to the Project at the Site with Owner's employees, or through other direct contracts therefor, or have other work performed by utility owners. If such other work is not noted in the Contract Documents, then:

1. written notice thereof will be given to Contractor prior to starting any such other work; and

2. if Owner and Contractor are unable to agree on entitlement to or on the amount or extent, if any, of any adjustment in the Contract Price or Contract Times that should be allowed as a result of such other work, a Claim may be made therefor as provided in Paragraph 10.05.

B. Contractor shall afford each other contractor who is a party to such a direct contract, each utility owner, and Owner, if Owner is performing other work with Owner's employees, proper and safe access to the Site, provide a reasonable opportunity for the introduction and storage of materials and equipment and the execution of such other work, and properly coordinate the Work with theirs. Contractor shall do all cutting, fitting, and patching of the Work that may be required to properly connect or otherwise make its several parts come together and properly integrate with such other work. Contractor shall not endanger any work of others by cutting, excavating, or otherwise altering such work; provided, however, that Contractor may cut or alter others' work with the written consent of Engineer and the others whose work will be affected. The duties and responsibilities of Contractor under this Paragraph are for the benefit of such utility owners and other contractors to the extent that there are comparable provisions for the benefit of Contractor in said direct contracts between Owner and such utility owners and other contractors.

C. If the proper execution or results of any part of Contractor's Work depends upon work performed by others under this Article 7, Contractor shall inspect such other work and promptly report to Engineer in writing any delays, defects, or deficiencies in such other work that render it unavailable or unsuitable for the proper execution and results of Contractor's Work. Contractor's failure to so report will constitute an acceptance of such other work as fit and proper for integration with Contractor's Work except for latent defects and deficiencies in such other work.

7.02 *Coordination*

A. If Owner intends to contract with others for the performance of other work on the Project at the Site, the following will be set forth in Supplementary Conditions:

1. the individual or entity who will have authority and responsibility for coordination of the activities among the various contractors will be identified;

2. the specific matters to be covered by such authority and responsibility will be itemized; and

3. the extent of such authority and responsibilities will be provided.

B. Unless otherwise provided in the Supplementary Conditions, Owner shall have sole authority and responsibility for such coordination.

7.03 *Legal Relationships*

A. Paragraphs 7.01.A and 7.02 are not applicable for utilities not under the control of Owner.

B. Each other direct contract of Owner under Paragraph 7.01.A shall provide that the other contractor is liable to Owner and Contractor for the reasonable direct delay and disruption costs incurred by Contractor as a result of the other contractor's wrongful actions or inactions.

C. Contractor shall be liable to Owner and any other contractor under direct contract to Owner for the reasonable direct delay and disruption costs incurred by such other contractor as a result of Contractor's wrongful action or inactions.

ARTICLE 8 – OWNER'S RESPONSIBILITIES

8.01 *Communications to Contractor*

A. Except as otherwise provided in these General Conditions, Owner shall issue all communications to Contractor through Engineer.

8.02 *Replacement of Engineer*

A. In case of termination of the employment of Engineer, Owner shall appoint an engineer to whom Contractor makes no reasonable objection, whose status under the Contract Documents shall be that of the former Engineer.

8.03 *Furnish Data*

A. Owner shall promptly furnish the data required of Owner under the Contract Documents.

8.04 *Pay When Due*

A. Owner shall make payments to Contractor when they are due as provided in Paragraphs 14.02.C and 14.07.C.

8.05 *Lands and Easements; Reports and Tests*

A. Owner's duties with respect to providing lands and easements and providing engineering surveys to establish reference points are set forth in Paragraphs 4.01 and 4.05. Paragraph 4.02 refers to Owner's identifying and making available to Contractor copies of reports of explorations and tests of subsurface conditions and drawings of physical conditions relating to existing surface or subsurface structures at the Site.

8.06 *Insurance*

A. Owner's responsibilities, if any, with respect to purchasing and maintaining liability and property insurance are set forth in Article 5.

8.07 *Change Orders*

A. Owner is obligated to execute Change Orders as indicated in Paragraph 10.03.

8.08 *Inspections, Tests, and Approvals*

A. Owner's responsibility with respect to certain inspections, tests, and approvals is set forth in Paragraph 13.03.B.

8.09 *Limitations on Owner's Responsibilities*

A. The Owner shall not supervise, direct, or have control or authority over, nor be responsible for, Contractor's means, methods, techniques, sequences, or procedures of construction, or the safety precautions and programs incident thereto, or for any failure of Contractor to comply with Laws and Regulations applicable to the performance of the Work. Owner will not be responsible for Contractor's failure to perform the Work in accordance with the Contract Documents.

8.10 *Undisclosed Hazardous Environmental Condition*

A. Owner's responsibility in respect to an undisclosed Hazardous Environmental Condition is set forth in Paragraph 4.06.

8.11 *Evidence of Financial Arrangements*

A. Upon request of Contractor, Owner shall furnish Contractor reasonable evidence that financial arrangements have been made to satisfy Owner's obligations under the Contract Documents.

8.12 *Compliance with Safety Program*

A. While at the Site, Owner's employees and representatives shall comply with the specific applicable requirements of Contractor's safety programs of which Owner has been informed pursuant to Paragraph 6.13.D.

ARTICLE 9 – ENGINEER'S STATUS DURING CONSTRUCTION

9.01 *Owner's Representative*

A. Engineer will be Owner's representative during the construction period. The duties and responsibilities and the limitations of authority of Engineer as Owner's representative during construction are set forth in the Contract Documents.

9.02 *Visits to Site*

A. Engineer will make visits to the Site at intervals appropriate to the various stages of construction as Engineer deems necessary in order to observe as an experienced and qualified design professional the progress that has been made and the quality of the various aspects of Contractor's executed Work. Based on information obtained during such visits and observations, Engineer, for the benefit of Owner, will determine, in general, if the Work is proceeding in accordance with the Contract Documents. Engineer will not be required to make exhaustive or continuous inspections on the Site to check the quality or quantity of the Work. Engineer's efforts will be directed toward providing for Owner a greater degree of confidence that the completed Work will conform generally to the Contract Documents. On the basis of such visits and observations, Engineer will keep Owner informed of the progress of the Work and will endeavor to guard Owner against defective Work.

B. Engineer's visits and observations are subject to all the limitations on Engineer's authority and responsibility set forth in Paragraph 9.09. Particularly, but without limitation, during or as a result of Engineer's visits or observations of Contractor's Work, Engineer will not supervise, direct, control, or have authority over or be responsible for Contractor's means, methods, techniques, sequences, or procedures of construction, or the safety precautions and programs incident thereto, or for any failure of Contractor to comply with Laws and Regulations applicable to the performance of the Work.

9.03 *Project Representative*

A. If Owner and Engineer agree, Engineer will furnish a Resident Project Representative to assist Engineer in providing more extensive observation of the Work. The authority and responsibilities of any such Resident Project Representative and assistants will be as provided in the Supplementary Conditions, and limitations on the responsibilities thereof will be as provided in Paragraph 9.09. If Owner designates another representative or agent to represent Owner at the Site who is not Engineer's consultant, agent or employee, the responsibilities and authority and limitations thereon of such other individual or entity will be as provided in the Supplementary Conditions.

9.04 *Authorized Variations in Work*

A. Engineer may authorize minor variations in the Work from the requirements of the Contract Documents which do not involve an adjustment in the Contract Price or the Contract Times and are compatible with the design concept of the completed Project as a functioning whole as indicated by the Contract Documents. These may be accomplished by a Field Order and will be binding on Owner and also on Contractor, who shall perform the Work involved promptly. If Owner or Contractor believes that a Field Order justifies an adjustment in the Contract Price or Contract Times, or both, and the parties are unable to agree on entitlement to or on the amount or extent, if any, of any such adjustment, a Claim may be made therefor as provided in Paragraph 10.05.

9.05 *Rejecting Defective Work*

A. Engineer will have authority to reject Work which Engineer believes to be defective, or that Engineer believes will not produce a completed Project that conforms to the Contract Documents or that will prejudice the integrity of the design concept of the completed Project as a functioning whole as indicated by the Contract Documents. Engineer will also have authority to require special inspection or testing of the Work as provided in Paragraph 13.04, whether or not the Work is fabricated, installed, or completed.

9.06 *Shop Drawings, Change Orders and Payments*

A. In connection with Engineer's authority, and limitations thereof, as to Shop Drawings and Samples, see Paragraph 6.17.

B. In connection with Engineer's authority, and limitations thereof, as to design calculations and design drawings submitted in response to a delegation of professional design services, if any, see Paragraph 6.21.

C. In connection with Engineer's authority as to Change Orders, see Articles 10, 11, and 12.

D. In connection with Engineer's authority as to Applications for Payment, see Article 14.

9.07 *Determinations for Unit Price Work*

A. Engineer will determine the actual quantities and classifications of Unit Price Work performed by Contractor. Engineer will review with Contractor the Engineer's preliminary determinations on such matters before rendering a written decision thereon (by recommendation of an Application for Payment or otherwise). Engineer's written decision thereon will be final and binding (except as modified by Engineer to reflect changed factual conditions or more accurate data) upon Owner and Contractor, subject to the provisions of Paragraph 10.05.

9.08 *Decisions on Requirements of Contract Documents and Acceptability of Work*

A. Engineer will be the initial interpreter of the requirements of the Contract Documents and judge of the acceptability of the Work thereunder. All matters in question and other matters between Owner and Contractor arising prior to the date final payment is due relating to the acceptability of the Work, and the interpretation of the requirements of the Contract Documents pertaining to the performance of the Work, will be referred initially to Engineer in writing within 30 days of the event giving rise to the question.

B. Engineer will, with reasonable promptness, render a written decision on the issue referred. If Owner or Contractor believes that any such decision entitles them to an adjustment in the Contract Price or Contract Times or both, a Claim may be made under Paragraph 10.05. The date of Engineer's decision shall be the date of the event giving rise to the issues referenced for the purposes of Paragraph 10.05.B.

C. Engineer's written decision on the issue referred will be final and binding on Owner and Contractor, subject to the provisions of Paragraph 10.05.

D. When functioning as interpreter and judge under this Paragraph 9.08, Engineer will not show partiality to Owner or Contractor and will not be liable in connection with any interpretation or decision rendered in good faith in such capacity.

9.09 *Limitations on Engineer's Authority and Responsibilities*

A. Neither Engineer's authority or responsibility under this Article 9 or under any other provision of the Contract Documents nor any decision made by Engineer in good faith either to exercise or not exercise such authority or responsibility or the undertaking, exercise, or performance of any authority or responsibility by Engineer shall create, impose, or give rise to any duty in contract, tort, or otherwise owed by Engineer to Contractor, any Subcontractor, any Supplier, any other individual or entity, or to any surety for or employee or agent of any of them.

B. Engineer will not supervise, direct, control, or have authority over or be responsible for Contractor's means, methods, techniques, sequences, or procedures of construction, or the safety precautions and programs incident thereto, or for any failure of Contractor to comply with Laws and Regulations applicable to the performance of the Work. Engineer will not be responsible for Contractor's failure to perform the Work in accordance with the Contract Documents.

C. Engineer will not be responsible for the acts or omissions of Contractor or of any Subcontractor, any Supplier, or of any other individual or entity performing any of the Work.

D. Engineer's review of the final Application for Payment and accompanying documentation and all maintenance and operating instructions, schedules, guarantees, bonds, certificates of inspection, tests and approvals, and other documentation required to be delivered by Paragraph 14.07.A will only be to determine generally that their content complies with the requirements of, and in the case of certificates of inspections, tests, and approvals that the results certified indicate compliance with, the Contract Documents.

E. The limitations upon authority and responsibility set forth in this Paragraph 9.09 shall also apply to the Resident Project Representative, if any, and assistants, if any.

9.10 *Compliance with Safety Program*

A. While at the Site, Engineer's employees and representatives shall comply with the specific applicable requirements of Contractor's safety programs of which Engineer has been informed pursuant to Paragraph 6.13.D.

ARTICLE 10 – CHANGES IN THE WORK; CLAIMS

10.01 *Authorized Changes in the Work*

A. Without invalidating the Contract and without notice to any surety, Owner may, at any time or from time to time, order additions, deletions, or revisions in the Work by a Change Order, or a Work Change Directive. Upon receipt of any such document, Contractor shall promptly proceed with the Work involved which will be performed under the applicable conditions of the Contract Documents (except as otherwise specifically provided).

B. If Owner and Contractor are unable to agree on entitlement to, or on the amount or extent, if any, of an adjustment in the Contract Price or Contract Times, or both, that should be allowed as a result of a Work Change Directive, a Claim may be made therefor as provided in Paragraph 10.05.

10.02 *Unauthorized Changes in the Work*

A. Contractor shall not be entitled to an increase in the Contract Price or an extension of the Contract Times with respect to any work performed that is not required by the Contract Documents as amended, modified, or supplemented as provided in Paragraph 3.04, except in the case of an emergency as provided in Paragraph 6.16 or in the case of uncovering Work as provided in Paragraph 13.04.D.

10.03 *Execution of Change Orders*

A. Owner and Contractor shall execute appropriate Change Orders recommended by Engineer covering:

1. changes in the Work which are: (i) ordered by Owner pursuant to Paragraph 10.01.A, (ii) required because of acceptance of defective Work under Paragraph 13.08.A or Owner's correction of defective Work under Paragraph 13.09, or (iii) agreed to by the parties;

2. changes in the Contract Price or Contract Times which are agreed to by the parties, including any undisputed sum or amount of time for Work actually performed in accordance with a Work Change Directive; and

3. changes in the Contract Price or Contract Times which embody the substance of any written decision rendered by Engineer pursuant to Paragraph 10.05; provided that, in lieu of executing any such Change Order, an appeal may be taken from any such decision in accordance with the provisions of the Contract Documents and applicable Laws and Regulations, but during any such appeal, Contractor shall carry on the Work and adhere to the Progress Schedule as provided in Paragraph 6.18.A.

10.04 *Notification to Surety*

A. If the provisions of any bond require notice to be given to a surety of any change affecting the general scope of the Work or the provisions of the Contract Documents (including, but not limited to, Contract Price or Contract Times), the giving of any such notice will be Contractor's responsibility. The amount of each applicable bond will be adjusted to reflect the effect of any such change.

10.05 *Claims*

A. *Engineer's Decision Required:* All Claims, except those waived pursuant to Paragraph 14.09, shall be referred to the Engineer for decision. A decision by Engineer shall be required as a condition precedent to any exercise by Owner or Contractor of any rights or remedies either may otherwise have under the Contract Documents or by Laws and Regulations in respect of such Claims.

B. *Notice:* Written notice stating the general nature of each Claim shall be delivered by the claimant to Engineer and the other party to the Contract promptly (but in no event later than 30 days) after the start of the event giving rise thereto. The responsibility to substantiate a Claim shall rest with the party making the Claim. Notice of the amount or extent of the Claim, with supporting data shall be delivered to the Engineer and the other party to the Contract within 60 days after the start of such event (unless Engineer allows additional time for claimant to submit additional or more accurate data in support of such Claim). A Claim for an adjustment in Contract Price shall be prepared in accordance with the provisions of Paragraph 12.01.B. A Claim for an adjustment in Contract Times shall be prepared in accordance with the provisions of Paragraph 12.02.B. Each Claim shall be accompanied by claimant's written statement that the adjustment claimed is the entire adjustment to which the claimant believes it is entitled as a result of said event. The opposing party shall submit any response to Engineer and the claimant within 30 days after receipt of the claimant's last submittal (unless Engineer allows additional time).

C. *Engineer's Action*: Engineer will review each Claim and, within 30 days after receipt of the last submittal of the claimant or the last submittal of the opposing party, if any, take one of the following actions in writing:

1. deny the Claim in whole or in part;

2. approve the Claim; or

3. notify the parties that the Engineer is unable to resolve the Claim if, in the Engineer's sole discretion, it would be inappropriate for the Engineer to do so. For purposes of further resolution of the Claim, such notice shall be deemed a denial.

D. In the event that Engineer does not take action on a Claim within said 30 days, the Claim shall be deemed denied.

E. Engineer's written action under Paragraph 10.05.C or denial pursuant to Paragraphs 10.05.C.3 or 10.05.D will be final and binding upon Owner and Contractor, unless Owner or Contractor invoke the dispute resolution procedure set forth in Article 16 within 30 days of such action or denial.

F. No Claim for an adjustment in Contract Price or Contract Times will be valid if not submitted in accordance with this Paragraph 10.05.

ARTICLE 11 – COST OF THE WORK; ALLOWANCES; UNIT PRICE WORK

11.01 *Cost of the Work*

A. *Costs Included:* The term Cost of the Work means the sum of all costs, except those excluded in Paragraph 11.01.B, necessarily incurred and paid by Contractor in the proper performance of the Work. When the value of any Work covered by a Change Order or when a Claim for an adjustment in Contract Price is determined on the basis of Cost of the Work, the costs to be reimbursed to Contractor will be only those additional or incremental costs required because of the change in the Work or because of the event giving rise to the Claim. Except as otherwise may be agreed to in writing by Owner, such costs shall be in amounts no higher than those prevailing in the locality of the Project, shall not include any of the costs itemized in Paragraph 11.01.B, and shall include only the following items:

1. Payroll costs for employees in the direct employ of Contractor in the performance of the Work under schedules of job classifications agreed upon by Owner and Contractor. Such employees shall include, without limitation, superintendents, foremen, and other personnel employed full time on the Work. Payroll costs for employees not employed full time on the Work shall be apportioned on the basis of their time spent on the Work. Payroll costs shall include, but not be limited to, salaries and wages plus the cost of fringe benefits, which shall include social security contributions, unemployment, excise, and payroll taxes, workers' compensation, health and retirement benefits, bonuses, sick leave, vacation and holiday pay applicable thereto. The expenses of performing Work outside of regular working hours, on

Saturday, Sunday, or legal holidays, shall be included in the above to the extent authorized by Owner.

2. Cost of all materials and equipment furnished and incorporated in the Work, including costs of transportation and storage thereof, and Suppliers' field services required in connection therewith. All cash discounts shall accrue to Contractor unless Owner deposits funds with Contractor with which to make payments, in which case the cash discounts shall accrue to Owner. All trade discounts, rebates and refunds and returns from sale of surplus materials and equipment shall accrue to Owner, and Contractor shall make provisions so that they may be obtained.

3. Payments made by Contractor to Subcontractors for Work performed by Subcontractors. If required by Owner, Contractor shall obtain competitive bids from subcontractors acceptable to Owner and Contractor and shall deliver such bids to Owner, who will then determine, with the advice of Engineer, which bids, if any, will be acceptable. If any subcontract provides that the Subcontractor is to be paid on the basis of Cost of the Work plus a fee, the Subcontractor's Cost of the Work and fee shall be determined in the same manner as Contractor's Cost of the Work and fee as provided in this Paragraph 11.01.

4. Costs of special consultants (including but not limited to engineers, architects, testing laboratories, surveyors, attorneys, and accountants) employed for services specifically related to the Work.

5. Supplemental costs including the following:

   a. The proportion of necessary transportation, travel, and subsistence expenses of Contractor's employees incurred in discharge of duties connected with the Work.

   b. Cost, including transportation and maintenance, of all materials, supplies, equipment, machinery, appliances, office, and temporary facilities at the Site, and hand tools not owned by the workers, which are consumed in the performance of the Work, and cost, less market value, of such items used but not consumed which remain the property of Contractor.

   c. Rentals of all construction equipment and machinery, and the parts thereof whether rented from Contractor or others in accordance with rental agreements approved by Owner with the advice of Engineer, and the costs of transportation, loading, unloading, assembly, dismantling, and removal thereof. All such costs shall be in accordance with the terms of said rental agreements. The rental of any such equipment, machinery, or parts shall cease when the use thereof is no longer necessary for the Work.

   d. Sales, consumer, use, and other similar taxes related to the Work, and for which Contractor is liable, as imposed by Laws and Regulations.

   e. Deposits lost for causes other than negligence of Contractor, any Subcontractor, or anyone directly or indirectly employed by any of them or for whose acts any of them may be liable, and royalty payments and fees for permits and licenses.

f. Losses and damages (and related expenses) caused by damage to the Work, not compensated by insurance or otherwise, sustained by Contractor in connection with the performance of the Work (except losses and damages within the deductible amounts of property insurance established in accordance with Paragraph 5.06.D), provided such losses and damages have resulted from causes other than the negligence of Contractor, any Subcontractor, or anyone directly or indirectly employed by any of them or for whose acts any of them may be liable. Such losses shall include settlements made with the written consent and approval of Owner. No such losses, damages, and expenses shall be included in the Cost of the Work for the purpose of determining Contractor's fee.

g. The cost of utilities, fuel, and sanitary facilities at the Site.

h. Minor expenses such as telegrams, long distance telephone calls, telephone service at the Site, express and courier services, and similar petty cash items in connection with the Work.

i. The costs of premiums for all bonds and insurance Contractor is required by the Contract Documents to purchase and maintain.

B. *Costs Excluded:* The term Cost of the Work shall not include any of the following items:

1. Payroll costs and other compensation of Contractor's officers, executives, principals (of partnerships and sole proprietorships), general managers, safety managers, engineers, architects, estimators, attorneys, auditors, accountants, purchasing and contracting agents, expediters, timekeepers, clerks, and other personnel employed by Contractor, whether at the Site or in Contractor's principal or branch office for general administration of the Work and not specifically included in the agreed upon schedule of job classifications referred to in Paragraph 11.01.A.1 or specifically covered by Paragraph 11.01.A.4, all of which are to be considered administrative costs covered by the Contractor's fee.

2. Expenses of Contractor's principal and branch offices other than Contractor's office at the Site.

3. Any part of Contractor's capital expenses, including interest on Contractor's capital employed for the Work and charges against Contractor for delinquent payments.

4. Costs due to the negligence of Contractor, any Subcontractor, or anyone directly or indirectly employed by any of them or for whose acts any of them may be liable, including but not limited to, the correction of defective Work, disposal of materials or equipment wrongly supplied, and making good any damage to property.

5. Other overhead or general expense costs of any kind and the costs of any item not specifically and expressly included in Paragraphs 11.01.A.

C. *Contractor's Fee:* When all the Work is performed on the basis of cost-plus, Contractor's fee shall be determined as set forth in the Agreement. When the value of any Work covered by a Change Order or when a Claim for an adjustment in Contract Price is determined on the basis of Cost of the Work, Contractor's fee shall be determined as set forth in Paragraph 12.01.C.

D. *Documentation:* Whenever the Cost of the Work for any purpose is to be determined pursuant to Paragraphs 11.01.A and 11.01.B, Contractor will establish and maintain records thereof in accordance with generally accepted accounting practices and submit in a form acceptable to Engineer an itemized cost breakdown together with supporting data.

11.02 *Allowances*

A. It is understood that Contractor has included in the Contract Price all allowances so named in the Contract Documents and shall cause the Work so covered to be performed for such sums and by such persons or entities as may be acceptable to Owner and Engineer.

B. *Cash Allowances:*

1. Contractor agrees that:

a. the cash allowances include the cost to Contractor (less any applicable trade discounts) of materials and equipment required by the allowances to be delivered at the Site, and all applicable taxes; and

b. Contractor's costs for unloading and handling on the Site, labor, installation, overhead, profit, and other expenses contemplated for the cash allowances have been included in the Contract Price and not in the allowances, and no demand for additional payment on account of any of the foregoing will be valid.

C. *Contingency Allowance:*

1. Contractor agrees that a contingency allowance, if any, is for the sole use of Owner to cover unanticipated costs.

D. Prior to final payment, an appropriate Change Order will be issued as recommended by Engineer to reflect actual amounts due Contractor on account of Work covered by allowances, and the Contract Price shall be correspondingly adjusted.

11.03 *Unit Price Work*

A. Where the Contract Documents provide that all or part of the Work is to be Unit Price Work, initially the Contract Price will be deemed to include for all Unit Price Work an amount equal to the sum of the unit price for each separately identified item of Unit Price Work times the estimated quantity of each item as indicated in the Agreement.

B. The estimated quantities of items of Unit Price Work are not guaranteed and are solely for the purpose of comparison of Bids and determining an initial Contract Price. Determinations of the actual quantities and classifications of Unit Price Work performed by Contractor will be made by Engineer subject to the provisions of Paragraph 9.07.

C. Each unit price will be deemed to include an amount considered by Contractor to be adequate to cover Contractor's overhead and profit for each separately identified item.

D. Owner or Contractor may make a Claim for an adjustment in the Contract Price in accordance with Paragraph 10.05 if:

1. the quantity of any item of Unit Price Work performed by Contractor differs materially and significantly from the estimated quantity of such item indicated in the Agreement; and

2. there is no corresponding adjustment with respect to any other item of Work; and

3. Contractor believes that Contractor is entitled to an increase in Contract Price as a result of having incurred additional expense or Owner believes that Owner is entitled to a decrease in Contract Price and the parties are unable to agree as to the amount of any such increase or decrease.

ARTICLE 12 – CHANGE OF CONTRACT PRICE; CHANGE OF CONTRACT TIMES

12.01 *Change of Contract Price*

A. The Contract Price may only be changed by a Change Order. Any Claim for an adjustment in the Contract Price shall be based on written notice submitted by the party making the Claim to the Engineer and the other party to the Contract in accordance with the provisions of Paragraph 10.05.

B. The value of any Work covered by a Change Order or of any Claim for an adjustment in the Contract Price will be determined as follows:

1. where the Work involved is covered by unit prices contained in the Contract Documents, by application of such unit prices to the quantities of the items involved (subject to the provisions of Paragraph 11.03); or

2. where the Work involved is not covered by unit prices contained in the Contract Documents, by a mutually agreed lump sum (which may include an allowance for overhead and profit not necessarily in accordance with Paragraph 12.01.C.2); or

3. where the Work involved is not covered by unit prices contained in the Contract Documents and agreement to a lump sum is not reached under Paragraph 12.01.B.2, on the basis of the Cost of the Work (determined as provided in Paragraph 11.01) plus a Contractor's fee for overhead and profit (determined as provided in Paragraph 12.01.C).

C. *Contractor's Fee:* The Contractor's fee for overhead and profit shall be determined as follows:

1. a mutually acceptable fixed fee; or

2. if a fixed fee is not agreed upon, then a fee based on the following percentages of the various portions of the Cost of the Work:

a. for costs incurred under Paragraphs 11.01.A.1 and 11.01.A.2, the Contractor's fee shall be 15 percent;

b. for costs incurred under Paragraph 11.01.A.3, the Contractor's fee shall be five percent;

c. where one or more tiers of subcontracts are on the basis of Cost of the Work plus a fee and no fixed fee is agreed upon, the intent of Paragraphs 12.01.C.2.a and 12.01.C.2.b is that the Subcontractor who actually performs the Work, at whatever tier, will be paid a fee of 15 percent of the costs incurred by such Subcontractor under Paragraphs 11.01.A.1 and 11.01.A.2 and that any higher tier Subcontractor and Contractor will each be paid a fee of five percent of the amount paid to the next lower tier Subcontractor;

d. no fee shall be payable on the basis of costs itemized under Paragraphs 11.01.A.4, 11.01.A.5, and 11.01.B;

e. the amount of credit to be allowed by Contractor to Owner for any change which results in a net decrease in cost will be the amount of the actual net decrease in cost plus a deduction in Contractor's fee by an amount equal to five percent of such net decrease; and

f. when both additions and credits are involved in any one change, the adjustment in Contractor's fee shall be computed on the basis of the net change in accordance with Paragraphs 12.01.C.2.a through 12.01.C.2.e, inclusive.

12.02 *Change of Contract Times*

A. The Contract Times may only be changed by a Change Order. Any Claim for an adjustment in the Contract Times shall be based on written notice submitted by the party making the Claim to the Engineer and the other party to the Contract in accordance with the provisions of Paragraph 10.05.

B. Any adjustment of the Contract Times covered by a Change Order or any Claim for an adjustment in the Contract Times will be determined in accordance with the provisions of this Article 12.

12.03 *Delays*

A. Where Contractor is prevented from completing any part of the Work within the Contract Times due to delay beyond the control of Contractor, the Contract Times will be extended in an amount equal to the time lost due to such delay if a Claim is made therefor as provided in Paragraph 12.02.A. Delays beyond the control of Contractor shall include, but not be limited to, acts or neglect by Owner, acts or neglect of utility owners or other contractors performing other work as contemplated by Article 7, fires, floods, epidemics, abnormal weather conditions, or acts of God.

B. If Owner, Engineer, or other contractors or utility owners performing other work for Owner as contemplated by Article 7, or anyone for whom Owner is responsible, delays, disrupts, or interferes with the performance or progress of the Work, then Contractor shall be entitled to an equitable adjustment in the Contract Price or the Contract Times, or both. Contractor's entitlement to an adjustment of the Contract Times is conditioned on such adjustment being essential to Contractor's ability to complete the Work within the Contract Times.

C. If Contractor is delayed in the performance or progress of the Work by fire, flood, epidemic, abnormal weather conditions, acts of God, acts or failures to act of utility owners not under the control of Owner, or other causes not the fault of and beyond control of Owner and Contractor, then Contractor shall be entitled to an equitable adjustment in Contract Times, if such adjustment is essential to Contractor's ability to complete the Work within the Contract Times. Such an adjustment shall be Contractor's sole and exclusive remedy for the delays described in this Paragraph 12.03.C.

D. Owner, Engineer, and their officers, directors, members, partners, employees, agents, consultants, or subcontractors shall not be liable to Contractor for any claims, costs, losses, or damages (including but not limited to all fees and charges of engineers, architects, attorneys, and other professionals and all court or arbitration or other dispute resolution costs) sustained by Contractor on or in connection with any other project or anticipated project.

E. Contractor shall not be entitled to an adjustment in Contract Price or Contract Times for delays within the control of Contractor. Delays attributable to and within the control of a Subcontractor or Supplier shall be deemed to be delays within the control of Contractor.

ARTICLE 13 – TESTS AND INSPECTIONS; CORRECTION, REMOVAL OR ACCEPTANCE OF DEFECTIVE WORK

13.01 *Notice of Defects*

A. Prompt notice of all defective Work of which Owner or Engineer has actual knowledge will be given to Contractor. Defective Work may be rejected, corrected, or accepted as provided in this Article 13.

13.02 *Access to Work*

A. Owner, Engineer, their consultants and other representatives and personnel of Owner, independent testing laboratories, and governmental agencies with jurisdictional interests will have access to the Site and the Work at reasonable times for their observation, inspection, and testing. Contractor shall provide them proper and safe conditions for such access and advise them of Contractor's safety procedures and programs so that they may comply therewith as applicable.

13.03 *Tests and Inspections*

A. Contractor shall give Engineer timely notice of readiness of the Work for all required inspections, tests, or approvals and shall cooperate with inspection and testing personnel to facilitate required inspections or tests.

B. Owner shall employ and pay for the services of an independent testing laboratory to perform all inspections, tests, or approvals required by the Contract Documents except:

1. for inspections, tests, or approvals covered by Paragraphs 13.03.C and 13.03.D below;

2. that costs incurred in connection with tests or inspections conducted pursuant to Paragraph 13.04.B shall be paid as provided in Paragraph 13.04.C; and

3. as otherwise specifically provided in the Contract Documents.

C. If Laws or Regulations of any public body having jurisdiction require any Work (or part thereof) specifically to be inspected, tested, or approved by an employee or other representative of such public body, Contractor shall assume full responsibility for arranging and obtaining such inspections, tests, or approvals, pay all costs in connection therewith, and furnish Engineer the required certificates of inspection or approval.

D. Contractor shall be responsible for arranging and obtaining and shall pay all costs in connection with any inspections, tests, or approvals required for Owner's and Engineer's acceptance of materials or equipment to be incorporated in the Work; or acceptance of materials, mix designs, or equipment submitted for approval prior to Contractor's purchase thereof for incorporation in the Work. Such inspections, tests, or approvals shall be performed by organizations acceptable to Owner and Engineer.

E. If any Work (or the work of others) that is to be inspected, tested, or approved is covered by Contractor without written concurrence of Engineer, Contractor shall, if requested by Engineer, uncover such Work for observation.

F. Uncovering Work as provided in Paragraph 13.03.E shall be at Contractor's expense unless Contractor has given Engineer timely notice of Contractor's intention to cover the same and Engineer has not acted with reasonable promptness in response to such notice.

13.04 *Uncovering Work*

A. If any Work is covered contrary to the written request of Engineer, it must, if requested by Engineer, be uncovered for Engineer's observation and replaced at Contractor's expense.

B. If Engineer considers it necessary or advisable that covered Work be observed by Engineer or inspected or tested by others, Contractor, at Engineer's request, shall uncover, expose, or otherwise make available for observation, inspection, or testing as Engineer may require, that portion of the Work in question, furnishing all necessary labor, material, and equipment.

C. If it is found that the uncovered Work is defective, Contractor shall pay all claims, costs, losses, and damages (including but not limited to all fees and charges of engineers, architects, attorneys, and other professionals and all court or arbitration or other dispute resolution costs) arising out of or relating to such uncovering, exposure, observation, inspection, and testing, and of satisfactory replacement or reconstruction (including but not limited to all costs of repair or replacement of work of others); and Owner shall be entitled to an appropriate decrease in the Contract Price. If the parties are unable to agree as to the amount thereof, Owner may make a Claim therefor as provided in Paragraph 10.05.

D. If the uncovered Work is not found to be defective, Contractor shall be allowed an increase in the Contract Price or an extension of the Contract Times, or both, directly attributable to such uncovering, exposure, observation, inspection, testing, replacement, and reconstruction. If the parties are unable to agree as to the amount or extent thereof, Contractor may make a Claim therefor as provided in Paragraph 10.05.

13.05 *Owner May Stop the Work*

A. If the Work is defective, or Contractor fails to supply sufficient skilled workers or suitable materials or equipment, or fails to perform the Work in such a way that the completed Work will conform to the Contract Documents, Owner may order Contractor to stop the Work, or any portion thereof, until the cause for such order has been eliminated; however, this right of Owner to stop the Work shall not give rise to any duty on the part of Owner to exercise this right for the benefit of Contractor, any Subcontractor, any Supplier, any other individual or entity, or any surety for, or employee or agent of any of them.

13.06 *Correction or Removal of Defective Work*

A. Promptly after receipt of written notice, Contractor shall correct all defective Work, whether or not fabricated, installed, or completed, or, if the Work has been rejected by Engineer, remove it from the Project and replace it with Work that is not defective. Contractor shall pay all claims, costs, losses, and damages (including but not limited to all fees and charges of engineers, architects, attorneys, and other professionals and all court or arbitration or other dispute resolution costs) arising out of or relating to such correction or removal (including but not limited to all costs of repair or replacement of work of others).

B. When correcting defective Work under the terms of this Paragraph 13.06 or Paragraph 13.07, Contractor shall take no action that would void or otherwise impair Owner's special warranty and guarantee, if any, on said Work.

13.07 *Correction Period*

A. If within one year after the date of Substantial Completion (or such longer period of time as may be prescribed by the terms of any applicable special guarantee required by the Contract Documents) or by any specific provision of the Contract Documents, any Work is found to be defective, or if the repair of any damages to the land or areas made available for Contractor's use by Owner or permitted by Laws and Regulations as contemplated in Paragraph 6.11.A is found to be defective, Contractor shall promptly, without cost to Owner and in accordance with Owner's written instructions:

1. repair such defective land or areas; or

2. correct such defective Work; or

3. if the defective Work has been rejected by Owner, remove it from the Project and replace it with Work that is not defective, and

4. satisfactorily correct or repair or remove and replace any damage to other Work, to the work of others or other land or areas resulting therefrom.

B. If Contractor does not promptly comply with the terms of Owner's written instructions, or in an emergency where delay would cause serious risk of loss or damage, Owner may have the defective Work corrected or repaired or may have the rejected Work removed and replaced. All claims, costs, losses, and damages (including but not limited to all fees and charges of engineers, architects, attorneys, and other professionals and all court or arbitration or other dispute resolution costs) arising out of or relating to such correction or repair or such removal and replacement (including but not limited to all costs of repair or replacement of work of others) will be paid by Contractor.

C. In special circumstances where a particular item of equipment is placed in continuous service before Substantial Completion of all the Work, the correction period for that item may start to run from an earlier date if so provided in the Specifications.

D. Where defective Work (and damage to other Work resulting therefrom) has been corrected or removed and replaced under this Paragraph 13.07, the correction period hereunder with respect to such Work will be extended for an additional period of one year after such correction or removal and replacement has been satisfactorily completed.

E. Contractor's obligations under this Paragraph 13.07 are in addition to any other obligation or warranty. The provisions of this Paragraph 13.07 shall not be construed as a substitute for, or a waiver of, the provisions of any applicable statute of limitation or repose.

13.08 *Acceptance of Defective Work*

A. If, instead of requiring correction or removal and replacement of defective Work, Owner (and, prior to Engineer's recommendation of final payment, Engineer) prefers to accept it, Owner may do so. Contractor shall pay all claims, costs, losses, and damages (including but not limited to all fees and charges of engineers, architects, attorneys, and other professionals and all court or arbitration or other dispute resolution costs) attributable to Owner's evaluation of and determination to accept such defective Work (such costs to be approved by Engineer as to reasonableness) and for the diminished value of the Work to the extent not otherwise paid by Contractor pursuant to this sentence. If any such acceptance occurs prior to Engineer's recommendation of final payment, a Change Order will be issued incorporating the necessary revisions in the Contract Documents with respect to the Work, and Owner shall be entitled to an appropriate decrease in the Contract Price, reflecting the diminished value of Work so accepted. If the parties are unable to agree as to the amount thereof, Owner may make a Claim therefor as provided in Paragraph 10.05. If the acceptance occurs after such recommendation, an appropriate amount will be paid by Contractor to Owner.

13.09 *Owner May Correct Defective Work*

A. If Contractor fails within a reasonable time after written notice from Engineer to correct defective Work, or to remove and replace rejected Work as required by Engineer in accordance with Paragraph 13.06.A, or if Contractor fails to perform the Work in accordance with the Contract Documents, or if Contractor fails to comply with any other provision of the Contract Documents, Owner may, after seven days written notice to Contractor, correct, or remedy any such deficiency.

B. In exercising the rights and remedies under this Paragraph 13.09, Owner shall proceed expeditiously. In connection with such corrective or remedial action, Owner may exclude Contractor from all or part of the Site, take possession of all or part of the Work and suspend Contractor's services related thereto, take possession of Contractor's tools, appliances, construction equipment and machinery at the Site, and incorporate in the Work all materials and

C. All claims, costs, losses, and damages (including but not limited to all fees and charges of engineers, architects, attorneys, and other professionals and all court or arbitration or other dispute resolution costs) incurred or sustained by Owner in exercising the rights and remedies under this Paragraph 13.09 will be charged against Contractor, and a Change Order will be issued incorporating the necessary revisions in the Contract Documents with respect to the Work; and Owner shall be entitled to an appropriate decrease in the Contract Price. If the parties are unable to agree as to the amount of the adjustment, Owner may make a Claim therefor as provided in Paragraph 10.05. Such claims, costs, losses and damages will include but not be limited to all costs of repair, or replacement of work of others destroyed or damaged by correction, removal, or replacement of Contractor's defective Work.

D. Contractor shall not be allowed an extension of the Contract Times because of any delay in the performance of the Work attributable to the exercise by Owner of Owner's rights and remedies under this Paragraph 13.09.

ARTICLE 14 – PAYMENTS TO CONTRACTOR AND COMPLETION

14.01 *Schedule of Values*

A. The Schedule of Values established as provided in Paragraph 2.07.A will serve as the basis for progress payments and will be incorporated into a form of Application for Payment acceptable to Engineer. Progress payments on account of Unit Price Work will be based on the number of units completed.

14.02 *Progress Payments*

A. *Applications for Payments:*

1. At least 20 days before the date established in the Agreement for each progress payment (but not more often than once a month), Contractor shall submit to Engineer for review an Application for Payment filled out and signed by Contractor covering the Work completed as of the date of the Application and accompanied by such supporting documentation as is required by the Contract Documents. If payment is requested on the basis of materials and equipment not incorporated in the Work but delivered and suitably stored at the Site or at another location agreed to in writing, the Application for Payment shall also be accompanied by a bill of sale, invoice, or other documentation warranting that Owner has received the materials and equipment free and clear of all Liens and evidence that the materials and equipment are covered by appropriate property insurance or other arrangements to protect Owner's interest therein, all of which must be satisfactory to Owner.

2. Beginning with the second Application for Payment, each Application shall include an affidavit of Contractor stating that all previous progress payments received on account of the

Work have been applied on account to discharge Contractor's legitimate obligations associated with prior Applications for Payment.

3. The amount of retainage with respect to progress payments will be as stipulated in the Agreement.

B. *Review of Applications:*

1. Engineer will, within 10 days after receipt of each Application for Payment, either indicate in writing a recommendation of payment and present the Application to Owner or return the Application to Contractor indicating in writing Engineer's reasons for refusing to recommend payment. In the latter case, Contractor may make the necessary corrections and resubmit the Application.

2. Engineer's recommendation of any payment requested in an Application for Payment will constitute a representation by Engineer to Owner, based on Engineer's observations of the executed Work as an experienced and qualified design professional, and on Engineer's review of the Application for Payment and the accompanying data and schedules, that to the best of Engineer's knowledge, information and belief:

   a. the Work has progressed to the point indicated;

   b. the quality of the Work is generally in accordance with the Contract Documents (subject to an evaluation of the Work as a functioning whole prior to or upon Substantial Completion, the results of any subsequent tests called for in the Contract Documents, a final determination of quantities and classifications for Unit Price Work under Paragraph 9.07, and any other qualifications stated in the recommendation); and

   c. the conditions precedent to Contractor's being entitled to such payment appear to have been fulfilled in so far as it is Engineer's responsibility to observe the Work.

3. By recommending any such payment Engineer will not thereby be deemed to have represented that:

   a. inspections made to check the quality or the quantity of the Work as it has been performed have been exhaustive, extended to every aspect of the Work in progress, or involved detailed inspections of the Work beyond the responsibilities specifically assigned to Engineer in the Contract Documents; or

   b. there may not be other matters or issues between the parties that might entitle Contractor to be paid additionally by Owner or entitle Owner to withhold payment to Contractor.

4. Neither Engineer's review of Contractor's Work for the purposes of recommending payments nor Engineer's recommendation of any payment, including final payment, will impose responsibility on Engineer:

   a. to supervise, direct, or control the Work, or

b. for the means, methods, techniques, sequences, or procedures of construction, or the safety precautions and programs incident thereto, or

c. for Contractor's failure to comply with Laws and Regulations applicable to Contractor's performance of the Work, or

d. to make any examination to ascertain how or for what purposes Contractor has used the moneys paid on account of the Contract Price, or

e. to determine that title to any of the Work, materials, or equipment has passed to Owner free and clear of any Liens.

5. Engineer may refuse to recommend the whole or any part of any payment if, in Engineer's opinion, it would be incorrect to make the representations to Owner stated in Paragraph 14.02.B.2. Engineer may also refuse to recommend any such payment or, because of subsequently discovered evidence or the results of subsequent inspections or tests, revise or revoke any such payment recommendation previously made, to such extent as may be necessary in Engineer's opinion to protect Owner from loss because:

a. the Work is defective, or completed Work has been damaged, requiring correction or replacement;

b. the Contract Price has been reduced by Change Orders;

c. Owner has been required to correct defective Work or complete Work in accordance with Paragraph 13.09; or

d. Engineer has actual knowledge of the occurrence of any of the events enumerated in Paragraph 15.02.A.

C. *Payment Becomes Due:*

1. Ten days after presentation of the Application for Payment to Owner with Engineer's recommendation, the amount recommended will (subject to the provisions of Paragraph 14.02.D) become due, and when due will be paid by Owner to Contractor.

D. *Reduction in Payment:*

1. Owner may refuse to make payment of the full amount recommended by Engineer because:

a. claims have been made against Owner on account of Contractor's performance or furnishing of the Work;

b. Liens have been filed in connection with the Work, except where Contractor has delivered a specific bond satisfactory to Owner to secure the satisfaction and discharge of such Liens;

c. there are other items entitling Owner to a set-off against the amount recommended; or

d. Owner has actual knowledge of the occurrence of any of the events enumerated in Paragraphs 14.02.B.5.a through 14.02.B.5.c or Paragraph 15.02.A.

2. If Owner refuses to make payment of the full amount recommended by Engineer, Owner will give Contractor immediate written notice (with a copy to Engineer) stating the reasons for such action and promptly pay Contractor any amount remaining after deduction of the amount so withheld. Owner shall promptly pay Contractor the amount so withheld, or any adjustment thereto agreed to by Owner and Contractor, when Contractor remedies the reasons for such action.

3. Upon a subsequent determination that Owner's refusal of payment was not justified, the amount wrongfully withheld shall be treated as an amount due as determined by Paragraph 14.02.C.1 and subject to interest as provided in the Agreement.

14.03 *Contractor's Warranty of Title*

A. Contractor warrants and guarantees that title to all Work, materials, and equipment covered by any Application for Payment, whether incorporated in the Project or not, will pass to Owner no later than the time of payment free and clear of all Liens.

14.04 *Substantial Completion*

A. When Contractor considers the entire Work ready for its intended use Contractor shall notify Owner and Engineer in writing that the entire Work is substantially complete (except for items specifically listed by Contractor as incomplete) and request that Engineer issue a certificate of Substantial Completion.

B. Promptly after Contractor's notification, Owner, Contractor, and Engineer shall make an inspection of the Work to determine the status of completion. If Engineer does not consider the Work substantially complete, Engineer will notify Contractor in writing giving the reasons therefor.

C. If Engineer considers the Work substantially complete, Engineer will deliver to Owner a tentative certificate of Substantial Completion which shall fix the date of Substantial Completion. There shall be attached to the certificate a tentative list of items to be completed or corrected before final payment. Owner shall have seven days after receipt of the tentative certificate during which to make written objection to Engineer as to any provisions of the certificate or attached list. If, after considering such objections, Engineer concludes that the Work is not substantially complete, Engineer will, within 14 days after submission of the tentative certificate to Owner, notify Contractor in writing, stating the reasons therefor. If, after consideration of Owner's objections, Engineer considers the Work substantially complete, Engineer will, within said 14 days, execute and deliver to Owner and Contractor a definitive certificate of Substantial Completion (with a revised tentative list of items to be completed or corrected) reflecting such changes from the tentative certificate as Engineer believes justified after consideration of any objections from Owner.

D. At the time of delivery of the tentative certificate of Substantial Completion, Engineer will deliver to Owner and Contractor a written recommendation as to division of responsibilities pending final payment between Owner and Contractor with respect to security, operation, safety, and protection of the Work, maintenance, heat, utilities, insurance, and warranties and guarantees. Unless Owner and Contractor agree otherwise in writing and so inform Engineer in writing prior to Engineer's issuing the definitive certificate of Substantial Completion, Engineer's aforesaid recommendation will be binding on Owner and Contractor until final payment.

E. Owner shall have the right to exclude Contractor from the Site after the date of Substantial Completion subject to allowing Contractor reasonable access to remove its property and complete or correct items on the tentative list.

14.05 *Partial Utilization*

A. Prior to Substantial Completion of all the Work, Owner may use or occupy any substantially completed part of the Work which has specifically been identified in the Contract Documents, or which Owner, Engineer, and Contractor agree constitutes a separately functioning and usable part of the Work that can be used by Owner for its intended purpose without significant interference with Contractor's performance of the remainder of the Work, subject to the following conditions:

1. Owner at any time may request Contractor in writing to permit Owner to use or occupy any such part of the Work which Owner believes to be ready for its intended use and substantially complete. If and when Contractor agrees that such part of the Work is substantially complete, Contractor, Owner, and Engineer will follow the procedures of Paragraph 14.04.A through D for that part of the Work.

2. Contractor at any time may notify Owner and Engineer in writing that Contractor considers any such part of the Work ready for its intended use and substantially complete and request Engineer to issue a certificate of Substantial Completion for that part of the Work.

3. Within a reasonable time after either such request, Owner, Contractor, and Engineer shall make an inspection of that part of the Work to determine its status of completion. If Engineer does not consider that part of the Work to be substantially complete, Engineer will notify Owner and Contractor in writing giving the reasons therefor. If Engineer considers that part of the Work to be substantially complete, the provisions of Paragraph 14.04 will apply with respect to certification of Substantial Completion of that part of the Work and the division of responsibility in respect thereof and access thereto.

4. No use or occupancy or separate operation of part of the Work may occur prior to compliance with the requirements of Paragraph 5.10 regarding property insurance.

14.06 *Final Inspection*

A. Upon written notice from Contractor that the entire Work or an agreed portion thereof is complete, Engineer will promptly make a final inspection with Owner and Contractor and will notify Contractor in writing of all particulars in which this inspection reveals that the Work is incomplete or defective. Contractor shall immediately take such measures as are necessary to complete such Work or remedy such deficiencies.

14.07 *Final Payment*

A. *Application for Payment:*

1. After Contractor has, in the opinion of Engineer, satisfactorily completed all corrections identified during the final inspection and has delivered, in accordance with the Contract Documents, all maintenance and operating instructions, schedules, guarantees, bonds, certificates or other evidence of insurance, certificates of inspection, marked-up record documents (as provided in Paragraph 6.12), and other documents, Contractor may make application for final payment following the procedure for progress payments.

2. The final Application for Payment shall be accompanied (except as previously delivered) by:

   a. all documentation called for in the Contract Documents, including but not limited to the evidence of insurance required by Paragraph 5.04.B.6;

   b. consent of the surety, if any, to final payment;

   c. a list of all Claims against Owner that Contractor believes are unsettled; and

   d. complete and legally effective releases or waivers (satisfactory to Owner) of all Lien rights arising out of or Liens filed in connection with the Work.

3. In lieu of the releases or waivers of Liens specified in Paragraph 14.07.A.2 and as approved by Owner, Contractor may furnish receipts or releases in full and an affidavit of Contractor that: (i) the releases and receipts include all labor, services, material, and equipment for which a Lien could be filed; and (ii) all payrolls, material and equipment bills, and other indebtedness connected with the Work for which Owner might in any way be responsible, or which might in any way result in liens or other burdens on Owner's property, have been paid or otherwise satisfied. If any Subcontractor or Supplier fails to furnish such a release or receipt in full, Contractor may furnish a bond or other collateral satisfactory to Owner to indemnify Owner against any Lien.

B. *Engineer's Review of Application and Acceptance:*

1. If, on the basis of Engineer's observation of the Work during construction and final inspection, and Engineer's review of the final Application for Payment and accompanying documentation as required by the Contract Documents, Engineer is satisfied that the Work has been completed and Contractor's other obligations under the Contract Documents have been fulfilled, Engineer will, within ten days after receipt of the final Application for Payment, indicate in writing Engineer's recommendation of payment and present the Application for Payment to Owner for payment. At the same time Engineer will also give written notice to Owner and Contractor that the Work is acceptable subject to the provisions of Paragraph 14.09. Otherwise, Engineer will return the Application for Payment to Contractor, indicating in writing the reasons for refusing to recommend final payment, in which case Contractor shall make the necessary corrections and resubmit the Application for Payment.

C. *Payment Becomes Due:*

1. Thirty days after the presentation to Owner of the Application for Payment and accompanying documentation, the amount recommended by Engineer, less any sum Owner is entitled to set off against Engineer's recommendation, including but not limited to liquidated damages, will become due and will be paid by Owner to Contractor.

14.08 *Final Completion Delayed*

A. If, through no fault of Contractor, final completion of the Work is significantly delayed, and if Engineer so confirms, Owner shall, upon receipt of Contractor's final Application for Payment (for Work fully completed and accepted) and recommendation of Engineer, and without terminating the Contract, make payment of the balance due for that portion of the Work fully completed and accepted. If the remaining balance to be held by Owner for Work not fully completed or corrected is less than the retainage stipulated in the Agreement, and if bonds have been furnished as required in Paragraph 5.01, the written consent of the surety to the payment of the balance due for that portion of the Work fully completed and accepted shall be submitted by Contractor to Engineer with the Application for such payment. Such payment shall be made under the terms and conditions governing final payment, except that it shall not constitute a waiver of Claims.

14.09 *Waiver of Claims*

A. The making and acceptance of final payment will constitute:

1. a waiver of all Claims by Owner against Contractor, except Claims arising from unsettled Liens, from defective Work appearing after final inspection pursuant to Paragraph 14.06, from failure to comply with the Contract Documents or the terms of any special guarantees specified therein, or from Contractor's continuing obligations under the Contract Documents; and

2. a waiver of all Claims by Contractor against Owner other than those previously made in accordance with the requirements herein and expressly acknowledged by Owner in writing as still unsettled.

ARTICLE 15 – SUSPENSION OF WORK AND TERMINATION

15.01 *Owner May Suspend Work*

A. At any time and without cause, Owner may suspend the Work or any portion thereof for a period of not more than 90 consecutive days by notice in writing to Contractor and Engineer which will fix the date on which Work will be resumed. Contractor shall resume the Work on the date so fixed. Contractor shall be granted an adjustment in the Contract Price or an extension of the Contract Times, or both, directly attributable to any such suspension if Contractor makes a Claim therefor as provided in Paragraph 10.05.

15.02 *Owner May Terminate for Cause*

A. The occurrence of any one or more of the following events will justify termination for cause:

1. Contractor's persistent failure to perform the Work in accordance with the Contract Documents (including, but not limited to, failure to supply sufficient skilled workers or suitable materials or equipment or failure to adhere to the Progress Schedule established under Paragraph 2.07 as adjusted from time to time pursuant to Paragraph 6.04);

2. Contractor's disregard of Laws or Regulations of any public body having jurisdiction;

3. Contractor's repeated disregard of the authority of Engineer; or

4. Contractor's violation in any substantial way of any provisions of the Contract Documents.

B. If one or more of the events identified in Paragraph 15.02.A occur, Owner may, after giving Contractor (and surety) seven days written notice of its intent to terminate the services of Contractor:

1. exclude Contractor from the Site, and take possession of the Work and of all Contractor's tools, appliances, construction equipment, and machinery at the Site, and use the same to the full extent they could be used by Contractor (without liability to Contractor for trespass or conversion);

2. incorporate in the Work all materials and equipment stored at the Site or for which Owner has paid Contractor but which are stored elsewhere; and

3. complete the Work as Owner may deem expedient.

C. If Owner proceeds as provided in Paragraph 15.02.B, Contractor shall not be entitled to receive any further payment until the Work is completed. If the unpaid balance of the Contract Price exceeds all claims, costs, losses, and damages (including but not limited to all fees and charges of engineers, architects, attorneys, and other professionals and all court or arbitration or other dispute resolution costs) sustained by Owner arising out of or relating to completing the Work, such excess will be paid to Contractor. If such claims, costs, losses, and damages exceed such unpaid balance, Contractor shall pay the difference to Owner. Such claims, costs, losses, and damages incurred by Owner will be reviewed by Engineer as to their reasonableness and, when so approved by Engineer, incorporated in a Change Order. When exercising any rights or remedies under this Paragraph, Owner shall not be required to obtain the lowest price for the Work performed.

D. Notwithstanding Paragraphs 15.02.B and 15.02.C, Contractor's services will not be terminated if Contractor begins within seven days of receipt of notice of intent to terminate to correct its failure to perform and proceeds diligently to cure such failure within no more than 30 days of receipt of said notice.

E. Where Contractor's services have been so terminated by Owner, the termination will not affect any rights or remedies of Owner against Contractor then existing or which may thereafter accrue. Any retention or payment of moneys due Contractor by Owner will not release Contractor from liability.

F. If and to the extent that Contractor has provided a performance bond under the provisions of Paragraph 5.01.A, the termination procedures of that bond shall supersede the provisions of Paragraphs 15.02.B and 15.02.C.

15.03 *Owner May Terminate For Convenience*

A. Upon seven days written notice to Contractor and Engineer, Owner may, without cause and without prejudice to any other right or remedy of Owner, terminate the Contract. In such case, Contractor shall be paid for (without duplication of any items):

1. completed and acceptable Work executed in accordance with the Contract Documents prior to the effective date of termination, including fair and reasonable sums for overhead and profit on such Work;

2. expenses sustained prior to the effective date of termination in performing services and furnishing labor, materials, or equipment as required by the Contract Documents in connection with uncompleted Work, plus fair and reasonable sums for overhead and profit on such expenses;

3. all claims, costs, losses, and damages (including but not limited to all fees and charges of engineers, architects, attorneys, and other professionals and all court or arbitration or other dispute resolution costs) incurred in settlement of terminated contracts with Subcontractors, Suppliers, and others; and

4. reasonable expenses directly attributable to termination.

B. Contractor shall not be paid on account of loss of anticipated profits or revenue or other economic loss arising out of or resulting from such termination.

15.04 *Contractor May Stop Work or Terminate*

A. If, through no act or fault of Contractor, (i) the Work is suspended for more than 90 consecutive days by Owner or under an order of court or other public authority, or (ii) Engineer fails to act on any Application for Payment within 30 days after it is submitted, or (iii) Owner fails for 30 days to pay Contractor any sum finally determined to be due, then Contractor may, upon seven days written notice to Owner and Engineer, and provided Owner or Engineer do not remedy such suspension or failure within that time, terminate the Contract and recover from Owner payment on the same terms as provided in Paragraph 15.03.

B. In lieu of terminating the Contract and without prejudice to any other right or remedy, if Engineer has failed to act on an Application for Payment within 30 days after it is submitted, or Owner has failed for 30 days to pay Contractor any sum finally determined to be due, Contractor may, seven days after written notice to Owner and Engineer, stop the Work until payment is made of all such amounts due Contractor, including interest thereon. The provisions of this Paragraph 15.04 are not intended to preclude Contractor from making a Claim under Paragraph 10.05 for an adjustment in Contract Price or Contract Times or otherwise for expenses or damage directly attributable to Contractor's stopping the Work as permitted by this Paragraph.

ARTICLE 16 – DISPUTE RESOLUTION

16.01 *Methods and Procedures*

A. Either Owner or Contractor may request mediation of any Claim submitted to Engineer for a decision under Paragraph 10.05 before such decision becomes final and binding. The mediation will be governed by the Construction Industry Mediation Rules of the American Arbitration Association in effect as of the Effective Date of the Agreement. The request for mediation shall be submitted in writing to the American Arbitration Association and the other party to the Contract. Timely submission of the request shall stay the effect of Paragraph 10.05.E.

B. Owner and Contractor shall participate in the mediation process in good faith. The process shall be concluded within 60 days of filing of the request. The date of termination of the mediation shall be determined by application of the mediation rules referenced above.

C. If the Claim is not resolved by mediation, Engineer's action under Paragraph 10.05.C or a denial pursuant to Paragraphs 10.05.C.3 or 10.05.D shall become final and binding 30 days after termination of the mediation unless, within that time period, Owner or Contractor:

1. elects in writing to invoke any dispute resolution process provided for in the Supplementary Conditions; or

2. agrees with the other party to submit the Claim to another dispute resolution process; or

3. gives written notice to the other party of the intent to submit the Claim to a court of competent jurisdiction.

ARTICLE 17 – MISCELLANEOUS

17.01 *Giving Notice*

A. Whenever any provision of the Contract Documents requires the giving of written notice, it will be deemed to have been validly given if:

1. delivered in person to the individual or to a member of the firm or to an officer of the corporation for whom it is intended; or

2. delivered at or sent by registered or certified mail, postage prepaid, to the last business address known to the giver of the notice.

17.02 *Computation of Times*

A. When any period of time is referred to in the Contract Documents by days, it will be computed to exclude the first and include the last day of such period. If the last day of any such period falls on a Saturday or Sunday or on a day made a legal holiday by the law of the applicable jurisdiction, such day will be omitted from the computation.

17.03 *Cumulative Remedies*

A. The duties and obligations imposed by these General Conditions and the rights and remedies available hereunder to the parties hereto are in addition to, and are not to be construed in any way as a limitation of, any rights and remedies available to any or all of them which are otherwise imposed or available by Laws or Regulations, by special warranty or guarantee, or by other provisions of the Contract Documents. The provisions of this Paragraph will be as effective as if repeated specifically in the Contract Documents in connection with each particular duty, obligation, right, and remedy to which they apply.

17.04 *Survival of Obligations*

A. All representations, indemnifications, warranties, and guarantees made in, required by, or given in accordance with the Contract Documents, as well as all continuing obligations indicated in the Contract Documents, will survive final payment, completion, and acceptance of the Work or termination or completion of the Contract or termination of the services of Contractor.

17.05 *Controlling Law*

A. This Contract is to be governed by the law of the state in which the Project is located.

17.06 *Headings*

A. Article and paragraph headings are inserted for convenience only and do not constitute parts of these General Conditions.

Engineers Joint Documents Committee
Design and Construction Related Documents
Instructions and License Agreement

Instructions

Before you use any EJCDC document:
1. Read the License Agreement. You agree to it and are bound by its terms when you use the EJCDC document.

2. Make sure that you have the correct version for your word processing software.

How to Use:
1. While EJCDC has expended considerable effort to make the software translations exact, it can be that a few document controls (e.g., bold, underline) did not carry over.

2. Similarly, your software may change the font specification if the font is not available in your system. It will choose a font that is close in appearance. In this event, the pagination may not match the control set.

3. If you modify the document, you must follow the instructions in the License Agreement about notification.

4. Also note the instruction in the License Agreement about the EJCDC copyright.

License Agreement

You should carefully read the following terms and conditions before using this document. Commencement of use of this document indicates your acceptance of these terms and conditions. If you do not agree to them, you should promptly return the materials to the vendor, and your money will be refunded.

The Engineers Joint Contract Documents Committee ("EJCDC") provides **EJCDC Design and Construction Related Documents** and licenses their use worldwide. You assume sole responsibility for the selection of specific documents or portions thereof to achieve your intended results, and for the installation, use, and results obtained from **EJCDC Design and Construction Related Documents**.

You acknowledge that you understand that the text of the contract documents of **EJCDC Design and Construction Related Documents** has important legal consequences and that consultation with an attorney is recommended with respect to use or modification of the text. You further acknowledge that EJCDC documents are protected by the copyright laws of the United States.

License:
You have a limited nonexclusive license to:

1. Use **EJCDC Design and Construction Related Documents** on any number of machines owned, leased or rented by your company or organization.

2. Use **EJCDC Design and Construction Related Documents** in printed form for bona fide contract documents.

3. Copy **EJCDC Design and Construction Related Documents** into any machine readable or printed form for backup or modification purposes in support of your use of **EJCDC Design and Construction Related Documents**.

You agree that you will:
1. Reproduce and include EJCDC's copyright notice on any printed or machine-readable copy, modification, or portion merged into another document or program. All proprietary rights in **EJCDC Design and Construction Related Documents** are and shall remain the property of EJCDC.

2. Not represent that any of the contract documents you generate from **EJCDC Design and Construction Related Documents** are EJCDC documents unless (i) the document text is used without alteration or (ii) all additions and changes to, and deletions from, the text are clearly shown.

You may not use, copy, modify, or transfer EJCDC Design and Construction Related Documents, or any copy, modification or merged portion, in whole or in part, except as expressly provided for in this license. Reproduction of EJCDC Design and Construction Related Documents in printed or machine-readable format for resale or educational purposes is expressly prohibited.

If you transfer possession of any copy, modification or merged portion of EJCDC Design and Construction Related Documents to another party, your license is automatically terminated.

Term:
The license is effective until terminated. You may terminate it at any time by destroying **EJCDC Design and Construction Related Documents** altogether with all copies, modifications and merged portions in any form. It will also terminate upon conditions set forth elsewhere in this Agreement or if you fail to comply with any term or

condition of this Agreement. You agree upon such termination to destroy **EJCDC Design and Construction Related Documents** along with all copies, modifications and merged portions in any form.

Limited Warranty:
EJCDC warrants the CDs and diskettes on which **EJCDC Design and Construction Related Documents** is furnished to be free from defects in materials and workmanship under normal use for a period of ninety (90) days from the date of delivery to you as evidenced by a copy of your receipt.

There is no other warranty of any kind, either expressed or implied, including, but not limited to the implied warranties of merchantability and fitness for a particular purpose. Some states do not allow the exclusion of implied warranties, so the above exclusion may not apply to you. This warranty gives you specific legal rights and you may also have other rights which vary from state to state.

EJCDC does not warrant that the functions contained in **EJCDC Design and Construction Related Documents** will meet your requirements or that the operation of **EJCDC Design and Construction Related Documents** will be uninterrupted or error free.

Limitations of Remedies:
EJCDC's entire liability and your exclusive remedy shall be:

1. the replacement of any document not meeting EJCDC's "Limited Warranty" which is returned to EJCDC's selling agent with a copy of your receipt, or

2. if EJCDC's selling agent is unable to deliver a replacement CD or diskette which is free of defects in materials and workmanship, you may terminate this Agreement by returning EJCDC Document and your money will be refunded.

In no event will EJCDC be liable to you for any damages, including any lost profits, lost savings or other incidental or consequential damages arising out of the use or inability to use **EJCDC Design and Construction Related Documents** even if EJCDC has been advised of the possibility of such damages, or for any claim by any other party.

Some states do not allow the limitation or exclusion of liability for incidental or consequential damages, so the above limitation or exclusion may not apply to you.

General:
You may not sublicense, assign, or transfer this license except as expressly provided in this Agreement. Any attempt otherwise to sublicense, assign, or transfer any of the rights, duties, or obligations hereunder is void.

This Agreement shall be governed by the laws of the State of Virginia. Should you have any questions concerning this Agreement, you may contact EJCDC by writing to:

Arthur Schwartz, Esq.
General Counsel
National Society of Professional Engineers
1420 King Street
Alexandria, VA 22314

Phone: (703) 684-2845
Fax: (703) 836-4875
e-mail: aschwartz@nspe.org

You acknowledge that you have read this agreement, understand it and agree to be bound by its terms and conditions. You further agree that it is the complete and exclusive statement of the agreement between us which supersedes any proposal or prior agreement, oral or written, and any other communications between us relating to the subject matter of this agreement.

This document has important legal consequences; consultation with an attorney is encouraged with respect to its use or modification. This document should be adapted to the particular circumstances of the contemplated Project and the controlling Laws and Regulations.

GUIDE TO THE PREPARATION OF SUPPLEMENTARY CONDITIONS

Prepared by

ENGINEERS JOINT CONTRACT DOCUMENTS COMMITTEE

and

Issued and Published Jointly by

AMERICAN COUNCIL OF ENGINEERING COMPANIES

ASSOCIATED GENERAL CONTRACTORS OF AMERICA

AMERICAN SOCIETY OF CIVIL ENGINEERS

PROFESSIONAL ENGINEERS IN PRIVATE PRACTICE
A Practice Division of the
NATIONAL SOCIETY OF PROFESSIONAL ENGINEERS

Endorsed by

CONSTRUCTION SPECIFICATIONS INSTITUTE

This Guide to the Preparation of Supplementary Conditions has been prepared for use with the Standard General Conditions of the Construction Contract (EJCDC C-700, 2007 Edition). Their provisions are interrelated and a change in one may necessitate a change in the other. The suggested language contained in the Guide to the Preparation of Instructions to Bidders (EJCDC C-200, 2007 Edition) is also carefully integrated with the suggested language of this document. Comments concerning their usage are contained in EJCDC guidance documents.

1420 King Street, Alexandria, VA 22314-2794
(703) 684-2882
www.nspe.org

American Council of Engineering Companies
1015 15th Street N.W., Washington, DC 20005
(202) 347-7474
www.acec.org

American Society of Civil Engineers
1801 Alexander Bell Drive, Reston, VA 20191-4400
(800) 548-2723
www.asce.org

Associated General Contractors of America
2300 Wilson Boulevard, Suite 400, Arlington, VA 22201-3308
(703) 548-3118
www.agc.org

TABLE OF CONTENTS

Page

I. INTRODUCTION 1
A. General 1
B. Mandatory Supplementary Conditions 2
C. Relationship of Supplementary Conditions to Other Contract Documents 2
D. Arrangement of Subject Matter 3
E. Use of this Guide 3

II. STANDARD PREFATORY LANGUAGE AND TRADITIONAL FORMAT FOR SUPPLEMENTARY CONDITIONS 4
A. Table of Contents 4
B. Pagination 4
C. Format for Complete Paragraph Change 4
D. Format for Change within a Paragraph 4
E. Format for Additional Language 5
F. Format for Additional Paragraph 5

III. ALTERNATE FORMAT FOR SUPPLEMENTARY CONDITIONS 5

IV. SUGGESTED SUPPLEMENTARY CONDITIONS 6
Caption and Introductory Statements 6
SC-2.02 Copies of Documents 6
SC-4.02 Subsurface and Physical Conditions 6
SC-4.06 Hazardous Environmental Conditions 8
SC-5.04 Contractor's Liability Insurance 9
SC-5.06 Property Insurance 11
SC-6.06 Concerning Subcontractors, Suppliers, and Others 13
SC-6.10 Taxes 13
SC-6.13 Safety and Protection 13
SC-6.17 Shop Drawings and Samples 14
SC-7.02 Coordination 14
SC-7.04 Claims Between Contractors 14
SC-8.11 Evidence of Financial Arrangements 16
SC-9.03 Project Representative 16
SC-11.01 Cost of the Work 20
SC-11.03 Unit Price Work 21
SC-12.01 Change of Contract Price 21
SC-16.01 Methods and Procedure 22

I. INTRODUCTION

A. *General*

The Engineers Joint Contract Documents Committee (EJCDC) has prepared and publishes forms for use as construction contract documents. The principal forms are listed in Table 1. EJCDC has also prepared other documents that may be useful in preparing construction contract documents. Some of the principal ones are listed in Table 2. For the most recent editions of these forms, guides, and other documents, please refer to EJCDC's website at www.ejcdc.org.

Table 1
Principal EJCDC Standard Forms and Related Guides for Construction Contracts

Name	Number	Short Title/Abbreviation
Suggested Instructions to Bidders for Construction Contracts	C-200	Instructions/I
Suggested Bid Form for Construction Contracts	C-410	Bid Form/BF
Suggested Form of Agreement between Owner and Contractor for Construction Contract (Stipulated Price)	C-520	Stipulated Price Agreement/A
Suggested Form of Agreement between Owner and Contractor for Construction Contract (Cost-Plus)	C-525	Cost-Plus Agreement/A
Standard General Conditions of the Construction Contract	C-700	General Conditions/GC
Guide to the Preparation of Supplementary Conditions	C-800	Supplementary Conditions/SC

Table 2
Principal EJCDC Documents Relating to Preparation of Construction Documents

Name	Number	Short Title
Narrative Guide to the 2007 EJCDC Construction Documents	C-001	Narrative Guide
Uniform Location of Subject Matter	N-122	Locator Guide
Bidding Procedures and Construction Contract Documents	C-050	Bidding Procedures
Engineer's Letter to Owner Requesting Instructions Concerning Bonds and Insurance	C-051	Engineer's Letter to Owner Concerning Bonds and Insurance
Owner's Instructions to Engineer Concerning Bonds and Insurance	C-052	Owner's Instructions Concerning Bonds and Insurance

EJCDC publications may be purchased from any of the organizations listed on the cover page of this document (see list of websites on the page immediately following the cover page).

B. *Mandatory Supplementary Conditions*

Several provisions of the General Conditions expressly indicate that essential project-specific information will be set out in a corresponding Supplementary Condition. For example, GC-5.04.B indicates that required insurance coverage limits will be specified in the Supplementary Conditions. Every EJCDC-based construction contract should include, at a minimum, the following Supplementary Conditions:

1. One of two suggested SC-4.02s, concerning reports and drawings of conditions at the Site, and any "technical data" in the reports and drawings on which the Contractor may rely;

2. one of two suggested SC-4.06s, concerning reports and drawings regarding Hazardous Environmental Conditions at the Site, and any "technical data" in those reports and drawings on which the Contractor may rely; and

3. SC-5.04, identifying specific insurance coverage requirements.

Other suggested Supplementary Conditions are mandatory under specific circumstances: for example, on projects in which the Engineer is providing the services of a Resident Project Representative (RPR), SC-9.03, concerning the authority and responsibilities of the RPR, would be mandatory; and on projects in which the Contractor will be responsible for compliance with Owner's safety program, SC-6.13 would be mandatory.

C. *Relationship of Supplementary Conditions to Other Contract Documents*

Supplementary Conditions are modifications to the General Conditions—additions, deletions, changes. This is as the term is defined by EJCDC and the Construction Specification Institute (CSI). Other organizations use their supplementary conditions to modify a broader range of contract documents, such as agreement forms and standard specifications.

This Guide and the other Construction-related documents prepared and issued by EJCDC assume use of the 1995 CSI MasterFormat™ concept, which provides an organizational format for location of all bound documentary information for a construction project: Bidding Requirements, contract forms (Agreement, Bonds, and certificates), General Conditions, Supplementary Conditions, and Specifications. Under the 1995 CSI MasterFormat™, the last grouping, Specifications, is divided into 16 Divisions, the first of which, Division 1, is entitled "General Requirements." (CSI issued a radically different form of MasterFormat™ in 2004. It is described in CSI's Project Resource Manual. EJCDC is expected to consider MasterFormat 2004 during the next revision cycle for the construction documents.)

The standard fundamental provisions affecting the rights and duties of the parties appear in the General Conditions. Language to modify the fundamental relationships between the parties, supplement the framework set forth in the General Conditions, or change the language of the General Conditions, should appear in the Supplementary Conditions. Examples of this are a change in the payment provisions and supplemental language specifying the details of insurance coverages and limits for the Project.

Price terms, monetary terms such as liquidated damages clauses, and completion dates should all be set forth in the Agreement (EJCDC C-520 or C-525), and should not be included in the Supplementary Conditions.

The substance of the General Requirements (Division 1 of the Specifications) falls generally into three categories: (1) administrative requirements, such as summary of work, allowances, coordination, alternatives (materials, equipment, or price), product options, project meetings, and project close-out; (2) work-related provisions, such as temporary facilities, field testing, and start-up; and (3) general provisions applicable to more than one section in Divisions 2 through 16.

D. *Arrangement of Subject Matter*

This Guide is arranged in the same order as the paragraphs in the 2007 edition of the General Conditions, and the paragraphs herein bear comparable addresses to those of the General Conditions but with the prefix "SC." A discussion of the purpose and function of these suggested Supplementary Conditions is included in EJCDC C-001, Narrative Guide.

E. *Use of this Guide*

The text presented in bold type in the remainder of this Guide is suggested language for some commonly used Supplementary Conditions. The drafter should bear in mind that most contractual provisions have important legal consequences. Consultation with legal counsel before finalization of any amendment or supplement is recommended.

Many sets of supplementary conditions examined by EJCDC contain typical or "boilerplate" provisions that have accumulated like moss over the years, appear to have no practical significance for the particular project, and may produce unintended and surprising legal consequences. Such provisions are usually there because someone saw similar terms in other contract documents and it "sounded good." Selecting contract terms in that manner is not recommended. Provisions of the Supplementary Conditions should address a particular point in the General Conditions or cover a particular topic. The Supplementary Conditions should not be a repository for general language of vague meaning for which another location cannot be readily found.

This Guide assumes a general familiarity with the other Contract Documents prepared by EJCDC and, when drafting language, specific attention to them is encouraged. Standard documents or prescribed forms issued by governmental bodies and other owners may differ materially from the documents of EJCDC so that careful correlation of any amending or supplementing language is essential. The loose practice of stating that any provision in one document that is inconsistent with another is superseded, or that one document always takes precedence over another in the event of a conflict in language or requirements, is discouraged. The resulting legal consequences of such provisions are frequently difficult to decipher and may be very different from what was anticipated.

The General Conditions use carefully chosen language and set forth the basic responsibilities of the parties with respect to fundamental matters and legal consequences. Their provisions should be altered only where mandated by the specific requirements of a given project and the consequences of any modification are thoroughly understood.

Caution should be exercised when making any change in the standard documents. They have been carefully prepared, terms are used uniformly throughout and are consistent with the terms in other EJCDC documents. Their provisions have been carefully integrated, and are dependent on one another. A change in one document may necessitate a change in another, and a change in one paragraph may necessitate a change in other language of the same document. No change should be made until its full effect on the rest of the General Conditions and other Contract Documents has been considered.

Lastly, remember that an engineer is neither qualified nor licensed to give advice to others on the legal consequences of contracts. All of the Contract Documents have important legal consequences. Owners should be encouraged to seek the advice of an attorney before accepting any modification of the printed forms, before the documents are sent out for bidding, and most assuredly before signing any agreement.

II. STANDARD PREFATORY LANGUAGE AND TRADITIONAL FORMAT FOR SUPPLEMENTARY CONDITIONS

Suggested format and wording conventions for Supplementary Conditions appear below.

A. *Table of Contents*

The inclusion of a table of contents will benefit the user of the Supplementary Conditions, especially if additional articles (beyond the 17 Articles of the General Conditions) are added for the purpose of including mandated or other provisions.

B. *Pagination*

The CSI MasterFormat™, October 1995, assigns Document Number 00800 to Supplementary Conditions. Unless another format is chosen, pages should be numbered 00800-1, 00800-2, 00800-n. If CSI's MasterFormat 04™ is being used for the Project Manual, consult MasterFormat 04 for the appropriate section number and number the pages accordingly.

C. *Format for Complete Paragraph Change*

When completely superseding a paragraph of the General Conditions, the following language may be used:

SC-5.09.B Delete Paragraph 5.09.B in its entirety and insert the following in its place:

____

____

D. *Format for Change within a Paragraph*

When changing language within a paragraph of the General Conditions, the following language may be used:

SC-6.21.A Amend the second sentence of Paragraph 6.21.A (to read as follows) *[or]* **(by striking out the following words):**

____
____
____

E. *Format for Additional Language*

When adding language to an existing paragraph of the General Conditions, the idea may be expressed as follows:

SC-9.03 Add the following language at the end of the second sentence of Paragraph 9.03:

____
____
____

F. *Format for Additional Paragraph*

If it is desired to add a new paragraph to the General Conditions, the thought may be expressed as follows:

SC-8.06 Add the following new paragraph immediately after Paragraph 8.06.B:

____
____
____

III. ALTERNATE FORMAT FOR SUPPLEMENTARY CONDITIONS

Electronic files are commonly used for transmittal and storage of the text of standard documents. In fact, EJCDC no longer prepares printed documents. Because it is easy to modify documents electronically, it is increasingly common for practitioners to integrate the text of desired Supplementary Conditions into the text of the General Conditions. Most word processing programs have line-out and underlining features which accurately show deletions, changes, and additions. Users of EJCDC's General Conditions are contractually obligated, through the terms of the purchase of the document, to clearly delineate all changes made to the standard text. It would be misleading to users to imply or represent that the General Conditions are EJCDC's General Conditions if changes are not properly and clearly identified.

IV. SUGGESTED SUPPLEMENTARY CONDITIONS

Caption and Introductory Statements

The following is a suggestion for use at the beginning of the Supplementary Conditions:

Supplementary Conditions

These Supplementary Conditions amend or supplement the Standard General Conditions of the Construction Contract, EJCDC C-700 (2007 Edition). All provisions which are not so amended or supplemented remain in full force and effect.

The terms used in these Supplementary Conditions have the meanings stated in the General Conditions. Additional terms used in these Supplementary Conditions have the meanings stated below, which are applicable to both the singular and plural thereof.

The address system used in these Supplementary Conditions is the same as the address system used in the General Conditions, with the prefix "SC" added thereto.

SC-2.02 *Copies of Documents*

As electronic documents become more widely used, electronic copies of the Drawings, Specifications, and other Contract Documents are often made available in place of multiple sets of hard copy. If electronic documents are to be made available, the following may be used:

SC-2.02 Delete Paragraph 2.02.A in its entirety and insert the following in its place:

A. Owner shall furnish to Contractor up to _____ printed or hard copies of the Drawings and Project Manual and one set in electronic format. Additional copies will be furnished upon request at the cost of reproduction.

SC-4.02 *Subsurface and Physical Conditions*

This is a mandatory Supplementary Condition. GC-4.02 requires the identification of all known documents regarding subsurface and physical conditions at the Site. Use the first version of SC-4.02, presented immediately below, for the purpose of identifying the known Site condition documents. If no such documents are known, then use the second version of SC-4.02, below.

SC-4.02 Add the following new paragraphs immediately after Paragraph 4.02.B:

C. The following reports of explorations and tests of subsurface conditions at or contiguous to the Site are known to Owner:

1. Report dated May 21, 2000, prepared by Aye and Bea, Consulting Engineers, Philadelphia, Pa., entitled: "Results of Investigation of Subsoil Conditions and Professional Recommendations for Foundations of Iron Foundry at South and Front Streets, Pembrig, NJ", consisting of 42 pages. The "technical data" contained in such report upon which

Contractor may rely is *[here indicate any such "technical data" or state "none."]*

2. **Report dated May 2, 2000, prepared by Ecks, Wye and Tszee, Inc., Baltimore, Md., entitled: "Tests of Water Quality in Mixter River at Pembrig, NJ", consisting of 26 pages. The "technical data" contained in such report upon which Contractor may rely are:** *[here indicate any such "technical data" or state "none."]*

D. The following drawings of physical conditions relating to existing surface or subsurface structures at the Site (except Underground Facilities) are known to Owner:

1. **Drawings dated March 2, 2000, of Route 24A Overpass Abutment, prepared by Dea & Associates, Inc., Wilmington, Del., entitled: "Record Drawings: Route No. 24A Overpass Abutment", consisting of 12 sheets numbered ___ to ___, inclusive.**

[Use one of the following two subparagraphs:]

a. **All of the information in such drawings constitutes "technical data" on which Contractor may rely, except for ______________________ appearing on Drawing No. _____ and ___________________ appearing on Drawing No. ______.**

[or]

a. **None of the contents of such drawings is "technical data" on which Contractor may rely.**

E. The reports and drawings identified above are not part of the Contract Documents, but the "technical data" contained therein upon which Contractor may rely, as expressly identified and established above, are incorporated in the Contract Documents by reference. Contractor is not entitled to rely upon any other information and data known to or identified by Owner or Engineer.

F. Copies of reports and drawings identified in SC-4.02.C and SC-4.02.D that are not included with the Bidding Documents may be examined at ___________________ *[insert location]* **during regular business hours.**

If there are no known Site-related reports or drawings, use the following version of SC-4.02:

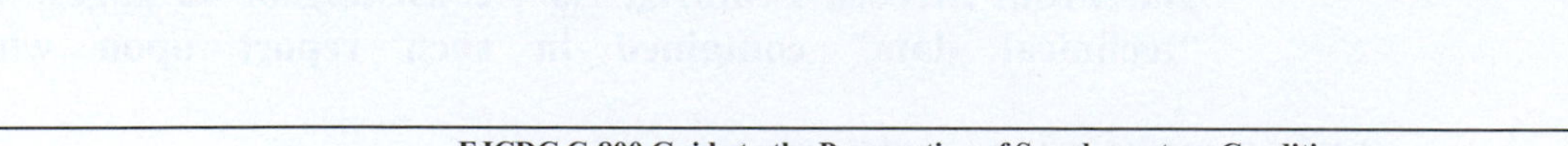

SC-4.02 Delete Paragraphs 4.02.A and 4.02.B in their entirety and insert the following:

A. No reports of explorations or tests of subsurface conditions at or contiguous to the Site, or drawings of physical conditions relating to existing surface or subsurface structures at the Site, are known to Owner.

SC-4.06 *Hazardous Environmental Conditions*

This is a mandatory Supplementary Condition. GC-4.06 contemplates the identification of all known documents regarding Hazardous Environmental Conditions (HEC) that have been identified at the Site. Use the first version of SC-4.06, presented immediately below, to identify the known HEC documents. If no HEC documents are known, then use the second version of SC-4.06, below.

SC-4.06 Add the following subparagraphs 4.06.A.1 and 4.06.A.2:

1. The following reports regarding Hazardous Environmental Conditions at the Site are known to Owner:

a. Report dated December 10, 2002, prepared by Eph Environmental Consultants, Princeton, N.J., entitled: "Results of Investigation of Conditions at Iron Foundry at South and Front Streets, Pembrig, NJ", consisting of 27 pages. The "technical data" contained in such report upon which Contractor may rely is *[here indicate any such "technical data" or state "none."]*

2. The following drawings regarding Hazardous Environmental Conditions at the Site are known to Owner:

a. Drawings dated November 27, 2002, prepared by Eph Environmental Consultants, Princeton, N.J., entitled: "Iron Foundry Site Conditions", consisting of 5 sheets numbered ___ to ___, inclusive.

[Use one of the following two subparagraphs:]

1) All of the information in such drawings constitutes "technical data" on which Contractor may rely, except for ___________________________ appearing on Drawing No. _____ and ____________________ appearing on Drawing No. ______.

[or]

1) None of the contents of such drawings is "technical data" on which Contractor may rely.

Use the following SC-4.06 if there are no known HEC reports or drawings:

SC-4.06 Delete Paragraphs 4.06.A and 4.06.B in their entirety and insert the following:

A. No reports or drawings related to Hazardous Environmental Conditions at the Site are known to Owner.

B. Not Used.

SC-5.04 *Contractor's Liability Insurance*

This is a mandatory Supplementary Condition, for it specifies the limits of the coverages for the insurance required in GC-5.04.

SC-5.04 Add the following new paragraph immediately after Paragraph 5.04.B:

C. The limits of liability for the insurance required by Paragraph 5.04 of the General Conditions shall provide coverage for not less than the following amounts or greater where required by Laws and Regulations:

1. Workers' Compensation, and related coverages under Paragraphs 5.04.A.1 and A.2 of the General Conditions:

a. State:	**Statutory**
b. Applicable Federal (e.g., Longshoreman's):	**Statutory**
c. Employer's Liability:	**$_____**

2. Contractor's General Liability under Paragraphs 5.04.A.3 through A.6 of the General Conditions which shall include completed operations and product liability coverages and eliminate the exclusion with respect to property under the care, custody and control of Contractor:

a. General Aggregate	**$_____**
b. Products - Completed Operations Aggregate	**$_____**
c. Personal and Advertising Injury	**$_____**
d. Each Occuarrence (Bodily Injury and Property Damage)	**$_____**

e. **Property Damage liability insurance will provide Explosion, Collapse, and Under-ground coverages where applicable.**

f. **Excess or Umbrella Liability**

☐ **General Aggregate** $_____
☐ **Each Occurrence** $_____

3. **Automobile Liability under Paragraph 5.04.A.6 of the General Conditions:**

a. **Bodily Injury:**
Each person $_____
Each Accident $_____

b. **Property Damage:**
Each Accident $_____

[or]

a. **Combined Single Limit of** $_____

4. **The Contractual Liability coverage required by Paragraph 5.04.B.4 of the General Conditions shall provide coverage for not less than the following amounts:**

a. **Bodily Injury:**
Each person $_____
Each Accident $_____

b. **Property Damage:**
Each Accident $_____
Annual Aggregate $_____

5. *[Here list additional types and amounts of insurance that may be required by Owner.]*

6. *[Here list by name (not genre) other persons or entities to be included on policy as additional insureds.]*

SC-5.06 *Property Insurance*

GC-5.06.A.1 refers to other individuals or entities that are to be identified in SCs as being entitled to protection as loss payees under the property insurance on the Work. In such cases use the following:

SC-5.06.A.1 Add the following new subparagraph after subparagraph GC-5.06.A.1:

a. **In addition to the individuals and entities specified, include as loss payees the following:**

[Here list by name (not genre) other persons or entities to be include on policy as loss payees.]

In the event that the Contractor, rather than the Owner, will purchase the Builder's Risk property insurance, use the following SC-5.06.A:

SC-5.06.A. Delete Paragraph 5.06.A in its entirety and insert the following in its place:

A. Contractor shall purchase and maintain property insurance upon the Work at the Site in the amount of the full replacement cost thereof. Contractor shall be responsible for any deductible or self-insured retention. This insurance shall:

1. **include the interests of Owner, Contractor, Subcontractors, Engineer, and** *[here identify by name (not genre) any other individuals or entities to be listed as loss payees]* **and the officers, directors, partners, employees, agents and other consultants and subcontractors of any of them, each of whom is deemed to have an insurable interest and shall be listed as an insured or loss payee;**

2. **be written on a Builder's Risk "all-risk" policy form that shall at least include insurance for physical loss and damage to the Work, temporary buildings, falsework, and materials and equipment in transit and shall insure against at least the following perils or causes of loss: fire, lightning, extended coverage, theft, vandalism and malicious mischief, earthquake, collapse, debris removal, demolition occasioned by enforcement of Laws and Regulations, water damage (other than that caused by flood), and such other perils or causes of loss as may be specifically required by these Supplementary Conditions.**

3. **include expenses incurred in the repair or replacement of any insured property (including but not limited to fees and charges of engineers and architects);**

4. **cover materials and equipment stored at the Site or at another location that was agreed to in writing by Owner prior to being incorporated in the Work, provided that such materials and equipment have been included in an Application for Payment recommended by Engineer;**

5. **allow for partial utilization of the Work by Owner;**

6. **include testing and startup;**

7. **be maintained in effect until final payment is made unless otherwise agreed to in writing by Owner, Contractor, and Engineer with 30 days written notice to each other loss payee to whom a certificate of insurance has been issued; and**

8. **comply with the requirements of Paragraph 5.06.C of the General Conditions.**

GC-5.06.B states that Owner will purchase "equipment breakdown" insurance (formerly referred to as "boiler and machinery" insurance) or other additional property insurance if required to do so by the Supplementary Conditions or Laws and Regulations. If there is a specific requirement that Owner purchase any such property insurance, include the following, selecting Owner as the purchaser. In the alternative, the Contract could require that Contractor purchase equipment breakdown or other additional property insurance, in which case the following may be used for that purpose, by selecting Contractor as purchaser:

SC-5.06 Delete Paragraph 5.06.B and replace with the following:

B. [Owner] [Contractor] *[select one, delete the other]* **shall purchase and maintain** *[here identify any specifically required equipment breakdown insurance or additional property insurance to be provided]*, **and any other additional property insurance required by Laws and Regulations, which insurance will include the interest of Owner, Contractor, Subcontractors, and Engineer, and** *[here identify any other individuals or entities to be included as loss payees]* **and the officers, directors, partners, employees, agents, consultants and subcontractors of each and any of them, each of whom is deemed to have an insurable interest and shall be listed as a loss payee.**

SC-6.06 *Concerning Subcontractors, Suppliers, and Others*

If the Owner wishes to release payment information, use the following:

SC-6.06 Add a new paragraph immediately after Paragraph 6.06.G:

H. Owner may furnish to any Subcontractor or Supplier, to the extent practicable, information about amounts paid to Contractor on account of Work performed for Contractor by a particular Subcontractor or Supplier.

SC-6.10 *Taxes*

If Owner qualifies for a state or local sales or use tax exemption in the purchase of certain materials and equipment, add the following:

SC-6.10 Add a new paragraph immediately after Paragraph 6.10.A:

B. Owner is exempt from payment of sales and compensating use taxes of the State of *[insert name of state where Project is located]* **and of cities and counties thereof on all materials to be incorporated into the Work.**

1. Owner will furnish the required certificates of tax exemption to Contractor for use in the purchase of supplies and materials to be incorporated into the Work.

2. Owner's exemption does not apply to construction tools, machinery, equipment, or other property purchased by or leased by Contractor, or to supplies or materials not incorporated into the Work.

SC-6.13 *Safety and Protection*

Some Owners have written safety programs with which construction contractors must comply. If such is the case, GC-6.13.C mandates that the safety program be identified in the Supplementary Conditions, which may be accomplished as follows:

SC-6.13 Delete the second sentence of Paragraph 6.13.C and insert the following:

The following Owner safety programs are applicable to the Work: *[here expressly identify by title and/or date, any such Owner safety programs]*

SC-6.17 *Shop Drawings and Samples*

Reviews of multiple resubmissions of Shop Drawings and other submittals may increase Project costs. To mitigate this, the following language may be used:

SC-6.17 Add the following new paragraphs immediately after Paragraph 6.17.E:

F. Contractor shall furnish required submittals with sufficient information and accuracy in order to obtain required approval of an item with no more than three submittals. Engineer will record Engineer's time for reviewing subsequent submittals of Shop Drawings, samples, or other items requiring approval and Contractor shall reimburse Owner for Engineer's charges for such time.

G. In the event that Contractor requests a change of a previously approved item, Contractor shall reimburse Owner for Engineer's charges for its review time unless the need for such change is beyond the control of Contractor.

SC-7.02 *Coordination*

GC-7.02 requires that if in addition to retaining Contractor, Owner will contract with others to perform work at the Site, Owner must provide to Contractor specified information. Use the following in that case:

SC-7.02 Delete Paragraph 7.02.A in its entirety and replace with the following:

A. Owner intends to contract with others for the performance of other work on the Project at the Site.

1. *[Here identify individual or entirety]* **shall have authority and responsibility for coordination of the various contractors at the Site;**

2. The following specific matters are to be covered by such authority and responsibility: *[here itemize such matters]***;**

3. The extent of such authority and responsibilities is: *[here provide the extent]*

SC-7.04 *Claims Between Contractors*

On projects involving multiple contractors, use the following:

SC-7.04 Add the following new paragraph immediately after paragraph GC-7.03:

SC-7.04 ***Claims Between Contractors***

A. Should Contractor cause damage to the work or property of any other contractor at the Site, or should any claim arising out of Contractor's performance of the Work at the Site be made by any other contractor against Contractor, Owner, Engineer, or the construction coordinator, then Contractor (without involving Owner, Engineer, or construction coordinator) shall either (1) remedy the damage, (2) agree to compensate the other contractor for remedy of the damage, or (3) remedy the damage and attempt to settle with such other contractor by agreement, or otherwise resolve the dispute by arbitration or at law.

B. Contractor shall, to the fullest extent permitted by Laws and Regulations, indemnify and hold harmless Owner, Engineer, the construction coordinator and the officers, directors, partners, employees, agents and other consultants and subcontractors of each and any of them from and against all claims, costs, losses and damages (including, but not limited to, fees and charges of engineers, architects, attorneys, and other professionals and court

and arbitration costs) arising directly, indirectly or consequentially out of any action, legal or equitable, brought by any other contractor against Owner, Engineer, consultants, or the construction coordinator to the extent said claim is based on or arises out of Contractor's performance of the Work. Should another contractor cause damage to the Work or property of Contractor or should the performance of work by any other contractor at the Site give rise to any other Claim, Contractor shall not institute any action, legal or equitable, against Owner, Engineer, or the construction coordinator or permit any action against any of them to be maintained and continued in its name or for its benefit in any court or before any arbiter which seeks to impose liability on or to recover damages from Owner, Engineer, or the construction coordinator on account of any such damage or Claim.

C. If Contractor is delayed at any time in performing or furnishing the Work by any act or neglect of another contractor, and Owner and Contractor are unable to agree as to the extent of any adjustment in Contract Times attributable thereto, Contractor may make a Claim for an extension of times in accordance with Article 12. An extension of the Contract Times shall be Contractor's exclusive remedy with respect to Owner, Engineer, and construction coordinator for any delay, disruption, interference, or hindrance caused by any other contractor. This paragraph does not prevent recovery from Owner, Engineer, or construction coordinator for activities that are their respective responsibilities.

SC-8.11 *Evidence of Financial Arrangements*

The following SC-8.11 is intended for use in contracts where Owner is a private entity. It is reasonable for Contractor to seek such information, particularly if Owner and Contractor do not have a continuing relationship.

SC-8.11 Add the following new paragraph immediately after Paragraph 8.11.A:

B. On request of Contractor prior to the execution of any Change Order involving a significant increase in the Contract Price, Owner shall furnish to Contractor reasonable evidence that adequate financial arrangements have been made by Owner to enable Owner to fulfill the increased financial obligations to be undertaken by Owner as a result of such Change Order.

SC-9.03 *Project Representative*

As indicated in GC-9.03, in those cases in which the Engineer will provide a Resident Project Representative (RPR) during construction, the authority and responsibilities of the RPR and any assistants must be specified in the Supplementary Conditions; thus this is a mandatory

Supplementary Condition in such cases. The following suggested language which parallels the working of Exhibit D to EJCDC E-500, the Standard Form of Agreement Between Owner and Engineer for Professional Services, should be edited to indicate the RPR authority and responsibilities that apply to this project:

SC-9.03 Add the following new paragraphs immediately after Paragraph 9.03.A:

B. The Resident Project Representative (RPR) will be Engineer's employee or agent at the Site, will act as directed by and under the supervision of Engineer, and will confer with Engineer regarding RPR's actions. RPR's dealings in matters pertaining to the Work in general shall be with Engineer and Contractor. RPR's dealings with Subcontractors shall be through or with the full knowledge and approval of Contractor. The RPR shall:

1. ***Schedules:*** **Review the progress schedule, schedule of Shop Drawing and Sample submittals, and schedule of values prepared by Contractor and consult with Engineer concerning acceptability.**

2. ***Conferences and Meetings:*** **Attend meetings with Contractor, such as preconstruction conferences, progress meetings, job conferences and other project-related meetings, and prepare and circulate copies of minutes thereof.**

3. ***Liaison:***

   a. **Serve as Engineer's liaison with Contractor, working principally through Contractor's authorized representative, assist in providing information regarding the intent of the Contract Documents.**

   b. **Assist Engineer in serving as Owner's liaison with Contractor when Contractor's operations affect Owner's on-Site operations.**

   c. **Assist in obtaining from Owner additional details or information, when required for proper execution of the Work.**

4. ***Interpretation of Contract Documents:*** **Report to Engineer when clarifications and interpretations of the Contract Documents are needed and transmit to Contractor clarifications and interpretations as issued by Engineer.**

5. ***Shop Drawings and Samples:***

a. Record date of receipt of Samples and approved Shop Drawings.

b. Receive Samples which are furnished at the Site by Contractor, and notify Engineer of availability of Samples for examination.

6. *Modifications:* Consider and evaluate Contractor's suggestions for modifications in Drawings or Specifications and report such suggestions, together with RPR's recommendations, to Engineer. Transmit to Contractor in writing decisions as issued by Engineer.

7. *Review of Work and Rejection of Defective Work*:

a. Conduct on-Site observations of Contractor's work in progress to assist Engineer in determining if the Work is in general proceeding in accordance with the Contract Documents.

b. Report to Engineer whenever RPR believes that any part of Contractor's work in progress will not produce a completed Project that conforms generally to the Contract Documents or will imperil the integrity of the design concept of the completed Project as a functioning whole as indicated in the Contract Documents, or has been damaged, or does not meet the requirements of any inspection, test or approval required to be made; and advise Engineer of that part of work in progress that RPR believes should be corrected or rejected or should be uncovered for observation, or requires special testing, inspection or approval.

8. *Inspections, Tests, and System Startups:*

a. Verify that tests, equipment, and systems start-ups and operating and maintenance training are conducted in the presence of appropriate Owner's personnel, and that Contractor maintains adequate records thereof.

b. Observe, record, and report to Engineer appropriate details relative to the test procedures and systems start-ups.

9. *Records:*

a. Record names, addresses, fax numbers, e-mail addresses, web site locations, and telephone numbers of all

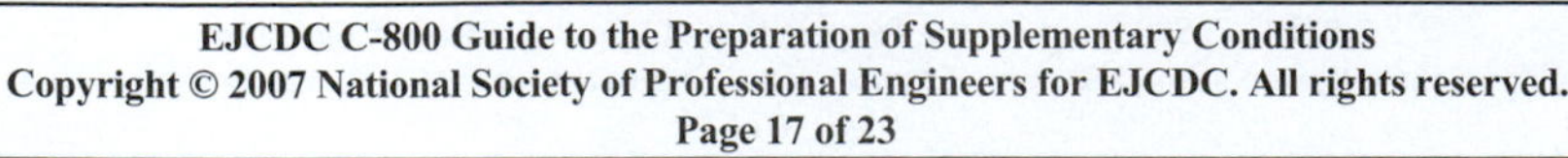

Contractors, Subcontractors, and major Suppliers of materials and equipment.

b. Maintain records for use in preparing Project documentation.

10. *Reports:*

a. Furnish to Engineer periodic reports as required of progress of the Work and of Contractor's compliance with the progress schedule and schedule of Shop Drawing and Sample submittals.

b. Draft and recommend to Engineer proposed Change Orders, Work Change Directives, and Field Orders. Obtain backup material from Contractor.

c. Immediately notify Engineer of the occurrence of any Site accidents, emergencies, acts of God endangering the Work, damage to property by fire or other causes, or the discovery of any Hazardous Environmental Condition.

11. *Payment Requests:* Review Applications for Payment with Contractor for compliance with the established procedure for their submission and forward with recommendations to Engineer, noting particularly the relationship of the payment requested to the schedule of values, Work completed, and materials and equipment delivered at the Site but not incorporated in the Work.

12. *Certificates, Operation and Maintenance Manuals:* During the course of the Work, verify that materials and equipment certificates, operation and maintenance manuals and other data required by the Specifications to be assembled and furnished by Contractor are applicable to the items actually installed and in accordance with the Contract Documents, and have these documents delivered to Engineer for review and forwarding to Owner prior to payment for that part of the Work.

13. *Completion*:

a. Participate in a Substantial Completion inspection, assist in the determination of Substantial Completion and the preparation of lists of items to be completed or corrected.

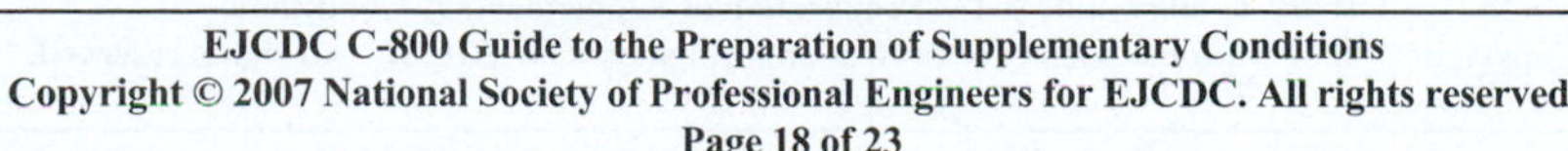

b. **Participate in a final inspection in the company of Engineer, Owner, and Contractor and prepare a final list of items to be completed and deficiencies to be remedied.**

c. **Observe whether all items on the final list have been completed or corrected and make recommendations to Engineer concerning acceptance and issuance of the Notice of Acceptability of the Work.**

C. **The RPR shall not:**

1. **Authorize any deviation from the Contract Documents or substitution of materials or equipment (including "or-equal" items).**

2. **Exceed limitations of Engineer's authority as set forth in the Contract Documents.**

3. **Undertake any of the responsibilities of Contractor, Subcontractors, Suppliers, or Contractor's superintendent.**

4. **Advise on, issue directions relative to, or assume control over any aspect of the means, methods, techniques, sequences or procedures of Contractor's work unless such advice or directions are specifically required by the Contract Documents.**

5. **Advise on, issue directions regarding, or assume control over safety practices, precautions, and programs in connection with the activities or operations of Owner or Contractor.**

6. **Participate in specialized field or laboratory tests or inspections conducted off-site by others except as specifically authorized by Engineer.**

7. **Accept Shop Drawing or Sample submittals from anyone other than Contractor.**

8. **Authorize Owner to occupy the Project in whole or in part.**

SC-11.01 *Cost of the Work*

Equipment rental charges, particularly with respect to Contractor-owned equipment, can sometimes lead to disagreements. To reduce the possibility of such disagreements, the following SC may be used. Note that it requires a published reference or method for determining the costs.

SC-11.01.A.5.c Delete Paragraph 11.01.A.5.c in its entirety and insert the following in its place:

c. *Construction Equipment and Machinery:*

1) **Rentals of all construction equipment and machinery, and the parts thereof in accordance with rental agreements approved by Owner with the advice of Engineer, and the costs of transportation, loading, unloading, assembly, dismantling, and removal thereof. All such costs shall be in accordance with the terms of said rental agreements. The rental of any such equipment, machinery, or parts shall cease when the use thereof is no longer necessary for the Work.**

2) **Costs for equipment and machinery owned by Contractor will be paid at a rate shown for such equipment in the** [*cite the rate book appropriate for the Project*]**. An hourly rate will be computed by dividing the monthly rates by 176. These computed rates will include all operating costs. Costs will include the time the equipment or machinery is in use on the changed Work and the costs of transportation, loading, unloading, assembly, dismantling, and removal when directly attributable to the changed Work. The cost of any such equipment or machinery, or parts thereof, shall cease to accrue when the use thereof is no longer necessary for the changed Work. Equipment or machinery with a value of less than $1,000 will be considered small tools.**

SC-11.03 *Unit Price Work*

The following SC is typically called a variation in estimated quantities clause and facilitates administrative resolution of situations where actual quantities differ materially from estimated quantities. Typically, the clause applies where the Bid price of an item of the Unit Price Work is more than 5 percent of the Contract Price and the actual quantity of the units of work performed varies by 15 to 25 percent.

SC-11.03.D Delete Paragraph 11.03.D in its entirety and insert the following in its place:

D. The unit price of an item of Unit Price Work shall be subject to reevaluation and adjustment under the following conditions:

1. if the Bid price of a particular item of Unit Price Work amounts to _____ percent or more of the Contract Price and the variation in the quantity of that particular item of Unit Price Work performed by Contractor differs by more than _____ percent

from the estimated quantity of such item indicated in the Agreement; and

2. **if there is no corresponding adjustment with respect to any other item of Work; and**

3. **if Contractor believes that Contractor has incurred additional expense as a result thereof or if Owner believes that the quantity variation entitles Owner to an adjustment in the unit price, either Owner or Contractor may make a Claim for an adjustment in the Contract Price in accordance with Article 10 if the parties are unable to agree as to the effect of any such variations in the quantity of Unit Price Work performed.**

SC-12.01 *Change of Contract Price*

In some cases a change in Contract Price will include a Contractor's fee for overhead and profit, which under GC-12.01.C.2 may be determined through application of prescribed percentages on portions of the Cost of the Work. The percentages to be applied to subcontract work of one or more tiers is set forth in GC-12.01.C.2.c, under which the Subcontractor actually performing the work earns a fee of 15 percent, and any higher tier Subcontractors and the Contractor will each earn a fee of 5 percent of the amount paid to the next lower tier Subcontractor. If the parties wish to cap the total amount of the fee that could result from application of this formula, EJCDC suggests the following Supplementary Condition establishing a maximum fee of 27 percent, based on a fee at three tiers (1.15 x 1.05 x 1.05=1.268):

SC-12.01.C ***Contractor's Fee.*** **Delete the semicolon at the end of GC 12.01.C.2.c, and add the following language:**

, provided, however, that on any subcontracted work the total maximum fee to be paid by Owner under this subparagraph shall be no greater than 27 percent of the costs incurred by the Subcontractor who actually performs the work;

SC-16.01 *Methods and Procedure*

As an alternative to the dispute resolution process set forth in the General Conditions (mediation followed by litigation), the contract could pair final and binding arbitration with mediation. A discussion of the pros and cons of the arbitration process (and there are many advocates on either side) is beyond the scope of this Guide. Consultation with the Owner's legal counsel is highly recommended. Users should also note that they will need to insert the name of an arbitration agency, such as the American Arbitration Association or the CPR Institute for Dispute Resolution, in SC-16.02.A. The mediation/arbitration option requires the following:

SC-16.01 **Delete Paragraph 16.01.C in its entirety and insert the following in its place:**

C. If the Claim is not resolved by mediation, Engineer's action under Paragraph 10.05.C or a denial pursuant to Paragraphs 10.05.C.3 or 10.05.D shall become final and binding 30 days after termination of the mediation unless, within that time period, Owner or Contractor:

1. elects in writing to demand arbitration of the Claim, pursuant to Paragraph SC-16.02; or

2. agrees with the other party to submit the Claim to another dispute resolution process.

SC-16.02 Add the following new paragraph immediately after Paragraph 16.01.

Arbitration

A. All Claims or counterclaims, disputes, or other matters in question between Owner and Contractor arising out of or relating to the Contract Documents or the breach thereof (except for Claims which have been waived by the making or acceptance of final payment as provided by Paragraph 14.09) including but not limited to those not resolved under the provisions of Paragraphs SC-16.01A and 16.01.B will be decided by arbitration in accordance with the rules of *[insert name of selected arbitration agency]*, **subject to the conditions and limitations of this Paragraph SC-16.02. This agreement to arbitrate and any other agreement or consent to arbitrate entered into will be specifically enforceable under the prevailing law of any court having jurisdiction.**

B. The demand for arbitration will be filed in writing with the other party to the Contract and with the selected arbitrator or arbitration provider, and a copy will be sent to Engineer for information. The demand for arbitration will be made within the 30 day period specified in Paragraph SC-16.01.C, and in all other cases within a reasonable time after the Claim or counterclaim, dispute, or other matter in question has arisen, and in no event shall any such demand be made after the date when institution of legal or equitable proceedings based on such Claim or other dispute or matter in question would be barred by the applicable statue of limitations.

C. No arbitration arising out of or relating to the Contract Documents shall include by consolidation, joinder, or in any other manner any other individual or entity (including Engineer, and Engineer's consultants and the officers, directors, partners, agents, employees or consultants of any of them) who is not a party to this Contract unless:

1. the inclusion of such other individual or entity is necessary if complete relief is to be afforded among those who are already parties to the arbitration; and

2. such other individual or entity is substantially involved in a question of law or fact which is common to those who are already parties to the arbitration and which will arise in such proceedings.

D. The award rendered by the arbitrator(s) shall be consistent with the agreement of the parties, in writing, and include: (i) a concise breakdown of the award; (ii) a written explanation of the award specifically citing the Contract Document provisions deemed applicable and relied on in making the award.

E. The award will be final. Judgment may be entered upon it in any court having jurisdiction thereof, and it will not be subject to modification or appeal, subject to provisions of the Controlling Law relating to vacating or modifying an arbitral award.

F. The fees and expenses of the arbitrators and any arbitration service shall be shared equally by Owner and Contractor.

CONSTRUCTION INDUSTRY ARBITRATION RULES AND MEDIATION PROCEDURES

(Including Procedures for Large, Complex Construction Disputes)
Amended and Effective September 1, 2007

TABLE OF CONTENTS

NATIONAL CONSTRUCTION DISPUTE RESOLUTION COMMITTEE
IMPORTANT NOTICE
INTRODUCTION
- Mediation
- Arbitration
- Track Procedures
- Fast Track Procedures
- Procedures for Large, Complex Construction Disputes
- The National Roster
- Administrative Fees
- ADR Clauses

CONSTRUCTION INDUSTRY MEDIATION PROCEDURES
- M-1. Agreement of Parties
- M-2. Initiation of Mediation
- M-3. Representation
- M-4. Appointment of the Mediator
- M-5. Mediator's Impartiality and Duty to Disclose
- M-6. Vacancies
- M-7. Date and Responsibilities of the Mediator
- M-8. Responsibilities of the Parties
- M-9. Privacy
- M-10. Confidentiality
- M-11. No Stenographic Record
- M-12. Termination of Mediation
- M-13. Exclusion of Liability
- M-14. Interpretation and Application of Procedures
- M-15. Deposits
- M-16. Expenses
- M-17. Cost of Mediation

CONSTRUCTION INDUSTRY ARBITRATION RULES-REGULAR TRACK PROCEDURES
- R-1. Agreement of Parties
- R-2. AAA and Delegation of Duties
- R-3. National Roster of Neutrals
- R-4. Initiation under an Arbitration Provision in a Contract
- R-5. Initiation under a Submission
- R-6. Changes of Claim
- R-7. Consolidation or Joinder
- R-8. Jurisdiction
- R-9. Mediation
- R-10. Administrative Conference
- R-11. Fixing of Locale
- R-12. Appointment from National Roster
- R-13. Direct Appointment by a Party
- R-14. Appointment of Chairperson by Party-Appointed Arbitrators or Parties
- R-15. Nationality of Arbitrator in International Arbitration
- R-16. Number of Arbitrators
- R-17. Disclosure
- R-18. Disqualification of Arbitrator
- R-19. Communication with Arbitrator
- R-20. Vacancies
- R-21. Preliminary Hearing
- R-22. Exchange of Information
- R-23. Date, Time, and Place of Hearing

R-24. Attendance at Hearings
R-25. Representation
R-26. Oaths
R-27. Stenographic Record
R-28. Interpreters
R-29. Postponements
R-30. Arbitration in the Absence of a Party or Representative
R-31. Conduct of Proceedings
R-32. Evidence
R-33. Evidence by Affidavit and Post-hearing Filing of Documents or Other Evidence
R-34. Inspection or Investigation
R-35. Interim Measures
R-36. Closing of Hearing
R-37. Reopening of Hearing
R-38. Waiver of Rules
R-39. Extensions of Time
R-40. Serving of Notice
R-41. Majority Decision
R-42. Time of Award
R-43. Form of Award
R-44. Scope of Award
R-45. Award upon Settlement
R-46. Delivery of Award to Parties
R-47. Modification of Award
R-48. Release of Documents for Judicial Proceedings
R-49. Applications to Court and Exclusion of Liability
R-50. Administrative Fees
R-51. Expenses
R-52. Neutral Arbitrator's Compensation
R-53. Deposits
R-54. Interpretation and Application of Rules
R-55. Suspension for Nonpayment

FAST TRACK PROCEDURES

F-1. Limitation on Extensions
F-2. Changes of Claim or Counterclaim
F-3. Serving of Notice
F-4. Appointment and Qualification of Arbitrator
F-5. Preliminary Telephone Conference
F-6. Exchange of Exhibits
F-7. Discovery
F-8. Proceedings on Documents
F-9. Date, Time, and Place of Hearing
F-10. The Hearing
F-11. Time of Award
F-12. Time Standards
F-13. Arbitrator's Compensation

PROCEDURES FOR LARGE, COMPLEX CONSTRUCTION DISPUTES

L-1. Administrative Conference
L-2. Arbitrators
L-3. Preliminary Hearing
L-4. Management of Proceedings

ADMINISTRATIVE FEES

Fees
Refund Schedule
Hearing Room Rental

NATIONAL CONSTRUCTION DISPUTE RESOLUTION COMMITTEE

Representatives of the more than 30 organizations listed below constitute the National Construction Dispute Resolution Committee (NCDRC). This Committee serves as an advisory body to the American Arbitration Association on mediation and arbitration procedures.

American Association of Airport Executives
American Bar Association—Construction Forum
American Bar Association—Construction Litigation Committee
American Bar Association—Public Contract Law Section
American College of Construction Lawyers
American Consulting Engineers Council
American Institute of Architects
American Public Works Association
American Road and Transportation Builders Association
American Society of Civil Engineers
American Subcontractors Association
Associated Builders & Contractors, Inc.
Associated General Contractors of America
American Specialty Contractors, Inc.
Buildings Future Council
Construction Specifications Institute
Construction Management Association of America
Design Build Institute of America
Engineers Joint Contract Documents Committee
National Association of Home Builders
National Association of Minority Contractors
National Association of Surety Bond Producers
National Society of Professional Engineers
National Utility Contractors Association
Victor O. Schinnerer
Women Construction Owners & Executives, USA

IMPORTANT NOTICE

These rules and any amendment of them shall apply in the form in effect at the time the administrative filing requirements are met for a demand for arbitration or submission agreement received by the AAA. To insure that you have the most current information, see our Web site at www.adr.org.

INTRODUCTION

Each year, many thousands of construction transactions take place. Occasionally, disagreements develop over these transactions. Many of these disputes are resolved by arbitration, the voluntary submission of a dispute to a disinterested person or persons for final and binding determination. Arbitration has proven to be an effective way to resolve disputes privately, promptly, and economically.

The American Arbitration Association (AAA) is a public service, not-for-profit organization offering a broad range of dispute resolution services to business executives, attorneys, individuals, trade associations, unions, management, consumers, families, communities, and all levels of government. Services are available through AAA headquarters in New York City and through offices located in major cities throughout the United States and Europe. Hearings may be held at locations convenient for the parties and are not limited to cities with AAA offices. In addition, the AAA serves as a center for education and training, issues specialized publications, and conducts research on all forms of out-of-court dispute settlement.

Mediation

Because of the increasing popularity of mediation, especially as a prelude to arbitration, the Association has combined its mediation procedures and arbitration rules into a single brochure.

By agreement, the parties may submit their dispute to mediation before arbitration under the mediation procedures in this brochure. Mediation involves the services of one or more individuals, to assist parties in settling a controversy or claim by direct negotiations between or among themselves. The mediator participates impartially in the negotiations, guiding and consulting the various parties involved. The result of the mediation should be an agreement that the parties find acceptable. The mediator cannot impose a settlement, but can only guide the parties toward achieving their own settlement.

The AAA will administer the mediation process to achieve orderly, economical, and expeditious mediation, utilizing to the greatest possible extent the competence and acceptability of the mediators on the AAA's Construction Mediation Panel. Depending on the expertise needed for a given dispute, the parties can obtain the services of one or more individuals who are willing to serve as mediators and who are trained by the AAA in the necessary mediation skills. In identifying those persons most qualified to mediate, the AAA is assisted by the NCDRC.

The AAA itself does not act as mediator. Its function is to administer the mediation process in accordance with the agreement of the parties, to teach mediation skills to members of the construction industry, and to maintain the National Roster from which topflight mediators can be chosen.

There is no additional administrative fee where parties to a pending arbitration attempt to mediate their dispute under the AAA's auspices. Procedures for mediation cases are described in Sections M-1 through M-17.

Arbitration

Regular Track Procedures

The rules contain Regular Track Procedures, which are applied to the administration of all arbitration cases, unless they conflict with any portion of the Fast Track Procedures or the Procedures for Large, Complex Construction Disputes whenever these apply. In the event of a conflict, either the Fast Track procedures or the Large, Complex Construction Disputes procedures apply.

The highlights of the Regular Track Procedures are:

- party input into the AAA's preparation of lists of proposed arbitrators;
- express arbitrator authority to control the discovery process;
- broad arbitrator authority to control the hearing;
- a concise written breakdown of the award and, if requested in writing by all parties prior to
- the appointment of the arbitrator or at the discretion of the arbitrator, a written explanation of the award;
- arbitrator compensation, with the AAA to provide the arbitrator's compensation policy with the biographical information sent to the parties;
- a demand form and an answer form, both of which seek more information from the parties to assist the AAA in better serving the parties.

Fast Track Procedures

The Fast Track Procedures were designed for cases involving claims of no more than $75,000. The highlights of this system are:

- a 60-day "time standard" for case completion;
- establishment of a special pool of arbitrators who are pre-qualified to serve on an expedited basis;
- an expedited arbitrator appointment process, with party input;
- presumption that cases involving $10,000 or less will be decided on a documents only basis;
- requirement of a hearing within 30 calendar days of the arbitrator's appointment;
- a single day of hearing in most cases;
- an award in no more than 14 calendar days after completion of the hearing.

Procedures for Large, Complex Construction Disputes

Unless the parties agree otherwise, the Procedures for Large, Complex Construction Disputes, which appear in this pamphlet, will be applied to all cases administered by the AAA under the Construction Arbitration Rules in which the disclosed claim or counterclaim of any party is at least $500,000 exclusive of claimed interest, arbitration fees and costs.

The key features of these procedures include:

- mandatory use of the procedures in cases involving claims of $500,000 or more;
- a highly qualified, trained Panel of Neutrals, compensated at their customary rates;
- a mandatory preliminary hearing with the arbitrators, which may be conducted by telephone;
- broad arbitrator authority to order and control discovery, including depositions;
- presumption that hearings will proceed on a consecutive or block basis.

The National Roster

The AAA has established and maintains as members of its National Roster individuals competent to hear and decide disputes administered under the Construction Industry Arbitration Rules. The AAA considers for appointment to the construction industry roster persons recommended by Regional Panel Advisory Committees as qualified to serve by virtue of their experience in the construction field. The majority of neutrals are actively engaged in the construction industry. Attorney neutrals generally devote at least half of their practice to construction matters. Neutrals serving under these rules must also attend periodic training.

The services of the AAA are generally concluded with the transmittal of the award. Although there is voluntary compliance with the majority of awards, judgment on the award can be entered in a court having appropriate jurisdiction if necessary.

Administrative Fees

The AAA charges a filing fee based on the amount of claim or counterclaim. This fee information, which is contained with these rules, allows the parties to exercise control over their administrative fees.

The fees cover AAA administrative services; they do not cover arbitrator compensation or expenses, if any, reporting services, hearing room rental or any post-award charges incurred by the parties in enforcing the award.

ADR Clauses

Mediation

If the parties elect to adopt mediation as a part of their contractual dispute settlement procedure, they can insert the following mediation clause into their contract in conjunction with a standard arbitration provision.

If a dispute arises out of or relates to this contract, or the breach thereof, and if the dispute cannot be settled through negotiation, the parties agree first to try in good faith to settle the dispute by mediation administered by the American Arbitration Association under its Construction Industry Mediation Procedures before resorting to arbitration, litigation, or some other dispute resolution procedure.

If the parties choose to use a mediator to resolve an existing dispute, they can enter into the following submission.

The parties hereby submit the following dispute to mediation administered by the American Arbitration Association under its Construction Industry Mediation Procedures. (The clause may also provide for the qualifications of the mediator(s), method of payment, locale of meetings, and any other item of concern to the parties.)

Arbitration

When an agreement to arbitrate is included in a construction contract, it might expedite peaceful settlement without the necessity of going to arbitration at all. Thus, an arbitration clause is a form of insurance against loss of good will. The parties can provide for arbitration of future disputes by inserting the following clause into their contracts.

Any controversy or claim arising out of or relating to this contract, or the breach thereof, shall be settled by arbitration administered by the American Arbitration Association under its Construction Industry Arbitration Rules, and judgment on the award rendered by the arbitrator(s) may be entered in any court having jurisdiction thereof.

Arbitration of existing disputes may be accomplished by use of the following.

We, the undersigned parties, hereby agree to submit to arbitration administered by the American Arbitration Association under its Construction Industry Arbitration Rules the following controversy: (cite briefly). We further agree that the above controversy be submitted to (one)(three) arbitrator(s). We further agree that we will faithfully observe this agreement and the rules, that we will abide by and perform any award rendered by the arbitrator(s), and that a judgment of the court having jurisdiction may be entered on the award.

For further information about the AAA's Construction Dispute Avoidance and Resolution Services, as well as the full range of other AAA services, contact the nearest AAA office or visit our Web site at www.adr.org.

CONSTRUCTION INDUSTRY MEDIATION PROCEDURES

M-1. Agreement of Parties

Whenever, by stipulation or in their contract, the parties have provided for mediation or conciliation of existing or future disputes under the auspices of the American Arbitration Association (AAA) or under these procedures, the parties and their representatives, unless agreed otherwise in writing, shall be deemed to have made these procedural guidelines, as amended and in effect as of the date of filing of a request for mediation, a part of their agreement and designate the AAA as the administrator of their mediation.

The parties by mutual agreement may vary any part of these procedures including, but not limited to, agreeing to conduct the mediation via telephone or other electronic or technical means.

M-2. Initiation of Mediation

Any party or parties to a dispute may initiate mediation under the AAA's auspices by making a request for mediation to any of the AAA's regional offices or case management centers via telephone, email, regular mail or fax. Requests for mediation may also be filed online via Web File at www.adr.org.

The party initiating the mediation shall simultaneously notify the other party or parties of the request. The initiating party shall provide the following information to the AAA and the other party or parties as applicable:

1. A copy of the mediation provision of the parties' contract or the parties' stipulation to mediate.
2. The names, regular mail addresses, email addresses, and telephone numbers of all parties to the dispute and representatives, if any, in the mediation.
3. A brief statement of the nature of the dispute and the relief requested.
4. Any specific qualifications the mediator should possess.

Where there is no preexisting stipulation or contract by which the parties have provided for mediation of existing or future disputes under the auspices of the AAA, a party may request the AAA to invite another party to participate in "mediation by voluntary submission". Upon receipt of such a request, the AAA will contact the other party or parties involved in the dispute and attempt to obtain a submission to mediation.

M-3. Representation

Subject to any applicable law, any party may be represented by persons of the party's choice. The names and addresses of such persons shall be communicated in writing to all parties and to the AAA.

M-4. Appointment of the Mediator

Parties may search the online profiles of the AAA's Panel of Mediators at www.aaamediation.com in an effort to agree on a mediator. If the parties have not agreed to the appointment of a mediator and have not provided any other method of appointment, the mediator shall be appointed in the following manner:

1. Upon receipt of a request for mediation, the AAA will send to each party a list of mediators from the AAA's Panel of Mediators. The parties are encouraged to agree to a mediator from the submitted list and to advise the AAA of their agreement.
2. If the parties are unable to agree upon a mediator, each party shall strike unacceptable names from the list, number the remaining names in order of preference, and return the list to the AAA. If a party does not return the list within the time specified, all mediators on the list shall be deemed acceptable. From among the mediators who have been mutually approved by the parties, and in accordance with the designated order of mutual preference, the AAA shall invite a mediator to serve.
3. If the parties fail to agree on any of the mediators listed, or if acceptable mediators are unable to serve, or if for any other reason the appointment cannot be made from the submitted list, the AAA shall have the authority to make the appointment from among other members of the Panel of Mediators without the submission of additional lists.

M-5. Mediator's Impartiality and Duty to Disclose

AAA mediators are required to abide by the Model Standards of Conduct for Mediators in effect at the time a mediator is appointed to a case. Where there is a conflict between the Model Standards and any provision of these Mediation Procedures, these Mediation Procedures shall govern. The Standards require mediators to (i) decline a mediation if the mediator cannot conduct it in an impartial manner, and (ii) disclose, as soon as practicable, all actual and potential conflicts of interest that are reasonably known to the mediator and could reasonably be seen as raising a question about the mediator's impartiality.

Prior to accepting an appointment, AAA mediators are required to make a reasonable inquiry to determine whether there are any facts that a reasonable individual would consider likely to create a potential or actual conflict of interest for the mediator. AAA mediators are required to disclose any circumstance likely to create a presumption of bias or prevent a resolution of the parties' dispute within the time frame desired by the parties. Upon receipt of such disclosures, the AAA shall immediately communicate the disclosures to the parties for their comments.

The parties may, upon receiving disclosure of actual or potential conflicts of interest of the mediator, waive such conflicts and proceed with the mediation. In the event that a party disagrees as to whether the mediator shall serve, or in the event that the mediator's conflict of interest might reasonably be viewed as undermining the integrity of the mediation, the mediator shall be replaced.

M-6. Vacancies

If any mediator shall become unwilling or unable to serve, the AAA will appoint another mediator, unless the parties agree otherwise, in accordance with section M-4.

M-7. Duties and Responsibilities of the Mediator

1. The mediator shall conduct the mediation based on the principle of party self-determination. Self-determination is the act of coming to a voluntary, uncoerced decision in which each party makes free and informed choices as to process and outcome.
2. The mediator is authorized to conduct separate or ex parte meetings and other communications with the parties and/or their representatives, before, during, and after any scheduled mediation conference. Such communications may be conducted via telephone, in writing, via email, online, in person or otherwise.
3. The parties are encouraged to exchange all documents pertinent to the relief requested. The mediator may request the exchange of memoranda on issues, including the underlying interests and the history of the parties' negotiations. Information that a party wishes to keep confidential may be sent to the mediator, as necessary, in a separate communication with the mediator.
4. The mediator does not have the authority to impose a settlement on the parties but will attempt to help them reach a satisfactory resolution of their dispute. Subject to the discretion of the mediator, the mediator may make oral or written recommendations for settlement to a party privately or, if the parties agree, to all parties jointly.
5. In the event a complete settlement of all or some issues in dispute is not achieved within the scheduled mediation session(s), the mediator may continue to communicate with the parties, for a period of time, in an ongoing effort to facilitate a complete settlement.
6. The mediator is not a legal representative of any party and has no fiduciary duty to any party.

M-8. Responsibilities of the Parties

The parties shall ensure that appropriate representatives of each party, having authority to consummate a settlement, attend the mediation conference.

Prior to and during the scheduled mediation conference session(s) the parties and their representatives shall, as appropriate to each party's circumstances, exercise their best efforts to prepare for and engage in a meaningful and productive mediation.

M-9. Privacy

Mediation sessions and related mediation communications are private proceedings. The parties and their representatives may attend mediation sessions. Other persons may attend only with the permission of the parties and with the consent of the mediator.

M-10. Confidentiality

Subject to applicable law or the parties' agreement, confidential information disclosed to a mediator by the parties or by other participants (witnesses) in the course of the mediation shall not be divulged by the mediator. The mediator shall maintain the confidentiality of all information obtained in the mediation, and all records, reports, or other documents received by a mediator while serving in that capacity shall be confidential.

The mediator shall not be compelled to divulge such records or to testify in regard to the mediation in any adversary proceeding or judicial forum.

The parties shall maintain the confidentiality of the mediation and shall not rely on, or introduce as evidence in any arbitral, judicial, or other proceeding the following, unless agreed to by the parties or required by applicable law:

1. Views expressed or suggestions made by a party or other participant with respect to a possible settlement of the dispute;
2. Admissions made by a party or other participant in the course of the mediation proceedings;
3. Proposals made or views expressed by the mediator; or
4. The fact that a party had or had not indicated willingness to accept a proposal for settlement made by the mediator.

M-11. No Stenographic Record

There shall be no stenographic record of the mediation process.

M-12. Termination of Mediation

The mediation shall be terminated:

1. By the execution of a settlement agreement by the parties; or
2. By a written or verbal declaration of the mediator to the effect that further efforts at mediation would not contribute to a resolution of the parties' dispute; or
3. By a written or verbal declaration of all parties to the effect that the mediation proceedings are terminated; or
4. When there has been no communication between the mediator and any party or party's representative for 21 days following the conclusion of the mediation conference.

M-13. Exclusion of Liability

Neither the AAA nor any mediator is a necessary party in judicial proceedings relating to the mediation. Neither the AAA nor any mediator shall be liable to any party for any error, act or omission in connection with any mediation conducted under these procedures.

M-14. Interpretation and Application of Procedures

The mediator shall interpret and apply these procedures insofar as they relate to the mediator's duties and responsibilities. All other procedures shall be interpreted and applied by the AAA.

M-15. Deposits

Unless otherwise directed by the mediator, the AAA will require the parties to deposit in advance of the mediation conference such sums of money as it, in consultation with the mediator, deems necessary to cover the costs and expenses of the mediation and shall render an accounting to the parties and return any unexpended balance at the conclusion of the mediation.

M-16. Expenses

All expenses of the mediation, including required traveling and other expenses or charges of the mediator, shall be borne equally by the parties unless they agree otherwise. The expenses of participants for either side shall be paid by the party requesting the attendance of such participants.

M-17. Cost of the Mediation

There is no filing fee to initiate a mediation or a fee to request the AAA to invite parties to mediate.

The cost of mediation is based on the hourly mediation rate published on the mediator's AAA profile. This rate covers both mediator compensation and an allocated portion for the AAA's services. There is a four-hour minimum charge for a mediation conference. Expenses referenced in Section M-16 may also apply.

If a matter submitted for mediation is withdrawn or cancelled or results in a settlement after the agreement to mediate is filed but prior to the mediation conference the cost is $250 plus any mediator time and charges incurred.

The parties will be billed equally for all costs unless they agree otherwise.

If you have questions about mediation costs or services visit our website at www.adr.org or contact your local AAA office.

Conference Room Rental

The costs described above do not include the use of AAA conference rooms. Conference rooms are available on a rental basis. Please contact your local AAA office for availability and rates.

CONSTRUCTION INDUSTRY ARBITRATION RULES—REGULAR TRACK PROCEDURES

R-1. Agreement of Parties

(a) The parties shall be deemed to have made these rules a part of their arbitration agreement whenever they have provided for arbitration by the American Arbitration Association (hereinafter AAA) under its Construction Industry Arbitration Rules. These rules and any amendment of them shall apply in the form in effect at the time the administrative requirements are met for a demand for arbitration or submission agreement received by the AAA. The parties, by written agreement, may vary the procedures set forth in these rules. After appointment of the arbitrator, such modifications may be made only with the consent of the arbitrator.

(b) Unless the parties or the AAA determines otherwise, the Fast Track Procedures shall apply in any case in which no disclosed claim or counterclaim exceeds $75,000, exclusive of interest and arbitration fees and costs. Parties may also agree to use these procedures in larger cases. Unless the parties agree otherwise, these procedures will not apply in cases involving more than two parties. The Fast Track Procedures shall be applied as described in Sections F-1 through F-13 of these rules, in addition to any other portion of these rules that is not in conflict with the Fast Track Procedures.

(c) Unless the parties agree otherwise, the Procedures for Large, Complex Construction Disputes shall apply to all cases in which the disclosed claim or counterclaim of any party is at least $500,000, exclusive of claimed interest, arbitration fees and costs. Parties may also agree to use these procedures in cases involving claims or counterclaims under $500,000, or in nonmonetary cases. The Procedures for Large, Complex Construction Disputes shall be applied as described in Sections L-1 through L-4 of these rules, in addition to any other portion of these rules that is not in conflict with the Procedures for Large, Complex Construction Disputes.

(d) All other cases shall be administered in accordance with Sections R-1 through R-55 of these rules.

R-2. AAA and Delegation of Duties

When parties agree to arbitrate under these rules, or when they provide for arbitration by the AAA and an arbitration is initiated under these rules, they thereby authorize the AAA to administer the arbitration. The authority and duties of the AAA are prescribed in the agreement of the parties and in these rules, and may be carried out through such of the AAA's representatives as it may direct. The AAA may, in its discretion, assign the administration of an arbitration to any of its offices.

R-3. National Roster of Neutrals

In cooperation with the National Construction Dispute Resolution Committee the AAA shall establish and maintain a National Roster of Construction Arbitrators ("National Roster") and shall appoint arbitrators as provided in these rules. The term "arbitrator" in these rules refers to the arbitration panel, constituted for a particular case, whether composed of one or more arbitrators, or to an individual arbitrator, as the context requires.

R-4. Initiation under an Arbitration Provision in a Contract

(a) Arbitration under an arbitration provision in a contract shall be initiated in the following manner.

  (i) The initiating party (the "claimant") shall, within the time period, if any, specified in the contract(s), give to the other party (the "respondent") written notice of its intention to arbitrate (the "demand"), which demand shall contain a statement setting forth the nature of the dispute, the names and addresses of all other parties, the amount involved, if any, the remedy sought, and the hearing locale requested.

(ii) The claimant shall file at any office of the AAA two copies of the demand and two copies of the arbitration provisions of the contract, together with the appropriate filing fee as provided in the schedule included with these rules.

(iii) The AAA shall confirm notice of such filing to the parties.

(b) A respondent may file an answering statement in duplicate with the AAA within 15 calendar days after confirmation of notice of filing of the demand is sent by the AAA. The respondent shall, at the time of any such filing, send a copy of the answering statement to the claimant. If a counterclaim is asserted, it shall contain a statement setting forth the nature of the counterclaim, the amount involved, if any, and the remedy sought. If a counterclaim is made, the party making the counterclaim shall forward to the AAA with the answering statement the appropriate fee provided in the schedule included with these rules.

(c) If no answering statement is filed within the stated time, respondent will be deemed to deny the claim. Failure to file an answering statement shall not operate to delay the arbitration.

(d) When filing any statement pursuant to this section, the parties are encouraged to provide descriptions of their claims in sufficient detail to make the circumstances of the dispute clear to the arbitrator.

R-5. Initiation under a Submission

Parties to any existing dispute may commence an arbitration under these rules by filing at any office of the AAA two copies of a written submission to arbitrate under these rules, signed by the parties. It shall contain a statement of the matter in dispute, the names and addresses of the parties, any claims and counterclaims, the amount involved, if any, the remedy sought, and the hearing locale requested, together with the appropriate filing fee as provided in the schedule included with these rules. Unless the parties state otherwise in the submission, all claims and counterclaims will be deemed to be denied by the other party.

R-6. Changes of Claim

A party may at any time prior to the close of the hearing increase or decrease the amount of its claim or counterclaim. Any new or different claim or counterclaim, as opposed to an increase or decrease in the amount of a pending claim or counterclaim, shall be made in writing and filed with the AAA, and a copy shall be mailed to the other party, who shall have a period of ten calendar days from the date of such mailing within which to file an answer with the AAA.

After the arbitrator is appointed no new or different claim or counterclaim may be submitted to the arbitrator except with the arbitrator's consent.

R-7. Consolidation or Joinder

If the parties' agreement or the law provides for consolidation or joinder of related arbitrations, all involved parties will endeavor to agree on a process to effectuate the consolidation or joinder.

If they are unable to agree, the Association shall directly appoint a single arbitrator for the limited purpose of deciding whether related arbitrations should be consolidated or joined and, if so, establishing a fair and appropriate process for consolidation or joinder. The AAA may take reasonable administrative action to accomplish the consolidation or joinder as directed by the arbitrator.

R-8. Jurisdiction

(a) The arbitrator shall have the power to rule on his or her own jurisdiction, including any objections with respect to the existence, scope or validity of the arbitration agreement.

(b) The arbitrator shall have the power to determine the existence or validity of a contract of which an arbitration clause forms a part. Such an arbitration clause shall be treated as an agreement independent of the other terms of the contract. A decision by the arbitrator that the contract is null and void shall not for that reason alone render invalid the arbitration clause.

(c) A party must object to the jurisdiction of the arbitrator or to the arbitrability of a claim or counterclaim no later than the filing of the answering statement to the claim or counterclaim that gives rise to the objection. The arbitrator may rule on such objections as a preliminary matter or as part of the final award.

R-9. Mediation

At any stage of the proceedings, the parties may agree to conduct a mediation conference under the Construction Industry Mediation Procedures in order to facilitate settlement. The mediator shall not be an arbitrator appointed to the case. Where the parties to a pending arbitration agree to mediate under the AAA's rules, no additional administrative fee is required to initiate the mediation.

R-10. Administrative Conference

At the request of any party or upon the AAA's own initiative, the AAA may conduct an administrative conference, in person or by telephone, with the parties and/or their representatives. The conference may address such issues as arbitrator selection, potential mediation of the dispute, potential exchange of information, a timetable for hearings and any other administrative matters.

R-11. Fixing of Locale

The parties may mutually agree on the locale where the arbitration is to be held. If any party requests that the hearing be held in a specific locale and the other party files no objection thereto within fifteen calendar days after notice of the request has been sent to it by the AAA, the locale shall be the one requested. If a party objects to the locale requested by the other party, the AAA shall have the power to determine the locale, and its decision shall be final and binding.

R-12. Appointment from National Roster

If the parties have not appointed an arbitrator and have not provided any other method of appointment, the arbitrator shall be appointed in the following manner:

(a) Immediately after the filing of the submission or the answering statement or the expiration of the time within which the answering statement is to be filed, the AAA shall send simultaneously to each party to the dispute an identical list of 10 (unless the AAA decides that a different number is appropriate) names of persons chosen from the National Roster, unless the AAA decides that a different number is appropriate. The parties are encouraged to agree to an arbitrator from the submitted list and to advise the AAA of their agreement. Absent agreement of the parties, the arbitrator shall not have served as the mediator in the mediation phase of the instant proceeding.

(b) If the parties are unable to agree upon an arbitrator, each party to the dispute shall have 15 calendar days from the transmittal date in which to strike names objected to, number the remaining names in order of preference, and return the list to the AAA. If a party does not return the list within the time specified, all persons named therein shall be deemed acceptable. From among the persons who have been approved on both lists, and in accordance with the designated order of mutual preference, the AAA shall invite the acceptance of an arbitrator to serve. If the parties fail to agree on any of the persons named, or if acceptable arbitrators are unable to act, or if for any other reason the appointment cannot be made from the submitted lists, the AAA shall have the power to make the appointment from among other members of the National Roster without the submission of additional lists.

(c) Unless the parties agree otherwise when there are two or more claimants or two or more respondents, the AAA may appoint all the arbitrators.

R-13. Direct Appointment by a Party

(a) If the agreement of the parties names an arbitrator or specifies a method of appointing an arbitrator, that designation or method shall be followed. The notice of appointment, with the name and address of the arbitrator, shall be filed with the AAA by the appointing party. Upon the request of any appointing party, the AAA shall submit a list of members of the National Roster from which the party may, if it so desires, make the appointment.

(b) Where the parties have agreed that each party is to name one arbitrator, the arbitrators so named must meet the standards of Section R-18 with respect to impartiality and independence unless the parties have specifically agreed pursuant to Section R-18(a) that the party-appointed arbitrators are to be non-neutral and need not meet those standards.

(c) If the agreement specifies a period of time within which an arbitrator shall be appointed and any party fails to make the appointment within that period, the AAA shall make the appointment.

(d) If no period of time is specified in the agreement, the AAA shall notify the party to make the appointment. If within 15 calendar days after such notice has been sent, an arbitrator has not been appointed by a party, the AAA shall make the appointment.

R-14. Appointment of Chairperson by Party-Appointed Arbitrators or Parties

(a) If, pursuant to Section R-13, either the parties have directly appointed arbitrators, or the arbitrators have been appointed by AAA, and the parties have authorized them to appoint a chairperson within a specified time and no appointment is made within that time or any agreed extension, the AAA may appoint the chairperson.

(b) If no period of time is specified for appointment of the chairperson and the party-appointed arbitrators or the parties do not make the appointment within 15 calendar days from the date of the appointment of the last party-appointed arbitrator, the AAA may appoint the chairperson.

(c) If the parties have agreed that their party-appointed arbitrators shall appoint the chairperson from the National Roster, the AAA shall furnish to the party-appointed arbitrators, in the manner provided in Section R-12, a list selected from the National Roster, and the appointment of the chairperson shall be made as provided in that Section.

R-15. Nationality of Arbitrator in International Arbitration

Where the parties are nationals of different countries, the AAA, at the request of any party or on its own initiative, may appoint as arbitrator a national of a country other than that of any of the parties. The request must be made before the time set for the appointment of the arbitrator as agreed by the parties or set by these rules.

R-16. Number of Arbitrators

If the arbitration agreement does not specify the number of arbitrators, the dispute shall be heard and determined by one arbitrator, unless the AAA, in its discretion, directs that three arbitrators be appointed. A party may request three arbitrators in the demand or answer, which request the AAA will consider in exercising its discretion regarding the number of arbitrators appointed to the dispute.

R-17. Disclosure

(a) Any person appointed or to be appointed as an arbitrator shall disclose to the AAA any circumstance likely to give rise to justifiable doubt as to the arbitrator's impartiality or independence, including any bias or any financial or personal interest in the result of the arbitration or any past or present relationship with the parties or their representatives. Such obligation shall remain in effect throughout the arbitration.

(b) Upon receipt of such information from the arbitrator or another source, the AAA shall communicate the information to the parties and, if it deems it appropriate to do so, to the arbitrator and others.

(c) In order to encourage disclosure by arbitrators, disclosure of information pursuant to this Section R-17 is not to be construed as an indication that the arbitrator considers that the disclosed circumstances is likely to affect impartiality or independence.

R-18. Disqualification of Arbitrator

(a) Any arbitrator shall be impartial and independent and shall perform his or her duties with diligence and in good faith, and shall be subject to disqualification for

(i) partiality or lack of independence,

(ii) inability or refusal to perform his or her duties with diligence and in good faith, and

(iii) any grounds for disqualification provided by applicable law. The parties may agree in writing, however, that arbitrators directly appointed by a party pursuant to Section R-13 shall be non-neutral, in which case such arbitrators need not be impartial or independent and shall not be subject to disqualification for partiality or lack of independence.

(b) Upon objection of a party to the continued service of an arbitrator, or on its own initiative, the AAA shall determine whether the arbitrator should be disqualified under the grounds set out above, and shall inform the parties of its decision, which decision shall be conclusive.

R-19. Communication with Arbitrator

(a) No party and no one acting on behalf of any party shall communicate ex parte with an arbitrator or a candidate for arbitrator concerning the arbitration, except that a party, or someone acting on behalf of a party, may communicate ex parte with a candidate for direct appointment pursuant to Section R-13 in order to advise the candidate of the general nature of the controversy and of the anticipated proceedings and to discuss the candidate's qualifications, availability or independence in relation to the parties or to discuss the suitability of the candidates for selection as a third arbitrator where the parties or party-designated arbitrators are to participate in that selection.

(b) Section R-19(a) does not apply to arbitrators directly appointed by the parties who, pursuant to Section R-18(a), the parties have agreed in writing are non-neutral. Where the parties have so agreed under Section R-18(a), the AAA shall as an administrative practice suggest to the parties that they agree further that Section R-19(a) should nonetheless apply prospectively.

R-20. Vacancies

(a) If for any reason an arbitrator is unable to perform the duties of the office, the AAA may, on proof satisfactory to it, declare the office vacant. Vacancies shall be filled in accordance with the applicable provisions of these rules.

(b) In the event of a vacancy in a panel of neutral arbitrators after the hearings have commenced, the remaining arbitrator or arbitrators may continue with the hearing and determination of the controversy, unless the parties agree otherwise.

(c) In the event of the appointment of a substitute arbitrator, the panel of arbitrators shall determine in its sole discretion whether it is necessary to repeat all or part of any prior hearings.

R-21. Preliminary Hearing

(a) At the request of any party or at the discretion of the arbitrator or the AAA, the arbitrator may schedule as soon as practicable a preliminary hearing with the parties and/or their representatives. The preliminary hearing may be conducted by telephone at the arbitrator's discretion.

(b) During the preliminary hearing, the parties and the arbitrator should discuss the future conduct of the case, including clarification of the issues and claims, a schedule for the hearings and any other preliminary matters.

R-22. Exchange of Information

(a) At the request of any party or at the discretion of the arbitrator, consistent with the expedited nature of arbitration, the arbitrator may direct

(i) the production of documents and other information, and

(ii) the identification of any witnesses to be called.

(b) At least five business days prior to the hearing, the parties shall exchange copies of all exhibits they intend to submit at the hearing.

(c) The arbitrator is authorized to resolve any disputes concerning the exchange of information.

(d) There shall be no other discovery, except as indicated herein or as ordered by the arbitrator in extraordinary cases when the demands of justice require it.

R-23. Date, Time, and Place of Hearing

The arbitrator shall set the date, time, and place for each hearing and/or conference. The parties shall respond to requests for hearing dates in a timely manner, be cooperative in scheduling the earliest practicable date, and adhere to the established hearing schedule. The AAA shall send a notice of hearing to the parties at least ten calendar days in advance of the hearing date, unless otherwise agreed by the parties.

R-24. Attendance at Hearings

The arbitrator and the AAA shall maintain the privacy of the hearings unless the law provides to the contrary. Any person having a direct interest in the arbitration is entitled to attend hearings. The arbitrator shall otherwise have the power to require the exclusion of any witness, other than a party or other essential person, during the testimony of any other witness. It shall be discretionary with the arbitrator to determine the propriety of the attendance of any person other than a party and its representative.

R-25. Representation

Any party may be represented by counsel or other authorized representative. A party intending to be so represented shall notify the other party and the AAA of the name and address of the representative at least three calendar days prior to the date set for the hearing at which that person is first to appear. When such a representative initiates an arbitration or responds for a party, notice is deemed to have been given.

R-26. Oaths

Before proceeding with the first hearing, each arbitrator may take an oath of office and, if required by law, shall do so. The arbitrator may require witnesses to testify under oath administered by any duly qualified person and, if it is required by law or requested by any party, shall do so.

R-27. Stenographic Record

Any party desiring a stenographic record shall make arrangements directly with a stenographer and shall notify the other parties of these arrangements at least three days in advance of the hearing. The requesting party or parties shall pay the cost of the record. If the transcript is agreed by the parties, or determined by the arbitrator to be the official record of the proceeding, it must be provided to the arbitrator and made available to the other parties for inspection, at a date, time, and place determined by the arbitrator.

R-28. Interpreters

Any party wishing an interpreter shall make all arrangements directly with the interpreter and shall assume the costs of the service.

R-29. Postponements

The arbitrator for good cause shown may postpone any hearing upon agreement of the parties, upon request of a party, or upon the arbitrator's own initiative.

R-30. Arbitration in the Absence of a Party or Representative

Unless the law provides to the contrary, the arbitration may proceed in the absence of any party or representative who, after due notice, fails to be present or fails to obtain a postponement. An award shall not be made solely on the default of a party. The arbitrator shall require the party who is present to submit such evidence as the arbitrator may require for the making of an award.

R-31. Conduct of Proceedings

(a) The claimant shall present evidence to support its claim. The respondent shall then present evidence supporting its defense. Witnesses for each party shall also submit to questions from the arbitrator and the adverse party. The arbitrator has the discretion to vary this procedure, provided that the parties are treated with equality and that each party has the right to be heard and is given a fair opportunity to present its case.

(b) The arbitrator, exercising his or her discretion, shall conduct the proceedings with a view to expediting the resolution of the dispute and may direct the order of proof, bifurcate proceedings, and direct the parties to focus their presentations on issues the decision of which could dispose of all or part of the case. The arbitrator shall entertain motions, including motions that dispose of all or part of a claim, or that may expedite the proceedings, and may also make preliminary rulings and enter interlocutory orders.

(c) The parties may agree to waive oral hearings in any case.

R-32. Evidence

(a) The parties may offer such evidence as is relevant and material to the dispute and shall produce such evidence as the arbitrator may deem necessary to an understanding and determination of the dispute. Conformity to legal rules of evidence shall not be necessary.

(b) The arbitrator shall determine the admissibility, relevance, and materiality of the evidence offered. The arbitrator may request offers of proof and may reject evidence deemed by the arbitrator to be cumulative, unreliable, unnecessary, or of slight value compared to the time and expense involved. All evidence shall be taken in the presence of all of the arbitrators and all of the parties, except where: 1) any of the parties is absent, in default, or has waived the right to be present, or 2) the parties and the arbitrators agree otherwise.

(c) The arbitrator shall take into account applicable principles of legal privilege, such as those involving the confidentiality of communications between a lawyer and client.

(d) An arbitrator or other person authorized by law to subpoena witnesses or documents may do so upon the request of any party or independently.

R-33. Evidence by Affidavit and Post-hearing Filing of Documents or Other Evidence

(a) The arbitrator may receive and consider the evidence of witnesses by declaration or affidavit, but shall give it only such weight as the arbitrator deems it entitled to after consideration of any objection made to its admission.

(b) If the parties agree or the arbitrator directs that documents or other evidence be submitted to the arbitrator after the hearing, the documents or other evidence, unless otherwise agreed by the parties and the arbitrator, shall be filed with the AAA for transmission to the arbitrator. All parties shall be afforded an opportunity to examine and respond to such documents or other evidence.

R-34. Inspection or Investigation

An arbitrator finding it necessary to make an inspection or investigation in connection with the arbitration shall direct the AAA to so advise the parties. The arbitrator shall set the date and time and the AAA shall notify the parties. Any party who so desires may be present at such an inspection or investigation. In the event that one or all parties are not present at the inspection or investigation, the arbitrator shall make an oral or written report to the parties and afford them an opportunity to comment.

R-35. Interim Measures

(a) The arbitrator may take whatever interim measures he or she deems necessary, including injunctive relief and measures for the protection or conservation of property and disposition of perishable goods.

(b) Such interim measures may be taken in the form of an interim award, and the arbitrator may require security for the costs of such measures.

(c) A request for interim measures addressed by a party to a judicial authority shall not be deemed incompatible with the agreement to arbitrate or a waiver of the right to arbitrate.

R-36. Closing of Hearing

When satisfied that the presentation of the parties is complete, the arbitrator shall declare the hearing closed.

If documents or responses are to be filed as provided in Section R-33, or if briefs are to be filed, the hearing shall be declared closed as of the final date set by the arbitrator for the receipt of documents, responses, or briefs. The time limit within which the arbitrator is required to make the award shall commence to run, in the absence of other agreements by the parties and the arbitrator, upon the closing of the hearing.

R-37. Reopening of Hearing

The hearing may be reopened on the arbitrator's initiative, or by direction of the arbitrator upon application of a party, at any time before the award is made. If reopening the hearing would prevent the making of the award within the specific time agreed to by the parties in the arbitration agreement, the matter may not be reopened unless the parties agree to an extension of time. When no specific date is fixed by agreement of the parties, the arbitrator shall have 30 calendar days from the closing of the reopened hearing within which to make an award.

R-38. Waiver of Rules

Any party who proceeds with the arbitration after knowledge that any provision or requirement of these rules has not been complied with and who fails to state an objection in writing shall be deemed to have waived the right to object.

R-39. Extensions of Time

The parties may modify any period of time by mutual agreement. The AAA or the arbitrator may for good cause extend any period of time established by these rules, except the time for making the award. The AAA shall notify the parties of any extension.

R-40. Serving of Notice

(a) Any papers, notices, or process necessary or proper for the initiation or continuation of an arbitration under these rules; for any court action in connection therewith, or for the entry of judgment on any award made under these rules, may be served on a party by mail addressed to the party or its representative at the last known address or by personal service, in or outside the state where the arbitration is to be held, provided that reasonable opportunity to be heard with regard thereto has been granted to the party.

(b) The AAA, the arbitrator and the parties may also use overnight delivery or electronic facsimile transmission (fax) to give the notices required by these rules. Where all parties and the arbitrator agree, notices may be transmitted by electronic mail (email), or other methods of communication.

(c) Unless otherwise instructed by the AAA or by the arbitrator, any documents submitted by any party to the AAA or to the arbitrator shall simultaneously be provided to the other party or parties to the arbitration.

R-41. Majority Decision

When the panel consists of more than one arbitrator, unless required by law or by the arbitration agreement, a majority of the arbitrators must make all decisions.

R-42. Time of Award

The award shall be made promptly by the arbitrator and, unless otherwise agreed by the parties or specified by law, no later than 30 calendar days from the date of closing the hearing, or, if oral hearings have been waived, from the date of the AAA's transmittal of the final statements and proofs to the arbitrator.

R-43. Form of Award

(a) Any award shall be in writing and signed by a majority of the arbitrators. It shall be executed in the manner required by law.

(b) The arbitrator shall provide a concise, written breakdown of the award. If requested in writing by all parties prior to the appointment of the arbitrator, or if the arbitrator believes it is appropriate to do so, the arbitrator shall provide a written explanation of the award.

R-44. Scope of Award

(a) The arbitrator may grant any remedy or relief that the arbitrator deems just and equitable and within the scope of the agreement of the parties, including, but not limited to, equitable relief and specific performance of a contract.

(b) In addition to the final award, the arbitrator may make other decisions, including interim, interlocutory, or partial rulings, orders, and awards. In any interim, interlocutory, or partial award, the arbitrator may assess and apportion the fees, expenses, and compensation related to such award as the arbitrator determines is appropriate.

(c) In the final award, the arbitrator shall assess fees, expenses, and compensation as provided in Sections R-50, R-51, and R-52. The arbitrator may apportion such fees, expenses, and compensation among the parties in such amounts as the arbitrator determines is appropriate.

(d) The award of the arbitrator may include interest at such rate and from such date as the arbitrator may deem appropriate; and an award of attorneys' fees if all parties have requested such an award or it is authorized by law or their arbitration agreement.

R-45. Award Upon Settlement

If the parties settle their dispute during the course of the arbitration and if the parties so request, the arbitrator may set forth the terms of the settlement in a "consent award." A consent award must include an allocation of arbitration costs, including administrative fees and expenses as well as arbitrator fees and expenses.

R-46. Delivery of Award to Parties

Parties shall accept as notice and delivery of the award the placing of the award or a true copy thereof in the mail addressed to the parties or their representatives at the last known address, personal or electronic service of the award, or the filing of the award in any other manner that is permitted by law.

R-47. Modification of Award

Within twenty calendar days after the transmittal of an award, the arbitrator on his or her initiative, or any party, upon notice to the other parties, may request that the arbitrator correct any clerical, typographical, technical or computational errors in the award. The arbitrator is not empowered to redetermine the merits of any claim already decided.

If the modification request is made by a party, the other parties shall be given ten calendar days to respond to the request. The arbitrator shall dispose of the request within twenty calendar days after transmittal by the AAA to the arbitrator of the request and any response thereto.

If applicable law provides a different procedural time frame, that procedure shall be followed.

R-48. Release of Documents for Judicial Proceedings

The AAA shall, upon the written request of a party, furnish to the party, at its expense, certified copies of any papers in the AAA's possession that may be required in judicial proceedings relating to the arbitration.

R-49. Applications to Court and Exclusion of Liability

(a) No judicial proceeding by a party relating to the subject matter of the arbitration shall be deemed a waiver of the party's right to arbitrate.

(b) Neither the AAA nor any arbitrator in a proceeding under these rules is a necessary or proper party in judicial proceedings relating to the arbitration.

(c) Parties to these rules shall be deemed to have consented that judgment upon the arbitration award may be entered in any federal or state court having jurisdiction thereof.

(d) Parties to an arbitration under these rules shall be deemed to have consented that neither the AAA nor any arbitrator shall be liable to any party in any action for damages or injunctive relief for any act or omission in connection with any arbitration under these rules.

R-50. Administrative Fees

As a not-for-profit organization, the AAA shall prescribe filing and other administrative fees and service charges to compensate it for the cost of providing administrative services. The fees in effect when the fee or charge is incurred shall be applicable.

The filing fee shall be advanced by the party or parties, subject to final apportionment by the arbitrator in the award.

The AAA may, in the event of extreme hardship on the part of any party, defer or reduce the administrative fees.

R-51. Expenses

The expenses of witnesses for either side shall be paid by the party producing such witnesses. All other expenses of the arbitration, including required travel and other expenses of the arbitrator, AAA representatives, and any witness and the cost of any proof produced at the direct request of the arbitrator, shall be borne equally by the parties, unless they agree otherwise or unless the arbitrator in the award assesses such expenses or any part thereof against any specified party or parties.

R-52. Neutral Arbitrator's Compensation

Arbitrators shall be compensated a rate consistent with the arbitrator's stated rate of compensation.

If there is disagreement concerning the terms of compensation, an appropriate rate shall be established with the arbitrator by the Association and confirmed to the parties.

Any arrangement for the compensation of a neutral arbitrator shall be made through the AAA and not directly between the parties and the arbitrator.

R-53. Deposits

The AAA may require the parties to deposit in advance of any hearings such sums of money as it deems necessary to cover the expense of the arbitration, including the arbitrator's fee, if any, and shall render an accounting to the parties and return any unexpended balance at the conclusion of the case.

R-54. Interpretation and Application of Rules

The arbitrator shall interpret and apply these rules insofar as they relate to the arbitrator's powers and duties. When there is more than one arbitrator and a difference arises among them concerning the meaning or application of these rules, it shall be decided by a majority vote. If that is not possible, either an arbitrator or a party may refer the question to the AAA for final decision. All other rules shall be interpreted and applied by the AAA.

R-55. Suspension for Nonpayment

If arbitrator compensation or administrative charges have not been paid in full, the AAA may so inform the parties in order that one of them may advance the required payment. If such payments are not made, the arbitrator may order the suspension or termination of the proceedings. If no arbitrator has yet been appointed, the AAA may suspend the proceedings.

FAST TRACK PROCEDURES

F-1. Limitation on Extensions

In the absence of extraordinary circumstances, the AAA or the arbitrator may grant a party no more than one seven-day extension of the time in which to respond to the demand for arbitration or counterclaim as provided in Section R-4.

F-2. Changes of Claim or Counterclaim

A party may at any time prior to the close of the hearing increase or decrease the amount of its claim or counterclaim. Any new or different claim or counterclaim, as opposed to an increase or decrease in the amount of a pending claim or counterclaim, shall be made in writing and filed with the AAA, and a copy shall be mailed to the other party, who shall have a period of five calendar days from the date of such mailing within which to file an answer with the AAA. After the arbitrator is appointed no new or different claim or counterclaim may be submitted to that arbitrator except with the arbitrator's consent.

If an increased claim or counterclaim exceeds $75,000, the case will be administered under the Regular procedures unless: 1) the party with the claim or counterclaim exceeding $75,000 agrees to waive any award exceeding that amount; or 2) all parties and the arbitrator agree that the case may continue to be processed under the Fast Track Procedures.

F-3. Serving of Notice

In addition to notice provided by Section R-40, the parties shall also accept notice by telephone. Telephonic notices by the AAA shall subsequently be confirmed in writing to the parties. Should there be a failure to confirm in writing any such oral notice, the proceeding shall nevertheless be valid if notice has, in fact, been given by telephone.

F-4. Appointment and Qualification of Arbitrator

Immediately after the filing of (a) the submission or (b) the answering statement or the expiration of the time within which the answering statement is to be filed, the AAA will simultaneously submit to each party a listing and biographical information from its panel of arbitrators knowledgeable in construction who are available for service in Fast Track cases. The parties are encouraged to agree to an arbitrator from this list, and to advise the Association of their agreement, or any factual objections to any of the listed arbitrators, within seven calendar days of the AAA's transmission of the list. The AAA will appoint the agreed-upon arbitrator, or in the event the parties cannot agree on an arbitrator, will designate the arbitrator from among those names not stricken for factual objections. Absent agreement of the parties, the arbitrator shall not have served as the mediator in the mediation phase of the instant proceeding.

The parties will be given notice by the AAA of the appointment of the arbitrator, who shall be subject to disqualification for the reasons specified in Section R-18. Within the time period established by the AAA, the parties shall notify the AAA of any objection to the arbitrator appointed. Any objection by a party to the arbitrator shall be for cause and shall be confirmed in writing to the AAA with a copy to the other party or parties.

F-5. Preliminary Telephone Conference

Unless otherwise agreed by the parties and the arbitrator, as promptly as practicable after the appointment of the arbitrator, a preliminary telephone conference shall be held among the parties or their attorneys or representatives, and the arbitrator.

F-6. Exchange of Exhibits

At least two business days prior to the hearing, the parties shall exchange copies of all exhibits they intend to submit at the hearing. The arbitrator is authorized to resolve any disputes concerning the exchange of exhibits.

F-7. Discovery

There shall be no discovery, except as provided in Section F-6 or as ordered by the arbitrator in extraordinary cases when the demands of justice require it.

F-8. Proceedings on Documents

Where no party's claim exceeds $10,000, exclusive of interest and arbitration costs, and other cases in which the parties agree, the dispute shall be resolved by submission of documents, unless any party requests an oral hearing or conference call, or the arbitrator determines that an oral hearing or conference call is necessary. The arbitrator shall establish a fair and equitable procedure for the submission of documents.

F-9. Date, Time, and Place of Hearing

In cases in which a hearing is to be held, the arbitrator shall set the date, time, and place of the hearing, to be scheduled to take place within 30 calendar days of confirmation of the arbitrator's appointment. The AAA will notify the parties in advance of the hearing date.

F-10. The Hearing

(a) Generally, the hearing shall not exceed one day. Each party shall have equal opportunity to submit its proofs and complete its case. The arbitrator shall determine the order of the hearing, and may require further submission of documents within two business days after the hearing. For good cause shown, the arbitrator may schedule one additional hearing day within seven business days after the initial day of hearing.

(b) Generally, there will be no stenographic record. Any party desiring a stenographic record may arrange for one pursuant to the provisions of Section R-27.

F-11. Time of Award

Unless otherwise agreed by the parties, the award shall be rendered not later than fourteen calendar days from the date of the closing of the hearing or, if oral hearings have been waived, from the date of the AAA's transmittal of the final statements and proofs to the arbitrator.

F-12. Time Standards

The arbitration shall be completed by settlement or award within 60 calendar days of confirmation of the arbitrator's appointment, unless all parties and the arbitrator agree otherwise or the arbitrator extends this time in extraordinary cases when the demands of justice require it. The Association will relax these time standards in the event the arbitration is stayed pending mediation.

F-13. Arbitrator's Compensation

Arbitrators will receive compensation at a rate to be suggested by the AAA regional office.

PROCEDURES FOR LARGE, COMPLEX CONSTRUCTION DISPUTES

L-1. Administrative Conference

Prior to the dissemination of a list of potential arbitrators, the AAA shall, unless the parties agree otherwise, conduct an administrative conference with the parties and/or their attorneys or other representatives by conference call. The conference call will take place within 14 days after the commencement of the arbitration. In the event the parties are unable to agree on a mutually acceptable time for the conference, the AAA may contact the parties individually to discuss the issues contemplated herein. Such administrative conference shall be conducted for the following purposes and for such additional purposed as the parties or the AAA may deem appropriate:

(a) to obtain additional information about the nature and magnitude of the dispute and the anticipated length of hearing and scheduling;

(b) to discuss the views of the parties about the technical and other qualifications of the arbitrators;

(c) to obtain conflicts statements from the parties; and

(d) to consider, with the parties, whether mediation or other non-adjudicative methods of dispute resolution might be appropriate.

L-2. Arbitrators

(a) Large, Complex Construction Cases shall be heard and determined by either one or three arbitrators, as may be agreed upon by the parties. If the parties are unable to agree upon the number of arbitrators and a claim or counterclaim involved at least $1,000,000, then three arbitrator(s) shall hear and determine the case. If the parties are unable to agree on the number of arbitrators and each claim and counterclaim is less than $1,000,000, then one arbitrator shall hear and determine the case.

(b) The AAA shall appoint arbitrator(s) as agreed by the parties. If they are unable to agree on a method of appointment, the AAA shall appoint arbitrator from the Large, Complex Construction Case Panel, in the manner provided in the Regular Construction Industry Arbitration Rules. Absent agreement of the parties, the arbitrator (s) shall not have served as the mediator in the mediation phase of the instant proceeding.

L-3. Preliminary Hearing

As promptly as practicable after the selection of the arbitrator(s), a preliminary hearing shall be held among the parties and/or their attorneys or other representatives and the arbitrator(s). Unless the parties agree otherwise, the preliminary hearing will be conducted by telephone conference call rather than in person.

At the preliminary hearing the matters to be considered shall include, without limitation:

(a) service of a detailed statement of claims, damages and defenses, a statement of the issues asserted by each party and positions with respect thereto, and any legal authorities the parties may wish to bring to the attention of the arbitrator(s);

(b) stipulations to uncontested facts;

(c) the extent to which discovery shall be conducted;

(d) exchange and premarking of those documents which each party believes may be offered at the hearing;

(e) the identification and availability of witnesses, including experts, and such matters with respect to witnesses including their biographies and expected testimony as may be appropriate;

(f) whether, and the extent to which, any sworn statements and/or depositions may be introduced;

(g) the extent to which hearings will proceed on consecutive days;

(h) whether a stenographic or other official record of the proceedings shall be maintained;

(i) the possibility of utilizing mediation or other non-adjudicative methods of dispute resolution; and

(j) the procedure for the issuance of subpoenas.

By agreement of the parties and/or order of the arbitrator(s), the pre-hearing activities and the hearing procedures that will govern the arbitration will be memorialized in a Scheduling and Procedure Order.

L-4. Management of Proceedings

(a) Arbitrator(s) shall take such steps as they may deem necessary or desirable to avoid delay and to achieve a just, speedy and cost-effective resolution of Large, Complex Construction Cases.

(b) Parties shall cooperate in the exchange of documents, exhibits and information within such party's control if the arbitrator(s) consider such production to be consistent with the goal of achieving a just, speedy and cost effective resolution of a Large, Complex Construction Case.

(c) The parties may conduct such discovery as may be agreed to by all the parties provided, however, that the arbitrator(s) may place such limitations on the conduct of such discovery as the arbitrator(s) shall deem appropriate. If the parties cannot agree on production of document and other information, the arbitrator(s), consistent with the expedited nature of arbitration, may establish the extent of the discovery.

(d) At the discretion of the arbitrator(s), upon good cause shown and consistent with the expedited nature of arbitration, the arbitrator(s) may order depositions of, or the propounding of interrogatories to such persons who may possess information determined by the arbitrator(s) to be necessary to a determination of the matter.

(e) The parties shall exchange copies of all exhibits they intend to submit at the hearing 10 business days prior to the hearing unless the arbitrator(s) determine otherwise.

(f) The exchange of information pursuant to this rule, as agreed by the parties and/or directed by the arbitrator(s), shall be included within the Scheduling and Procedure Order.

(g) The arbitrator is authorized to resolve any disputes concerning the exchange of information.

(h) Generally hearings will be scheduled on consecutive days or in blocks of consecutive days in order to maximize efficiency and minimize costs.

ADMINISTRATIVE FEES

The administrative fees of the AAA are based on the amount of the claim or counterclaim. Arbitrator compensation is not included in this schedule. Unless the parties agree otherwise, arbitrator compensation and administrative fees are subject to allocation by the arbitrator in the award.

Fees

An initial filing fee is payable in full by a filing party when a claim, counterclaim or additional claim is filed.

A case service fee will be incurred for all cases that proceed to their first hearing. This fee will be payable in advance at the time that the first hearing is scheduled. This fee will be refunded at the conclusion of the case if no hearings have occurred.

However, if the Association is not notified at least 24 hours before the time of the scheduled hearing, the case service fee will remain due and will not be refunded.

These fees will be billed in accordance with the following schedule:

Amount of Claim

Initial Filing Fee

Case Service Fee

Above $0 to $10,000

$750

$200

Above $10,000 to $75,000

$950

$300

Above $75,000 to $150,000

$1,800

$750

Above $150,000 to $300,000

$2,750

$1,250

Above $300,000 to $500,000

$4,250

$1,750

Above $500,000 to $1,000,000

$6,000

$2,500

Above $1,000,000 to $5,000,000

$8,000

$3,250

Above $5,000,000 to $10,000,000

$10,000

$4,000

Above $10,000,000

Nonmonetary Claims*

$3,250

$1,250

Fee Schedule for Claims in Excess of $10 Million

The following is the fee schedule for use in disputes involving claims in excess of $10 million. If you have any questions, please consult your local AAA office or case management center.

Claim Size

Fee

Case Service Fee

$10 million and above

Base fee of $ 12,500 plus .01% of the amount of claim above $ 10 million.

$6,000

Filing fees capped at $65,000

Refund Schedule

The AAA offers a refund schedule on filing fees. For cases with claims up to $75,000 a minimum filing fee of $300 will not be refunded. For all other cases, a minimum fee of $500.00 will not be refunded. Subject to the minimum fee requirements, refunds will be calculated as follows:

- 100% of the filing fee, above the minimum fee, will be refunded if the case is settled or withdrawn with five calendar days of filing.
- 50% of the filing fee will be refunded if the case is settled or withdrawn between six and 30 calendar days of filing.
- 25% of the filing fee will be refunded if the case is settled or withdrawn between 31 and 60 calendar days of filing.

No refund will be made once an arbitrator has been appointed (this includes one arbitrator on a three arbitrator panel). No refunds will be granted on awarded cases.

Note: The date of receipt of the demand for arbitration with the AAA will be used to calculate refunds of filing fees for both claims and counterclaims.

Hearing Room Rental

The fees described above do not cover the rental of hearing rooms, which are available on a rental basis. Check with the AAA for availability and rates.

AAA047-10M 7/03

*This fee is applicable when a claim or counterclaim is not for a monetary amount. Where a monetary claim is not known, parties will be required to state a range of claims or be subject to the highest possible filing fee. Fees are subject to increase if the amount of a claim or counterclaim is modified after the initial filing date. Fees are subject to decrease if the amount of a claim or counterclaim is modified before the first hearing. The minimum fees for any case having three or more arbitrators are $2,750 for the filing fee, plus a $1,250 case service fee. Fast Track Procedures are applied in any case where no disclosed claim or counterclaim exceeds $75,000, exclusive of interest and arbitration costs. Parties on cases held in abeyance for one year by agreement, will be assessed an annual abeyance fee of $300. If a party refuses to pay the assessed fee, the other party or parties may pay the entire fee on behalf of all parties, otherwise the matter will be closed.

INDEX

15 USC 2601 et seq, 35
31 USC 3729 3733, 54
40 CFR Part 211 Subpart B, 34
43 USC 6901 et seq, 35
60 minutes, 54
106th Congress of the United States, 173
1890 Planned Model Community, 211

acceleration of work, 172
acceptance, 94, 120
  final, 123
access to the work site, 118
Accreditation Board for Engineering and Technology (ABET), 59
acid deposition control, 33
Acret, J., 118, 168
acts of God, 118
actual authority, 66
ad hoc arbitration, 194
addenda, 161
addendum, 162
additional work, 122
adequate consideration, 94
adjustable rate mortgages, 2
adjustment
  for alternatives, 118
  of price, 192
administration of government contracts, 151
administrative
  agencies, 10-11
  law, 7
admiralty, 11
admitted to practice (professional engineer), 60
adversarial
  relationship, 88
  processes, 172
adverse possession, 76
Affirmative Procurement Program (U.S. Government), 40
American Federation of Labor-Congress of International Organizations (AFL-CIO)
  Building Trades Department, 28
Africa (Northern), 9
African Development Bank, 188
Agency
  disputes, 67
  termination, 67
Agreement
  Between Owner and Engineer for Professional Services, 238
  to arbitrate, 193
agreements, 91
aggregate model viewers, 174
Air
  Pollution Control Act of 1955, 33
  Quality Act of 1967, 34
  traffic controllers, 23
Age Discrimination in Employment Act of 1967, 24
agency, 66
  shop, 27
Al-Jarallah, M., 10
al-khitabat, 9
Al-Sadar, M., 9
alternative dispute resolution, 179
American Arbitration Association, 176, 178, 194
  Construction Industry Arbitration Rules and Mediation Procedures, vi, 437
  Guidelines on Construction Industry Arbitration Rules and Mediation Procedures, 176
Association of Cost Engineers, ix
American Concrete Institute, 132
  Council of Engineering Companies Guidelines to Practice, 175
  Federation of Labor, 26
Congress of International Organizations
  Building Trades Department, 28
Institute of Architects, 3, 49, 84, 103, 114, 115, 167, 174, 188
Document A201, 174
Document E202-2008 Building Information Model Exhibit and Protocol, 174
Institute of Chemical Engineers, 49
  code of ethics, 219
Institute of Steel Construction, 132
Institute of Timber Construction, 132
Motorcycle Association versus Superior Court of Los Angeles County, 71
National Standards Institute, 132
Society
  of Agricultural and Biological Engineers, 49
    code of ethics, 219
  of Civil Engineers, ix, 49, 52, 194
    code of ethics, 52, 216
    International Activities Committee, 194
  of Mechanical Engineers, 49
    code of ethics, 220
  for Testing and Materials, 132
Americans with Disabilities Act of 1990, 24
  Standards for Accessible Design, 24
America's Climate Security Act of 2007, 40
Amin, S., 195
anticorruption legislation, 194
Anti-kickback Act of 1948, 22
antimonopoly law, 21
Antiracketeering Act of 1951, 23
apparent
  authority, 67
  low bidder, 97, 102

appeal as a right, 15
appeals
  process, 19
  to arbitration, 177
appellant, 19
appellate court, 16
appellee, 19
application service providers, 174
apprenticeship programs, 27
approval of contractor's drawings, 121
arbitral tribunal, 193, 194
arbitrate (agreement to), 193
arbitration, 119, 177
  binding, 176, 179
  nonbinding, 176
archeological remains, 123
ArchiCAD 12's Virtual Builder Explorer, 174
architect/engineer responsibilities of, 117
Architectural and Constructors Estimating Service versus Smith, 93
Army
    Corps of Engineers (U.S.), 174
    Building Information Modeling Road Map, 174
  Services Board of Contract Appeals, 167
around the clock engineering, 141
arrow on arrow schedule notation, 111
asbestos, 36
Asbestos.com, 36
Asian
  Development Bank, 188
  law, 6
  legal systems, 7
assignment, 96
Associated
  Building Contractors, 101
  General Contractors, 3, 84, 101, 103
  Guide to Building Information Modeling, 174
Association of Consulting Engineers, 188
assumption of risk, 70
attorneys, 7
  contingent fee, 18
  cost of, 18
attractive nuisance, 184
auction, 3
Autodesk, 174

Ball, H., 15, 17
bankruptcy, 11
  of contracts, 189
bar charts, 112
  samples, 113
barrister, 7
Bartholomew, S., 5, 118
Basel Convention, 43
basis of bids, 103
Beijing Arbitration Commission's Arbitration Rules, 193
Bennett, J., 165
Bentley Systems, 174
Bentley versus State of Wisconsin, 131
bicameral parliament, 8
bid
  advertisement, 102
  bonds, 147
  depository, 102
  forms, 101
  opening, 97
  proposals, 101
  shopping, 101, 103
  submission process, 102
  substitution, 102
bidding
  project award phase, 86
bilateral contracts, 98
bill of materials, 123
Bill of Rights (U.S. Constitution), 10
Board of Contract Appeals (U.S.), 152
  decisions, 171
Bockroth, J., 10, 18, 66, 67, 70, 71, 83, 96, 118, 158
boilerplate, 103
bona fide occupational qualification, 24
bonding company, 9
bonds, 101, 145, 148
  advantages, 151
  bid, 146
  disadvantages, 146
  maintenance, 145
  payment, 149
  performance, 148
Bondzi-Simpson, P., 191
Bonnell, M., 189
Bontang, ix
borings, 123
Borneo, Indonesia, ix
Bramble, B. and Callahan, M., 163, 167
breach of contract, 3, 4, 124, 177
  anticipatory, 127
  by other party, 128
  nature of, 127
  partial, 127
  remedies for, 127
  total, 127
British
  Broadcasting Corporation (BBC), 196
  legal system, 7
  Standards Institute, 134
broker contracts, 98
Brooks Act, 113
Broward Business Review, 194
Brown, R., ix
Buddhism, 10
builders risk insurance, 184
Building Information Modeling, v, 115, 174
Building Trades Department (of AFL-CIO), 26
business agents, 27
Business License and Qualification Scheme (People's Republic of China), 8
Businessman's Assurance Company of America versus Graham, Representative for Skidmore, Owings, and Merrell et al, 73
but for, 167
Buy American clause, 123

cadastral survey, 75
California Civil Code 333.2, 73
Calspan Corporation, 213
Calvi, J. and Coleman, S., 7, 9, 15, 16, 90
Calvo clause, 191
Camp, Dresser, and McKee, Inc., 214
Canada, 7
Canadian Standards Association, 134
Cannon, A., 50
canvassing bids, 102
cap and trade system, 41
carbon
  dioxide, 38
  market efficiency board, 41
  sinks, 43
carcinogen, 38
cardinal change, 85, 161, 193
Caribbean Development Bank, 188
cartels, 21
case law, 7
census (U.S.), 11
Center for Disease Control, 213-214
Central Intelligence Agency, 11
certificate of substantial completion, 124
certification, 15
CH_2M Hill Investment versus Alexis Herman, 26
change order, 119, 159-161
  international, 192
  proposal request form, 161
  sample form, 160
changed conditions, 119
character of a contract, 84
Ch'in
  law codes of, 8
China (People's Republic of), 201
choice
  of forum, 189
    clauses, 191
  of law, 189
    clauses, 191
Cibinic, Jr. J., Nash, Jr. R., and Nagle, J., 153
circuit courts, 16
civil law, 6, 201
  proceedings, 7
Civil Rights Act
  of 1964, 10, 24
  of 1991, 102-166
  Title VII, 23
C&K Construction versus Amber Steel Company, 95
Chukwumerije, O., 194
Civil Engineering Contractors Association, 188
claim, 17, 119
  construction, 162
  contract, 172
  delay, 163, 168

claim (*Continued*)
  government, 152
  international, 192
  legal cases, 170
  negotiations, 175
  resolution process, 158
clash detection, 174
class action lawsuits, 16
  Florida, 34
Clayton Antitrust Act, 21
Clean
  Air Act, 33
  Development Mechanism, 43
  Water Act, 35
Climate Change Legislation Design, 41
closed shop, 22
Clough, Jr. J, and Nash, R., 26
Clough, R., Sears, G., and Sears, K., 104
    118
Code
  of Federal Regulations, 24, 152
  of Justinian, 7
codes of ethics for engineers, 219
collaborative agreements, 172
collateral, 77
collateralized debt obligations, 2
collective bargaining agreements, 27
collusion, 102
comity, 61
Commerce Clearing House Asian PTE
    Limited, 8
commercial insurance, 183
Committee of Action of the International
    Engineering Council, 137
common
  law, 6, 7, 201
  proceedings, 7
  situs picketing, 28
communication
  Decency Act of 1996, 184
  technologies, 173
Commission on the European
    Communities, 188
comparative negligence, 71
compensable excusable delays, 164
compensatory damages, 24
competitiveness, 201
compliance with laws, 119
comprehensive
  Environmental Response Compensation
    and Liability Act, 36
  general liability insurance, 184
  Procurement Guidelines (U.S.
    government), 40
computer aided design (CAD), v, 174
concealing information, 93
conciliation, 8, 23
concurrent delays, 164
Conditions
  of Contract for Building and Engineering
    Works Designed by the Employer,
    188
  of the Contract for Engineering
    Procurement and Construction/
    Turnkey Projects, 188
  of the Contract for Client and Design-
    Build for Electrical and
    Mechanical Plant and for Building
    and Engineering Works, 188
conduct of the work, 122
confidentiality, 189
  agreements, 190
  statements, 54
Confucian ideal, 8
Congress (U.S.), 41, 173
  of the Parties, 43
Congressional
  Digest, 21, 22
  Quarterly, 16
consensus
  decisions, 8
  standards, 134
consequential damages, 118
ConsensusDOCS, 172, 300
  Standard Form of Tri-Party Agreements for
  Collaborative Project Delivery, 174
contingent liability insurance, 184
constitution (U.S.), 1, 10
  articles of, 14, 202
construction
  change authorization form American
    Institute of Architects, 160
    sample, 161
  contracts, 89, 96, 115
  documentation, 163
  Industry Arbitration Rules and
    Mediation Procedures, 437
  Management Association of America,
    84
  managers, 8
  phase, 85, 86
  reports, 123
  Specification Institute, 84, 142
    Manual of Practice, 84, 104, 141
    Master Format, 141
constructive changes, 161
contract, 84
  acceptance, 94
  administration, 4
  agreements, 91
  Appeals Board, 165-168
  claims, 172
  clauses (international), 187, 188
  contract documents (AIA), 115
  Disputes Act, 152, 168
  drawings, 105
  interpreting, 3
  law, 12
  lump sum, 97
  mediation, 176
  negotiations, 175
  offer, 94
  plans, 105
  short form, 188
  termination, 127
  unit price, 97
contractor
  Code of Business Ethics and Conduct,
    53
  guarantee, 196
  required to furnish, 11
contractors, 85
  Business Ethics Compliance Programs
    and Disclosure Requirements, 53
  licensing requirements, 62
contracts, 1, 10
  bilateral, 98
  broken, 98
  construction, 96
  construction management, 115
  cost plus
    a fixed fee, 98
    a fixed sum, 98
    a percentage of the cost, 98
  engineering, 96, 112
  express, 98
  forming, 94
  general topics, 118
  government, 145
    construction, 151
  implied, 98
  joint and severable, 98
  joint and several, 98
  main components, 103
  topics
    avoided, 118
    covered, 118
  under seal, 95
  unilateral, 98
  void, 98
  voidable, 98
contributory negligence, 71
Convention
  on Combating Bribery of Foreign Public
    Officials, 194
  on the Recognition and Enhancement of
    Foreign Arbitral Awards, 187, 193
coordination
  with the work of others, 118
cooling-off period, 23
Copeland Act, 22
Copeland, L. and Griggs, L., 141
corporate
  average fuel economy, 41
  social responsibility, 37
  sustainability, 37
  taxes (Middle Eastern), 9
corporations, 82, 182
Corpus Juris Civilis, 7
cost plus
  a fixed fee contracts, 98
  a fixed sum contracts, 98
  a percentage of the cost contracts, 98
copyrights, 11
Council on Environmental Quality, 32
Counterclaim, 17

court
annexed arbitration, 180
docket, 177
of Appeals (U.S.), 15
of Claims (U.S.), 15, 152, 168
of Customs and Patent Appeals (U.S.), 15
of Federal Claims (U.S.), 152
of limited jurisdiction, 16
proceedings, 177
Crain versus Sestak, 71
criminal
law, 7
sanctions, 61
critical path method schedules, 110, 165, 167
cross-complaint, 17
Crow Construction Company, Inc. versus State, 167
Cullowhee North Carolina, i, ix
currency
clause, 189, 191
issues, 11
custom duties, 190

damages
compensatory, 127
consequential, 162
delays, 118
liquidated, 127, 168-169, 189, 191
nominal, 127
pencitory, 127
D'Angelo versus State of New York, 167
Davis Bacon Act, 22
decisions
consensus, 8
deed, 76
deep pockets, 71
default, 2, 119
defective
drawings, 121
work, 121
defendant, 17
delays, 163, 172
claims, 163, 167, 168
compensable, 164
concurrent, 164
excusable, 164, 169
noncompensable excusable, 163
nonexcusable, 164
owner damages for, 168
partial, 165
deleted work, 172
demolition, 87
denial of justice, 192
demotion, 87
demolition phase, 88
demurrer, 17
Department
of Defense (U.S.), 134
Technical Standard Program, 133
of Energy (U.S.), 134
of Health and Human Services (U.S), 214
of Justice (U.S.), 213
of Labor (U.S.), 25, 28
Whistleblower Program, 36
depositions, 17, 63, 175
destruction of subject matter, 124
design/build contractor, 90
design
errors, 172
phase, 87
specifications, 131
designers, 85
Deutsche
Institute, 134
Weller, 196
digital notary service, 174
dimensioning, 139
disadvantaged business entity, 93
discovery, 17, 175, 177
discrimination, 10, 23
intentional, 24
Diet (Japanese), 8
dispute
avoiding and resolving during construction,178
protests, 119
resolution techniques, 172
Review Boards, 178
Guide to Specifications, 178
Manual, 178
Weaknesses, 179
settlements, 179
district courts, 16
diversity of citizenship, 11, 15
doctrine of support, 78
domestic
corporations, 82
materials, 123
double breasting, 29
Dow Jones Global Sustainability Index, 37
Draft International Standard, 135
drawings, 105
dredging, 123
documentation, 162
due process of law, 11
Duran, J., iiv
duty of care, 71

early
completion claims, 167
finish schedule submissions, 167
neutral evaluation, 179
E. C. Ernst Inc. versus General Motors, 167
E.C. Jordan Company, 215
easements, 77
East Kalimantan, Indonesia, ix
eastern bloc countries, 6
economic duress, 91
ecotoxic waste, 43
electronic
marketplace, 173
signatures, 173, 174
in Global and National Commerce Act of 2000, 173
eLEGAL, 174
European Commission Research, 173
El-Nagar, H., and Yates, J., 163
emails, v
as evidence, 173
eminent domain, 78
emergencies, 123
employment, 123
encumbrances, 77
enemy alien, 7
energy
efficient equipment standards, 41
Independence and Security Act of 2007, 41
enjoin, 23
Engineer-in-Training (EIT) examination, 59
engineering, 86
Advancement Association of Japan, 188
codes of ethics, 53
contracts, 96, 112
Joint Construction Documents Committee, 84, 103, 114-115
Design and Construction Related Documentation, 222
News Record, ix
October 10, 2007, 92
January 9, 2008, 55
January 23, 2008, 72
February 14, 2008, 93
February 25, 2008, 26
March 24, 2008, 40
May 12, 2008, 36
May 16, 2008, 115
July 14, 2008, 185
September 24, 2008, 63
December 1, 2008
February 23, 2009, 57
March 3, 2009, 34
May, 15, 2009, 35
professional
ethics, 49
registration, 59
services fee schedule, 113
environmental
compliance, 45
impact statements, 32-33
protection, 119
Protection Agency, 33-34, 37, 40-41, 213, 214, 215
June 10, 2009, 34
strategy and goals, 38
Emergency Planning and Community Right-to-Know Act, 37
emission
targets. 43
trading, 43
entitlement, 153
Epstein, A., vii
Equal Employment Opportunity Commission, 24

equitable decrees, 18
equivalent uniform annual cost, 87
Erickson, M, vii
Erecto Corporation versus State, 168
escalation, 120
estoppel, 66
ethical dilemmas, 54
ethics, 51
Europe, 6
European
 Bank for Reconstruction and Development, 188
 Commission, 174
 Community for Standardization, 135
 Union, 43
evidence of land ownership, 76
exclusive jurisdiction clauses, 191
exculpatory clause, 164
executive branch, 10
exemption, 189
exemption clause, 191
expert witness, 63, 175
express contracts, 98
expropriation, 192
extras, 159

facilities, 123
Fair
 Employment Practices Agencies, 24
 Labor Standards Act, 22
feasibility
 phase, 86
 studies, 86, 87
family
 matters courts, 12
 Medical Leave Act of 1993, 25
fault doctrine, 70
Federal
 Acquisition Regulations, 32, 53, 54, 151
 agencies, 101
 Arbitration Act, 176
 Aviation Administration, 34
 and state licensing commissions (boards), 11
 Bureau of Investigation, 11
 Coordinating Officer, 214
 court system, 11
 Emergency Management Administration, 11, 213
 Highway Administration, 6, 11
 Insecticide, Fungicide, and Rodenticide Act, 35
 labor laws, 21
 judicial system, 15
 register, 21, 22, 33, 214
 specifications, 132
 and standards, 134
 supreme court, 10
 trial courts, 14
Federation Internationale des-Ingenreurs-Censeils (International Federation of Consulting Engineers), 188
fee simple, 77
felony, 16
field order change form, 162
Filbert versus Philadelphia, 131
final
 acceptance, 123
 inspection, 123
 payment, 120
fire protection, 119
first aid, 119
Fishbach and Moore International Corporation ASBCA, 166
Fleischer-Seeger Construction Company versus Hellmuth, Obat, and Kassabaum, 69
float, 165
Florida
 Homestead Act, 18
force majeure, 118, 165, 195
foreclosures, 2
foreign
 corporations, 82
 Corrupt Practices Act, 194
 Invested Construction Enterprises (People's Republic of China), 8
 materials, 123
 ministry, 9
Form of Agreement Between Owner and Contractor for Construction Contract (Stipulated Price), 224
forming contracts, 94
Fourth World Trade Organization Ministerial Conference, 9
fragnets, 167
France, 7
Frank Brisco Company, Inc. versus Clark County, 120
fraud, 92
 contract, 126
free float, 111
Freedom of Information Act, 152
French
 civil law, 7
 colonies (former), 6
front end loading, 98
frustration of object, 126
Fundamentals of Engineering examination, 59
functional specification, 132

Geneva Convention, 194
geotechnical reports, 102
Gerilla, G., Teknomo, K., and Kokoa, K., 41
Germanic code, 6
Germany, 197
Glass-Steagall Act, 2
global financial crisis, 2
general conditions, 103
Gold, H., 167
governing laws, 120
government
 Accountability Office Comptroller General (U.S.), 152
 claims, 153
 constitution (U.S.), 14
 construction contracts, 151
 contracting officer, 145
 contractor, 145
 power sharing, 10
 services, 55
 Administration (U.S.), 6, 134,
 standards, 134
 Tender Law of Saudi Arabia, 10
governors, 10
gratuitous offer, 94
Gray book 28
Great Britain, 7
 asbestos, 36
Green
 book, 28
 builders, 167
 Building Council, 38
 building rating system, 38
 building technologies (liability), 185
 house gas emissions, 190
 house gases, 41, 42
Gross, J., 134
gross receipts tax, 195
guarantee
 clauses, 120
 contractor, 196
Guide to the Preparation of Supplementary Conditions, 103, 411

Harbor Court Association versus Leo A. Daly Company, 73
hard laws, 45
hardship clause, 189, 191
Hazard Commission, 37
hazardous
 and Solid Waste Act, 35
 waste mitigation, 123
 wastes, 43
Heath et al, 213
Hewlett Packard Foundation, ix
Hickenlooper Amendment, 192
highlighting of interferences, 174
Hill, J., 7
Hinduism, 10
hiring halls (union), 27
Hobbs Act, 23
Hohns, M., 168
hold harmless clause, 73
Homestead Act, 18
Hooker Electrochemical Company, 211, 213
host
 country, 190
 nation laws, 189
house of representatives, 11
Housing and Urban Development Administration, 11
HR 4986 Public Law 110-181 Section 585, 25
Huang versus Garner, 72
Hughes versus Board of Architectural Examiners, 61

intellectual property rights, 172
impeachment of judges, 17
implied contracts, 98
impossibility
  objective, 124
  subjective, 124
incentive clause, 118
indefinite delivery/indefinite quantity, 151
indemnification clauses, 73
indemnity, 120, 182
index clause, 189, 191
incidence rates (accident), 26
India, 7, 201
information
  department (People's Republic of China), 9
  technologies, 173
initial decision maker, 115
inspection, 119, 122, 190
  agencies, 6
  of site and local conditions, 119
  of work in place, 119
Institute
  of Electronic and Electrical Engineers
    code of ethics, 220
  of Industrial Engineers
    code of ethics, 221
Institution
  of Civil Engineers, 49, 188
    Conditions of Contracts, 199
    New Engineering Contract, 188
  of Mechanical Engineers, 49
institutional memory, 175
instructions to bidders, 101
insurance, 119, 195
  policies, 182
  professional liability and structural defects, 196
integrated project delivery, 114, 174
InterAmerican
  Bank for Reconstruction and Development 188
  Commission of Commercial Arbitration, 194
interest, 110
intergovernmental treaties, 190
Intergovernmental Panel on Climate Change, 43
intermediate appellate court, 16
international
  Arbitration Association, 193, 194
    Centre for the Settlement of Investment Disputes, 194
    Chamber of Commerce, 187, 194
    Dispute Board Clauses, 178
    Dispute Board Rules, 178
    Model Dispute Board Member Agreements, 178
  Commercial Terms (INCOTERMS), 187
  construction contract clauses, 188
  conventions, 187, 189
  customery laws, 45
  Electrotechnical Commission, 135
  engineering and construction contracts, 187
  Federation of Consulting Engineers, 8, 188, 194
  joint ventures, 192
  Network for Environmental Compliance and Enforcement, 45
  Organization for Standardization, 42, 134–135
    Central Committee, 136
    liaisons, 136
  standards, 134
  system of units (metric), 139
  technical standards, 138
interpretation differences, 172
interrogations, 17
interrogatories, 17
interstate highway system, 33
intergovernmental treaties, 190
Islamic countries, 9
IS USC 2601 et seq, 35
Iowa
  Homestead Act, 18
  State University, ix
invitation to negotiate an offer, 94

Japan, 8
  Economic News, 195
Japanese
  law, 7
  Society of Civil Engineers, 194
Jewish faith, 10
John Martin Company versus Morse/Diesel, Inc., 95
John Wiley and Sons, 33
joint
  and several contact, 98
  and severable
    contract, 98
    liability, 81
  Engineering Construction Documents Committee, 3
  implementation, 43
  tenancy, 76
  venture (People's Republic of China), 9
  ventures, 82
Jenkins, J. and Stubbings, S., 180
judgments, 18, 177
judicial
  branch of government, 10, 11
  conduct board (commission), 17
judge-made laws, 7
jurisdictional
  disputes, 27
  venue, 120
jurors, 7
jury trial, 18
justifiable reliance, 94
Justinain
  Emperor of Constanople, 6
Kangari, R., and Lucas, C., 196
Kansas
  Homestead Act, 18
Katz, A., 7, 8
Kelo versus New London, 79
key performance indicators, 37
kidnapping insurance, 194
Kimmel, J., I, ix
Knutson, R., 7, 194
Koran, 6, 9
Krassow, E., vii
Kurtner, S., 168
Kyoto Protocol, 42
  Treaty, 190

Labeling of Hearing Protection Devices Regulations, 34
labor
  considerations, 122
  Department (U.S.), 26
  force, 86
  local, 197
  Management
    Relations Act of 1947, 23
    Reporting Disclosure Act of 1959, 23
    standards, 119
land, 123
  locked, 77
landlord, 78
Landrum-Griffin Act, 23
Lantis, M., 188
laminated wood, 200
lawsuit, 11, 17, 177
lawyers, 7
leachate, 213
Leadership in Energy and Environmental Design (LEED), v, 38, 185
lease, 78
left on the table, 146
legal aid, 18
Legal Guide for Drawing Up International Contracts for Construction of Industrial Work, 187
legal
  system (U.S.), 10
  theory of indemnification, 17
legislative
  address, 17
  branch, 10, 11
Lewis and Queen versus N.M. Ball and Sons, 62
liability
  insurance, 182, 184
  issues, 174
    global, 195
    green building technology, 185
  owner's protective, 185
  umbrella excess, 185
Liberal Democratic Party (Japan), 8
licenses, 78, 122, 199
licensing boards, 54
Lieberman et al 2007, 41
liens, 78, 189

life
  cycle environmental cost analysis, 38
  estate, 77
limited partnership, 81
liquidated damages, 7, 127, 168, 169, 189, 191
litigation, 7, 176, 177
live loads, 50
logical precedence, 111
living trust, 82
Lloyd's
  List - December 19, 2003, 195
  of London, 183
London Court of International Arbitration, 194
loss time accidents, 183
Louisiana, 6
Love, W., 211
Love Canal, 211, 212, 214,
  Emergency Declaration Area, 214
  Revitalization Agency, 214
lump sum contacts, 97

MacKnight Flintic Stone Company versus the Mayor, 131
maintenance
  after construction, 189
  bonds, 150
Malaysia, 7
Marshall versus Karl F. Schultz, 168
Martinale-Hubbell, 187
  International Law Digest, 187
material
  data safety sheets, 37
  specifications, 132
mayors, 10
McKechinie, J., 52, 133
mechanic's liens, 78, 148, 153
mediation, 23
  /arbitration, 179
  contract, 176
Medicare, 184
meeting of the minds, 91
metes and bounds, 75
method and material specifications, 131
metric system 139
Meyers, L., 56
Meyninger, R., viii, 36, 211, 214
Microsoft Project, 111
middle ages, 7
Military Court of Appeals (U.S.), 15
Miller Act, 145
minimum
  attractive rate of return, 86
  wage, 32
  wage rates, 123
minitrials, 179
Ministry
  of Construction (People's Republic of China), 9
  and National Commerce and Administrative Bureau (People's Republic of China), 8
  of International Trade and Industry (Japan), 8
minority owned business, 93
misdemeanor, 16
misrepresentation, 93
Model
  Form International Contract for Process Plant Construction (Japanese), 188
  Registration Law, 75
modifications, 159
moonlighting, 61
morals, 52
Moran, T., 196
Moransais versus Heatherman et al, 70
mortgages, 2, 77
Motor Vehicle Air Pollution Control Act of 1965, 34
Murphy, O., 187, 195
mutual
  agency, 81
  agreement, 127
  mistake,

Nadar, R., 56
Napoleanic Code, 9
National
  Ambient Air Quality Standards, 33
  Bureau of Standards, 214
  Council of Examiners for Engineering and Surveying, 60, 75
  Defense Authorization Act, 25
  Economic and Trade Commission, 8
  Environmental Policy Act, 32, 34
  Geographic, 35
  Institute for Engineering Ethics, 194
  Joint Board, 28
  Law Journal, 194
  Labor Relations Act, 22
  Labor Relations Board, 22
  Pollutant Discharge Elimination System, 35
  People's Congress (People's Republic of China), 8
  Security Administration, 11
  Society of Professional Engineers, 52
  unions, 27
Navisworks (Autodesk), 174
negligence, 183
  laws, 3
  lawsuits, 69
negotiations, 175
  contract, 175
negotiating claims, 175
Nielsen, Y., Hassan, T., and Cifti, C., 174
nemawashi (root-binding), 8
New York
  convention, 187, 194
  Department of Transportation, 212
  public works contracts, 164
  representatives, 11
  state, 213
  Stock Exchange, 82
  Times, 194, 215
  University, ix
new work item proposal, 136
Niagra Falls, 211
  Board of Education, 211
  Power and Development Company, 211
no damages for delay clauses, 164
nonadversarial processes, 172
nongovernment standards, 134
noncollussion, 102
noncompensable excusable delays, 163
nonexcusable delays, 164
nonunion shop, 28
Norman conquest, 6
Norris-LaGuardia Act of 1932, 22
North
  Africa, 200
  Carolina, i, ix
  Dakota
    Homestead Act, 18
    state representatives, 11
    State University, ix
Northbridge Company versus W.R. Grace and Company, 96
notary republic, 95
notice of occurrence of delay, 167

obligee, 146
occupational illnesses, 25
Occupational Health and Safety
  Administration, 6, 26, 36,
    Act of 1970, 25
    Communication Standard, 37
    Public Law 91-596, 25
offer, 94
Office of International Affairs - Environmental Protection Agency (OIA-EPA), 45
Office of Research and Development Strategic Plan, 38
Office of Technology Assessment, 214
O'Hare, C., 187
Ohio University, ix
Oklahoma Homestead Act, 18
open-ended liability, 17, 72
open shop, 28
operation and maintenance phase, 88
operations, 87
order of completion, 122
Organization of Economic Cooperation and Development, 194
Ortolono, L., 32, 33
Owner, 85
  damages for delays, 168
  protective liability insurance, 185
  right to supplement or complete work, 118
  will furnish, 118
  work done by, 122
overflow bathtub, 212
overseas private investment corporation, 196

ozone
  layer, 42
  standard, 40

Pacific Rim nations, 200
Pan American Federation of Consultants Institution of Engineers, 194
partnerships, 81
partial
  performance, 124
  suspensions, 165
partitioning, 76
patents, 11, 123, 189, 190
Pathman Construction Company versus the United States, 152
payment bonds, 119
payroll, 123
PCBS, 213
Pennor Installation Corporation versus the United States, 167
People's Republic of China, 8
  Construction Law of, 8
performance, 123
  bonds, 9, 148
  specifications, 131
permit departments, 6
permits, 119, 122
  transportation, 190
persistent organic pollutants, 45
personal torts, 69
Pew Charitable Trust, 40
picketing, 23
plan approval departments, 6
plaintiff, 17
plans, 105
Ploskonka, D., vii
political insurance, 196
Postner, W. and Rubin, R., 164, 165
Potts versus Halsted Financial Corporation, 70
Poynter, D., 63
Polytechnic University, ix
precedent
  cases, 19
  law, 7, 177
Pregnancy Discrimination Act of 1978, 24
prejudgment remedies, 18
preliminary
  approvals, 119
  engineering design phase, 86
  injunction, 18
prequalification, 97, 103
present worth analysis, 87
president, 10
  Carter, 214
  Clinton, ix
  Dwight D. Eisenhower, 33
  Ronald Reagan, 23
pretrial
  activities, 17
prevention by other, 126
price adjustments, 119
primary trustee, 82
Primavera Project Planner, 111
Principals and Practice Examination, 60
privity, 95
process server, 17
procurement, 86
  phase, 88
product specifications, 132
professional engineer, 60
  examination, 60
  registration, 59
  liability insurance, 182, 196
Project Wise Navigator, 174
profit a prendre, 78
project
  baseline, 84
  contract administration models, 88
  development, 86
  Management Institute, ix, 52
    oversight constructor, 178
    Network March 2008, 56
  schedules, 110, 165, 167
promissory estoppel, 95
proper subject matter, 91
property
  surveys, 102
  torts, 69
proprietary specifications, 132
proximate cause, 70
Public Law 91-190, 32, 34 93-288, 34
public liability and property damage insurance, 184
purchase specifications, 132

qualified
  certificate, 174
  electronic signature, 174
quality
  assurance, 130
  control, 130
quantum, 153
Quantity Take-Offs (TQO – Autodesk), 174
quasi-criminal sanctions, 61
Quebec, 6, 7
Quran, 6, 9

Ramono-Germanic, 6
ransom insurance, 194
real estate, 75
  contracts, 94
  property, 75
recall of judges, 17
reciprocity, 61
Redfern, A. and Hunter, M., 8
regime change, 195
registered professional engineers, 11
registration of ownership, 76
Regulation
  on Administration of Foreign-Invested Construction Enterprises (People's Republic of China), 8
  on Examining Occupational Safety and Health Management System, 8
Ren, Z. and Hasan, T., 174
renewable fuel standard, 41
rent-a-judge, 180
repeal of oil and gas tax incentives, 41
reports
  construction, 123
Republic of Korea (South Korea), 9
request for equitable adjustment, 152
Resource and Conservation Recovery Act, 35
respondent, 19
  superior, 71
responsibility for safeguarding existing structures, 119
responsive, 101
  bidder, 97
restitution, 127, 162
restrictive specifications, 132
retainage, 118
Revised Code of Montana, 119
revision
  index clause, 192
  of price, 192
revisions, 151
Revit (Autodesk), 174
right of way, 79
right principles, 8
right-to-work laws, 23, 28
Rio Declaration, 43
risk, 195
  management, 119, 182
Robillard, W., Wilson, D., and Brown, C., 75, 76
Roman legal system, 6
  precepts, 6
Royal Institute
  of British Architects, 99
  Chartered Surveyors, 8
Rubin, R., Gray, S., Maevis, A., and Fairweather, 163
Rubino-Samanartano, M., 191, 195
rule of law, 1
rulings of general obligations (Koran, Quran), 9

safety, 119
Salman, J., 174
Samaras, C., 38
San Jose
  City of, 131
  State University, ix
Sarbanes-Oxley Act of 2002, 25, 194
Saudi Industrial Development Loan Fund, 9
S.C. Anderson Investment versus Bank of America NT and SA, 147
scabs, 28
schedule of equipment, 123
scope of work, 84, 118
Scotland, 6

Scott, S., 192
seal, 95
Secretary
  of Health (United States), 33
  of Labor (United States), 26
secondary
  boycotts, 28
  trustee, 82
secure signature creation device, 174
security, 119
Security and Exchange Commission, 41, 82
seig (right principles), 8
senate, 11
separate
  gates, 28
  prime contractor, 89
settlements, 175, 177
service of process, 17
servitudes, 77
severable contract, 98
several contract, 98
severe weather, 169
Shari'a Law, 6, 9, 201
Shen, L., Li, D., Drew, D., and Shen, Q., 8, 9
Sherman Antitrust Act, 21
shop-stewards, 27
Short Form of Agreement Between Owner and Engineer for Professional Services, 329
single prime contractor, 89
sinoforeign (Asian-foreign), 8
site conditions (differing), 172
small claims court, 16
Smith-Connelly Act, 22
Smith
  Currie, and Hancock, 118, 124
  Gary, viii
  Joseph, vii
Social Security Administration, 184
software incompatibility, 174
sole ownership, 76
solicitors, 7
Social Responsibility Investment, 37
soft laws, 45
sole propriertorship, 81
Solid Waste Disposal Act, 35
Sotelo, C., vii
South
  America, 6
  Korea, 6, 9
spare parts, 189
special conditions, 104
specifications, 105, 130
  design, 131
  federal, 132
  functional, 132
  material, 132
  methods and materials, 131
  performance, 131
  product, 132
  proprietary, 132
  purchase, 132
  restrictive, 132
  standard, 132
  types of, 130
standard
  contracts, 3
  form 23-A, 151
  of Agreement Between Owner and Engineer, 91
  of Agreement Between Owner and Engineer for Professional Services, 114
  General Conditions of the Contract, 104, 341
  of care, 70
  of conduct, 69
  specifications, 132
standards, 133
  consensus, 134
  government, 134
  international, 134
    technical, 138
  nongovernmental, 134
  technical, 133
standby letter of credit, 196
state
  agencies, 101
  Board of Architects versus Clark, 62
  boards of professional registration, 59
  council, 9
  court systems, 12, 16
  Department of Transportation, 6, 11
  judicial systems, 16
  supreme court, 16
statutes
  of fraud, 94
  of limitations, 17, 72
statutory law, 7
Stockholm
  Chamber of Commerce, 194
  Convention, 45
Stokes, M., 7, 9, 10, 191, 192, 195, 196
stop notices, 153
stored materials, 118
strikes, 23, 28, 126
structural drawings, 110
Subcommittee on Energy and Air Quality, 41
subcontracts, 99
subcontractor, 189, 190
subcontracting requirements, 119
sublease, 78
substantial performance, 124
Sun Shipbuilding and Dry Dock Company versus United States Lines, Inc., 167
sunnah, 9
summary jury trials, 180
Superfund
  Amendment and Reauthorization Act, 36
  National Priority List, 36
  program, 36
superior courts, 16
supplementary conditions, 3, 104
suppliers, 189
Supreme Court (United States), 11, 14
  judges, 15
surety, 85, 146
  bonds, 146
  company, 145
suspension
  of work, 118, 172
  partial, 165
sustainable development, 37
  industrial construction, 38
surveying departments, 6
swales, 212
Sweet, J. and Schneier, M., 17, 18, 66, 69, 71, 81, 82, 90, 113, 118, 131
symbols in drawings, 140

Taft-Hartley Act, 23
tax liens, 78
taxes host country, 95
technical
  advisory group, 135
  committee drafts, 138
  Committee on Contracting Practices of the Underground Technology Research Council, 178
  standards, 133, 189, 190
technological dependences, 111
Telka structures, 174
tenancy in common, 76
tenants, 78
Termaat, K., 137
terminating contracts, 127
termination, 189
  clause, 191
  for convenience, 119
Test Oil Company versus La Tourette, 76
testing, 87
testimony, 17
Texas
  Agricultural and Mining (A&M) University, ix
  Homestead Act, 18
three-dimensional models, 174
  software, 175
time
  is of the essence, 118, 163
  of completion, 103
Title
  18 USC 874, 22
  42 Ch 126 U. S. Code, 24
Tommy L. Guggin Plumbing and Heating Company versus Jordan, Jones, and Goulding, Inc., 70
Torrens registration, 76
tort
  laws, 3
  lawsuits, 69
  liability, 5
    compensation, 73

total organic carbon, 213
toxic
  Substance Act, 35
  wastes, 43
trademarks, 198, 190
transfer of ownership of property, 195
transit taxes, 190
transportation
  of equipment, 189, 190
  permits, 190
Treasury Department (United States), 146
trial courts, 16
Trial Court of Original and General Jurisdiction (U.S.), 16
Truth in Negotiations Act, 152
Turkish Electronic Signature Law, 174
twenty-four hour engineering, 141
two-dimensional drawings, 175

umbrella excess liability insurance, 185
unbalanced bids, 97
unemployment insurance, 184
unilateral
  contracts, 98
  mistakes, 93
unjust enrichment, 196
under oath, 17
underpinning, 122
undue influence, 91
Uniform Commercial Code (United States), 90
union
  contracts, 27
  hiring halls, 27
  locals, 27
  shop, 22
  structure, 27
unions, 26
unit
  price contract, 96
  price bid, 97
United Kingdom, 7
  Institution of Civil Engineers, 194
United Nations
  Commission on International Trade Law (UNICITRAL), 187, 189, 190, 191, 192, 193, 194, 195, 196
  Compensation Commission, 195
  Convention Against Corruption, 194
  Convention on Contracts for the International Sale of Goods, 187
  Development Program, 188
  Framework Convention on Climate Change, 42, 45
United States, 7
  government, 152
  Green Building Council, 38-39
  Mexican General Claims Commission, 191
  News and World Report - July 2009, 184
  versus Blair, 167
  versus Spearin, 131
  Telecommunication, Inc. versus American Television and Communication *Corporation*, 168
University
  of Colorado, ix
  of Washington, ix
utilities, 119

vague terms, 3, 139
value engineering, 85
Veterans
  Administration 1988 Master Specifications, 167
  Affairs Board of Contract Appeals, 166
VICO (virtual construction software) contractor, 174
voidable contracts, 18
voiding contracts, 98
Vorster, M., 178

Wagner Act, 22
waivers, 159
War Dispute Act of 1943, 22
Water Pollution Act, 35
waiver of liability, 70
warranty clauses, 120
Wearne, S., 195
weather severe, 169
West Warwick Rhode Island, 49
Western
  Carolina University, i, ix
  nations, 201
Whalen, R., 213
whistle-blowing, 55
whistle-blower protection laws, 57
White, N., 118
Wholly Foreign Invested Construction Enterprise (People's Republic of China), 8
Wickwire, J., Hurburt, S., Lerman,. L., 165, 166
Wigal, G., 167
Will, G., 79
Wolf, R., 192
women business entity, 93
work
  additional, 172
  breakdown structure, 111
  deleted, 172
  Directive Change Order Form, 159, 161
  done by the owner, 122
  in place, 118
worker's compensation insurance, 183, 196
World Federation of Engineering Organizations, 194
Writ of Centiurari, 15

XCU exclusions, 195
Xichuan, D., and Lingyuan, Z., 8

Yates, J., i, ix, 38, 40
  and Epstein, 164
yellow dog contracts, 26
Yohara, C., Turange, C., and Gatlin, F., 148
Yturralde, D., 168

Zack, J., 165
Zegarowski, G., 25